Botany

A Brief Introduction To Plant Biology

Botany

A Brief Introduction
To Plant Biology

**Thomas L. Rost • Michael G. Barbour
Robert M. Thornton • T. Elliot Weier
C. Ralph Stocking**

University of California • Davis, California

John Wiley & Sons

New York • Santa Barbara • Chichester • Brisbane
Toronto

Library of Congress Cataloging in Publication Data:

Main entry under title: Botany.

   Based on Botany, An Introduction to Plant
Biology, 5th ed., by T. E. Weir, C. R. Stocking,
and M. G. Barbour.

   Includes index.

   1. Botany. I. Rost, Thomas L. II. Weir,
Thomas Elliot, 1903—Botany, An Introduction
To Plant Biology 5th ed.

QK47.B775   581   78-5433
ISBN 0-471-02114-8

Printed in the United States of America

10 9 8 7 6 5 4 3 2 1

preface

Our objective was to provide an abridged and shortened version of *Botany: An Introduction to Plant Biology,* Fifth Edition, by Weier, Stocking, and Barbour. However, this text is significantly different from that text and is more than a simple abridgement.

Like the parent text, this is intended for introductory courses at the university, college, or community college level. Prior courses in biology, mathematics, or physics are not required, but some acquaintance with elementary inorganic chemistry is helpful for understanding Chapters 2, 5, and 6. Appendix A introduces the basic ideas of chemistry that are needed, and should be valuable for students who have no background in chemistry. Much of the parent text has been rewritten and many new illustrations have been added. Topics in the fifth edition of *Botany* by Weier et al. that have been extensively revised include: metabolism, absorption and transport, photosynthesis, growth, algae, fungi, and angiosperms. Many detailed new drawings by Alice B. Addicott and Jacqueline L. Lockwood accompany these and other chapters. All the drawings convey important information in a dramatic manner, but we think that some are original enough in themselves to be contributions to botany. The chapters on bacteria and viruses in the larger text have been deleted in this adaptation, and much of the material on genetics has been condensed and placed in Chapters 3, 7, 10, and Appendix B.

We believe that this text has several features not shared by any other botany books of comparable size: (1) A complete, unbiased coverage of botanical topics with equally detailed sections on anatomy, cytology, economic botany, ecology, evolution, morphology, physiology, taxonomy, and a survey of the plant kingdom. (2) Many original drawings and photographs that are large, detailed, and numerous enough to be true learning aids. (3) A traditional pedagogic organization written in a clear and direct style. (4) An extensive glossary that defines frequently used terms in the text and that shows their etymological derivation.

We take pleasure in acknowledging the following individuals for their help in the development of the manuscript in whole or in part or for providing the materials that are in this book: Alice B. Addicott, Dorothy Brandon, Dr. Edward Butler, Robin Camp, Dr. Norma Lang, Jacqueline Lockwood, Walter Russell, Lorna R. Thornton, and Dr. John Tucker. We also wish to thank the following individuals who reviewed the manuscript: Charles H. Field, Cochise College; Jerry Davis, University of Wisconsin, La Crosse; David Dallas, Northeastern Oklahoma A & M College; and Mary McLanathan, Foothill College. We are especially grateful to Ted Barkley of Kansas State University for his meticulous and detailed suggestions on the entire manuscript. Others who provided illustrations are cited at the end of the book. The many students who have provided suggestions for the improvement of the general botany course taught at the University of California, Davis, cannot be acknowledged individually, yet they should be aware of our appreciation for them collectively. We apologize for the unintentional omission here of others who have contributed to the text.

Thomas L. Rost
Michael G. Barbour
Robert M. Thornton
T. Elliot Weier
C. Ralph Stocking

contents

Contents

Contents

Contents

introduction

The Scope of Botany

Botany began with tribal lore about edible, medicinal, and poisonous plants. From this narrow focus on familiar leafy plants and mushrooms, curiosity spread to diverse forms until today more than 550,000 kinds, or **species,** of organisms are identified as part of the plant kingdom. New species are found continuously because there are still regions of the world that have not been thoroughly explored: the tropics, with their lush rain forests; the arctic; and the microscopic worlds of soils, oceans, and sediments.

Perhaps in the earliest days of botany the field could easily be defined as the study of life forms that are rooted and essentially immobile. But the identification of additional species and more detailed study have erased any clear boundaries. Thus for example the mosses, or Bryophytes (Fig. 1.1), have always been considered ''plants'' and appropriate subjects for botanists to study. But in its early development the moss plant consists of green, threadlike filaments that resemble certain species of aquatic organisms, the **algae** (Fig. 1.2). Furthermore, both the moss and the filamentous alga have a phase of the life cycle in which they produce free-living reproductive cells (Fig. 1.3). These cells swim by means of flagella resembling those of animal sperm cells. Still other algae spend their whole lives as actively swimming, flagellated single cells. These discoveries confirm the fact that true natural boundaries between groups of organisms are hard to find.

Is there any constant feature that is characteristic of all the organisms that botanists study and not of other forms of life? The answer is, ''not quite.'' But two features—the presence of cell walls and the ability to perform photosynthesis—almost serve that purpose and are worth special comment.

Whenever large, complex forms of life are closely inspected, they are found to be composed of numerous microscopic units of living material called **cells.** In all but one kind of organism that botanists study, each cell is surrounded by a tough, fibrous **cell wall.** The walls of adjacent cells are cemented together, giving the plant as a whole a rigid shape and preventing individual cells from moving. The one exception is a small group of organisms known as **slime molds** (Fig. 1.4), which do not have walls during most of their life cycle. In this feature the slime molds are like animals. Animal cells do not have walls,

and, thus, possess the flexibility needed for cooperative movements such as muscle contraction.

Even though most plant cells have walls, there are major differences in wall structure and composition among the organisms of the plant kingdom. In green plants the strength of the walls results from a network of **cellulose** fibers. In the fungi **chitin** is usually found instead of cellulose, while the bacteria and blue-green algae have walls with a fishnet structure built from polymers of another, more complex set of subunits. These major differences in wall structure create a suspicion that the fungi, the bacteria, and the rooted green plants may be only remotely related.

The ability to perform photosynthesis is an extremely important property of plants (Fig. 1.5). This process enables the organism to trap radiant energy from sunlight in order to construct organic materials. The foods produced by photosynthetic organisms are essential not only for the organisms themselves but also for life forms such as animals (including human beings) that cannot trap sunlight. However, some of the ''plants'' discussed in this book do not perform photosynthesis. An example is ''Indian pipe,'' a parasitic plant that has roots, stems, and flowers (Fig. 1.6). Most bacteria do not photosynthesize; nor do any of the 200,000 species of fungi (Fig. 1.7). We have no reason to suspect that the fungi ever had any photosynthetic ancestors. It is clear that botanists study these life forms because they have cell walls and some of the life cycle characteristics of the green, photosynthetic plants.

Ancestry and Classification of Plants

Because of the difficulties just described, not all botanists agree on the proper way to sort and group the organisms included in the Plant Kingdom. Nevertheless, classification is both a practical necessity and an important intellectual goal of botanists. In this regard the highest goal is to organize a true or *natural* classification of the organisms.

The idea of a natural system depends on the belief that present-day organisms are related by common ancestry and that the differences we observe between organisms

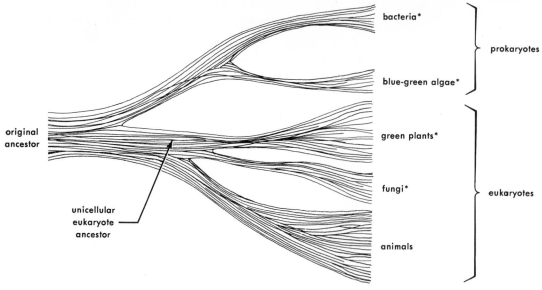

Figure 1.8 Ancestral relationships between modern organisms as deduced from protein structure and the structure and function of cells. The starred groups are all studied in botany.

are a result of extensive changes in heredity (the process of **organic evolution**) over the three billion years that plants have been on Earth. In a perfect natural classification organisms would be grouped according to the closeness of their ancestry. Of course the construction of such a classification must depend on indirect evidence, since human observers did not trace events earlier than a few thousand years ago. This evidence includes fossils of ancient plants as well as observations of similarities and differences between present-day plants (Chapter 10).

The most fundamental dividing line among life forms separates the **prokaryotes** from the **eukaryotes** (Fig. 1.8). The differences can be seen in the structure of the cells. One of several fundamental differences is that prokaryotes have their hereditary material (DNA) floating free in the same fluid mass as the rest of the cellular material, whereas the eukaryotes have their DNA separated from the rest of the cell contents by a surrounding membranous envelope. All the common plants are eukaryotes whereas the bacteria and the blue-green algae represent the world's only known prokaryotes. The entire animal kingdom consists of eukaryotes.

One of the most recent and promising tools for judging the hereditary relations between species depends on examining and comparing protein molecules. The hereditary information that is passed from generation to generation consists largely of instructions for building protein molecules. Since the number of possible different proteins is astronomical, a high degree of similarity between the proteins in two organisms indicates a

common heredity. Two organisms are not likely to have arrived at similarly constructed proteins by chance. The proteins built by eukaryotes and prokaryotes are similar enough to indicate that they arose from a common ancestor, but different enough to suggest that the eukaryotes and prokaryotes diverged long before there were any organisms higher than unicells and before there were any distinctions between plants and animals.

Probably the most significant way to divide up the large group of eukaryotes is according to whether or not they have chloroplasts. These are bodies in the cell that perform photosynthesis. Animal and fungal cells lack them. They are present in all the trees and shrubs and nearly all the herbs that make up our familiar landscape. They are also present in a multitude of less familiar forms, including many microscopic unicells.

There is evidence to suggest that chloroplasts may be the descendants of free-living bacteria that entered the cells of early eukaryotes, forming a symbiosis that became permanent. Comparable symbioses can be seen today between bacteria and plants in the root nodules of legumes (Chapter 5), but in these, the partners are still separable. If chloroplasts did arise from symbiosis, there may have been many points in evolution where various species acquired chloroplasts. Alternatively, chloroplasts may have originated within a single line of ancient cells, without symbiosis. These questions remain to be resolved, and they leave us uncertain about how far back we must reach to find ancestral connections between the bacteria, the fungi, and the green plants.

Figure 1.1 A common moss.

Figure 1.2 A filamentous green alga of the genus *Stigeoclonium.*

Figure 1.3 A later stage of development in the alga *Stigeoclonium.* The small free green cells are swimming reproductive cells.

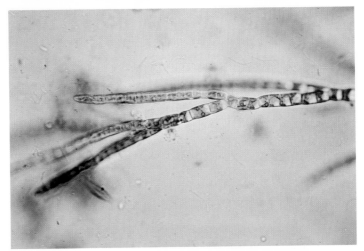

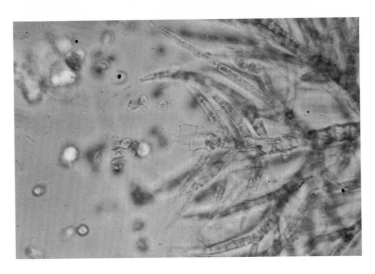

Figure 1.4 Plasmodium of the slime mold *Physarum polycephalum.*

Figure 1.5 A species of *Mimulus,* a green flowering plant.

Figure 1.6 The nonphotosynthetic flowering plant *Monotropa uniflora.*

Figure 1.7 The fruiting body (basidiocarp) of *Amanita muscaria,* a true fungus.

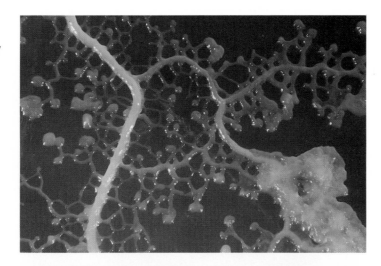

metabolism

The life of the plant cannot be understood without discussing its chemistry.[1] Since muscular movements and nervous responses are lacking, the visible signs of life in most plants are limited to slow changes such as the growth of organs. But if we consider the units of matter called molecules, of which plants are composed, the plant body proves to be a place of incessant, complex activity. This chemical activity is known collectively as **metabolism**. It occurs throughout the plant body, at all stages of life except the state known as dormancy. (We shall meet the subject of dormancy later in discussing seeds and buds.) With its thousands of different chemical reactions, metabolism makes the plant body millions of times more active chemically than the surrounding nonliving environment.

Principal Materials

Raw Materials

The plant extracts raw materials from the environment and converts them into a great variety of more complex products.

Water is the substance that plants take up in the greatest quantity. The plant acts as a wick between the moist soil and dry air; about 90% of the water that enters the plant is later evaporated away. The remaining 10% remains in the plant, where it provides bulk, serves as a medium for storing and transporting dissolved materials, and contributes atoms to the metabolic system. Water is directly consumed or produced in many reactions. Having the formula H_2O, water is a major source of the elements hydrogen (H) and oxygen (O).

Carbon dioxide (CO_2) is taken up by the land plant from the air. This compound is the plant's chief source of carbon (C) and oxygen.

Mineral elements are also taken up by the plant from the environment. These are discussed extensively in Chapter 5. They include nitrogen (N), phosphorus (P), sulfur (S), and several other elements. They are usually taken up from the soil, where they occur as ions in solution with water. Some of them (e.g., magnesium and potassium) are present as single charged atoms, whereas others occur as ionic compounds with oxygen and hydrogen (e.g., NO_3^-, NH_4^+, SO_4^{-2}, and HPO_4^{-2}).

[1] Readers who have not previously studied chemistry may find it useful to read the Appendix before pursuing this chapter.

Common Metabolic Products

There is a noticeable difference between the simple raw materials that the plant takes in, such as CO_2, H_2O, and mineral elements, and the complex final products of metabolism. For example, the compound NAD^+, which we will meet again later, has the formula $C_{21}H_{28}O_{14}N_7P_2^+$ and is formed from raw materials by a complex series of reactions.

Nearly all the molecules formed in metabolism contain carbon and hydrogen; other elements may or may not be present. Such molecules are called **organic compounds** because they are rarely found in nature except as the products of metabolic activity by living organisms. By contrast, the simple molecules that plants take in as raw materials are termed **inorganic compounds**.

Organic compounds are produced in such great variety that only a few types can be discussed here. But there are several classes of organic compounds that have universal importance and can be used as the basis of an introduction to metabolism.

Carbohydrates are composed of the elements C, H, and O; they have the general formula $(CH_2O)_n$, where *n* may be any number. The simplest carbohydrates are the **sugars**, which are important both as sources of energy and as building units in the construction of many other kinds of compounds.

Let us consider briefly the structure of a sugar molecule. In one group of simple sugars, the **hexoses**, each molecule contains six carbon atoms and has the general formula $C_6H_{12}O_6$. For example, **glucose** has the straight-chain structure shown in Fig. 2.1*A*. The —CHO end is an **aldehyde** group. The sugar molecule may occur in two forms. When not in solution the carbon atoms form a straight chain. When the sugar is in solution, four or five of the carbon atoms (depending on the kind of sugar) and an oxygen atom form a closed ring (Fig. 2.1*C*). In the straight-chain form, note the difference in the end carbon atom of glucose and fructose. As noted, the —CHO of glucose is an aldehyde. The C=O of fructose (Fig. 2.1*B*) characterizes a **ketone**.

The —OH and —H groups of the sugar molecule may be arranged in different positions in the ring without changing the relative numbers of carbon, hydrogen, and oxygen in the formula. In fact, a shifting of these atoms, as in the examples of glucose and fructose, results in sugars with different chemical properties. Sixteen different hexoses are possible, but only a few occur naturally in plants. Molecules such as these sugars with the same

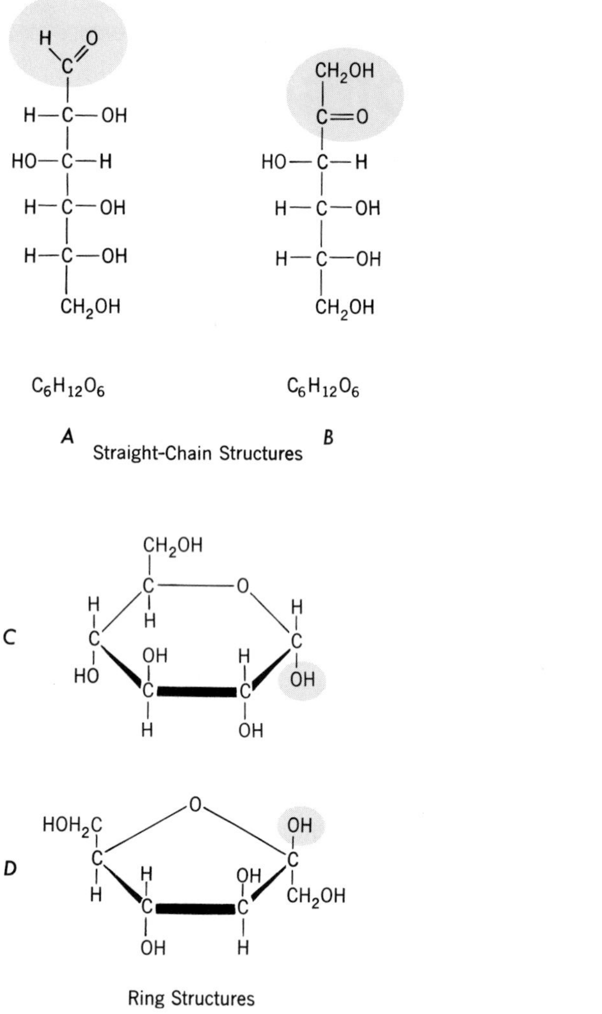

$C_6H_{12}O_6$

$C_6H_{12}O_6$

A **B**

Straight-Chain Structures

C

D

Ring Structures

Figure 2.1 Glucose and fructose, two hexose sugars, have the same formula but different structures. Both can exist as a straight chain or ring. A and C, glucose; B and D, fructose.

chemical composition but a different internal arrangement are called **isomers**.

The most common hexoses in plants are glucose and fructose. Sugars with three carbon atoms are called **trioses**; with five and seven carbon atoms, **pentoses** and **heptoses**. The pentose sugar **ribose** is a constituent of the giant molecules that carry hereditary information (the nucleic acids). Ribose is also part of several energy-carrying molecules such as ATP (Fig. 2.8).

Simple sugars are called **monosaccharides**. The union of two of these molecules produces a **disaccharide** and a

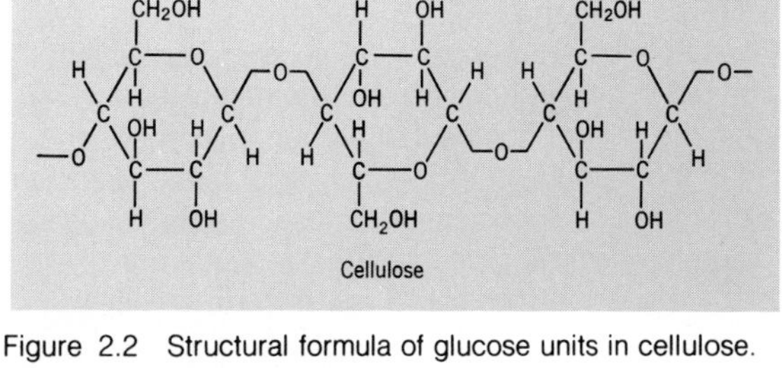

Cellulose

Figure 2.2 Structural formula of glucose units in cellulose.

water molecule. For example, the disaccharide produced from glucose and fructose is the commonest of all sugars, **sucrose**:

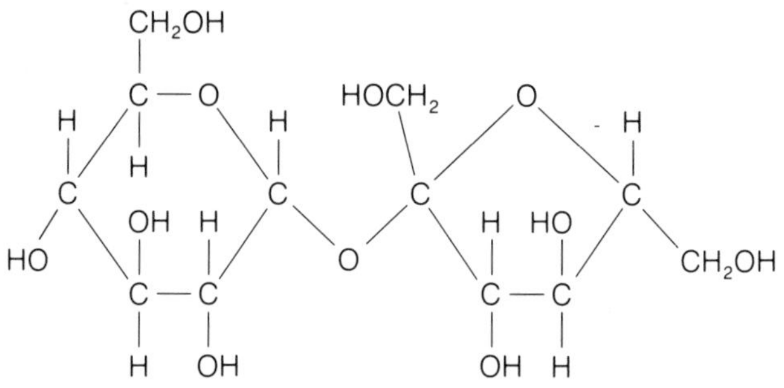

Sucrose can easily be split into fructose and glucose. One water molecule is consumed in the process. Reactions such as this, where water splits another compound, are called **hydrolysis** reactions. They are important steps in breaking down stored food molecules. Sucrose, our common table sugar, is very important as a mobile food storage compound in most plants.

Three or more monosaccharide molecules may join to form tri-, tetra-, or **polysaccharides**. The latter are composed of the union of many simple sugar molecules. As with the formation of a disaccharide a water molecule is given off for each pair of simple sugar molecules united. Polysaccharides are not generally soluble in water, nor are they sweet. **Starch** and **cellulose** are the two most abundant polysaccharides in plants. Each is composed of a long chain of glucose molecules. In starch, the chain may be coiled because of the way the glucose units are linked together and some chains are branched, while in cellulose the chains are unbranched and more or less straight. Cellulose (Fig. 2.2) is a major structural material in the plant, while starch is a reserve, water-insoluble food.

The union of relatively simple molecules, like sugar, into long-chain gigantic molecules composed of the repetition of simple units is a common chemical process known as **polymerization**. We shall meet it again in our discussion of proteins and nucleic acids.

Lipids form a very diverse collection of compounds; the chief similarity among all of them is a tendency to be insoluble in water (that is, molecules of lipids do not readily mingle with water molecules).

Cutin and **suberin** are two waxy lipids that often coat the surfaces of plant organs and serve to limit water loss. Some plant waxes (e.g., carnauba) are widely used in furniture and automobile waxing compounds.

Fats are simple and abundant lipids that are composed of **fatty acids** united with the three-carbon alcohol, **glycerol**. A fatty acid is a molecule that has an acidic group at one end; the rest of the molecule is a long carbon chain to which little other than hydrogen is attached. In **lauric acid** (Fig. 2.3) the chain is composed of 12 carbon atoms.

Figure 2.3 shows the structure of glycerol and fatty acids, and the way in which they react to produce a fat. Three molecules of water are produced in the process. The fat molecule itself is nonpolar and does not mingle readily with water. This means that fat molecules tend to

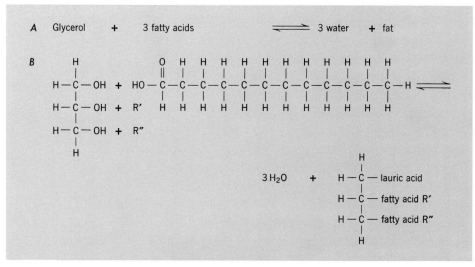

Figure 2.3 Formation of a fat by condensing fatty acids with glycerol. *A*, in words; *B*, structural formulas. The fatty acid shown in detail is lauric acid.

accumulate in droplets. They are rich in energy and their insolubility makes them good food storage compounds.

Phospholipids are similar to fats, but here one of the fatty acids is replaced by a **phosphoryl group:**

$$-O-\overset{\overset{\displaystyle O}{\|}}{\underset{\underset{\displaystyle OH}{|}}{P}}-O-$$

One end of the phosphoryl group binds to the glycerol unit, while the other end of the phosphoryl may be attached to any of several different organic groups. The phospholipids are unusual in that they are water-soluble at one end (the phosphoryl part, which is polar) while insoluble at the other end (the nonpolar fatty acid tails). These molecules tend to line up at the boundary of any water mass in which they happen to be immersed; the insoluble parts jut out of the water. Phospholipids are essential to the structure of **membranes** in the plant cell (Chapter 3), and their importance is a result of their semisoluble property.

Amino acids are important as the molecular units from which **proteins** are built. Some of the amino acids also serve as carriers for temporarily storing nitrogenous groups. There are 20 common kinds of amino acids. With one exception they are alike in their basic structure:

$$H_2N-\overset{\overset{\displaystyle R}{|}}{\underset{\underset{\displaystyle H}{|}}{C}}-\overset{\overset{\displaystyle O}{/\!/}}{C}\overset{}{\underset{\displaystyle OH}{\diagdown}}$$

amino group   carboxyl group

The **carboxyl group** readily donates its hydrogen nucleus to water; hence it acts as an acid. Thus the amino and carboxyl groups give the compound its name, amino acid. The symbol R signifies a special group of atoms called a

side chain. The side chain is not shown in detail here because it differs from one amino acid to another. The 20 kinds of amino acids differ from one another according to the side chains they possess. The exception to the picture shown above is the amino acid **proline**, in which the R group bends over and attaches to the N of the amino group. Several amino acids are shown in Fig. 2.4.

Proteins are polymers built by attaching amino acids together, end to end (Fig. 2.5). The bond is made between the amino group of one molecule and the carboxyl group of another. A molecule of water is released in making this bond.

The bond between the amino acids is termed a **peptide bond** and therefore proteins are sometimes called **polypeptides.** Some proteins have as few as 16 amino acids while others may contain hundreds. An average protein might contain 150 to 200 amino acids. Proteins also differ in the kinds of amino acids they contain and in their sequence along the chain. With these differences an immense variety of proteins is possible. No single organism can manufacture more than a small number of the possibilities.

It is impossible to exaggerate the importance of proteins. They serve structural roles, storage roles, and regulatory roles; in addition they are the agents **(enzymes)** that govern chemical reactions in metabolism. Their ability to perform such a variety of functions is a result of their structural variety. Each kind of protein, with its unique amino acid sequence, performs just one function. Proteins can perform hundreds of different functions because there are hundreds of different proteins in the organism.

The precise shape of the protein determines its function. The shape is determined by the sequence of amino acids. But that sequence alone is not the only factor responsible for the definite shape of the protein. Since rotation is possible around some of the bonds in the protein polymer, the protein has a potential for assuming many different patterns of coiling or folding. Of these possibilities, there is usually only one pattern of folding that is associated with a biological function (Fig. 2.6). An improperly folded protein has no function except as a store of amino acids.

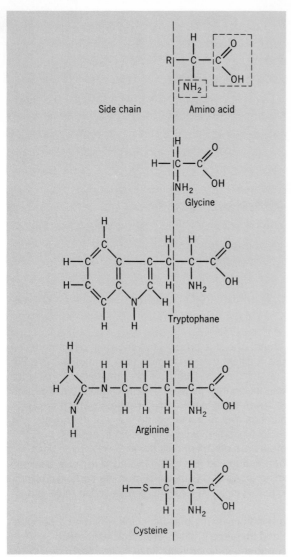

Figure 2.4 Some amino acids.

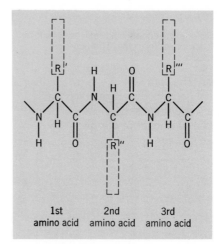

Figure 2.5 A polypeptide chain is a part of a protein molecule. The backbone of the protein molecule is formed by many amino acids joined by the union of the amino group (NH_2) of one amino acid to the acid group (COOH) of another amino acid by the removal of a water molecule. The R groups represent side chains of the different amino acids.

In the living organism each protein ordinarily maintains the pattern of folding that has functional value. Several forces contribute to the stability of the folding. Most important, the surrounding water forms a cage, with polar water molecules attracted to one another by hydrogen bonds. Some of the side chains of the protein molecule are also polar and can interact with the surrounding water, ionizing or forming hydrogen bonds. But other side chains are nonpolar. These cannot break through the web of forces that unites the surrounding water mass. Therefore the protein folds in a shape that places the nonpolar side chains out of contact with water. Most proteins assume a globular or rounded shape with the interior occupied by nonpolar parts and the surface, in contact with water, carrying the parts that are attracted to water.

Dehydration (removal of water) allows the protein to unfold. This is one reason why an abundance of water is vital to the normal operation of the organism.

The folding of the protein molecule may also be influenced by interactions between the side chains. For instance, the groups

$$\overset{O}{\underset{||}{-C-}} \quad \text{and} \quad \overset{H}{\underset{|}{-N-}}$$

occur regularly along the polymer; if the folding brings them close together, they can form a hydrogen bond (dotted line):

$$\diagdown C = O ----- H - N \diagup$$

Many such hydrogen bonds form within the folds of a typical protein, often giving rise to coiled or helical foldings in part of the protein chain. In addition, attractions and repulsions can occur between ionized side chains. Strong electron-sharing bonds known as **disulfide bridges** can also occur to firmly cross-link two folds of the protein when the two side chains of the amino acid **cysteine** (Fig. 2.4) come together.

Most of the forces that maintain the folding of proteins are relatively weak. For this reason the folding, hence the function, can be disrupted by such factors as elevated temperatures (which leave the amino acid sequence intact but tangle or **denature** the protein); changes in acidity; and the addition of specific small molecules that may bind to the protein and modify the balance of internal forces. The environmental sensitivity of proteins goes far toward explaining the physical limits of life.

Figure 2.6 Proposed folded configuration of a molecule of cytochrome c, a plant protein.

Enzymes and Catalysis

Enzymes are vital in two ways. First, in their absence the molecules in the body react so slowly that life would be nonexistent at the pace at which we know it, even at the slow pace of plant life. *Enzymes speed reactions.* Second, the molecules of life are capable of an almost infinite variety of reactions, involving different combinations of raw materials and products. The development of an organized plant body with a shape determined by heredity implies a careful selection of the chemical reactions that occur and when they occur. *Enzymes are selective* in speeding reactions. Each kind of enzyme affects only a narrow range of reactions.

The property of selectivity or **specificity** mentioned

above is widely attributed to a "lock and key" relationship between the enzyme and the molecules (**substrates**) that react with it. Enzymes are protein molecules. The folding of the protein includes a furrow or pocket (the **active site**) which has a shape complementary to the shape of the substrate molecule. This allows the enzyme and substrate to bind together; in contrast molecules with other shapes cannot bind to that particular enzyme.

The enzyme functions as a broker or **catalyst**, opening an avenue for rapid reaction but giving no energy for it. Reactions are driven by the kinetic energy of molecular collisions and sometimes by light energy that the molecules absorb. Without catalysts, biological molecules rarely react because their existing structure is maintained by internal forces too strong to be overcome by the forces that are generated in most molecular collisions. Enzymes often serve to weaken the stabilizing forces within the molecule so that the energy of collisions may be enough to bring about the reaction. The enzyme may act by (a) warping the substrate molecule; (b) temporarily reacting with the molecule so that its internal organization is altered; or (c) changing the electrochemical environment, which affects the pattern of electronic movements in the molecule. Sometimes an enzyme may also speed a reaction by holding two substrate molecules together longer and with more exact orientation than would have been possible in free solution. This gives the molecules more opportunities to collide effectively.

Phases of Metabolism

The action of each enzyme is limited to the performance of only one simple catalytic task. A single reaction between two substrates usually involves changes in only about two bonds. Therefore the construction of a complex substance such as a protein from simple materials requires an extensive series of reactions with almost every step catalyzed by a different enzyme. A series of such reactions is termed a **metabolic pathway**. Most of the reactions in a pathway require two substrates, one that was made in the previous step of the pathway, and the other made by a separate pathway. Thus metabolism is actually a web or network of intersecting and branching pathways. The enzymes determine the pathways that are available, thereby shaping metabolism as a whole.

Photosynthesis, Anabolism, and Catabolism

The whole of metabolism is far too complex to discuss here. However, we can distinguish three general phases of metabolism (Fig. 2.7). One phase is **photosynthesis**. This is a complex system of pathways by which green plants use light energy to build sugar molecules. It is the primary way in which energy is brought into the living system to power chemical constructions, growth, and repair. Its occurrence in plants is especially important to man because we cannot use light energy ourselves but must derive food (i.e., energy-rich molecules such as sugars) from other organisms and ultimately from plants.

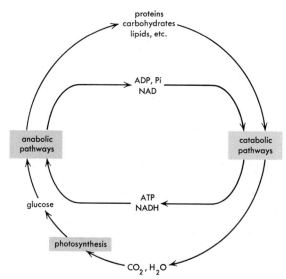

Figure 2.7   Phases of metabolism. Anabolism produces complex, high-energy products from simpler raw materials. Catabolism breaks down fuels to form compounds such as ATP and NADH that are needed for anabolism. Photosynthesis completes the cycle.

The second phase of metabolism is termed **anabolism**. This heading refers to pathways that build more complex molecules from simpler ones. Photosynthesis, which makes sugar from carbon dioxide and water, could have been put in this category but was discussed separately for special emphasis. The construction of proteins from amino acids is an example of an anabolic process. Anabolism is the basis of all developmental processes and reproductive events.

The third phase of metabolism is known as **catabolism**. Catabolic pathways extract energy from fuel (food) molecules, for use in powering anabolism and for driving the transport of materials from one location to another. Other catabolic pathways can dismantle all the kinds of molecules that are built during anabolism (though a given plant may lack some of these pathways). Proteins, lipids, and carbohydrates as well as less common compounds can serve in the plant as fuels because of these pathways.

Anabolism and catabolism are linked by several kinds of molecules that act as carriers of energy and building materials. The most prominent and universal of these are **ATP** (adenosine triphosphate) and the pyridine nucleotides ($NADP^+$ and NAD^+).

ATP is produced from ADP (adenosine diphosphate) and phosphate ions (Fig. 2.8). Energy is required to couple the extra phosphate group onto ADP; molecules with three phosphate groups joined in series are unstable,

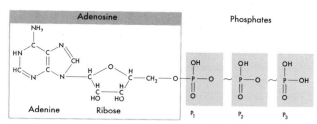

Figure 2.8   The structure of ATP (adenosine triphosphate). ADP has only two phosphate groups and AMP has only one. AMP is an example of a *nucleotide*.

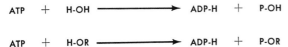

$$ATP \; + \; H\text{-}OH \; \longrightarrow \; ADP\text{-}H \; + \; P\text{-}OH$$

$$ATP \; + \; H\text{-}OR \; \longrightarrow \; ADP\text{-}H \; + \; P\text{-}OR$$

Figure 2.9 Two reactions of ATP. The first is hydrolysis; the energy stored in ATP is released as heat. In the second, an organic compound replaces water and accepts phosphate from ATP, storing part of the energy and becoming more reactive as a result. R can be any of a wide array of organic groups.

reactive, and rich in energy. If molecules of ATP and water are put in contact and supplied with a suitable catalyst, hydrolysis will spontaneously occur and heat will be released, resulting in a final equilibrium that very strongly favors ADP and inorganic phosphate (Fig. 2.9). The role of catabolism is to drive this hydrolysis reaction in reverse, using energy from fuel molecules. By means of suitable enzymes, the anabolic system allows ATP to react with organic molecules but not with water (Fig. 2.9). With a portion of the ATP molecule attached, the organic product inherits some of the reactivity (energy) of the ATP and can participate more readily in reactions that yield complex products. Note that the organic reactant in Fig. 2.9 is exchanging an H atom for a phosphate-containing group. Evidently, attachment to phosphorus-containing groups tends to make organic compounds more reactive and energy-rich. In most reaction pathways the phosphate-containing group is eventually released, to be recycled back to catabolism and built into new ATP molecules (Fig.

2.7). Thus we can characterize ADP as a phosphate carrier as well as an energy carrier, which travels back and forth between catabolism and anabolism. These relationships help to explain why phosphorus is an essential element in the life of the plant.

Just as ADP picks up phosphate and (in the form of ATP) carries it to anabolic pathways, so also NAD^+ picks up electrons and hydrogen ions at several points in catabolism. Carbon dioxide, the starting material from which plants build organic materials, contains no hydrogen, but all the complex functional molecules of metabolism are rich in hydrogen. Carriers such as NAD^+ get hydrogen from the fuel molecules (e.g., sugars) that are being broken down in catabolism. The hydrogen is then supplied to anabolic pathways.

Respiration

The principal events in catabolism form a series of reactions known as **respiration**. Dozens of reactions are involved in this process, using many enzymes. All of these reactions cannot be presented in detail here, but we can outline some of their most important aspects in general terms. Three distinct phases in the respiration system can be distinguished.

The first phase is known as **glycolysis** (Fig. 2.10). This is a series of some 11 reactions in which molecules of the sugar glucose are trimmed and modified for entry into the

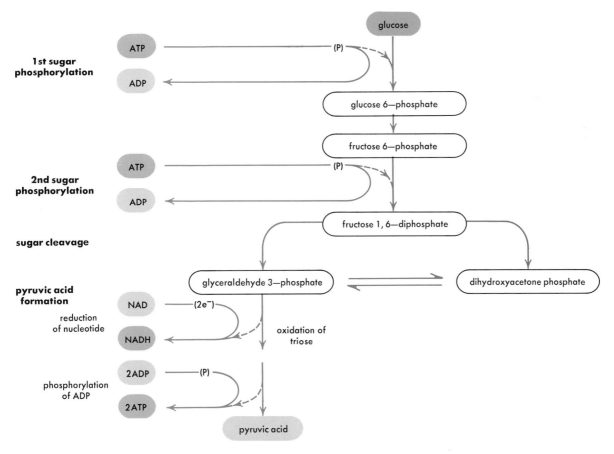

Figure 2.10 Steps in the process of glycolysis.

subsequent phase of respiration. The effect of glycolysis may be compared to trimming logs to a size that will fit into a wood stove.

Though glucose is rich in energy, it is too stable to start the process of breakdown without help. Thus at two points in glycolysis the sugar molecules are reacted with ATP so that the sugar acquires a reactive phosphate group. This process is termed **phosphorylation**. Glucose-phosphate undergoes a series of internal rearrangements, acquires another phosphate from ATP, and is split into 2 three-carbon compounds (sugar cleavage). Another reaction makes these three-carbon compounds identical. Another long series of internal reorganizations then converts each three-carbon unit into a molecule of the final product of glycolysis, **pyruvic acid** or **pyruvate**. In the course of these changes a molecule of NAD^+ picks up hydrogen

and electrons from each three-carbon unit; and at two points in the process the three-carbon compounds give up phosphate to ADP, making ATP.

Glycolysis is important to the plant partly because some of its intermediate products are useful in anabolism. Only about 17% of the energy of glucose is transferred to energy-carrying compounds. A little energy (3%) is lost as heat. And almost 80% of the energy still lies trapped in the final product, pyruvate.

The second phase of respiration, termed either the **citric acid cycle** or the **Krebs cycle** (after its discoverer, Sir Hans Krebs), breaks pyruvate down to carbon dioxide (Fig. 2.11). In this process the hydrogen that was present in the pyruvate molecules is transferred to hydrogen-carriers, chiefly NAD^+. Water molecules also enter, donating oxygen to help form CO_2 and giving up their

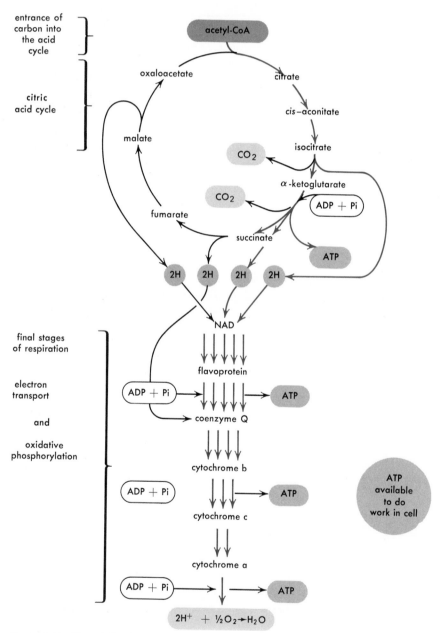

Figure 2.11 The relation between the citric acid cycle, electron transport, and oxidative phosphorylation.

hydrogen to NAD^+. There are two points in the process where organic intermediates pick up phosphate from the surrounding solution and transfer it to ADP, making ATP.

Pyruvic acid enters this phase of metabolism by attaching to a complex of enzymes. One of the three carbon atoms of pyruvate is cleaved off and released as a molecule of CO_2. The two-carbon remainder, still bound to the enzyme, is an **acetyl group**:

$$H - \overset{\displaystyle H}{\underset{\displaystyle H}{C}} - \overset{\displaystyle O}{C} - \quad \text{acetyl group}$$

The acetyl group is transferred to a mobile carrier compound that is called **coenzyme A**. The combination is highly reactive, and coenzyme A can transfer its acetyl group to several different compounds. Thus in anabolism acetyl-coenzyme A donates two-carbon units, which help to build more complex molecules. But its most frequent role is to feed acetyl groups into the citric acid cycle.

The cycle begins when an enzyme transfers the acetyl group from coenzyme A to a four-carbon acid called **oxaloacetate**. The product, with six-carbon atoms is **citric acid**. The rest of the citric acid cycle consists of a series of reactions that gradually remove all the atoms that were originally brought in as the acetyl group: the carbon and oxygen are released as CO_2; the H is transferred to carriers. The final reaction produces another oxaloacetate molecule. This makes the whole process appear as a cycle, though it also represents an open-ended flow of materials from pyruvate to other products.

Among the intermediates of the citric acid cycle are several compounds that are starting materials for pathways of anabolism. This phase of catabolism therefore is a material interface between catabolism and anabolism. It is commonly described as a "hub" of metabolism. Since pathways ending with lipids and proteins meet here, this cycle is the location where the breakdown of lipids and proteins feed into respiration.

The principal result of glycolysis and the citric acid cycle is the transfer of hydrogen from sugar to carriers such as NAD^+. The loaded hydrogen carriers are extremely reactive (rich in energy) and may donate H to many molecules that are being built during anabolism. But most of the H is consumed in the third phase of respiration, the **electron transport system**.

The electron transport system acts as an ATP generator, which harvests the energy of reactive hydrogen by letting it combine with oxygen. The system consists of catalysts that are grouped into many short chains. Each chain draws H from the loaded hydrogen carriers. The hydrogen nuclei are released early in the chain; they are picked up by water molecules to form H_3O^+. The electrons remain and pass along the chain of catalysts as if they were on a bucket brigade. Several of the catalysts in the electron transport chains have iron ions that are converted to the Fe^{+2} form when they take up an electron, and then return to the Fe^{+3} form as they pass the electron on to the next catalyst in the chain. The final catalyst is an enzyme that combines its electrons with H^+ (from the surrounding water) and O_2 to form water molecules. Oxygen has a very high affinity for electrons. Its position as the final electron acceptor explains why oxygen is necessary for most life.

In the electron transport process, two or three molecules of ATP are made from ADP and phosphate for each pair of electrons that travel through the chain. Since many charged hydrogen carriers were formed during earlier phases of respiration and each donates a pair of electrons to electron transport, this last phase of respiration is the place where the greatest abundance of ATP is produced: about 32 molecules produced for every glucose molecule consumed in respiration. In comparison, glycolysis and the citric acid cycle each produces only two molecules of ATP per glucose molecule consumed.

Overall, the complete breakdown of glucose yields six CO_2 and six H_2O molecules, absorbing six O_2 molecules and trapping about 40% of the energy of glucose in the form of some 36 molecules of ATP. The rest of the energy is released as heat during the various steps of respiration.

Alcoholic Fermentation

Most higher plants are not able to live for long periods of time in the absence of O_2. Without O_2, electron transport stops and ATP production is much reduced. In contrast, some fruits, notably apples, may live and produce CO_2 for long periods even if stored in an atmosphere with little O_2. Yeast may live and reproduce with very little O_2, producing large amounts of CO_2 and ethyl alcohol as wastes.

But yeasts have only a limited tolerance for alcohol. When the alcohol concentration in the environment reaches about 12%, the yeast cells become inactive. This is why natural wines do not have alcohol content above 12%. Yeast actually grows faster with an O_2 supply, but then it does not produce alcohol. Only a few bacteria grow better without O_2 present.

In higher plants that are deprived of O_2, the pyruvic acid formed during glycolysis is converted into carbon dioxide and ethyl alcohol. This also happens in yeast. This process, called **alcoholic fermentation**, occurs in two steps. First, an enzyme separates a CO_2 molecule from the pyruvic acid. This leaves a two-carbon compound, **acetaldehyde**. Next, the NADH that was formed in glycolysis donates its H to the acetaldehyde, forming alcohol. NAD^+ is released to take part in glycolysis again.

Thus, during growth without O_2, the energy trapped in NADH is used to form alcohol. For each glucose molecule fermented to alcohol a net of only two ATP molecules are formed. This is less than 3% of the energy available in glucose. About 6% of this energy is lost as heat, while almost 84% is still locked in the alcohol and is unavailable to the plant. In addition, the alcohol itself may be toxic. Fermentation is not an efficient way to extract food energy, compared to complete respiration.

The Control of Metabolism

In preceding sections we have outlined the basic structure of metabolism. We now turn to the control systems that balance the reaction rates within the pathways and that direct the pattern of productions.

Local Control Within Metabolism

Since metabolism comprises many pathways, its control is also complex. But several kinds of controls occur repeatedly at different places in the metabolic system and, taken together, they constitute the total control system.

Compartmentation is one important aspect of control: the enzymes that govern various pathways are often confined in organelles so that they are separated from the enzymes of other pathways by differentially permeable membranes. This confinement permits the membrane to exert control over the flow of molecules between pathways. Compartmentation is further discussed in Chapter 3.

Regulatory enzymes are found at many points in the network of metabolism, particularly where pathways branch off. Regulatory enzymes differ from other enzymes in that their rate of catalytic activity can be controlled by special small molecules known as **modulators**. Modulators act by attaching to the enzyme at a particular location known as the **allosteric site**. The attachment changes the shape of the enzyme and its ability to work with its normal substrate. Depending on the enzyme and the modulator, the attachment may either speed or slow the enzyme's action. A single regulatory enzyme may have sites to attach as many as a dozen different modulators. This gives the regulatory enzyme a fine degree of sensitivity to conditions in the surrounding plant body.

Often the final product of a pathway acts as a modulator to inhibit a regulatory enzyme at the start of the pathway. This relationship is called **feedback inhibition**. It provides a supply-and-demand control of production: if the product is formed faster than it is used, its concentration rises and the regulatory enzymes governing the pathway are inhibited. The attachment between modulator and enzyme is weak and reversible, so a later demand for more product, signaled by a drop in concentration of the modulator, releases the enzymes from inhibition.

To understand the operation of regulatory enzymes it is important to realize that many copies of each kind of enzyme are present in the system, and that some enzyme molecules may be free of modulators at the same time that other molecules of the same enzyme are bound and inhibited (or promoted). This makes the rate of the reaction smoothly adjustable in proportion to the levels of modulator concentrations.

Regulatory enzymes balance the rate of catabolism against the needs of anabolism. The anabolic processes consume ATP, converting it to ADP and phosphate ions. It has been found that one of the enzymes in glycolysis is regulatory, with ATP as an inhibitory modulator and ADP as a promotory modulator. If the rate of anabolism increases, the cell soon contains less ATP and more ADP. This change increases the rate at which the regulatory enzyme in glycolysis operates. The result is that glycolysis runs faster, feeding the other phases so that respiration as a whole is accelerated. ATP is therefore produced at a higher rate.

The Control of Enzyme Quantities

The rate of activity along a pathway may also be controlled by adding or subtracting enzyme molecules. Indeed, whole phases of metabolism may be completely turned on or off by this means. More than any other factor, the timed and controlled production of enzymes guides the transformation of a seed into an adult plant. It has been mentioned before that an almost unlimited variety of different proteins could theoretically be made, whereas real plants actually build only a few hundreds or thousands. The ability to build a particular set of proteins is a property that the plant has acquired from its ancestors. Biologists currently believe that the hereditary control of protein synthesis—both the kinds of proteins and the timing of their production—form the essential basis of all heredity. Therefore the subject of protein (including enzyme) synthesis and its control have primary importance in any study of life.

To control enzyme quantity, systems are needed for removing old enzyme molecules as well as building new ones. Plants often contain relatively nonspecific protein-degrading enzymes known as **proteases**. Proteases break down proteins more or less indiscriminately, so that accessible proteins must be continually replaced if the quantity present is to remain unchanged.

The control of enzyme synthesis is complex. The control system needs (1) a set of recipes, which spell out the amino acid sequences of each type of protein that the plant can build; (2) a reading system that can translate the recipes into protein molecules; and (3) a control system that instructs the reading apparatus on the specific recipes to use at each moment.

The Mechanism of Protein Synthesis

The recipes for proteins are technically known as **genes**. Recipes are stored in molecules of **deoxyribonucleic acid (DNA)**. Like proteins, DNA is a polymer. Its subunits are called **nucleotides** (Fig. 2.8). Each nucleotide of DNA consists of a molecule of the sugar deoxyribose to which is attached a phosphate group and a complex organic group known as a base. In the polymer, the phosphate of one nucleotide is linked to the sugar of the next. Thus the sugar and phosphate groups comprise a "backbone" of the polymer with the bases jutting out to the side. There are four kinds of bases in DNA: **adenine, guanine, cytosine,** and **thymine** (Fig. 2.12). These are often symbolized as A, G, C, and T. Just as an English sentence carries information in its sequence of letters, so also the DNA polymer carries information about proteins in its sequence of bases.

The structure of DNA is a little more complex because the DNA polymers combine in pairs, wound together as a **double helix** (Fig. 2.13). This pairing is maintained by hydrogen bonds that occur between the bases of the two polymers. For such a helix to be stable, the two DNA polymers must have a **complementary** sequence of bases. The required relationship is that wherever one polymer has an A, the other must have a T; and wherever one has a G, the other must have a C. Thus for a polymer

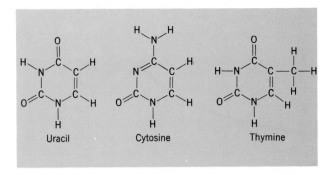

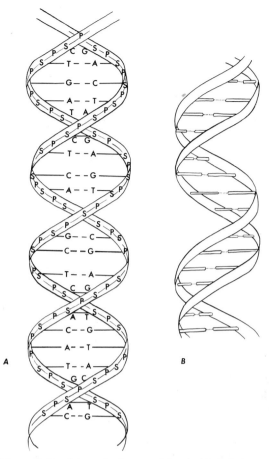

Figure 2.12   The bases in DNA and RNA.

Figure 2.13   Schematic diagram of the double helix in a portion of a DNA molecule. Alternating sugar, S (deoxyribose); and phosphate, P, groups make up the backbones of the strands. Attached to each sugar unit is one of the purine or pyrimidine bases, adenine, A; thymine, T; guanine, G; cytosine, C. The strands of the helix are held together through hydrogen bonds between the base pairs: adenine to thymine and guanine to cytosine. *A,* one strand is displaced along the axis of the helix for the convenience of diagramming; *B,* a diagram of the parallel strands in the helix.

with the sequence ACCTG, the complementary polymer would have the sequence TGGAC. The reason for this requirement is that the bases differ in size and shape; only the A–T and C–G pairs have the proper match of size and shape to fit comfortably within the helically folded arrangement. Note that when the two polymers are combined in the double helix, the important base sequence is hidden in the interior of the helix. In this form the genes cannot be read and the information in the DNA is not available for protein synthesis. To permit reading of the recipes, the DNA helix must unwind. Possibly the primary importance of the helical arrangement is that it protects the DNA and prevents untimely reading.

The *reading system* is a two-phase process, consisting of *transcription* followed by *translation*.

Transcription is the transfer of information from DNA to a similar substance, **RNA (ribonucleic acid)**. RNA is also built of nucleotides, but its sugar is ribose instead of deoxyribose; and in place of the base thymine, RNA carries the base **uridine**. Otherwise the two kinds of polymer are similar. Transcription works in the following way (Fig. 2.14). Under the influence of an enzyme called **RNA polymerase,** the DNA double helix uncoils at the region containing the recipe of the protein to be manufactured. The enzyme attaches to one of the two DNA polymers. Free subunits of RNA, which have been built by the metabolic system, attach to the bases of the DNA polymer according to the base pairing rules: where the DNA has an A, an RNA subunit bearing the base uracil (U) attaches by hydrogen bonds; where DNA has C, a subunit with G will attach; and so on. The enzyme moves along the DNA polymer and connects the RNA subunits to one another so that an RNA polymer results. This RNA polymer has a base sequence complementary to the DNA molecule, with the substitution of U for T. When the

enzyme reaches the end of the DNA segment that was to be read, the RNA molecule is released and the enzyme may detach and return for another act of transcription. The RNA molecule that results from this act of transcription may now be read by the protein-synthesizing machinery, as instructions for building a protein. Because of its intermediary role, this kind of RNA is termed **messenger RNA** or *m*RNA. The use of an intermediary information carrier presumably allows the organism to control its information more effectively; when no longer needed, the *m*RNA molecule may be destroyed, leaving the original DNA intact for possible later transcription.

Translation is a more complex process than transcription. Here the *m*RNA polymer is "read" by a complex unit called a **ribosome**, with a protein molecule as the product. The ribosome consists of some 50 different protein molecules together with three large segments of a special type of RNA known as ribosomal RNA. (Ribosomal RNA is produced by transcription from special regions of DNA by the same process that yields *m*RNA. However, ribosomal RNA is embedded in the ribosome where it presumably serves a structural role.)

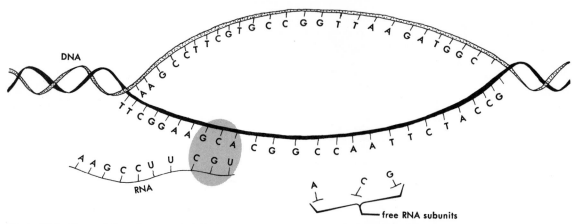

Figure 2.14 Transcription. A segment of DNA about to be read has uncoiled from the helix. The segment may contain genes for one or several proteins. The RNA polymerase enzyme, shown as a circle, has built part of an RNA molecule and is waiting for the next free subunit (which should carry the base guanine) to arrive.

Ribosomes need help to read the information carried by *m*RNA. Specifically, there is required a set of enzymes called amino acid activating enzymes and a set of RNA molecules of a third type, known as transfer RNA (*t*RNA). To understand how these components cooperate let us consider the nature of the "language" or "code" in which DNA and RNA carry information.

In English, a word consists of a group of letters that may be anywhere from one to perhaps 20 letters in length. The RNA language also has words, technically known as **codons**. All the codons have the same length, however. Each consists of three bases along the RNA molecule. The *m*RNA molecule acts as a string of three-base codons. Most of these codons represent the names of particular amino acids. With three bases per codon and four kinds of bases (A, C, G and U), there are 64 possible different codons. This is more than enough to name all the 20 kinds of amino acids. Some of the extras codons serve as punctuation, instructing the ribosome where to start and stop along the RNA message. Most of the extras are synonyms, however, for amino acids that have already been named by other codons. For example the amino acid isoleucine is represented by six different codons.

For each codon that names an amino acid, there is a unique, matching kind of **transfer RNA (*t*RNA)**. There are 61 kinds of *t*RNA. The *t*RNA's are all alike in that they are small compared to *m*RNA: each contains about 80 nucleotides each, whereas *m*RNA may contain thousands. Also, all the *t*RNA's fold into cloverleaf shapes (Fig. 2.15) that are stabilized by the same kind of base pairing used in the DNA double helix. Each *t*RNA molecule can carry its appropriate amino acid at one end. The other end of the *t*RNA molecule is specialized to bind with the appropriate *m*RNA codon. This binding to *m*RNA is achieved by means of a set of three bases, forming an **anticodon**, on the *t*RNA molecule. The anticodon is complementary to the codon. Thus a *t*RNA with the anticodon CAG would bind to the *m*RNA codon GUC.

The *t*RNA cannot pick up amino acids by itself. To do this, **amino acid activating enzymes** are needed. There are 20 of these enzymes, one for each kind of amino acid. Each enzyme has an active site that is shaped to accept only the correct amino acid and *t*RNA combination (Fig.

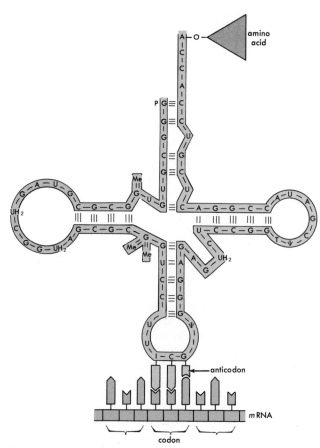

Figure 2.15 The structure of a *t*RNA molecule.

2.16). The enzyme uses energy from ATP to join the amino acids to the *t*RNA molecules. Once joined, the amino acids are much more reactive than free amino acids.

The next step in building proteins is for the "loaded" *t*RNA's to attach to the right *m*RNA codons. The ribosome seems to serve as a kind of "workbench" on which the reading of *m*RNA is completed (Fig. 2.17). The ribosome attaches to the *m*RNA molecule at a "start" codon, and attaches to a loaded *t*RNA that matches this codon. The

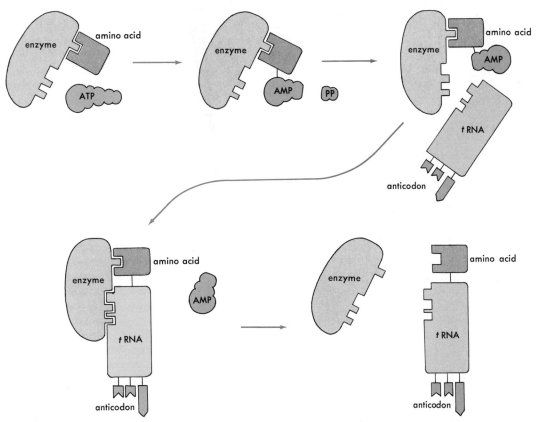

Figure 2.16 Diagram showing one manner in which *t*RNA could select a specific amino acid from an amino acid pool. Note that ATP and an enzyme are involved. The binding sites on the enzyme are specific for a certain amino acid and for the molecule of *t*RNA bearing the anticodon for that amino acid.

ribosome shifts its grip on the *m*RNA, moving to the next codon. Eventually another loaded *t*RNA arrives, matching this codon. The ribosome detaches the first amino acid from its *t*RNA and attaches it by a peptide bond to the newly arrived amino acid. The ribosome shifts its grip to the next codon and is ready to receive another loaded *t*RNA. This cycle of steps is repeated again and again with one amino acid after another being added to the growing protein until the ribosome reaches a stop codon. At this point the ribosome releases both the *m*RNA and the completed protein. The ribosome, the *m*RNA, and the *t*RNA may be reused many times to produce more protein molecules.

The Control of Protein Synthesis

The system that controls the reading of the genes is not well understood in plants. However, we have some knowledge of the controls in bacteria and these may give us an idea of the possible systems.

Control requires accuracy in selecting the right gene to be read at the right time. In bacteria, the task of locating the right gene falls to the RNA polymerase enzyme. The DNA molecule contains many genes arranged in a series. The points where transcription may begin are marked by special regions of DNA known as **promoter sites**. These sites presumably have special base sequences that permit the enzyme to bind to the intact DNA helix.

Whether an enzyme that has found a promoter site may produce *m*RNA depends on the presence or absence of a **repressor** protein that blocks the way. The repressor binds to the DNA at a site near the promoter called the **operator site**. Repressors are analogous to the locks on filing cabinets. They are similar to regulatory enzymes in that their shapes can be changed when special small signal molecules are attached to them. Depending on the particular repressor and signal molecule, attachment may either improve or decrease the repressor's ability to bind with DNA. Therefore the reading of genes is controlled by supplying or removing the signal molecules. Such molecules may be products of local metabolism, they may enter from the environment, or they may migrate from a distant point of origin elsewhere in the organism. These possibilities may give a basis for keying the production of enzymes to prevailing conditions, as well as coordinating development in various parts of the organism.

The control of gene reading by repressors and signal molecules is known as the **operon** mechanism. The extent to which this mechanism operates in higher plants is uncertain, and additional modes of control undoubtedly remain to be discovered.

Summary

1. The plant body is built and governed by a complex system of reactions called metabolism.
2. The plant metabolic system builds organic compounds from carbon dioxide, water, and minerals absorbed from the environment.

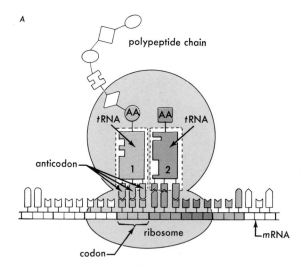

A

polypeptide chain

tRNA   AA   AA   tRNA

anticodon

1   2

ribosome

codon

mRNA

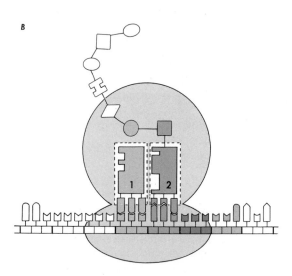

B

1   2

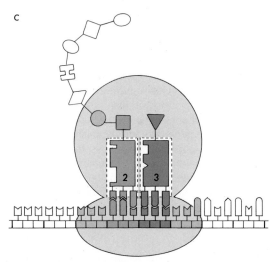

C

2   3

Figure 2.17 Steps in translation. In *A*, the ribosome binds an *m*RNA molecule with two codons in working positions. A partially completed protein is bound to *t*RNA 1, and another *t*RNA with its amino acid has just arrived at the second working site. *B*, as the next step, the ribosome shifts the protein from *t*RNA 1 to the newly arrived amino acid. *t*RNA 1 is now free to depart. *C*, the ribosome has moved along the *m*RNA molecule a distance of one codon, so that *t*RNA 2 is now in the left-hand working site. A third *t*RNA (3) with its amino acid has bound to the next codon. Step *B* can now take place again. This cycle of events occurs each time a new amino acid is added to the protein.

The catalysts are special proteins called enzymes.

5. Each enzyme catalyzes only a specific kind of reaction, so that the course of metabolism is determined by the selection of enzymes on hand.

6. Sequences of reactions are called metabolic pathways. The pathways branch and join as a network.

7. The three main phases of metabolism are photosynthesis, anabolism, and catabolism. Anabolic pathways build complex molecules; catabolic pathways break down fuels to produce building materials and provide energy for anabolism and transport processes.

8. Energy and hydrogen are carried from catabolism to anabolic reactions by recyclable carrier compounds such as ATP and NADH.

9. The principal events in catabolism comprise respiration, which is the oxidation of fuels such as sugar, fats, and amino acids to produce ATP.

10. Respiration consists of three blocks of reactions: glycolysis, the citric acid cycle, and the electron transport system. Many intermediate products formed in the first two phases are used as building blocks in anabolism.

11. In the absence of O_2, respiration cannot be completed, but glycolysis may continue with the terminal product, pyruvic acid, being converted to ethyl alcohol instead of feeding into the citric acid cycle. This is the process of alcoholic fermentation. It yields much less ATP than respiration.

12. The choice of metabolic pathways and their rates are controlled by mechanisms that include separation of enzymes in organelles, the action of regulatory enzymes that change their rate of action on signal, and the production and destruction of enzyme molecules. Some of these controls use feedback signals to pace reactions.

13. Protein (enzyme) synthesis uses the primary hereditary information of the organism as recipes for building particular proteins.

14. The hereditary information is permanently stored in the base sequences of DNA molecules. The first step in protein synthesis, termed transcription, consists in building disposable molecules of *m*RNA that have base sequences matching that of DNA.

15. Protein synthesis is completed by ribosomes (units built of RNA and protein) that assemble amino acids into proteins, using *m*RNA molecules as templates to determine the proper amino acid sequence. This process, termed translation, also involves many enzymes, *t*RNA molecules, and energy sources.

16. The control of protein synthesis by the operon mechanism, found mainly in bacteria, is based on blocking the process of transcription through the reversible attachment of proteins to the DNA molecule.

3. The most abundant compounds formed by plants are carbohydrates, lipids, amino acids, proteins, and nucleic acids.

4. Nearly all reactions in metabolism require catalysis.

the plant cell

Galen, the last of the great Greek doctors, who lived in Asia Minor during the second century A.D., thought that all organs, such as spleen, brain, and kidney of animals, leaves, stems and roots of plants, were a "sensible element, of similar parts all through, simple and uncompounded." Others before him had thought that animal tissues were simple coagulated "juices" seeping through the walls of the intestine. No one dreamed that these tissues and their plant counterparts had an astonishing and complicated structure, a structure that could not be seen until the invention of a microscope, which Zacharias Jansen, a spectacle maker of Holland, accomplished in 1590. This instrument was later improved upon by Robert Hooke, an Englishman, who was not only interested in optics but was also an architect and an experimenter with flying machines. Hooke, who lived from 1636 to 1703, examined all sorts of natural objects with his improved microscope. Among these were thin slices of cork (the dead outer bark of an oak). Figure 3.1 shows cork tissue as Hooke saw it under his microscope. This illustration was published in 1664 in an article entitled *Micrographia,* or *Some Physiological Descriptions of Minute Bodies Made by Magnifying Glasses.* The term **cell** was first used by Hooke to denote in cork the "little boxes or cells distinct from one another . . . that perfectly

enclosed air." He estimated that a cubic inch of cork would contain about 1259 million such cells.

Because of the prominence of cell walls in plant tissue, the cell was soon considered the unit of structure and of life in plants. However, during these early years, zoologists considered the tissues as the true centers of life in the animal body. There were supposed to be 21 different tissues, able to change within themselves, depending upon the organ in which they were located. In these tissues life was thought to reside. However, numerous observations upon protozoa and animal tissues led to a gradual accumulation of evidence that cells also existed in animals. In 1838 the German zoologist Theodor Schwann and the botanist Matthias Jacob Schleiden collaborated in a paper entitled, *Microscope Investigations on the Similarity of the Structure and Growth in Animals and Plants.*

This paper established on a firm basis the theory that the cell is a basic unit of structure in both plants and animals. However, another 30 years of research were necessary for the general acceptance of such a new idea: that all organisms are composed of cells, and that cells are indeed the basic units or building blocks of life.

This chapter is divided into three parts. The first examines plant cell structure, the second summarizes cell division and the cell cycle, and the third reviews the process of DNA replication.

Cell Structure

There are two large groups of organisms that differ in basic cellular structure. Bacteria and blue-green algae (perhaps more correctly, blue-green bacteria, Chapter 11) constitute one group called *prokaryotes.* Prokaryotic cells are considered to be primitive and are characterized by their lack of a membrane-enclosed nucleus and of membrane-bound organelles. The second group comprises all other plants and animals, the *eukaryotes.* Eukaryotic cells separate many of the metabolic processes of the cell into discrete membrane-bound, protoplasmic particles called **organelles.** Each organelle has a characteristic structure and function wherever it occurs. The organelles are the **nuclei** (Fig. 3.4), the **mitochondria** (Fig. 3.7), the **endoplasmic reticulum** (Fig. 3.11), the **dictyosomes** (Fig. 3.15), the **microbodies** (Fig. 3.18), and the **plastids** (Figs. 3.12 and 3.14). Although the **ribosome** is a small particle rather than a membrane-bound organelle, it does play an important role in protein

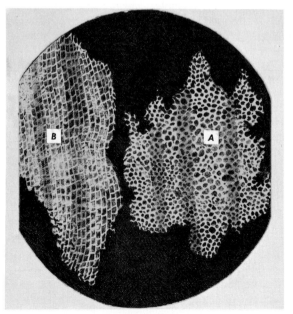

Figure 3.1 Cork tissue as Robert Hooke observed it under his microscope.

Table 3.1
Some Dimensions

	inch	cm	mm	μm	nm	Å
1 inch	1	2.54	25.4	25.4×10^3	25.4×10^6	25.4×10^7
1 cm	0.393	1	10	10^4	10^7	10^8
1 mm	0.393×10^{-1}	10^{-1}	1	10^3	10^6	10^7
1 μm	0.393×10^{-4}	10^{-4}	10^{-3}	1	10^3	10^4
1 nm	0.393×10^{-7}	10^{-7}	10^{-6}	10^{-3}	1	10
1 Å	0.393×10^{-8}	10^{-8}	10^{-7}	10^{-4}	10^{-1}	1

The symbol for micrometer (μm) now replaces the micron (μ); nanometer(nm) replaces millimicron (mμ).

synthesis and is included in this list. The organelles are bathed in ground **cytoplasm** through which diffusion of materials occurs, in which certain phases of metabolism take place, and in which the activities of the organelles are coordinated.

Eukaryotic plant and animal cells differ in several basic ways: (a) Plant cells specialized to carry out photosynthesis are provided with one to several organelles, the **chloroplast** (Fig. 3.12), in which reside most of the enzymes, energy compounds, and intermediate substances required for photosynthesis. (b) With few exceptions, plant cells are provided with a **cell wall** composed of **cellulose** and secondarily deposited materials that gives greater rigidity and lessened mobility to plants. (c) Most plant cells are provided with a large central aqueous **vacuole** (Fig. 3.6).

As you recall, Robert Hooke calculated that there would be 1259 million cells in a cubic inch of cork. This figure is based on his estimate of about 1100 cork cells along a line 1 inch long. One thousand cells along 1 inch would give us about 40 cells for the length of 1 mm; therefore a single cork cell is 1/40 mm in diameter. It now becomes convenient to use a smaller unit of measurement, the **micrometer** (micron) (μm). There are 1000 μm in 1 mm, and a cell 1/40 mm in diameter would be 25 μm in diameter. This is about correct for a cork cell, although many plant cells are larger. Table 3.1 lists other units of measure commonly used.

Technique

Living cells may be observed for many hours on the light microscope by maintaining them in a proper medium. Because the parts of the protoplast have an index of refraction close to that of water (i.e., the protoplasm bends light rays to the same degree that water does), many details of cell structure are not apparent in living cells. More detail can be observed by killing cells in various mixtures of alcohol, acetic acid, chromic acid, formaldehyde, or osmium tetroxide. The latter two compounds are of particular interest, because when they kill cells they do not coagulate the proteins. The killed tissue is dehydrated, embedded in paraffin, and cut into sections from 5 to 20 μm thick. When this tissue is mounted on a glass slide and properly stained, much detail becomes apparent.

Plants or plant tissues may be examined with a beam of electrons very much as they are examined with a light beam. However, electrons are not visible to the eye, and some device, such as activation of phosphors on a fluorescent screen, must be used to enable us to see the beam of electrons. There are two types of electron microscopes: transmission electron microscopes and scanning electron microscopes. In the first type, the beam passes through the specimen as in the compound microscope. In the second, the beam is reflected as with the stereodissecting microscope.

For transmission electron microscopy, small pieces of tissue are killed in glutaraldehyde and osmium tetroxide, then dehydrated in a graded series of alcohols and embedded in a plastic that must be held at 70°C for 24 hours to harden it properly. Sections of tissue and plastic, about 70 nm thick (0.07 μm), are stained with a heavy-metal stain such as lead citrate, which binds to membranes. The tissues are then observed with the electron microscope in a high vacuum. A beam of electrons is passed through the sections, and an image is formed on a fluorescent screen. Magnetic fields bend the beam of electrons just as glass bends light rays. Thus, the lenses in the electron microscope are electromagnets. The dark masses seen on the screen of the transmission electron microscope will not be images of organic matter but shadows of the high-atomic-weight metals, osmium from the fixative and lead from the stain, which have become preferentially bound to some components of the protoplasm.

For viewing with the scanning electron microscope, an object is mounted on a small metal plate and placed in a vacuum in the path of a beam of electrons. Living tissue may be examined directly; however, it rapidly desiccates in the vacuum and does not provide a good reflecting surface for the electrons. This means that plant parts to be studied with the scanning microscope must usually be killed, dehydrated carefully, with a final drying in liquid CO_2, and then shadowed or coated with a thin layer of gold. Thus prepared, the tissue has a normal appearance, does not deteriorate further, and has a better reflecting surface in the electron beam.

Structure and Function

Cells may be considered by referring to their (a) structure, (b) activity, or (c) chemical organization. It is necessary to know the interrelationships of all three levels to understand the cell as a whole. Within a cell, certain activities are confined to definite structures or organelles. Photosynthesis, for example, takes place in chloroplasts, while respiration is confined to mitochondria.

Cells are dynamic living units that are maintained in balance with their surroundings only through the expenditure of energy. The maintenance of a steady living state through the expenditure of energy is known as **homeostasis**. Disruption of the source of energy results in the death of the cell.

Cells are continually renewing their substance. To keep pace with the increasing number of cells, the total mass of protoplasm must also increase. Cells also accumulate materials to be stored and used in their own metabolism. For instance, leaf cells synthesize sugar from simpler substances, and store it in large amounts as starch.

The products, or accumulations, of protoplasmic activity (water, sugar, cellulose, hormones, and similar items) are found in the dynamic cooperating unit of protoplasm, called the **protoplast** (Figs. 3.2, 3.3*A,B*). The protoplast secretes about itself a **cell wall** which frequently later receives a deposition of secondary products. All of the protoplast, plus the cell wall, make up the **plant cell**.

The Living Cell

In a living leaf, such as that in the aquatic plant *Elodea*, the chloroplasts are the most prominent organelles. They are embedded in the **cytoplasm**, which is confined to the periphery of the cell, or to thin strands crossing the clear central vacuole (Figs. 3.2, 3.3*A,B*). The central vacuole contains water with dissolved salts, various organic solutes, and sometimes water-soluble pigments. The interfaces of the cytoplasm–vacuole and the cytoplasm–cell wall are characterized by special cytoplasmic membranes that exercise some control over the passage of substances into and out of the cytoplasm. These membranes are not visible with the light microscope but, because of their activity, their existence has long been recognized. The membrane separating cytoplasm and vacuole is known as the **tonoplast**; that bounding the outer surface of the cytoplasm is the **plasmalemma**. The cytoplasm and plasmalemma may be removed, leaving the tonoplast as a sac enclosing the vacuole. A single **nucleus** (Figs. 3.2, 3.3*B*) is embedded in the cytoplasm. In the living cell it generally appears structureless, although with careful focusing, one to four small denser bodies, the **nucleoli** (nucleolus, singular) can be seen (Fig. 3.2). There is also an envelope (membrane) around the nucleus.

The cytoplasm in a leaf cell, especially when warmed by the light of a microscope, will show active streaming, in which the organelles are carried. In time the velocity reaches 5 to 10 mm per min. This rapid flowing of the cytoplasm is a dramatic indication of its plasticity, and

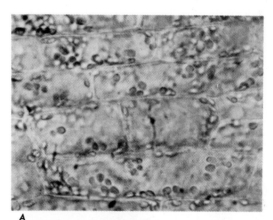

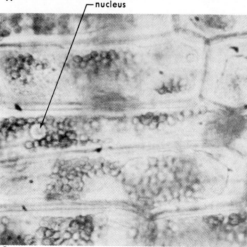

Figure 3.3   Leaf cells from the aquatic plant *Elodea*. *A*, living untreated cells; note the distribution of chloroplasts around sides, top, and bottom of the cell, with a vacuole occupying the rest of the cell. *B*, in cells recently killed with formaldehyde, the protoplast may withdraw slightly from the cell wall, chloroplasts may form irregular clumps, and the nucleus frequently becomes visible.

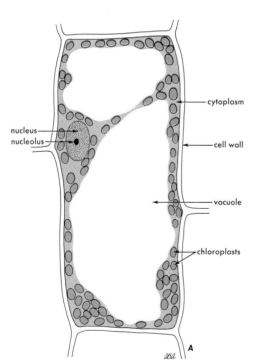

Figure 3.2   Diagram of a living cell from an *Elodea* leaf. When viewed on the light microscope, organelles in living cells, other than chloroplasts, lack contrast. The nucleus and nucleolus are generally visible. Mitochrondria are seen only with perfect lighting conditions, ×700.

should be recalled when studying microscope slides and electron micrographs.

Cell Fine Structure

A thin slice of a cell from a young leaf, as it appears in the transmission electron microscope, is shown in Fig. 3.4. Cells at this stage of development are not conducting photosynthesis at their potential capacity. These young cells tend to have relatively dense protoplasm, small vacuoles, and a large nucleus. The cells shown in Fig. 3.4 have several small chloroplasts, mitochondria, microbodies, dictyosomes, endoplasmic reticulum, and many small vesicles and ribosomes. This abundance of organelles indicates a high level of metabolic activity. Compare the "texture" of these cells to that of the young root cell in Fig. 3.5. Both of the cells in these two figures, from a young leaf and a young root, are highly active. The primary difference between them is that the root cell has colorless starch-storing plastids (amyloplasts) while the leaf cell contains green chloroplasts. Now examine Fig. 3.6, which was taken from a mature leaf of alfalfa. In comparison to the young leaf and root cells these mature cells are less "busy" in terms of their organelle complement; they also contain a large dominant central vacuole. Note that the nucleus and other organelles are pressed against the cell wall. The chloroplasts are also quite large and numerous.

Membranes

Perhaps the most notable thing about cells at this magnification is their compartmentalization into **membrane-bound organelles**. The protoplasm itself is bounded externally by the plasmalemma (Figs. 3.4, 3.5, 3.7) and internally by the tonoplast (Figs. 3.4, 3.4, 3.6). The plasmalemma separates the protoplast from the external cell wall or other nonliving systems. Within the protoplast, the tonoplast separates the protoplasm from the vacuole. In any organized living system membranes are needed to separate high-energy from low-energy regions.

Note that each of these membranes consists of a light line sandwiched between two dark lines. Such membranes are known as **unit membranes** (Fig. 3.7). They have been thought to be composed of a central bimolecular layer of lipid sandwiched between two monomolecular layers of fully extended protein (Figs. 3.7, 3.8A,B).

Two other membrane models have been recently proposed—the **subunit** model (Fig. 3.8C), and the **fluid mosaic** model (Fig. 3.8D). The two models are in some ways similar, both consisting of globular protein molecules partially or completely embedded in a lipid bilayer. The presence of protein actually within the membrane and not just on its surface helps to explain certain aspects of membrane permeability (Chapter 5).

Ribosomes

The cytoplasm, after glutaraldehyde and osmium fixation, is moderately electron transparent and has a soft gray granular appearance. In it always appears at least one type of densely stained granule, roughly angular and from 17 to 20 nm in diameter (Figs. 3.4, 3.7). The enzyme **RNAase** specifically degrades RNA. Treating tissue with RNAase results in the disappearance of these granules, with no change taking place in the other organelles. This is good evidence that RNA is present in these particles, the **ribosomes**. Ribosomes are present in plastids and

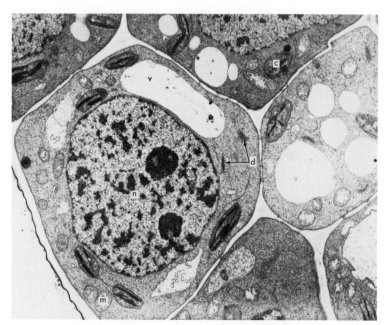

Figure 3.4 Immature cells from apex of sunflower **(Hellanthus annuus).** Note the small vacuoles (v), large nucleus (n) and other organelles—mitochondria (m), chloroplasts (c) and dictyosomes (d), ×6416.

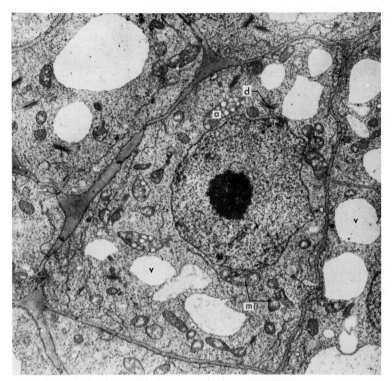

Figure 3.5 Immature cells from young root tip of *Gladiolus* sp.
Notice the similarities to the immature leaf cells in Fig. 3.4; small
vacuoles (v), amyloplasts (a), mitochondria (m), and dictyosomes
(d) are apparent, ×4583.

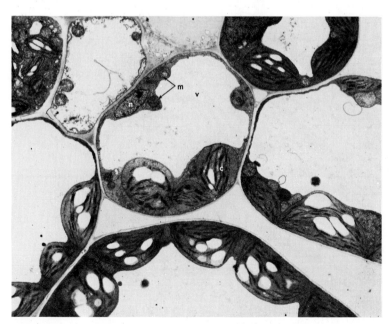

Figure 3.6 Mature cells from an alfalfa (*Medicago sativa*) leaf.
The vacuoles *(v)* are much larger in these old cells. Note also that
the nucleus (n) is proportionally smaller than in immature cells.
Chloroplasts (c) and mitochondria (m) are also abundant, ×4191.

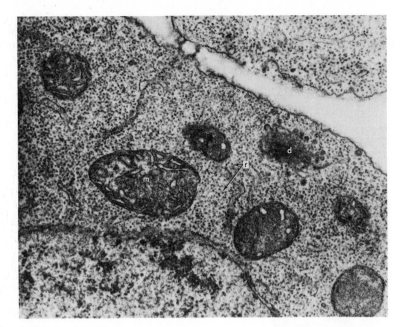

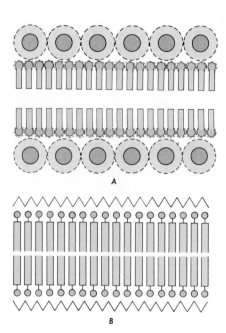

Figure 3.7 Enlarged view of mitochondria (m) in a corn root cell (*Zea mays*). The nucleus (n), a dictyosome (d), and numerous ribosomes (r) are also visible, ×25,333.

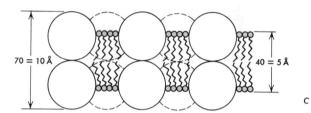

Figure 3.8 Membrane models. *A, B,* unit membrane; *C,* subunit model; *D,* fluid mosaic model.

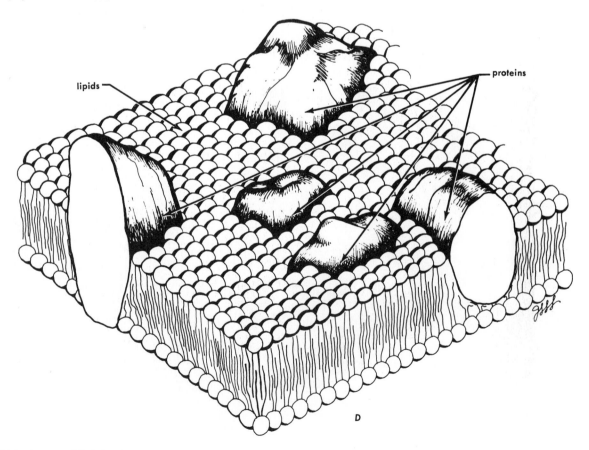

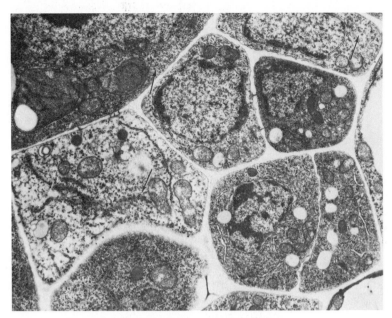

Figure 3.9 Electron microscope photograph of an immature wheat *(Triticum aestivum)* leaf. Numerous organelles are present. Note especially the coiled polyribosomes (arrows), ×6231.

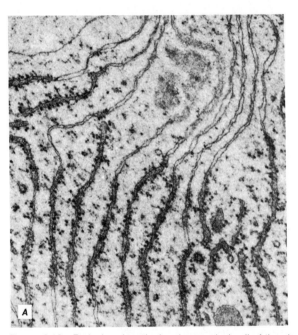

Figure 3.10 Endoplasmic reticulum in an apical cell of the alga *Chara*. Both rough and smooth endoplasmic reticulum are present and are continuous with each other. A microbody is also visible. ×15,000.

mitochondria, as well as in the cytoplasm. They appear to be lacking in nuclei. Some micrographs show them attached to the cytoplasmic side of the nuclear envelope. If they do occur within the nucleus, they are probably confined to the nucleolus (Fig. 3.5). Ribosomes are frequently associated with the cytoplasmic-membrane system, the endoplasmic reticulum (Figs. 3.9, 3.10). Under certain circumstances they appear in groups, frequently as helical aggregations. These arrays are designated as **polyribosomes** or simply **polysomes** (Fig. 3.9).

Organelles

Endoplasmic Reticulum

In cross section, the endoplasmic reticulum is represented by profiles of two parallel membranes separated by a narrow light space about 4 nm wide (Figs. 3.9, 3.10, 3.11). These profiles are actually cross sections through extensive, flattened vesicles. Note that these membranes form a closed system; their ends are never open to the ground cytoplasm. The endoplasmic reticulum frequently has ribosomes pressed to its outer, or cytoplasmic, surface. This is known as **rough endoplasmic reticulum**. **Smooth endoplasmic reticulum** lacks ribosomes and is not as common as rough endoplasmic reticulum (Fig. 3.10). There is considerable evidence suggesting that protein synthesis may be associated with the rough endoplasmic reticulum. For example, endoplasmic reticulum is most highly developed in cells active in protein synthesis. It is poorly developed in mature leaf cells (Fig. 3.6). The endoplasmic reticulum is associated with **plasmodesmata**, or cytoplasmic connections from cell to cell across cell walls (Figs. 3.4, 3.7, 3.9).

Mitochondria

The mitochondrial envelope separates internal mitochondrial space from the cytoplasm (Figs. 3.4, 3.7). The mitochondrial envelope is double and each component is a typical unit membrane. The outer component forms a continuous barrier around the mitochondrion. The inner membrane is thrown into folds called **cristae** (Figs. 3.7, 3.11). Internal to the cristae is a fluid called the **mitochondrial matrix**. Cristae in plant mitochondria frequently are sparse, irregularly arranged, and not always seemingly connected with the inner membrane of the envelope.

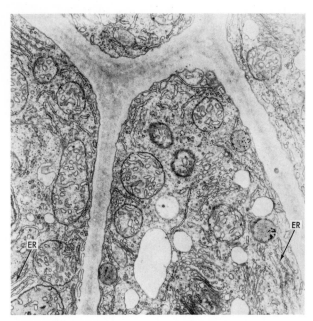

Figure 3.11   Meiocytes of *Selaginella* to show the endoplasmic reticulum (ER) and other membranes surrounding cellular organelles, ×16,250.

Plastids

The chloroplasts (Fig. 3.12) are bounded by an envelope of two membranes, just as are the mitochondria. The plastids have a granular background material, the **stroma**. Particles having the properties of cytoplasmic ribosomes may be seen in the stroma. Clear areas are also present, and in favorable sections, such clear areas contain an array of slender fibrils that represent chloroplastic DNA. Also present are closed membranous sacs called

thylakoids. Some of these occur in cylindrical stacks (Fig. 3.12) called **grana**. The grana are connected at irregular intervals by membranes called **frets**.

Kinds of Plastids.   Plastids respond to environmental conditions by reversibly changing their structure, contents, and function. Dark-grown seedlings, for example, are long and spindly and contain plastids with an elaborate crystallinelike structure usually called a prolamellar body (Fig. 3.13). In nongreen embryonic tissue and in roots are

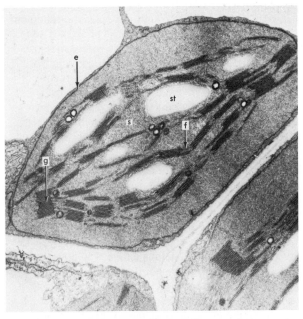

Figure 3.12   Chloroplast from alfalfa *(Medicago sativa)* leaf. The parts of the chloroplast, the grana (g), frets (f), stroma (s), stored starch (st) and surrounding envelope (e) are clearly shown, ×16,696.

chapter **3** | **The Plant Cell**

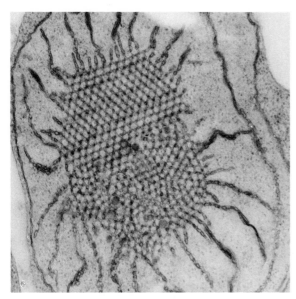

Figure 3.13 An etioplast from a yellow leaf of a dark-grown bean seedling *(Phaseolus vulgaris)*, ×20,000.

found small plastids that will later develop into normal plastids; these are called **proplastids**. Other colorless plastids, **leucoplasts**, are found in the epidermal cells of leaves, in onion, in storage tissue of apples, and in other white tissues. Some of these plastids store large amounts of starch and may be called **amyloplasts** (Figs. 3.12, 3.14).

As fruits ripen on trees and leaves prepare to fall in the end of summer, they change from green to red, orange, or yellow. This results from the destruction of chlorophyll, accompanied by an accumulation of yellow or red pigments known as the **carotenoids**. The plastids with a dominance of red and yellow pigments are **chromoplasts**. A chromoplast from ripe tomato is shown in Fig. 3.14 (right).

Dictyosomes

Dictyosomes are composed of from 5 to 15 circular flattened vesicles, or **cisternae**, aligned in stacks (Fig. 3.15). In cross sections, many small vesicles appear at the margins of the cisternae. However, face views reveal that the margins of each cisterna form a coarse net with true vesicles only at the extreme outer regions (Figs. 3.15, 3.16). The peripheral vesicles, in most cases, contain granules of secretory material elaborated in the dictyosome cisternae, possibly in collaboration with the endoplasmic reticulum. There is considerable evidence that the precursors of cell wall material are synthesized in the dictyosomes. The stack of cisternae appears to be polarized; that is, there is a forming face and a concave disappearing face that is somehow used up in the formation of the vesicles (Fig. 3.16).

Microtubules

Microtubules are hollow tubes of indefinite length and are approximately 28 nm in diameter (Fig. 3.17). Their function is not exactly understood, but it is connected to certain cellular events. The spindle apparatus, for example, is composed of bundles of microtubules and evidence indicates that they are involved in chromosome movements (Figs. 3.27, 3.28). Microtubules are also found in bands that spiral around cells and somehow regulate the pattern of secondary and primary cell wall formation (Fig. 3.30). They also occur in the cilia, or flagella, of motile cells of all eukaryotes and may, therefore, be involved in motility.

Microbodies

Preparations of different kinds of plant and animal tissues show a variety of spherical organelles that vary from about mitochondrial size to a size very much smaller. These organelles can be distinguished morphologically from mitochondria, since they have a single outer membrane rather than a double membrane envelope that is characteristic of the mitochondria. Cisternae and cristae are absent from microbodies, and the central area frequently appears rather dense under the electron microscope and may contain a variety of crystals (Fig. 3.18A).

Considerable research is being carried out to determine the exact chemical and physiological significance of these organelles. Evidence is accumulating that many different enzymes may be found in single-membrane-bound spherical organelles in both plant and animal cells. To some extent, the types of enzymes found depend on the tissue and kind of cells being studied. Thus, it appears that there is not just one kind of microbody but that the term "microbody" as defined above includes a fairly large number of different kinds of organelles that contain different enzymes and perform different functions in the cell.

Beginning students can appreciate some of the confusion in terminology when they realize that microbodies isolated from leaves and containing a complex of oxidative enzymes have been called peroxisomes (Fig. 3.18A); microbodies isolated from castor bean seeds and containing other oxidative enzymes have been called glyoxysomes (Fig. 3.18B).

Nucleus

Electron micrographs of nuclei show few details not already recognized by the light microscopist (Figs. 3.4, 3.5). There are one or more nucleoli and some denser areas embedded in a granular nucleoplasm. The denser areas may represent sections of chromosomes, but such a relationship has not yet been definitely demonstrated. High magnification sometimes suggests fibrils in sectioned nuclei.

The nuclear envelope is double and is provided with pores (Figs. 3.4, 3.6). Nuclear pores are evident in special preparations of frozen nuclei, when fractured and viewed through the electron microscope (Fig. 3.19). These pores appear as distinct holes, and although there is some evidence that rather large molecules can pass through them, not all investigators agree that they do serve as passageways.

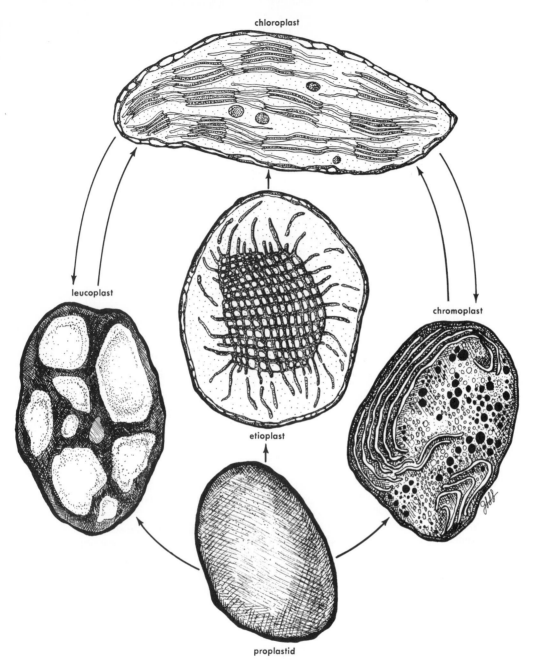

chloroplast

leucoplast

etioplast

chromoplast

proplastid

Figure 3.14 Diagram to show the interconversions of plastids; bottom center (proplastid), center (etioplast), top center (chloroplast), left (leucoplast) and right (chromoplast).

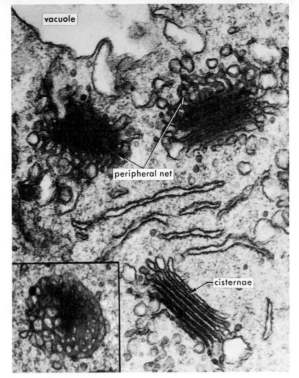

Figure 3.15 Four dictyosomes in a cell from the root tip of a water plant *(Hydrocharis)*. One dictyosome has been sectioned at right angles to its cisternae, and another one almost parallel to its cisternae. Note the peripheral net around this second dictyosome. The two other dictyosomes have been sectioned obliquely to the cisternae and some material appears to be passing from them to the vacuole, ×15,000.

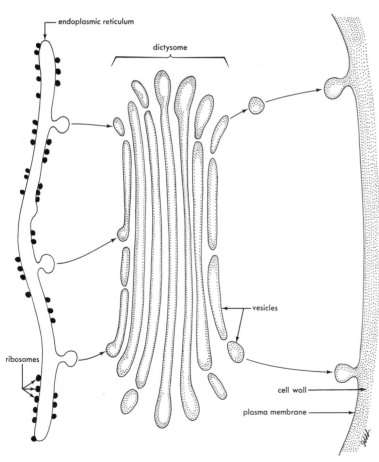

Figure 3.16 Diagram to show membrane flow from ER through a dictyosome to the plasmalemma.

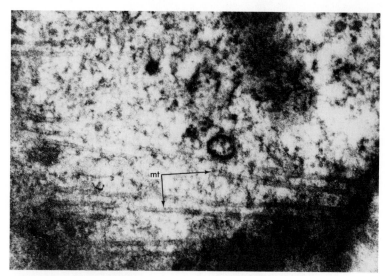

Figure 3.17 Microtubules (mt) are shown in a parallel array near the cell wall, ×62,480.

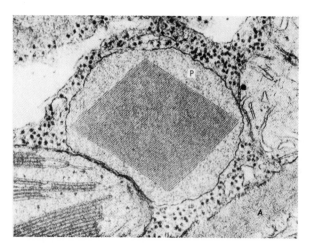

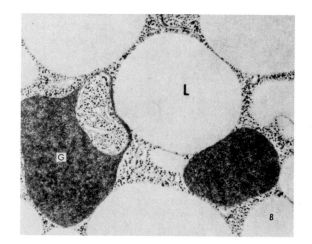

Figure 3.18 *A,* electronmicrograph of a type of microbody, the peroxisome (p), which is found in leaf cells, ×54,500. *B,* A second type of microbody, the glyoxysome (G), is found in seeds and is associated with lipid (L) digestion during seed germination, ×20,714.

Figure 3.19   Freeze-etch preparation of an onion root tip cell,
showing the organelles as they appear after being frozen, rather
than chemically fixed. The similarity between the chemical fixation
image and the frozen image is striking, ×10,000.

Nonprotoplasmic Portions of the Cell

Vacuoles

Vacuoles are definitely a part of the protoplast, but are involved in passive, or active, mechanisms to remove organic or inorganic materials from the protoplasm.

In many young cells the cytoplasm occupies much of the space in the cell; small vacuoles are, however, present (Fig. 3.20*A*). As the cell grows larger, the small vacuoles within the cytoplasm increase in size, coalesce, and become fewer in number (Figs. 3.20*B,C*). Finally, when the cell has attained its mature size, only a few large vacuoles, or even only one, may remain (Fig. 3.20*D*). The protoplasmic contents (nucleus and cytoplasm) of the cell lie compressed against the cell wall. The nucleus may occupy a position near the center of the cell, where it is connected with the cytoplasm around the cell wall by strands of cytoplasm (Fig. 3.20*D*).

Vacuoles do not contain pure water, but rather a dilute solution of many substances. This aqueous solution in the cell is termed **cell sap**. Among the substances dissolved in the cell sap are (a) atmospheric gases, including nitrogen, oxygen, and carbon dioxide; (b) inorganic salts, such as nitrates, sulfates, phosphates, and chlorides of potassium, sodium, calcium, iron, and magnesium; (c) organic acids, such as oxalic, citric, malic, tartaric; (d) salts of organic acids; (e) sugars, such as grape sugar (glucose) and cane sugar (sucrose); (f) water-soluble proteins, alkaloids, and certain pigments, such as anthocyanin. Generally, the cell sap is slightly acid. The concentration of cell sap varies from cell to cell, and it may vary in the same cell during the course of the cell's life.

The most common pigments of the vacuolar sap are the **anthocyanins**. These pigments are responsible for the red, purple, or blue of the petals of many flowers or of other parts of plants. The yellow coloration of poppy flowers is due to yellow vacuolar anthocyaninlike pigments called anthoxanthins. The red color of roots and leaves of garden beets is due to another vacuolar pigment called betacyanin.

The Cell Wall

The protoplast is surrounded by a plasmalemma or cytoplasmic membrane. Outside this membrane, and surrounding the entire protoplast, is a rigid wall synthesized by the protoplast that it encloses. Walls of adjacent cells are cemented together by an intercellular substance, the **middle lamella** (Fig. 3.21*A*), which is composed of pectin (polymers of various sugars, especially rich in partly oxidized galactose) and certain other substances. The first wall formed by the protoplast is the **primary wall**, made of cellulose and other carbohydrates. When the cell ages, the protoplast may deposit more wall material on the primary wall. Thus, a **secondary wall** (Fig. 3.21*D,E*) is formed, and the completely mature cell wall may have a thickness many times greater than the primary wall. In some tissues, the secondary wall is stratified and composed of several layers. In others, the cells do not lay down any secondary wall material. The secondary wall may be of cellulose or of

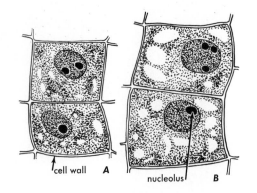

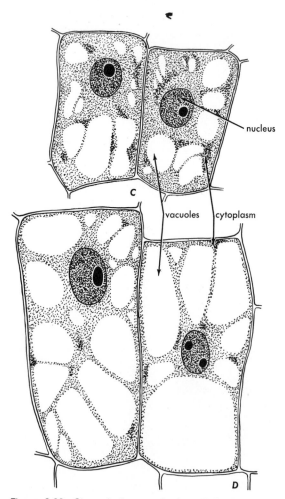

Figure 3.20 Stages in the growth of a cell. Progressively older cells shown from *A* through *D*, ×2000.

cellulose impregnated with other substances. Some of these substances, notably **lignin**, slows decay; others, like **suberin** and **cutin** (Figs. 3.21*B,C*), are waxy and protect leaves and stems against water loss. In addition, certain other materials may enter into the composition of the cell wall—gums, tannins, minerals, pigments, proteins, fats, and oils. It should be emphasized that in mature hard tissues, such as wood, lignin may be deposited not only in the secondary wall but also in the primary wall and middle lamella.

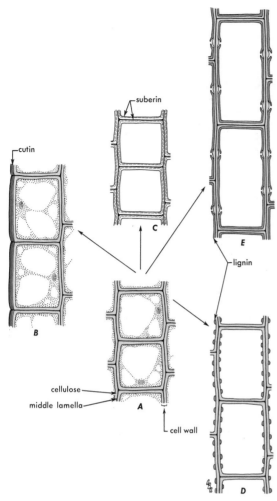

Figure 3.21 Development of cell walls during primary growth while stem is elongating. *A*, cell with cellulose walls. *B*, cutin may be laid down on the outside of an epidermal cell. *C*, in another location, suberin may be deposited within the cellulose primary cell wall to form a cork cell. Lignin is deposited within the primary wall and between the primary cell wall and the protoplast. *D*, during rapid elongation, it is laid down as rings or as a spiral band. *E*, when elongation stops (except for pits), the lignin forms a complete secondary wall around the cell.

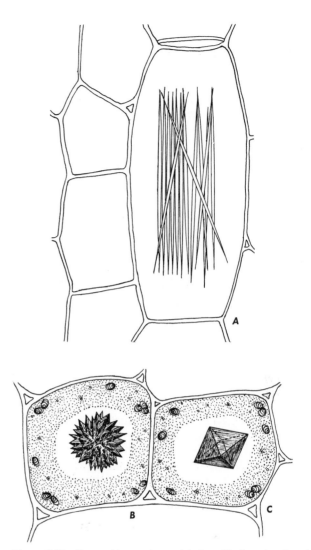

Figure 3.22 Types of inorganic crystals found in the vacuoles of living cells and in the cell walls of older nonliving cells. *A*, raphides; *B*, a cluster of crystals; *C*, a single crystal, ×2000.

Although the walls of cells vary considerably in composition in different species, and from one part to another in the same individual plant, **cellulose** constitutes the greatest percentage of the material of which most cell walls are made. It is synthesized by the protoplast and deposited across the plasmalemma by some mechanism.

Living cells are interconnected with each other through cytoplasmic channels called **plasmodesmata** (Fig. 3.7). The plasmalemma lines these small channels, and endoplasmic reticulum can often be observed within them.

Inorganic Crystals

Cells with crystals are found in almost all plants and in many different plant tissues. Crystals form within vacuoles and vary in chemical composition and in form (Fig. 3.22). The most common crystals are of calcium oxalate; it is generally held that they are an excretory product of the protoplast formed by the union of calcium and oxalic acid. This acid is soluble in cell sap and is toxic to the protoplasm if it attains a high concentration in the cell. By its union with calcium, the soluble oxalic acid is converted into the highly insoluble calcium oxalate, which will not injure the protoplasm. In addition to calcium oxalate, crystals of calcium sulfate, calcium carbonate, or protein are sometimes found.

Summary

1. A cell may be divided into protoplast and cell wall.
2. Protoplasm is divided into an array of particles, membranes, and organelles.
3. The plasmalemma is the membrane separating the protoplast from the cell wall, and the tonoplast is the membrane separating the vacuole from the cytoplasm.
4. At least three major theories have been proposed to explain membrane structure. The unit membrane model consists of a bilayer of protein with a middle layer of lipid. The subunit and fluid-mosaic models

propose that globular proteins are embedded in a lipid bilayer.

5. Ribosomes are small particles having a high RNA content. Polyribosomes are aggregates of ribosomes that are involved in protein synthesis.

6. The endoplasmic reticulum consists of an extensive array of flattened vesicles. It may or may not have ribosomes associated with the outer surface of the vesicles and appears to be involved in protein synthesis.

7. Mitochondria are small organelles about 0.5 μm in width and from 1 to 3 μm in length. They are bounded by a double membrane envelope. The inner component invaginates to form cristae, which are surrounded by a homogeneous matrix. Respiration is localized in the mitochondria.

8. Chloroplasts are approximately 5 $\times$ 10 μm in size. They are bounded by a double membrane envelope. The internal chloroplast lamellae aggregate to form cylindrical grana, which are connected by intergranal lamellae (frets). The photochemical reactions of photosynthesis take place on the membranes; the enzymatic reactions of photosynthesis are located in the stroma.

9. In addition to chloroplasts, cells may contain leucoplasts, amyloplasts, proplastids, and chromoplasts.

10. Dictyosomes are stacks of from 3 to 10 flattened cisternae. Each cisterna is surrounded by a peripheral net. Dictyosomes appear to be involved in the synthesis of various cellular products.

11. Microtubules are of indefinite length; they are about 28 nm in diameter, and have an internal core about 8 nm in diameter. They are known to be involved in chromosome movements, to possibly regulate the pattern of cell wall fibril deposition, and to somehow control cell motility.

12. Microbodies are all bounded by a single membrane. They are variable in size and in morphology. They are closely associated with various types of intracellular enzyme activities.

13. The nucleus is bounded by a double membrane envelope provided with pores. The nucleoplasm appears to be characterized by a closely packed array of unit fibers about 22.5 nm in diameter and of indefinite length. Cells may live and even differentiate for a short time without a nucleus; however a nucleus is required for the continued life of a cell and for cell division.

14. Vacuoles contain aqueous solutions within the protoplast, and are separated from the cytoplasm by the tonoplast. Inorganic crystals, when present, are located in vacuoles.

15. The cell wall, which bounds the protoplast, is formed of cellulose fibrils embedded in an amorphous matrix. Other compounds (suberin, pectin, cutin, and lignin) may be present. Plasmodesmata are cytoplasmic connections between cells through the primary cell wall.

16. The type of cell just summarized is highly compartmentalized. It is known as a eukaryotic cell. More primitive cells are not compartmentalized; DNA, photosynthetic processes, and respiratory activity all share a common cytoplasm. These are prokaryotic cells.

The Cell Cycle

Cells don't simply divide, but instead must progress through a sequence of four precise steps known as the cell cycle (Fig. 3.23). G1 refers to the period preceding DNA synthesis. During **G1** the cell accumulates the chemical energy and proteins needed before DNA synthesis can occur. For example, if an inhibitor of protein synthesis were added to a cell in G1, that cell would be also inhibited from entering DNA synthesis, which is the second step **(S)** in the cell cycle. G2 is a third step; it precedes nuclear division **(M)**. During G2 energy is stored that enables chromosome movement to take place. In addition, proteins such as those used to construct the microtubules, which make the spindle fibers, are accumulated. Inhibition of any of these events will mean that cells will be prevented from dividing.

Each plant species has a characteristic average amount of DNA per nucleus. Pea (*Pisum sativum*) plants, for example, have 7.9 pg (1 picogram = 10^{-9}g) DNA per nucleus; longpod bean (*Vicia faba*) plants have 24.3 pg per nucleus. Cytologists (scientists who study cells) have established the generalization that the more DNA a cell has in its nucleus, the longer it will take for that cell to progress through the cell cycle. As an example, pea cells take 14 hours to progress once through the cell cycle, and longpod beans take approximately 18 hours. Each stage of the cell cycle generally takes a characteristic proportion of the total cell cycle time. G1 usually is the most variable in length; mitosis is usually the shortest period. Table 3.2 gives the durations of different stages of the cell cycle in three plants and total cycle time (CT).

M, the last of the four stages of the cell cycle, refers to nuclear division, the actual separation of chromosome replicates to form derivative nuclei. This division may be

Table 3.2
Cell Cycle Durations

	DNA Amount (pg)	CT (hours)	G1	S	G2	M
Helianthus annus (sunflower)	6.3	7.8	1.2	4.5	1.5	0.6
Pisum sativum (pea)	7.9	14	5	4.5	3	1.2
Vicia faba (longpod bean)	24.3	18	4	9	3.5	1.9

Data from J. Van't Hof. 1974. The duration of chromosomal DNA synthesis, the mitotic cycle and meiosis in higher plants. In *Handbook of Genetics,* Vol. II, R. C. King (Ed.). Plenum, New York.

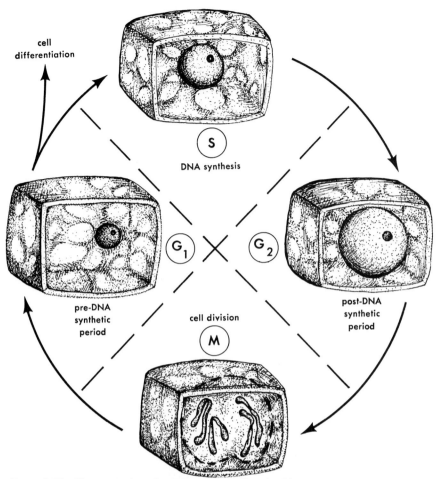

cell differentiation

S
DNA synthesis

G₁
pre-DNA synthetic period

G₂
post-DNA synthetic period

cell division

M

Figure 3.23   Diagram to show the stages of the cell cycle: G1 (pre-DNA synthesis phase), S (DNA synthesis phase), G2 (post-DNA synthesis phase), and M (mitosis).

either by **mitosis,** which results in two genetically identical derivative nuclei, or by **meiosis,** which results in the formation of four nuclei that have a reduced chromosome complement. Each body cell contains two identical sets of **homologous** chromosomes. One set is originally contributed by each parent; hence a cell before division is called a **diploid** cell and is said to have a **2n** chromosome number. A cell derived by meiosis, however, has a reduced chromosome complement and is said to be **haploid.** Haploid cells have a **1n** chromosome complement. Whether it is mitosis or meiosis that occurs depends on the cell type or tissue (group of cells) that is dividing. Each type of cell division will now be examined in detail.

Mitosis

Cell division is apparently required because a single nucleus with a full complement of genes can control only a small amount of cytoplasm. Increase in size and complexity of the plant body requires cell division. The resulting increase in the number of cells makes possible the specialization of tissues for the different structural and

physiological activities including absorption, conduction, reproduction, photosynthesis, and support.

The formation of new tissue cannot take place without cell division. During the period of division, both nucleus and cytoplasm divide. It is customary to designate the period of nuclear division as **mitosis** and to divide it into four phases: (a) **prophase,** (b) **metaphase,** (c) **anaphase,** and (d) **telophase.** The division of the cytoplasm is known as **cytokinesis.** Mitosis gives rise to derivative nuclei having identical gene complements. Cytokinesis gives rise to two new parcels of cytoplasm that are similar, but probably never identical (Fig. 3.24). The period of preparation for division is known as **interphase** and consists of the stages G1, S, and G2. The sequence of events taking place during interphase is equal in importance, in the complete process of the cell cycle, to the events that take place during mitosis and cytokinesis.

Prophase

The DNA strands in the interphase nucleus are long, slender, and seem tangled. The onset of mitosis is heralded by the presence of definite chromatin threads (Figs. 3.24A, 3.25A). These threads gradually shorten and thicken and become easier to see (Figs. 3.24B,C). They

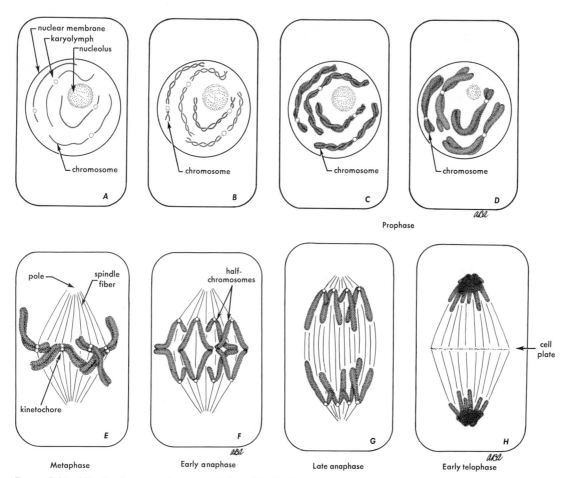

Prophase

Metaphase   Early anaphase   Late anaphase   Early telophase

Figure 3.24 Mitosis, diagrammatic representation; A to D, stages in prophase. Two pairs of chromosomes are represented; the "green" pair have a median kinetochore, while in the "gray" pair the kinetochore is close to one end. The chromosomes shorten and thicken, and chromatids become apparent, E, metaphase; F, early anaphase; G, late anaphase; H, telophase.

stain more heavily with certain dyes. For this reason they are called colored (chromo) bodies (soma), or **chromosomes**. Each chromosome consists of an individual strand of DNA. The nucleolus slowly decreases in size and finally disappears during this stage (Figs. 3.24A to D).

It becomes apparent that at late prophase each chromosome is composed not of one but of two threads coiled about each other (Fig. 3.24C). The nuclear membrane disappears toward the end of prophase.

Metaphase

Forces active within the cell now arrange the chromosomes, or at least a specialized portion of each chromosome (the **kinetochore** or **centromere**), in the equatorial plane of the cell (Figs. 3.24E, 3.25C,D). **Spindle fibers** are attached to the kinetochores. As the nucleoplasm elongates, the chromosomes, or at least the kinetochores, are moved to the equatorial plane of the cell (Figs. 3.24E, 3.25D, 3.26). Spindle fibers extend from the chromosomes to the opposite poles of the cells. Other fibers apparently reach to the poles but are not attached to the chromosomes. The electron microscope demonstrates that these spindle fibers are bundles of

microtubules (Fig. 3.26). These structures, the microtubules, or spindle fibers, plus any adherent nucleoplasm is called the **mitotic spindle**.

The chromosomes are now visibly composed of two closely associated halves, each half being known as a **chromatid** (Figs. 3.24E,F). In plants, such as corn, which have been intensively studied, each chromosome can be recognized and numbered. There are 20 chromosomes in corn, but only 10 different types that can be distinguished by their size and form. There are thus two chromosomes of each type. The 20 chromosomes of corn may be arranged in 10 pairs. Maps have been prepared of corn chromosomes showing the relative positions of the genes along them. Since the chromosomes split longitudinally, each gene is replicated and each chromatid contains a full set of genes.

Anaphase

The chromosomes do not remain long in the equatorial plane. The chromatids soon separate from each other and move to opposite poles of the cell. This period of separation of chromatids is **anaphase** (Figs. 3.24F,G, 3.25D,E).

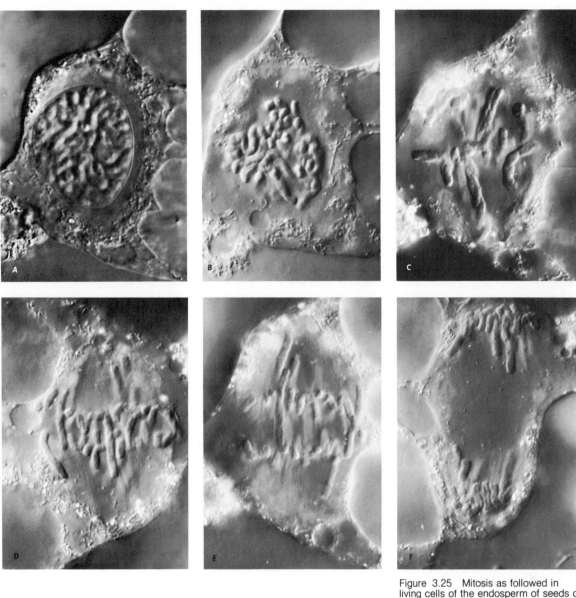

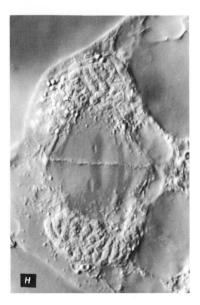

Figure 3.25 Mitosis as followed in living cells of the endosperm of seeds of the blood lily (Haemanthus katherinae). (Photographs taken with Nomarski optics.) A, early prophase; the nuclear membrane is still present. Note the clear zone of cytoplasm surrounding the nucleus. B, the nuclear envelope has disappeared, and the clear zone still surrounds the chromosomes. A few short spindle fibers are already present. C, full metaphase; the coiled chromatids are distinct. The two chromatids moving to the upper pole have separated and a few spindle fibers may be seen. D, early anaphase; the chromatids are moving to opposite poles of the cell. E, midanaphase; spindle fibers are in evidence between the chromosomes and the poles of the cell. F, telophase; the chromatids have aggregated at the opposite poles of the cell. A nuclear envelope has not yet formed. Time intervals after A, B, 14 min; C, 64 min; D, 74 min; E, 93 min; F, 107 min, ×2000. G, telophase; the chromosomes are tightly clustered at opposite poles of the cell, and the cell plate has started to form; ×4000, H, late telophase; the cell plate is almost continuous, and fibrils may be seen extending poleward a short distance from the plate; ×4000.

Telophase

When the divided chromosomes have reached the opposite poles of the cell, they group together and the nuclear membrane and the nucleolus again become apparent. This period is known as **telophase** (Figs. 3.24*H*, 3.25*F* to *H*). Cells in telophase will now begin to recycle through the cell cycle (G1–S–G2) (Fig. 3.23).

Polarity of Cell Division

If the spindles in a meristematic tissue were oriented at random, an irregular mass of tissue would result. This does occur when a single cell or small group of cells from a carrot root in tissue culture produces an irregular mass of cells called a **callus** (Fig. 3.27*A*). For orderly growth there must be a precise orientation of the mitotic spindles. A polarity is established so that, in general, the axes of spindles are parallel with each other and with the axis of the shoot or root (Fig. 3.27*B*). When the spindle axes are oriented parallel to the root-shoot axis, the divisions are called **anticlinal**. When the axes are perpendicular to the root-shoot axis, the divisions are called **periclinal** (Fig. 3.27*B*).

Polarity seems to be inherent in cells and in tissues of which they are a part. Changes in polarity may be induced by hormones. Occasionally, the orientation to be assumed by the spindle can be detected in plants at the cellular level before metaphase. In some instances, the cellular organelles will pass largely to one end of the interphase cell. According to some observations microtubules form a circular band at the equatorial plane of the cell during interphase, just before the onset of mitosis. These microtubules may be involved in establishing the location of cytokinesis, but we do not yet know how the placement of the microtubules is controlled.

Cytokinesis

In the great majority of cases, the division of the nucleus is followed by the division of the cytoplasm. Light microscopy has demonstrated that a cell plate forms at the equatorial plane of the cell at right angles to the spindle fibers (Figs. 3.25*G,H*, 3.28). From electron microscopy we know that the spindle fibers are groups of microtubules. At the equatorial region, the spindle fibers are surrounded by an amorphous material (Fig. 3.28*A*). Vesicles appear in this region and eventually fuse to form the first barrier dividing the derivative protoplasts (Fig. 3.28*B*). This structure, formed of pectin, is known as the **middle lamella**. The new cell wall is formed by the deposition of cellulose by each protoplast on its side of the middle lamella. The resulting structure is the **primary cell wall**.

The formation of the primary wall and, subsequently, the **secondary wall** poses an interesting problem of genetic control. The cellulose is laid down outside the protoplast, apparently resulting from the polymerization of many sugar molecules to form long, unbranched molecules of cellulose. The molecules are organized into a crystalline array by hydrogen bonds to form long unbranched **fibrils** that are visible in the electron microscope.

After removal of the amorphous materials from the wall, the array of fibrils may be studied with the electron microscope. When first deposited by the protoplast, the fibrils are in parallel array and form a band around the protoplast (Figs. 3.29*A*, 3.30). The mechanism bringing about this precise arrangement is not understood. However, the microtubules in the cytoplasm are also present in a similar parallel array. It is logical to postulate that the cytoplasmic microtubules may be involved in the orientation of the cellulose fibrils.

As the cell lengthens, the deposition of fibrils by the

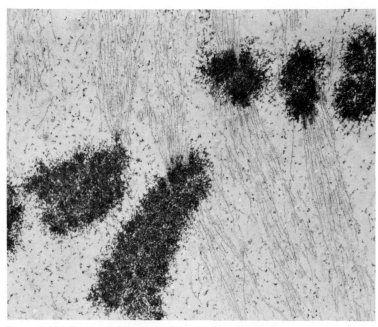

Figure 3.26   Electron micrograph of thin section of metaphase chromosomes in the endosperm of *Haemanthus katherinae*, showing the aligned chromosomes with microtubules (spindle fibers) attached to the kinetochore. Magnification ×40,000.

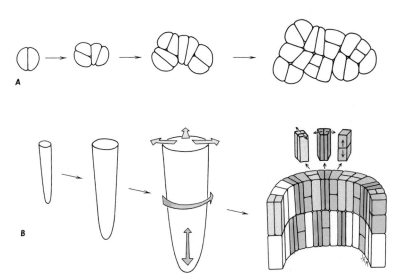

A

B

Figure 3.27   Influence of polarity on cell divisions and development of form. *A,* nonpolarized cell divisions result in an irregular mass of cells; *B,* cell division parallel with the axis of the root (gray arrow and cells) results in growth in length; cell division parallel with the circumference of the root (dark green arrow and cells) increases the circumference of the root; cell division at right angles to the circumference of the root (light green arrows and cells) increases the diameter of the root.

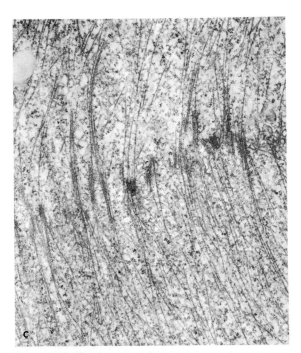

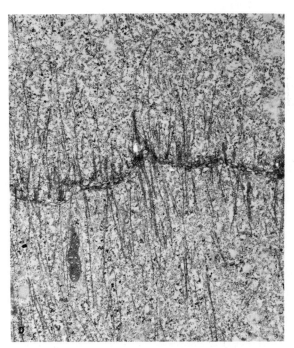

Figure 3.28   The formation of the cell plate in the endosperm of *Haemanthus katherinae. A,* early stage in plate formation; the microtubules are clustered into groups that are associated with a denser material, apparently in the plane of the future cell plate, ×26,000. *B,* late stage; the microtubules are fewer in number and are not clustered; vesicles are present and apparently fusing to form a continuous separation phase between the sister cells, ×26,000.

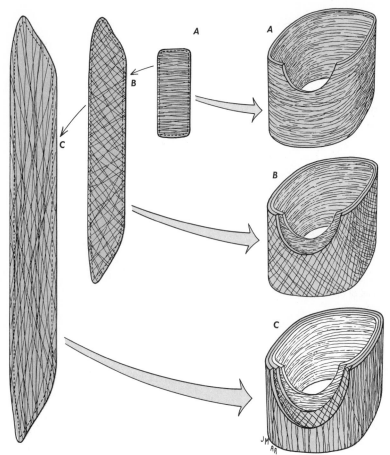

Figure 3.29 Diagram to show deposition and change in orientation of the cellulose fibrils in an elongating primary cell wall. *A, B,* and *C,* represent increasing age and length of the same cell. The cellulose fibrils are first deposited parallel to the circumference of the protoplast. They are then pulled out of this orientation as the cell wall elongates. (In the actual wall, the different stages of fibril orientation would grade into each other.) Green represents the earliest fibrils deposited, and light gray represents those most recently deposited.

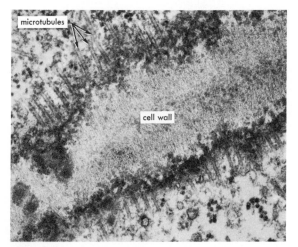

Figure 3.30 Microtubules in the root tip of *Arabidopsis thaliana.* The microtubules are oriented parallel to the circumference of the cell wall and thus parallel to the cellulose fibrils, ×50,000.

protoplast continues, with no change in the orientation of the newly deposited fibrils. As elongation continues, the fibrils, now further removed from the protoplast, assume a position more in line with the axis of the elongating cell.

A section through a primary cell wall after completion of elongation shows the oldest cellulose fibrils to be parallel to the long axis of the cell; the most recently deposited cellulose fibrils encircle the cell at right angles to the long axis (Fig. 3.29*A* to 3.29*C*). Intermediate fibrils have intermediate positions.

Meiosis

Meiosis is a type of nuclear division that is important during the life cycle of plants, because it involves a reduction in the chromosome number by one half. This process is significant in sexual life cycles because it offers a mechanism for genetic exchange and recombination during reproduction. The role of meiosis in plant life cycles is discussed further in Chapter 7 and in the chapters surveying the plant kingdom.

The division stages of chromosomes during meiosis are similar to the division stages of chromosomes undergoing mitosis. Meiosis begins with the chromosomes present as long, slender uncoiled DNA strands (Figs. 3.31*A,* 3.32*A*). They proceed to shorten and thicken and split into chromatids as in mitosis (Fig. 3.32*C*). The chromatids separate at anaphase. In meiosis (but not in mitosis), there is a *pairing* of *homologous chromosomes* so that four

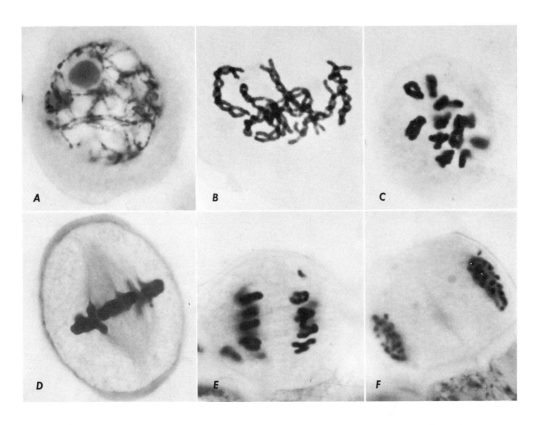

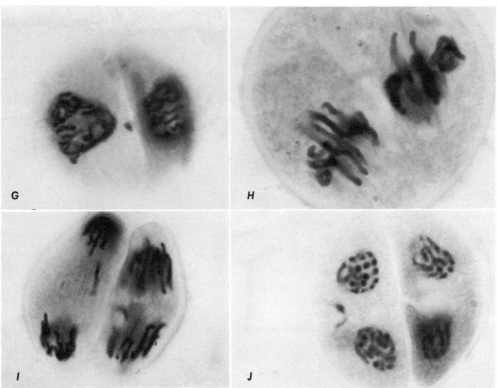

Figure 3.31 Meiosis in lily anther. Photomicrographs showing:
A, early prophase I. Note paired threads. B, late prophase I, each
body represents two paired chromosomes; note chiasmata. C,
late prophase I. Paired chromosomes; two chiasmata present.
D, metaphase I. Note spindle fibers extending from chromosomes.
E, anaphase I. F, telophase I. G, prophase II. H, metaphase II. I,
anaphase II. J, telophase II.

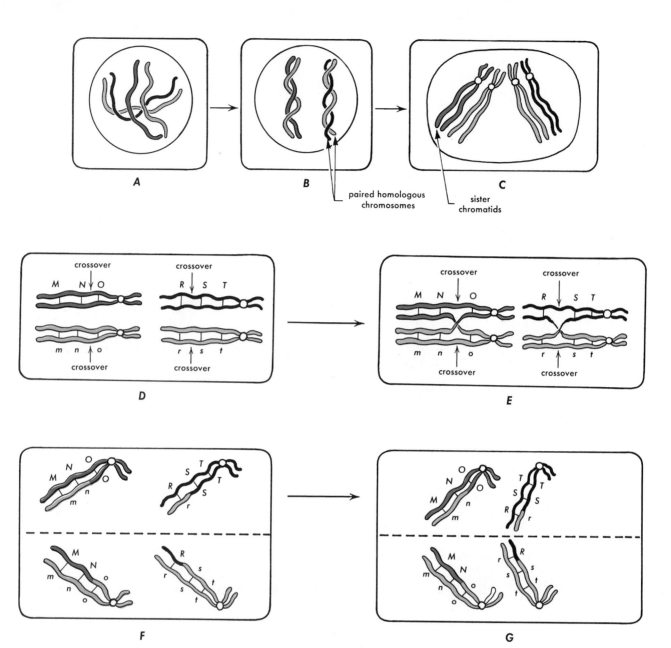

Figure 3.32 Diagram of meiosis, two pairs of homologous chromosomes are shown. A, early prophase with unpaired chromosomes. B, the homologous chromosomes have paired. C, the chromatids have formed. Those of a single chromosome are sister chromatids. Here and in the following diagrams they are represented as paired parallel lines (this is not the normal situation). D, markers on the chromosome, future location of crossover indicated by arrows. E, metaphase I. Nonsister chromatids of homologous chromosomes have broken and, at the crossover region, have exchanged pieces with each other. F, early anaphase I. Kinetochores separate; sister chromatids are still associated at the kinetochore but, because of the crossover, nonsister chromatids are associated distally to the kinetochore. G, early telophase I. The chromosomes start to regroup into a nucleus; note that there is only one member of a homologous pair represented. H, interphase. Two sister chromatids attached at the kinetochores. I, early anaphase II. The chromosomes have moved to the metaphase II plate and the kinetochores have split. Here chromatids are moving to opposite poles of the cell. Note the disposition of the markers. J, early telophase II. K, four meiospores are formed. L, meiospores that would arise if the crossover had occurred distally to the third marker or in the short arm of the chromosome; the three markers in each chromosome would have remained linked in their original order.

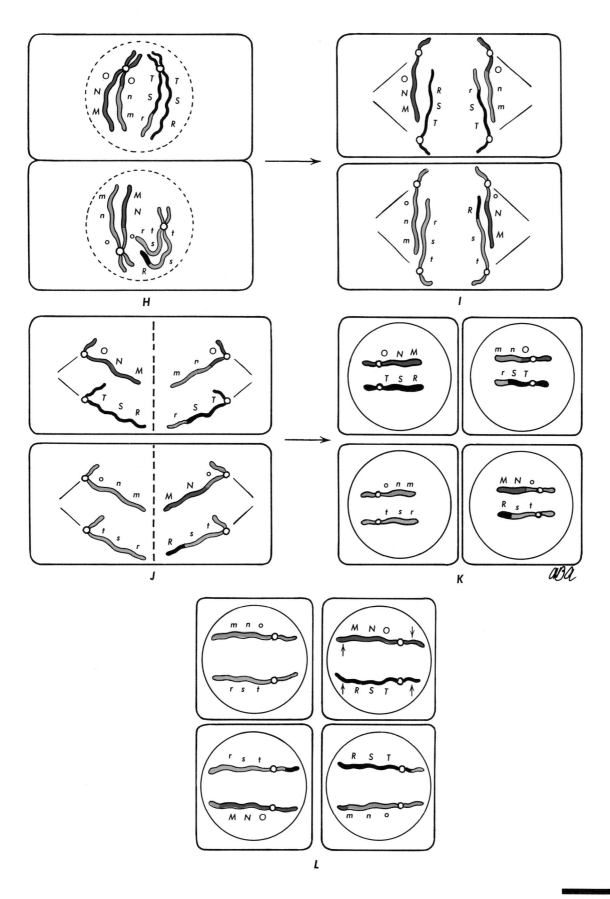

H

I

J

K

L

chromatids, two derived from each of the pairing homologous chromosomes, are associated at mid-prophase (Figs. 3.32B,C). This introduces a complication, because the chromatids of paired homologous chromosomes are able to exchange partners with each other by tangling their chromosome arms and exchanging segments (Fig. 3.32E). Such exchanges between chromatids, derived from opposite members of a homologous pair of chromosomes (nonsister chromatids) (Fig. 3.32F), may later be expressed as visible changes in the offspring.

In order for meiosis to take place, there are two sequential divisions that result in four cells from a single initial cell. However, DNA replication occurs only during the interphase preceding the first division. Each of the four resulting cells, therefore, has only one complete set of chromosomes. Thus, during meiosis the two sets of homologous chromosomes found in a diploid parent are reduced to a single set of chromosomes, and the cell or plant bearing this single set is said to be haploid.

Since two divisions are required for the completion of meiosis, it is common to designate the two divisions by the numerals I and II and the phases as prophase I, metaphase I, anaphase I, telophase I, prophase II, metaphase II, anaphase II, and telophase II.

Meiosis—First Division

Prophase I

During prophase I of meiosis (Figs. 3.31A, 3.32A,B,C) the chromatin contracts to form chromosomes, each with two chromatids; in this respect the process resembles mitosis. The process is complicated, however, because before the chromatids become apparent, the homologous chromosomes pair.

In so doing, they approach and coil about each other (Figs. 3.31B, 3.32B). The two chromatids of a chromosome are visible only after the pairing of the homologous chromosomes is well-advanced (Fig. 3.32C). The resulting figure is composed of two paired homologous chromosomes, with four chromatids (Figs. 3.31B, 3.32C). While they are paired, chromatids frequently break at several points and rejoin in such a way that a given reconstituted chromatid may be composed of parts of four chromatids (Fig. 3.32C). This breaking and rejoining of the chromatids is called **crossing over**, and the crossover formed by the chromatids involved in the interchange is known as a **chiasma** (chiasmata, plural) (Fig. 3.32E). Chiasmata may occur at any point along the paired chromosomes. Figure 3.32D represents two pairs of homologous chromosomes, each with three pairs of genes to serve as markers; *Mm, Nn,* and *Oo* on the green homologs and *Rr, Ss,* and *Tt* on the gray homologs. The symbols *Mm, Nn,* etc. are used to show the hypothetical positions, on homologous chromosomes, of gene loci. Each of these letters refers to a gene on one homologous chromosome; for instance, *M* is an **allele** for the corresponding gene *m* on the other homolog. The two together, *Mm,* are called an **allelic pair.** If chiasmata occurred between *Oo* and the kinetochore, or to the left of

Mm, the relationship between these three pairs would not change: *M, N,* and *O* would remain linked together as would *m, n,* and *o* on the chromatids of the original chromosomes. However, in Fig. 3.32E chiasmata are shown between *Nn* and *Oo* on the green chromosome and between *Rr* and *Ss* on the gray chromosome.

Thus, a third step in the prophase I of meiosis is (c) the formation of chiasmata resulting from the breaking and rejoining of chromatids from homologous chromosomes. It is normal for some chiasmata to form; however, some chromatids produce no chiasmata.

As in mitosis, the nucleolus disappears. The nuclear membrane also disappears during the later stages of prophase I (Figs. 3.31B, 3.32C).

Metaphase I

At metaphase I (Figs. 3.32C,D), the kinetochores of the paired homologous chromosomes pass to the equator of the cell.

Anaphase I

In anaphase I, the kinetochores of *whole chromosomes* separate from each other and, with their associated chromosomes, move to opposite poles of the cell. However, because of chiasmata formation, whole, complete chromosomes are not separated from each other, for on the side of the chiasma away from the kinetochore, sister chromatids will separate just as in mitosis (Fig. 3.32F). Anaphase I of meiosis is thus the separation of *two chromatids.* Since the chromatids have become variously modified, the separation involves both a separation of whole chromosomes (at the kinetochores) and a separation of sister chromatids as in mitosis (across the chiasma, Fig. 3.32F).

Telophase I

Following anaphase I, the chromatids group together at opposite poles of the cell (Figs. 3.31F, 3.32G) and immediately prepare for the second meiotic division. Each telophase chromosome consists, as usual, of two chromatids. In this sense, the **1n,** or haploid number of chromosomes, is present. However, it must be remembered that some of these chromosomes consist of two chromatids derived from the same parent (sister chromatids), and others consist of two chromatids derived from each of the two parents (nonsister chromatids).

Meiosis—Second Division

There now occurs a second meiotic division that separates these rearranged chromatids. During prophase II (Figs. 3.31G, 3.32H), chromosomes again form, each with two composite chromatids. The kinetochores approach the equatorial plate, forming metaphase II (Figs. 3.31H, 3.32I). As in mitosis, the kinetochores now split and separate. In anaphase II, single chromatids move to opposite poles of the cell (Figs. 3.31I, 3.32J) and are reconstituted in telophase II into nuclei (Fig. 3.32K). Each nucleus formed in this way contains one of the four chromatids derived from the pairing of homologous chromosomes in prophase (Fig. 3.32L).

Walls develop about each new nucleus and associated cytoplasm, thus forming cells with the **1n** or haploid number of chromosomes. Since the chromatids within each of these four cells have been variously modified by crossing over, each of the four cells may be genetically different. In what specific way do they differ and of what importance is this variance in the life cycle of plants?

Recombination of Genes

The distribution of genes during meiosis depends on the separation of the chromatids, and on the location and the amount of interchange taking place between the chromatids. If, for instance, as in Fig. 3.32D, the interchange took place to the left of *Mm*, then *M, N,* and *O* would move with the kinetochore at anaphase I, and the two telophase I nuclei would contain the markers *M, N, O* and *m, n, o*; these nuclei would remain linked together so that only two types of meiospores *M, N, O* and *m, n, o* (Fig. 3.32L) would result. The same would happen if the exchange occurred to the left of *Rr. T, S,* and *R* would remain linked, as would *t, s,* and *r*. The number of different kinds of meiospores to be derived from one to several linked allelic pairs is two or four, depending on whether crossing over has occurred between the alleles. Without crossing over, two of the meiospores would be similar; with *crossing over*, all four meiospores would be different. The proportion of meiospores having crossover chromosomes in a large number of meiotic divisions will depend on the number and location of interchanges.

Now, consider the distribution of the genes, *Oo* and *Tt*, which are on nonhomologous chromosomes. Each of the four meiospores are different in the example shown; they are *OT, Ot, oT,* and *ot*.

Combining segregation of chromatids during meiosis with chromatid interchanges, or crossing over, brings about variation of the genetic constitution of the resulting meiospores or gametes. In either case, gametes of different chromosomal makeup will ultimately fuse to produce a new diploid plant that may be quite genetically different from the parent. Thus, meiosis greatly enhances the possibility for variation, and results in *genetic recombination* from generation to generation.

Summary of the Cell Cycle

1. The cell cycle consists of G1–S–G2–M. G1 and G2 are important, because it is during this time that cells prepare for DNA synthesis (S) and nuclear division (M).
2. Cell division increases the numbers of cells.
3. All cells have the potentiality to divide but are normally blocked from doing so.
4. Nuclear division is called mitosis; cytoplasmic division is called cytokinesis.
5. The derivative nuclei are genetically identical.
6. The derivative parcels of cytoplasm may be unlike each other, but are usually similar.
7. In general, mitotic spindles are oriented parallel to each other and to the axis of the shoot. A definite, predetermined polarity produces this orientation.
8. The stages of mitosis are prophase, metaphase, anaphase, and telophase.
9. The spindle, which somehow directs chromosome movement, is constructed from many microtubules.
10. The middle lamella is the first layer to separate the two derivative protoplasts.
11. Vesicles collect around the microtubules in the equatorial plane of the cell. They fuse to form the middle lamella. Cellulose formed by the two daughter protoplasts and deposited on the middle lamella forms the primary wall.
12. Meiosis involves two divisions with only one period of DNA replication.
13. Meiosis is different from mitosis in that during prophase I homologous pairs of chromosomes join together. After both divisions each new haploid cell contains one half the number of chromosomes of its parent cell.
14. Meiosis is important because its mechanisms for recombination of genes allow variation in the genes of resulting progeny.

DNA and its Replication

Deoxyribose nucleic acid, DNA, is the relatively simple molecule in which, it appears, is stored all the information needed for the development of the several hundred thousand plant species, past and present.

There may be variations in the type and distribution of the nitrogen bases in DNA from different species, but there is no reason to suppose that there has ever been any fundamentally different kind of carrier of genetic information than the DNA known today.

Enzymes regulate metabolic reactions in the cell. Each different kind of enzyme is a different protein and regulates a different reaction in the metabolism, growth, and development of the cell. Since each kind of protein has its own exact sequence of amino acids, the way in which the cell regulates the sequence of amino acid in proteins is very important (Chapter 2). We now know that the basis of genetic information is the sequence of bases that each part of the DNA molecule contains. Precise duplication of DNA insures that each new cell contains the genetic information needed for all of its potential functions.

Replication of DNA

We shall now consider the manner in which DNA itself is replicated. From each double helix (Fig. 2.13) two new double helices must form, without at the same time, disturbing the precise ordering of the base pairs (Fig. 2.12). The dividing double helix may be several million nanometers long and, since a single base pair occupies the space of only a few nanometers, there are many base pairs along a double helix.

Precise information telling us the exact time and location of the synthesis of new DNA may be obtained by using radioactive precursors of nucleic acid. A commonly used substance of this type is radioactive thymidine. It seems to enter cells and to be directly incorporated into nuclei of dividing cells of both plants and animals. Root tips of

Tradescantia (a common house plant) grown in a culture solution containing radioactive thymidine (a form of the DNA base thymine) show that interphase may be divided into three periods (the cell cycle). During the first stages the nucleus increases in volume, but there seems to be no incorporation of radioactive thymidine; this is G1. Later, radioactive material is incorporated into newly synthesized DNA (s). A period follows in which the nucleus maintains a constant size without any further incorporation of labeled thymidine (G2). Refer back to Table 3.2 to reexamine the time it takes for a cell to transverse each stage of the cell cycle.

But these observations do not tell us the manner of replication of the double-stranded DNA helix. It is possible that a whole new double-stranded helix might be replicated, thus totally conserving the old double helix intact. Or the coiled double strands could somehow separate and a new strand could be formed by each old strand. This is semiconservative; half of the new helix contains old DNA. Or the replication could take place at random throughout the DNA molecule.

Careful experiments conducted by Dr. J. Herbert Taylor when he was at Columbia University, demonstrated that DNA replication was, in fact, semiconservative. The mechanism for this replication can be explained in the following way.

Assume that a double helix is present in early interphase at the onset of replication. This helix may untwist, separating the two strands (Fig. 3.33). Each strand now serves as a template for the formation of a new strand. Thus, following replication with radioactive thymidine available, each chromatid will contain a strand bearing the radioactive material. In a subsequent division in the absence of radioactive thymidine, the radioactive strand will serve as a template for a new nonradioactive strand as well as the old nonradioactive strand.

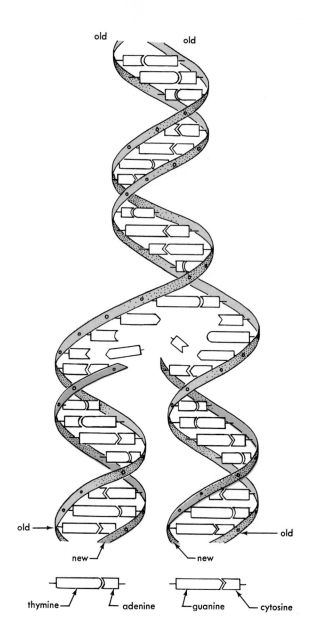

Figure 3.33 Schematic diagram showing how one "old" double-stranded molecule of DNA could be replicated into two new double-stranded molecules of DNA. Each new molecule has one old and one newly synthesized strand.

4 CHAPTER

the plant body

The vegetative plant body consists of three organs—stems, leaves and roots. In this chapter we will examine their external form (**morphology**) and internal structure (**anatomy**). One serious problem in presenting this material is that photographs and drawings are two-dimensional, and words are often inadequate to explain the pattern and symmetry of plant parts. So, while reading these sections, you should visualize the plant's internal structure in three-dimensions. Try to use your "mind's eye" as a kind of X-ray vision to see the length, width, and breadth of cells and tissues and their interconnections. It will become apparent, as you study plant structure, that all organs are integrated functionally and structurally. This means that the entire plant acts as one unit, and is not simply an aggregate of isolated parts.

PART ONE

Stems

Stems provide *mechanical support* for leaves in erect plants, and are an axis for attached leaves in horizontal plants. Flowers and fruits are also produced in positions on stems that allow for pollination and seed dispersal.

Stems provide a pathway for the *conduction* of water and mineral nutrients from roots to leaves, and for transfer of foods, hormones, and other metabolites from one part of the plant to another.

The normal life span of plant cells is from 1 to 3 years. Water and mineral salts in dilute solution move in dead cells, but this movement depends upon the activity of living cells in leaves and roots that are generally less than three years old. Stems in herbaceous perennials (Fig. 4.1A) and in 2000-year-old redwoods, or bristlecone pines (Fig. 4.1B), annually *provide new living tissue* for normal metabolism of the plant. Other stems are modified for the *storage* of plant products.

Stems thus have four major functions: (a) support; (b) conduction; (c) the production of new living tissue; and (d) storage.

Stem Morphology

The organs of plants can usually be distinguished from one another by their external shape (morphology) or by the distribution of internal cells and tissues (anatomy). Stems are elongated organs that form the axis on which the lateral appendages, **leaves** and **buds,** are attached. The place on a stem where leaves and buds are attached is called a **node,** the portion between nodes is an **internode.** Some plants have erect stems, others have horizontal, creeping stems (Fig. 4.35), and some have stems that do not elongate except at flowering time. (These are rosette plants, Fig. 4.1C.) All stems whether short or long, horizontal or erect, are distinguishable as stems by the presence of nodes and internodes.

Arrangement of Leaves and Buds

A three-year-old twig of walnut in winter condition is shown in Fig. 4.2. The tip of the twig generally bears a large **terminal leaf bud.** At regular intervals along the stem, other buds may be seen; they are called **lateral buds.** Note that below the base of each lateral bud there is a scar that was made when a leaf fell from the twig; this is a **leaf scar.** Vascular bundle scars (Fig. 4.2) may be seen within each leaf scar; strands of food- and water-conducting tissues passing from the stem into the leaf were broken when the leaf fell, leaving these scars. Buds and leaves are usually borne in this relationship to each other; buds form in the angle made by the stem and the leaf stalk. This angle is termed the **leaf axil,** and consequently these buds may also be called axillary buds.

Protecting the young immature leaves and cells within the bud is a series of overlapping scales, **bud scales.** They are usually shed when the bud develops into a new shoot, and they also leave scars, **bud-scale scars.** The part of a stem or twig between sets of terminal-bud-scale scars is generally formed during one growing season. For instance, growth made by the twig this year is set off from growth made last year by means of a ring or girdle or terminal-bud-scale scars (Fig. 4.2). When scales of a terminal bud fall off in spring, they leave a number of closely crowded scars that form a distinct ring. Examination of twigs several years old shows that growth in length may vary from year to year, as revealed by the different spacings between the terminal-bud-scale scars. There is no increase or decrease in the length of any portion of a stem after that portion is one year old.

The slightly raised areas on the bark are **lenticels.** They are composed of cells that fit loosely together, with air spaces between, which permit passage of gases inward and outward.

Figure 4.1 *A,* bristlecone pine *(Pinus aristata);* gymnosperms and woody angiosperms achieve longevity through secondary growth, which annually results in a cylinder of new tissue around the trunk. The older tissues die. *B,* Iris; many other seed plants attain longevity by the continued growth of new primary tissue at the apex of the shoot. Older parts of the stem die. New leaves form at the apex, roots arise in nodes behind the apex, and older portions of rhizome die. *C, Echeveria* sp. rosette plant with shortened internodes. Note the elongated internodes of the floral branches to the left and right.

Position of Buds on a Woody Twig

In the walnut twig there is just one leaf bud and one leaf at each node. This arrangement of buds and leaves on the stem is spoken of as **alternate** (Figs. 4.3*A,* 4.4*A,C,E*). It is the most common type of bud and leaf arrangement. Ash, maple, lilac, and many other plants have two leaves opposite each other at each node, and a bud in the axil of each leaf (Figs. 4.3*B,* 4.4*D,F*). This arrangement of leaves and buds is spoken of as **opposite**. When three or more leaves and buds occur at each node, as in *Catalpa,* leaf arrangement and bud arrangement are said to be **whorled** (Fig. 4.3*G*).

Some plants have several buds in or near the leaf axil. For example, apricot (*Prunus* sp.) often has a group of three buds in the leaf axil: a central bud, which develops into a side branch, and two lateral ones, which are called **accessory buds**. Walnut (*Juglans* sp.) may also have more than one bud in the leaf axil (Fig. 4.3*A*).

Not infrequently, buds may arise on the plant at places other than leaf axils. They may appear on stems, roots, or even leaves, and give rise to new shoots. Such buds are called **adventitious buds**. Their formation may be stimulated by injury, such as occurs in pruning.

Dormant or **latent buds** arise in a regular fashion in the leaf axil, but their development is usually inhibited by the dominance of the terminal bud. This mechanism, **apical dominance**, is fully discussed in Chapter 8.

From the above discussion it is seen that buds may be classified by their arrangement on the stem, which may be (a) alternate, (b) opposite, or (c) whorled; by their position on the stem, which may by (a) terminal, (b) lateral

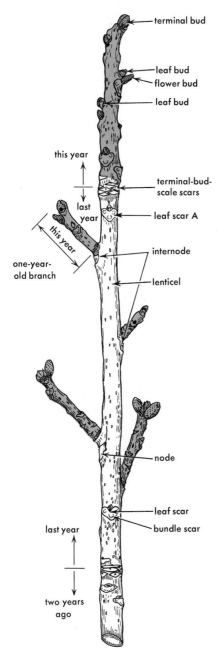

Figure 4.2   Three-year-old twig of walnut *(Juglans regia),* ×1.

The figure contains the following labels:

- terminal bud
- leaf bud
- flower bud
- leaf bud
- this year
- last year
- terminal-bud-scale scars
- leaf scar A
- this year
- one-year-old branch
- internode
- lenticel
- node
- leaf scar
- bundle scar
- last year
- two years ago

(axillary), (c) accessory, or (d) adventitious; and by the nature of the organs into which they develop, which may be (a) leaf, (b) flower, or (c) mixed.

As a rule, the terminal bud of a stem is the most active and grows more vigorously than any of the axillary buds. Usually, the lowest lateral buds on a year's growth of the shoot remain dormant and do not develop into branches. If the terminal bud is removed, however, as may be done in pruning, lateral buds, otherwise dormant, may become active.

Development of Tissues of the Primary Plant Body

The Primary Meristems

Plant organs (leaves, stems, roots, and flower parts) are obviously different from each other morphologically (i.e., based on their external form). But, if we examine their internal anatomy it is apparent that all organs are composed of similar structural units—cells and tissues—that are dissimilarly arranged. Each cell *(cell type)* is modified to make it ideally suited to perform one or more specific functions. **Tissues** are organizations of cells of one or more types that have a common origin and a common collective function.

Cell types and tissues develop by a process called **differentiation.** A differentiating cell progresses through a series of steps that result in the cell becoming mature and functional. Each different cell type "experiences" slightly different differentiation steps. The formation of new cells, and the initiation of differentiation in plants takes place in specific regions called **meristems.** Meristems can also be categorized. The **apical meristems** occur in the **shoot** and **root tips.** Apical meristems are the source of all other meristems. They form the three primary meristematic tissues (or **primary meristems**): protoderm, ground meristem, and procambium. These three primary meristems will differentiate into the three **primary** tissues: epidermis, ground tissues ("pith and cortex") and vascular tissues (xylem and phloem). Now we will consider in detail each of the primary meristems and tissues in the plant body.

Protoderm

This is the outermost layer of cells (Figs. 4.5*A,D,* 4.6). It develops into **epidermis**—the special primary tissue that covers and protects all underlying primary tissues. The epidermis prevents excessive water loss and yet allows for the exchange of gases necessary for respiration and photosynthesis.

Ground Meristem

The ground meristem comprises the greater portion of meristematic tissue of the shoot tip (Figs. 4.5*B,D,* 4.6). Its cells are relatively large, thin-walled, and isodiametric. The regions forming from the ground meristem are (a) **pith,** in the very center of the stem, and (b) **cortex,** in a cylinder just beneath the epidermis and surrounding the vascular

Figure 4.3 Twigs showing three methods of bud and leaf arrangement. The position of leaves is shown by the leaf bases and scars. *A*, alternate, walnut *(Juglans regia); B,* opposite, lilac *(Syringa vulgaris); C,* whorled, *Catalpa,* × ½.

tissue. A note of clarification must be added here. Traditionally speaking, the pith and cortex are not tissues, but instead are stem and root regions that are composed of the ground tissues, parenchyma, sclerenchyma, and collenchyma. These will be discussed in the following section.

Procambium

These cells usually appear first as strands among ground meristem cells (Fig. 4.6). In cross section (Fig. 4.5*C,D*) strands appear as isolated groups of cells arranged in a circle. Sometimes a continuous **procambium cylinder** is formed. As seen in a transverse section of a single procambium strand, procambium cells are smaller than those of the surrounding ground meristem; in longitudinal section, they are much longer, and some of them may be pointed at the ends. Procambium cells give rise to **primary vascular tissues** (Figs. 4.5*E,F,* 4.7). These primary tissues carry out several functions and are divided rather rigidly into two groups according to these functions. Food synthesized by photosynthesis is conducted in the **phloem**. Water and mineral salts are conducted in **xylem**.

Primary Tissues

Primary tissues of the stem are differentiated from the three primary meristematic tissues—protoderm, ground meristem, and procambium—and that these three are derived from the apical meristem of the shoot tip. In woody plants, we must look for primary tissues of the stem a very short distance behind the stem tip. Even before the end of the first season's growth, differentiation of these primary tissues from primary meristematic tissues is completed, and **secondary tissues** may be formed in abundance. Whereas primary tissues are derived from primary meristematic tissues of the shoot, secondary tissues are the result of production of new cells by the **vascular cambium** and by the **cork cambium**. The origins and nature of these two types of cambia are discussed later.

The Epidermis

The epidermis is usually a single superficial layer of cells that covers all other primary tissues, protecting them from drying out and, to some extent, from mechanical injury. It is the limiting layer of cells between the plant and its environment. In surface views (Fig. 4.7, 4.8), epidermal

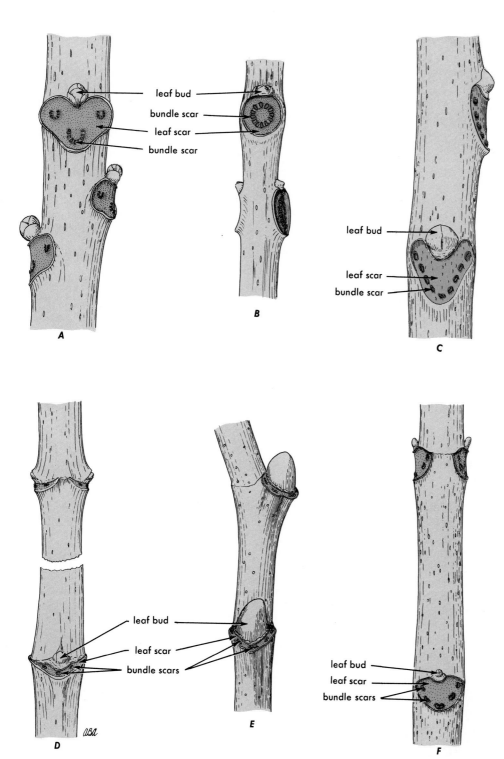

Figure 4.4 Leaf scars and bud arrangement of different species of woody plants. A, walnut (Juglans regia), ×2; B, catalpa (Catalpa bignonioides), ×2½; C, tree of heaven (Ailanthus altissma) ×2; D, box elder (Acer negundo), ×2; E, European plane (Platanus acerifolia), ×3; F, buckeye (Aesculus californica), ×2.

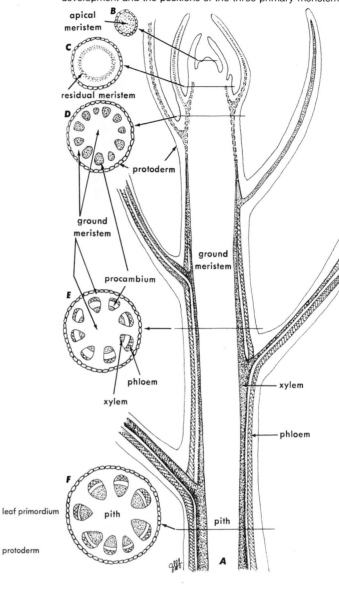

Figure 4.5 *A to F:* diagrams showing the pattern of vascular development and the positions of the three primary meristems.

apical meristem

B

C

residual meristem

D

protoderm

ground meristem

procambium

E

phloem

xylem

ground meristem

xylem

phloem

F

pith

pith

A

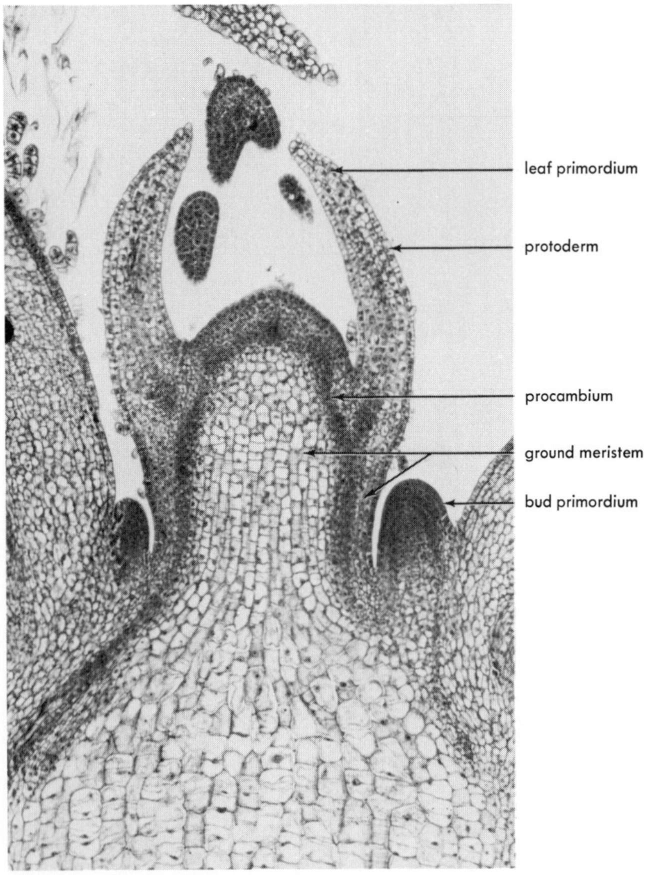

leaf primordium

protoderm

procambium

ground meristem

bud primordium

Figure 4.6 Longitudinal section of the shoot apex of the bean *(Phaseolus vulgaris),* ×90.

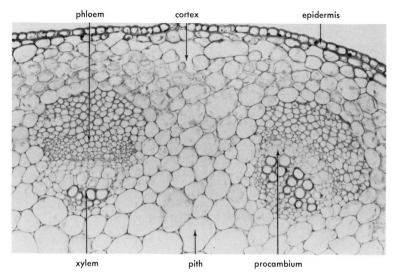

phloem   cortex   epidermis

xylem   pith   procambium

Figure 4.7 Cross sectional view of clover (*Trifolium* sp.) stem showing the primary vascular tissues, xylem and phloem. The epidermis, and the stem regions, the cortex and pith are also illustrated, ×339.

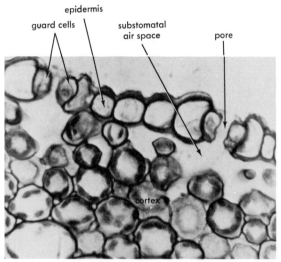

epidermis
guard cells   substomatal
                          air space   pore

cortex

Figure 4.8 Epidermis; cross section of alfalfa (*Medicago sativa*) stem showing epidermal cells, guard cells, and cortex with a substomatal chamber, ×400.

cells are elongated in the direction of the stem's length; in transverse section, they are usually isodiametric. Its protoplasm forms a thin layer that lines cell cavities and normally retains its living properties for a long period.

The outer tangential wall of cells exposed to air is usually thicker than the other walls, and its surface layer is usually coated with a waxy substance called **cutin**. This superficial layer is the **cuticle**, which is quite impermeable to water and gases. There may be cracks or other imperfections in the cuticle, however, and water vapor may pass out of the plant at those points. The inner walls, parallel to the stem surface, are thinnest, and radial walls, at right angles to the surface, often taper in thickness toward the inner wall (Fig. 4.8).

Young stems usually possess specialized epidermal cells

called **guard cells**. Between each pair of guard cells is a small opening, or **pore**, through which gases enter and leave the underlying stem tissues. Two guard cells plus the pore make one **stoma (pl. stomata)** (Fig. 4.8A). Guard cells differ from other epidermal cells by their crescent shape and the fact that they contain chloroplasts. Stomata are common in the epidermis of leaves, floral structures, and fruits as well as of stems. Stomata are discussed more thoroughly in Part II of this chapter.

Epidermal appendages, such as hairs, may occur on young stems (Fig. 4.44). These structures will also be discussed in the section on leaves.

The Cortex

This complex region, derived from ground meristem, forms beneath the epidermis as a cylindrical zone that extends inward to the primary phloem (Figs. 4.7A,B, 4.15). The following tissues or cell types may be found within it: **parenchyma, collenchyma, sclerenchyma,** and **secretory tissue.**

Parenchyma. The principal tissue of the cortex is parenchyma (Figs. 4.7, 4.9G,I). It consists of isodiametric cells with thin walls, made mostly of cellulose, and with protoplasts that remain alive for a long time. We may speak either of a parenchyma tissue, where many parenchyma cells are found together, or of individual parenchyma cells.

Parenchyma tissue is characterized by the presence of intercellular air spaces, which vary greatly in size; in some parenchyma tissues they are difficult to find, while in others they are very apparent (Fig. 4.7). Because parenchyma cells retain active protoplasts, they function in the storage of water and food, in photosynthesis, and sometimes in secretion. When the parenchyma cells contain chloroplasts, they are collectively referred to as **chlorenchyma.** Parenchyma cells can be reprogrammed to differentiate into different cell types. One example would be the changes parenchyma cells undergo in response to

Development of Tissues of the Primary Plant Body

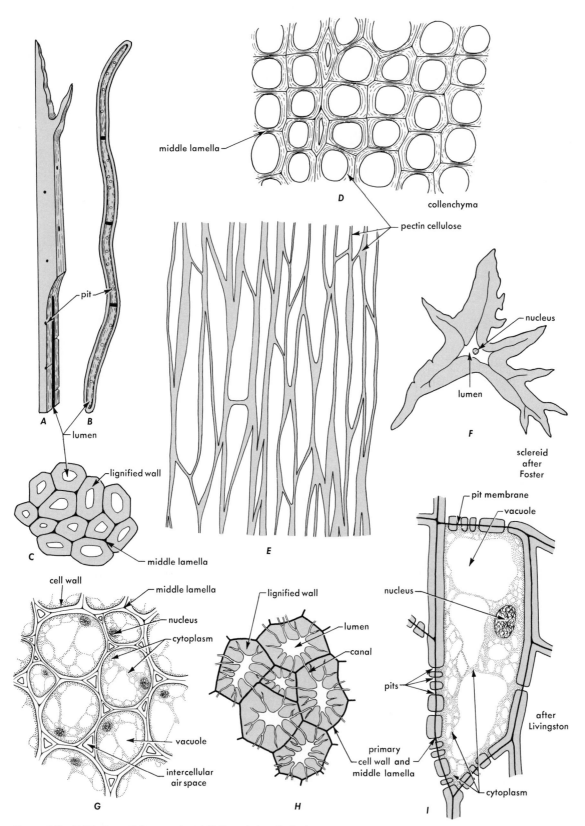

Figure 4.9 Cell types and tissues. *A* and *B,* fibers in longitudinal
view; *C,* fibers in cross section; *D,* collenchyma in cross section;
E, collenchyma in lontitudinal view; *F,* sclereid; *G,* parenchyma;
H, stone cells; *I,* woody parenchyma.

being wounded. In this case, the wounded cells on the surface simply die, and the inner cells divide and form a layer that is quite similar to the bark on the outside of stems.

Parenchyma cells are not confined only to the cortex and pith of the stem, but also occur in practically all other types of tissue and in other organs of the plant.

Collenchyma. The outermost cells of the cortex of young stems, lying just beneath the epidermis, often constitute a tissue known as **collenchyma**. This tissue may form a complete cylinder, or it may occur in separate strands. Collenchyma cells are elongated, often contain chloroplasts, and are living at maturity. The walls of collenchyma cells are composed of alternating layers of pectin and cellulose. In the most common type of collenchyma, the cell walls are thickened at the corners (Figs. 4.9D,E). These thickenings are quite flexible and will stretch without resistance, in much the same way as heated plastic. Collenchyma is, therefore, an ideal strengthening tissue because it strengthens and at the same time allows normal tissue growth. Collenchyma serves as a strengthening tissue in young expanding stems and also in the petioles of leaves.

Sclerenchyma. The main functions of **sclerenchyma** cells are support and, in many cases, protection. Their shape and the thickness and toughness of their walls contribute to the ability of these cells to support and protect the stem. Thickness and toughness of walls are increased by deposition, within the original cellulose wall, of a substance known as **lignin**. Lignin is made by the protoplast and deposited on the wall. When this secondary wall is completely formed, the protoplast usually dies. There are two types of sclerenchyma cells: (a) **sclereids** and (b) **fibers** (Figs. 4.9A,B,F,H).

Of various types of sclereids, the most common are stone cells. Other types of sclereids are branched, resembling very irregular stars (Fig. 4.9F). Some sclereids are derived from parenchyma cells by a pronounced thickening of cell walls; others arise from separate meristematic cells.

Sclereids occur not only in the cortex of stems but also in the hard shells of fruits, seed coats, and bark, in pith of some stems, and in certain leaves.

Fibers are elongated, strengthening cells that are thick-walled, and usually pointed at the ends (Figs. 4.9A,B). Their walls may or may not be lignified. The walls may become so thick that the cell cavity, the lumen, almost disappears. Simple pits form in their thick walls (Fig. 4.9A). Fibers are sometimes very elastic and can be stretched to a great degree without losing their ability to return to their original length. Protoplasts of fibers often disappear as the cells attain maturity.

Secretory Cells. Secretory cells are parenchymalike and contain dense protoplasm. They secrete various substances, such as resinous materials (Fig. 4.27) and nectar in many flowers. Many epidermal hairs are secretory cells (Fig. 4.44C).

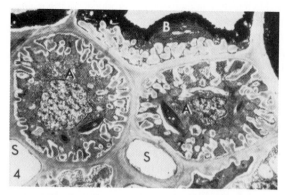

Figure 4.10 Transfer cells (A) around vascular bundle in leaf of *Armeria corsica*, ×1887.

Transfer Cells. Frequently, cells located in positions of active solute transfer will show an irregular extension of the cell wall into the protoplast. Since the plasmalemma follows the contour of the wall, its surface is greatly extended. Mitochondria appear to aggregate adjacent to these areas of increased membrane surface. The morphological picture thus presented (Fig. 4.10) suggests an adaptation that facilitates transport from one cell to another, or from the interior to the exterior of the plant. These cells have been called **transfer cells**.

The Pith

The pith makes up the central core of many stems. It is composed mostly of parenchyma cells that store food products such as starch. Sclerenchyma cells (especially sclereids) can also be found in the pith. The cellular region between vascular bundles is sometimes called a pith ray (Fig. 4.5F).

The Primary Vascular Tissues

The term "vascular" pertains to tissues that conduct various substances in liquid form.

In vascular plants, water and different water-soluble inorganic salts from soil, as well as food substances, are conducted throughout the plant in well-defined vascular tissues.

In a young stem, very near the tip (Figs. 4.5, 4.6), vascular tissues occur as separate bundles, **primary vascular bundles**. Each primary vascular bundle is differentiated from a procambium strand and consists of primary xylem and primary phloem. If secondary growth is to occur, a thin band of meristematic tissue destined to become vascular cambium remains between primary xylem and primary phloem (Fig. 4.17).

The Primary Phloem

Phloem in angiosperms may possess several types of cells: **sieve-tube members, companion cells, fibers, sclereids,** and **parenchyma**. A **sieve tube** is a vertical row of elongated cells; each cell is known as a **sieve-tube member**. These are the conducting elements of phloem

Development of Tissues of the Primary Plant Body

(Fig. 4.11) that translocate sugars, produced in the leaves by photosynthesis, to other plant parts (see Chapter 5). Among angiosperms, a sieve-tube member and a companion cell are sister cells; that is, they originate by division from the same procambial cell. Young sieve elements have the usual complement of organelles: nucleus, plastids, mitochondria, and dictyosomes. As the element matures, its protoplast becomes greatly modified. Its nucleus is thought to disintegrate. Plastids lose most of their internal membranes, but usually retain starch.

The mitochondria become small. The cytoplasm, much reduced in amount, becomes reduced to a thin peripheral layer. The central part of the cell is occupied by a mass of strands or tubules. This mass may be seen with the light microscope, and has been called **slime**. Since it is now known to be a protein, it is more correctly referred to as **P-protein** (Fig. 4.11). At maturity one or more companion cells lie adjacent to each sieve-tube member. Since

companion cells have a normal protoplast with a full complement of organelles (Fig. 4.11), these cells possibly regulate the metabolic activity of the sieve-tube members that have no nucleus. Plasmodesmata connect the protoplasts of companion cells and sieve-tube members.

A characteristic structural feature of mature sieve-tube members is the **sieve plate**. It may occur in the end or side walls (Fig. 4.11). The end wall between two adjacent sieve-tube members is thickened and strands of cytoplasm pass through pores adjoining them. With the exception of a few trees such as palms, where they live longer, sieve-tube members live and function about 1 to 3 years. New ones are formed annually.

In many studies on the structure of mature sieve-tube members, a carbohydrate known as **callose** is seen around the margins of pores in the sieve plate (Fig. 4.12). In some instances, protein may also collect at the sieve plate. Obviously, this would block the pores of the sieve

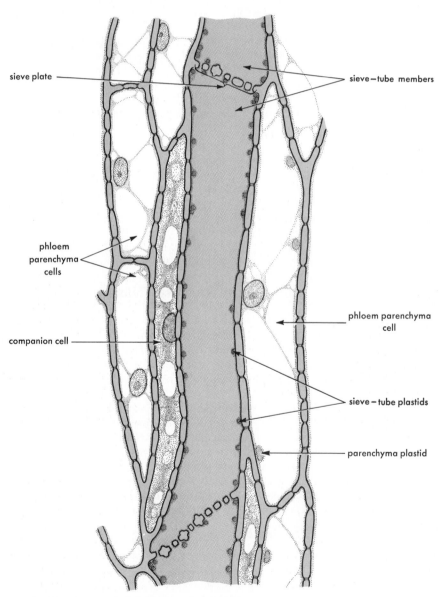

Figure 4.11   Phloem tissue from the stem of tobacco *(Nicotiana),* ×400.

plate and obstruct movement of food materials. It has been demonstrated in other tissues that callose forms very rapidly in response to wounding.

In gymnosperms there are **sieve cells** rather than sieve-tube members; these have tapered end walls without sieve plates and they do not connect to form sieve tubes.

The Primary Xylem

The conducting cells that occur in primary xylem of vascular plants are **tracheids** and **vessel elements.** These cells conduct water and mineral salts. Associated with them may be fibers (xylem fibers) and **parenchyma** (xylem parenchyma).

A tracheid is a single elongated cell more or less pointed at its ends (Fig. 4.13). Functioning tracheids are

not alive. The tracheid wall may not be the same thickness throughout. All the wall may be thickened except for numerous small, circular, or oval areas called **pits** (Fig. 4.13C). There are two types of pits in xylem cells: **simple pits** and **bordered pits.** Simple pits as shown in Figs. 4.9A and I, occur in fibers, sclereids, and in parenchyma cells when they have secondary walls. Pits form opposite each other in the secondary walls of adjacent cells. A pit-pair is not a hole in the wall, since the primary wall and the middle lamella of the two communicating cells remain intact. These primary layers, however, are penetrated by plasmodesmata while the cells are living. The type of pit known as a bordered pit (Figs. 4.13A,E) occurs in tracheids, vessel elements, and some xylem fibers. This type of pit is more structurally complex. It consists of an expanded border of the secondary cell wall that extends

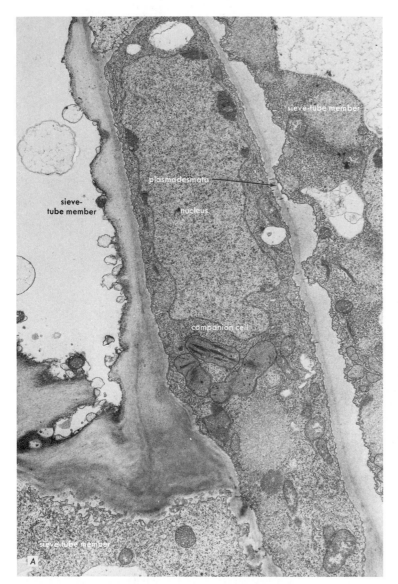

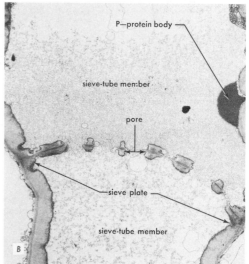

Figure 4.12 *A,* longitudinal section of three sieve-tube members and one companion cell. The nucleus and various organelles are present in the central companion cell. Most organelles are missing in the left sieve-tube member, but are present in the one on the right, *Cucurbita maxima,* ×8300. *B,* micrograph of sieve plate between two sieve-tube members from *Cucurbita maxima,* ×4600.

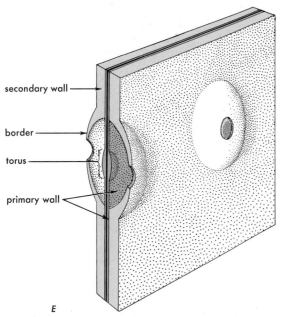

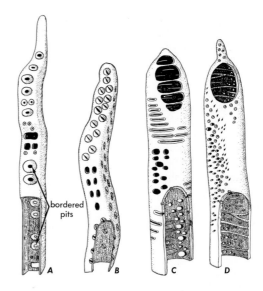

Figure 4.13 Tracheids and vessel elements from secondary wood. *A*, tracheid from spring wood of white pine *(Pinus)*; *B*, tip of tracheid from wood of oak *(Quercus)*; *C*, tip of vessel element from wood of Magnolia; *D*, tip of vessel element from wood of basswood *(Tilia)*; *E*, diagram of bordered pit.

over a small pit chamber. Within the chamber is a diaphragmlike primary cell wall. In gymnosperms, the center portion, the torus, is thickened and impregnated with a waxy material. The torus apparently acts as a valve that plugs the pit opening during times of drought (Chapter 5).

A **vessel element** is a single cell with oblique, pointed, or transverse ends. A **vessel** is a series of vessel elements differentiating end to end, with perforated end walls. Vessels are often several centimeters long, and in some vines and trees they may be many meters in length. Before the protoplasts disappear, vessel walls become thickened, forming a secondary wall (Fig. 4.14); the thickening material is laid down on primary walls in various patterns so that in some places the secondary walls are thick and in others thin. The material deposited is cellulose; later, the layers of cellulose become lignified. The end walls of the vessel elements also dissolves before the protoplasts disappear. Thus, the deposition of thickening material forming the secondary walls and the dissolution of end walls are functions of living cells. After these events have taken place, the protoplast dies.

The secondary walls of vessels in angiosperm stems are deposited in several different patterns (Figs. 4.14*A,B*): **annular, scalariform, reticulate,** and **pitted.** The ends of the vessel elements (Figs. 4.13*A,D*), are generally on a slant and, although open, they may have bars of wall material across them. The shape of vessel elements may indicate evolutionary relationships among plants (Chapter 10). Vessel elements do not occur in small veins in leaves, and are lacking in most gymnosperms and in the lower vascular plants. In these forms, tracheids occur in elongating regions, and they may have annular and spiral types of secondary walls.

Xylem parenchyma cells outlive vessel elements,

tracheids, and most xylem fibers. They function in the storage of water and foods, which, as we have learned, is one of the principal functions of parenchyma wherever it occurs in the plant. Parenchyma may also conduct materials for short distances.

Xylem fibers are similar to the fibers described elsewhere.

Summary of Primary Tissues

1. Buds are characteristic of woody stems. Bud scales enclose and protect rudimentary leaves surrounding an apex. The apex may be either a vegetative shoot or a floral apex. A floral apex terminates growth. Woody plants each year require new vegetative growth, which is normally accompanied by flowers.

2. Primary growth brings about the elongation of stems and establishes the basic pattern of cells and primary tissues characteristic of the particular stem and upon which the functioning and future growth of the stem depend.

3. Primary meristematic tissues are apical meristem, protoderm, ground meristem, and procambium.

4. The primary plant body is composed of (a) the epidermis, which may be differentiated into three cell types: epidermal, guard cells, and epidermal hairs; (b) cortex, which is composed of collenchyma, sclerenchyma (fibers and sclereids), and parenchyma; (c) vascular tissue, composed of xylem (fibers, tracheids, vessel elements, and parenchyma) and phloem (fibers, sieve-tube members, companion cells, and parenchyma); (d) pith, composed largely of parenchyma and sometimes accompanied by sclereids; (e) pith rays, composed of parenchyma cells.

5. The functions of these cells and tissues are as follows.

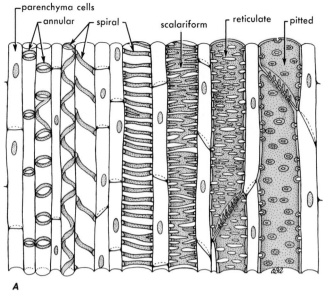
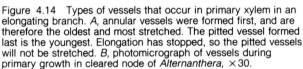
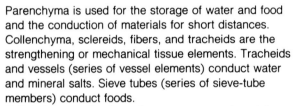

parenchyma cells
annular — spiral — scalariform — reticulate — pitted

A

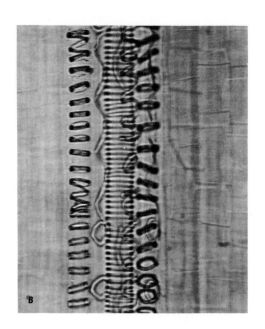

B

Figure 4.14 Types of vessels that occur in primary xylem in an elongating branch. *A*, annular vessels were formed first, and are therefore the oldest and most stretched. The pitted vessel formed last is the youngest. Elongation has stopped, so the pitted vessels will not be stretched. *B*, photomicrograph of vessels during primary growth in cleared node of *Alternanthera*, ×30.

Parenchyma is used for the storage of water and food and the conduction of materials for short distances. Collenchyma, sclereids, fibers, and tracheids are the strengthening or mechanical tissue elements. Tracheids and vessels (series of vessel elements) conduct water and mineral salts. Sieve tubes (series of sieve-tube members) conduct foods.

6. Collenchyma and sclerenchyma may occur in patches or completely surround the stem just underneath the epidermis. Sclerenchyma is frequently associated with vascular bundles.

The Dicotyledonous Stem

Stem Primary Growth

The tip of the stem consists of small immature leaves enclosing a dome-shaped apical meristem. The apical meristem is composed of dividing cells, arranged in various ways, that give rise to the leaves, buds, and the primary meristematic tissues (Figs. 4.6, 4.5). The apical meristem is said to be indeterminate, that is, if conditions were ideal the apex could grow continuously. We know, however, that this doesn't happen in nature either because environmental factors are limiting or because the onset of flowering usually causes vegetative growth to stop. In addition, each plant species is genetically programmed to develop within a certain size/age range.

Beneath the apical meristem are the regions of the three primary meristematic tissues—protoderm, ground meristem, and procambium. The ground meristem starts to differentiate first. These tissues form a cylinder near the outside of the stem and a core in the inside (Fig. 4.5C),

which will continue to differentiate into the cortex and pith, respectively.

Between these two tissues, near the apex, is a ring of **residual meristem** cells that retain the cellular characteristics of the apical meristem (Fig. 4.5C). This ring will become a cylinder of discrete procambium strands that later differentiate into primary xylem and phloem (vascular bundles). Each bundle is separated from others by regions of parenchyma cells (Figs. 4.5D,E, 4.15). The formation of vascular bundles from the residual meristem is apparently in response to leaf development. The mechanism is not exactly understood, but it may involve the production of some substance by the leaf primordia (immature leaves on the shoot tip) and its transport downward to the residual meristem ring. When the substance reaches a certain critical concentration, it apparently induces the residual meristem cells to form bundles of procambium and then xylem and phloem. Each bundle is called a **leaf trace** and, as it matures, will lead from the stem into a leaf, connecting it to the axis of the stem (Fig. 4.47). The vascular system of the stem actually consists, for the most part, of interconnected leaf traces. Other traces, however, connect to buds (bud traces), and some traces end at the apical meristem without connecting to either a leaf or a bud.

The role that leaves play in vascular differentiation and in the pattern of vascular bundles has been determined by elegant experiments. In one of these experiments, the leaves and leaf primordia were removed around the apex (Fig. 4.16A). As the stem elongated the new leaf primordia which formed were destroyed. Anatomical examination of the stem which differentiated during the weeks of the experiment revealed that the "vascular tissue" remained as an unbroken cylinder (Figs. 4.16B,C). The cells making up the cylinder did not differentiate completely, and

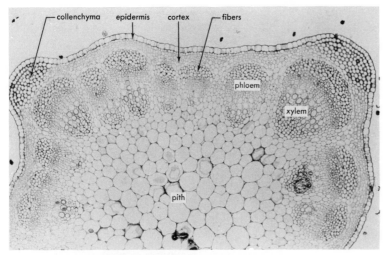

Figure 4.15 Young alfalfa *(Medicago sativa)* stem showing primary tissues. Notice the vascular bundles that are composed of xylem and phloem, the epidermis, and the cortex and pith regions. The regions between vascular bundles consist of only parenchyma cells, ×14.

vascular bundles did not form. If leaves were later permitted to develop, the newly formed vascular tissue did have bundles and did differentiate into xylem and phloem. This experiment demonstrated that leaves are needed for vascular differentiation and for the formation of procambial strands and vascular bundles.

Stem Secondary Growth

Initiation

During primary growth, stems increase in length. Perennial stems and some annual stems such as those of tomato

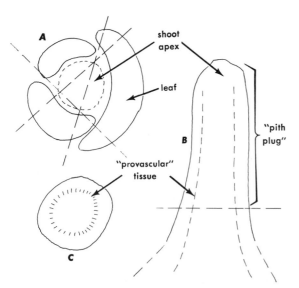

Figure 4.16 Diagram of an experiment to show the effect of removing leaves on vascular development in the stem of *Geum* sp. *A,* top view of shoot tip, dotted lines indicate cuts to remove young leaves and leaf primordia that are not yet visible. *B,* "pith plug" that forms several days after removing leaf primordia. *C,* cross section of pith plug to show the "provascular" tissue that forms in a complete circle or cylinder when leaves are removed.

(Lycopersicon), sunflower *(Helianthus),* and alfalfa *(Medicago),* also increase in diameter. This lateral thickening involves the activation of a secondary meristem, the **vascular cambium** (Fig. 4.17, 4.18). The vascular cambium is composed of two parts—the **fascicular cambium,** which forms from within the vascular bundles, and the **interfascicular cambium,** which originates from parenchyma cells that lie between vascular bundles (Figs. 4.17*A,D*).

Recall that the first stage of development of vascular tissues in stems was the formation of a cylinder of residual meristem. Next, the newly formed leaves acted on this cylinder to induce portions of it (vascular bundles) to become procambium strands and then vascular bundles of xylem and phloem. After development of the primary xylem and phloem, in some plants a portion of the procambium remains undifferentiated. At some time during the plant's life cycle, very early in the case of woody plants and later in the case of herbaceous plants that develop secondary growth, this residual procambium reinitiates divisions to become the fascicular cambium (Figs. 4.17*A,B*). Simultaneously, parenchyma cells adjacent to the fascicular cambium, but between vascular bundles, are also stimulated to divide, to become the interfascicular cambium. Together, both components form the vascular cambium that will make secondary phloem to the outside and secondary xylem to the inside of the stem (Fig. 4.19).

Organization

Close examination of a woody stem reveals that the cells that make up wood are actually oriented in two ways (Figs. 4.20*A,B*). Some cells are elongated parallel with the axis of the stem, making the **axial system.** The axial system is composed of vessel members, tracheids, fibers, and parenchyma. In the phloem, sieve-tube members, fibers, companion cells, and parenchyma make up the axial system. The source of these cells in the vascular cambium are **fusiform initials.** A second system, the **ray system,** is composed of cells oriented at right angles to

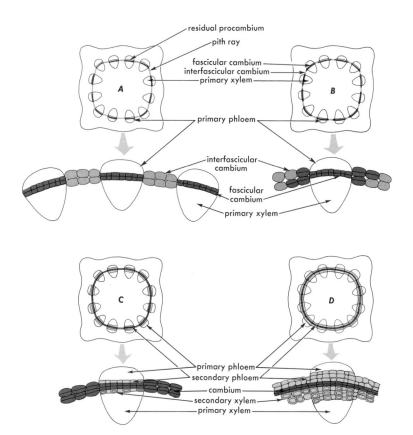

Figure 4.17 Formation of a complete cylinder of vascular cambium. *A*, at completion of primary growth, some meristematic cells remain between primary xylem and primary phloem. (Residual or "leftover" procambium shown in dark green.) Parenchyma cells appear in pith rays between vascular bundles (light green). *B*, the residual procambium becomes reactivated to form the fascicular cambium; some of the parenchyma cells of the pith ray become meristematic to form the interfascicular cambium (dark green). Together, the fascicular cambium and interfascicular cambium = vascular cambium. *C*, parenchyma cells of pith between the meristematic cells of the bundles have returned to a meristematic state, forming a cylinder of cambium (dark green). *D*, complete cylinder of secondary xylem and secondary phloem (light gray) is formed by the vascular cambium. Parenchyma cells are shown in light green, vascular cambium in dark green, and secondary tissues in light gray.

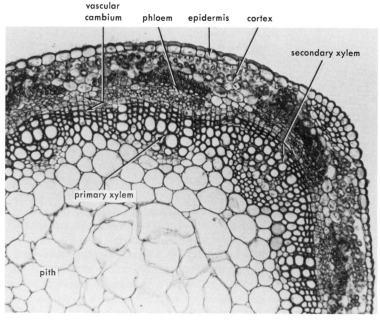

Figure 4.18 Older alfalfa *(Medicago sativa)* stem showing secondary growth. The vascular cambium is a continuous cylinder that forms progeny cells to the inside, the secondary xylem, and to the outside, the secondary phloem. The primary xylem, pith, cortex, and epidermis are also shown, ×25.

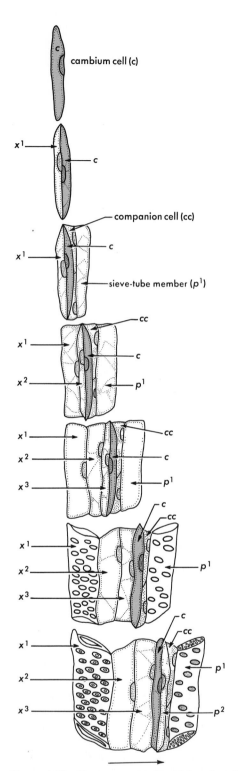

Figure 4.19 Diagram as seen in radial section showing stages in differentiation of vascular cambium cells (c, cambium; cc, companion cell; p^1, p^2, phloem; x^1, x^2, x^3, xylem).

the axis of the stem (Figs. 4.20*A*,*B*). The **rays** that make up this system develop from **ray initials** in the vascular cambium. Rays are living channels through which nutrients and water move laterally in stems with secondary growth. Xylem rays are made up of ray tracheids, and ray parenchyma and phloem rays are made of phloem parenchyma. Ray cells may remain alive for several years, perhaps 10 or more.

Anyone who has examined the stump of a cut tree has observed **annual rings** (Figs. 4.21, 4.22, 4.23). One annual ring represents the amount of secondary xylem growth for one season. Thick rings mean maximum growth has occurred during a season, and thin rings mean minimal growth. The density (width) of these rings give an indication of past climate conditions.

Microscopic examination of an individual growth ring shows that during the early part of the growing season, cells are large and have relatively thin walls. This part of the growth ring is called the **early wood** (springwood). Later in the season the cells become smaller in diameter and have thicker walls; this is the **late wood** (summer wood) (Figs. 4.23*A*, 4.25*A*). In certain trees, large vessel members form only in the early wood and only small vessel members occur in the late wood; this organization is called **ring porous** (Fig. 4.23). In other trees the occurrence of vessel members is uniform throughout the growing season; this is **diffuse porous** wood (Fig. 4.24). Many trees in tropical areas, where growing conditions are uniform throughout the year, show no annual rings in their wood.

After a tree or branch is several years old, the inner part of the stem usually becomes inactive and filled with resins. This dense, central core is called the **heartwood**. Good barrels and wooden tubs are made from heartwood, because the resin makes this wood impermeable to water. Outside the heartwood is a light-colored, less dense region of active wood called **sapwood** (Fig. 4.22).

Two things contribute to the formation of heartwood. One is the curious formation of structures called **tyloses**. Tyloses are ingrowths of the primary wall of parenchyma cells that grow through the pits of vessel members. The walls of tyloses eventually become enlarged and may form secondary walls that completely plug up the xylem. Species of trees that have abundant parenchyma cells adjacent to vessel members readily form tyloses; however other species with more scattered parenchyma do not (Fig. 4.26). The second contributing factor to forming heartwood is the activity of rays. Rays apparently play an important role in transporting resins from the active sapwood into the central heartwood. Heartwood is a repository for metabolic by-products, serving a function parallel to the excretory system of animals.

Gymnosperm wood is anatomically simpler than angiosperm wood. It is composed almost entirely of tracheids in the spring wood and of fiber-tracheids (cells that are intermediate between fibers and tracheids) in the summer wood. The rays in gymnosperms are usually only one cell layer thick and axial parenchyma is rare. Vessels are absent, but there are many **resin ducts**. These secretory ducts are long hollow tubes containing an internal layer of cells that produce resin (Fig. 4.27).

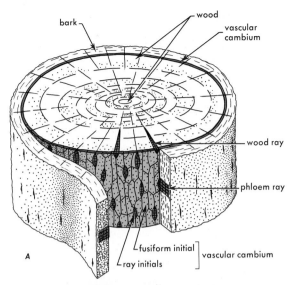

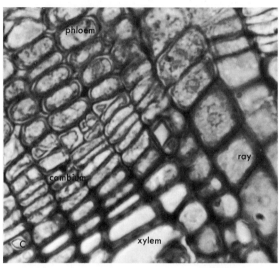

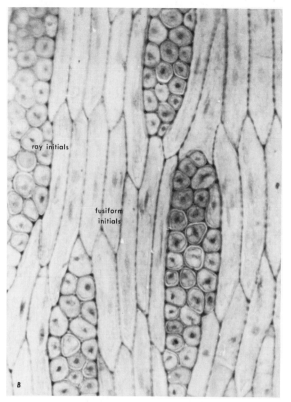

Figure 4.20 Vascular cambium. *A*, diagram showing relationship of cambium and cambial initial to stem; note also the cell orientations. The cells of the axial system are elongated vertically, those of the ray system are elongated horizontally; *B*, tangential section; *C*, cross section of quiescent cambium of locust (*Robinia*), ×300.

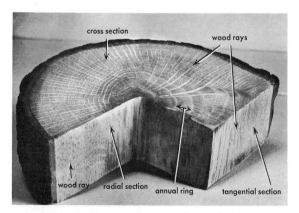

Figure 4.21 Portion of stem of oak *(Quercus)*, showing cross, radial, and tangential sections and their gross characteristics, ×½.

Figure 4.22 Cross section of a branch of mulberry *(Morus)*, ×½.

The Dicotyledonous Stem

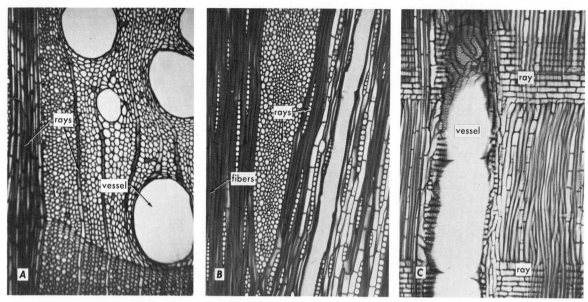

Figure 4.23 Sections of wood of oak *(Quercus borealis)*. *A*, cross section; *B*, tangential section; *C*, radial section, ×40.

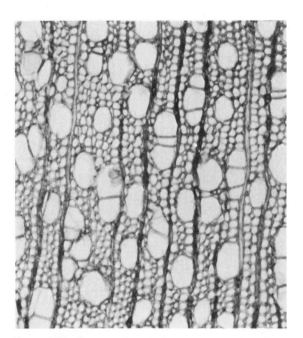

Figure 4.24 Cross section of diffuse-porous wood from popular *(Populus deltoides)*, ×150.

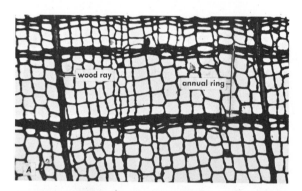

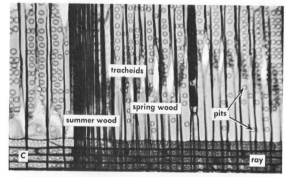

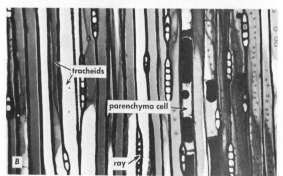

Figure 4.25 Sections of wood of redwood *(Sequoia sempervirens)*. *A*, cross section; *B*, tangential section; *C*, radial section; *D*, scanning electron micrograph of radial section, ×100.

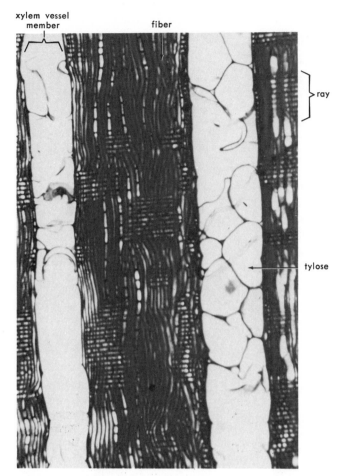

Figure 4.26 Radial section of wood from white oak *(Quercus alba)* that shows tyloses. Tyloses are ingrowths of parenchyma cells that form secondary cell walls and may plug up xylem vessel members, ×35.

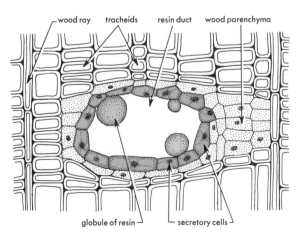

Figure 4.27 Resin duct in pine wood *(Pinus)* as seen in cross section.

Formation of Cork (Periderm)

Cork Cambium.   Figure 4.28 shows the origin of cork cambium and the development of secondary tissues from it. Cork cambium in most plants arises from outer cortical cells. The outer derivative cells from the cork cambium generally differentiate into cork cells, thus forming a layer of **cork** beneath the epidermis. The inner cells are known as **phelloderm**, and they are parenchymalike cells. Cork cambium may originate also from epidermal or phloem cells.

Cork tissue (Figs. 4.28, 4.29), is composed of flattened, thin-walled cells with no, or small, intercellular spaces. A fatty substance called **suberin** is deposited in the walls, rendering the cells almost impermeable to water and gases. Hence, this tissue provides protection for the stem against excessive loss of water and also against mechanical injury. The protoplasts of cork cells are short-lived.

The Dicotyledonous Stem

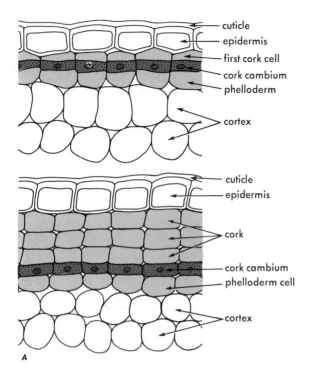

cuticle
epidermis
first cork cell
cork cambium
phelloderm
cortex

cuticle
epidermis
cork
cork cambium
phelloderm cell
cortex

A

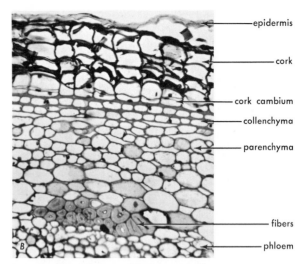

epidermis
cork
cork cambium
collenchyma
parenchyma
fibers
phloem

B

Figure 4.28 *A*, diagram showing origin of the first cork cambium and the first layer of cork. New cork cambia and new layers of cork form in a similar manner each spring. *B*, light micrograph showing cork cambium and newly formed cork in an elderberry stem *(Sambucus)*; ×200.

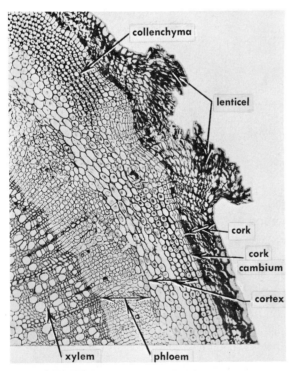

collenchyma
lenticel
cork
cork cambium
cortex
xylem
phloem

Figure 4.29 Cross section of a portion of a young stem of elderberry *(Sambucus)*. From the outside to the inside of the stem, the following tissues may be seen: cork, cork cambium, cortex, secondary phloem, and xylem. The vascular cambium is probably the two layers of thin-walled cells on the phloem side of the xylem. Note the lenticel in the cork, ×100.

Bark. In a young stem, the bark is made up of the following tissues, in this order, from the *outside* to the *inside*: **cork, cork cambium, phelloderm** (if present), **cortex,** and **phloem** (Figs. 4.28, 4.29). Microscopic examination may reveal the presence of epidermal cells still clinging to the cork. In old stems, the epidermis, cortex, and primary phloem become separated from the adjacent inner tissues by successively deeper layers of cork formation. As this happens, tissues outside newly formed cork die for lack of water and nutrients. They dry up and eventually wither away.

Lenticels

An impervious layer of cork would effectively cut off the oxygen supply of the living tissues beneath it if groups of parenchyma cells called **lenticels** did not develop in various places (Figs. 4.29, 4.2). They frequently originate beneath the epidermal stomata of young stems. When a cork cambium is formed, it produces ordinary parenchyma cells below these stomata in an outward direction. The resulting loose aggregation of parenchyma tissue bursts through the epidermis to form the lenticel. The air spaces between the parenchyma cells permit gaseous interchange.

The Monocotyledonous Stem

Primary Growth

The angiosperms can be conveniently divided into two major groups—**dicotyledonous** and **monocotyledonous** plants (Chapter 16). The primary difference between these two groups is that seeds of monocotyledonous plants, for

Figure 4.30 Arborescent monocotyledons. *A*, California fan palm (*Washingtonia filifera*) in a native stand; *B*, palms, and presumably *Pandanus*, have long-lived phloem; *C*, the Joshua tree *(Yucca brevifolia)* has secondary growth but produces typical closed vascular bundles from meristematic tissue resembling procambium strands.

example, grasses, contain only one cotyledon (seed leaf) while dicotyledonous seeds contain two. In addition, they have certain other characteristics in their plant bodies that are sometimes different.

Monocot stems are usually uniform in thickness from the top to the bottom of the plant, as opposed to being tapered like most dicot stems (Fig. 4.30).

The apical meristem of a typical monocot looks like a

The Monocotyledonous Stem

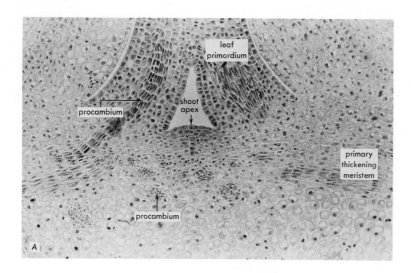

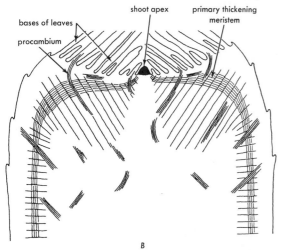

Figure 4.31 *A,* longitudinal section of *Iris* shoot apex to show the primary thickening meristem, ×88. *B,* longitudinal diagram of the shoot apex of corn *(Zea mays).*

small dome in a broad depression. Young leaves grow out of the depression and extend over the apical meristem. Beneath it is the inverted saucer-shaped primary thickening meristem (Figs. 4.31*A,B*). This meristem may extend down the stem a short way, so that it forms new cells upward to allow for an increase in length and outward to allow for an increase in girth. Leaf traces traverse the primary thickening meristem to connect it to already existing bundles.

The vascular bundles in monocots are usually scattered in a random pattern throughout the ground tissue (Fig. 4.32*A*), but some monocot stems have a hollow core and have bundles arranged in a ring similar to some dicot stems (Fig. 4.32*C*). The terms pith and cortex are often not applicable in monocot stems. The tangled vascular bundle pattern in the nodes of monocot stems is complex and is called a nodal plate (Fig. 4.32*B*).

Secondary Growth

Most monocot stems lack secondary growth. Palm trees increase their thickness in the apex by the activity of the

primary thickening meristem (Fig. 4.30*A*). Further down the stem, parenchyma cells continue to divide and enlarge, allowing for continued lateral stem enlargement. This process is called **diffuse secondary growth** since it does not involve an actual lateral meristem. In some palms, the stem looks much thicker at the base than at the top of the stem, but this thickening is caused partly by the overlapping leaf bases of old leaves, and partly by the presence, in some palms, of great masses of adventitious aerial roots.

Other monocots like *Agave, Cordyline, Sansevieria,* and *Dracaena* have true secondary growth that involves a secondary meristem (Fig. 4.33). This secondary meristem forms below the primary thickening meristem and extends to the base of the plant. It divides and forms only parenchyma cells to the outside (secondary cortex). To the inside, it forms parenchyma cells and secondary vascular bundles (Fig. 4.33*A*). The secondary vascular bundles consist of a ring of xylem surrounding the phloem, whereas the primary bundles consist of xylem and phloem side by side.

Secondary growth occurs in the roots of only one genus of monocot—*Dracaena.* A barklike layer is present in some monocots. It is derived from cortical parenchyma cells that divide to form files of suberin-filled parenchyma cells.

Stem Modifications

Stems may become adapted for functions other than support, conduction, and production of new growth. They may, for instance, become attachment organs for vines; they may carry on photosynthesis, store food or water, and develop protecting devices.

Rhizomes

A rhizome is a horizontal, underground stem. In most *Iris* species, leaves and flowering stalks are produced at the growing rhizome tip (Fig. 4.1*B*). The leaves die a relatively short distance back from the growing tip, so that many iris

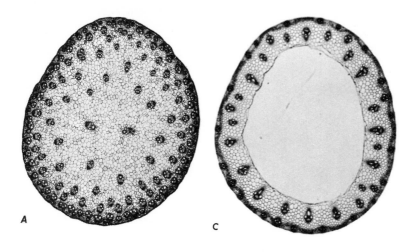

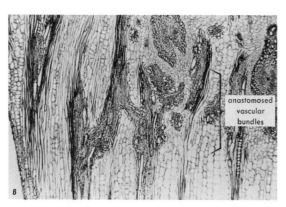

Figure 4.32   Internal anatomy of a monocotyledonous stem. *A*, cross section of corn stem *(Zea mays)* showing the scattered distribution of vascular bundles ×30. *B*, longitudinal section through a node in corn stem ×40. *C*, cross section of wheat stem *(Triticum aestivum)*, note that in this plant the core of the stem is hollow, ×30.

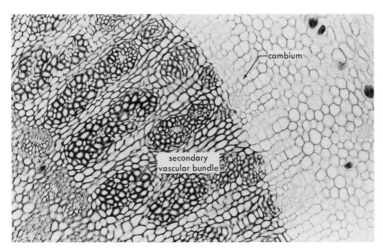

Figure 4.33   Cross section of a portion of monocotyledonous stem, *Dracaena* sp., that has true secondary growth. Note that the cambium forms discrete secondary vascular bundles and parenchyma cells to the inside of the stem, and parenchyma cells only toward the outside, ×13.

Stem Modifications

plants bear senescent leaves. Roots are also formed at nodes, and they may remain for the life of the rhizome. Other rhizomes, like *Canna* (Fig. 4.37A), produce upright leafy stems with terminal flowers at every third node. The intervening nodes are marked by only small sheath leaves.

Corms

A shortened vertical, thickened underground stem is a corm (Fig. 4.37). In *Gladiolus,* it consists of a short stem with much stored food. Nodes are, as usual, indicated by leaves; some bases are shown in Fig. 4.37B. Small buds occur in axils of some of these leaves. In a median section of a corm (Fig. 4.37C) one can distinguish between stored food and the central portion containing a single bud that will produce a single leafy, flowering shoot. Food stored in a dormant corm is used in the production of the leafy shoot. New corms will develop from axillary buds. In addition, short underground stems may form, each giving rise, at its tip, to a single small corm.

Bulbs

A bulb differs from a corm in that food is stored in leafy scales. The stem portion is small and has at least one central terminal bud that will produce a single upright leafy stem. In addition, there is at least one axillary bud that will produce a bulb for the subsequent year. In the longitudinal section of the sprouting daffodil bulb shown in Fig. 4.37D, the stem is producing three leafy stalks, one of which is forming a new bulb.

Food stored in the leafy scales of a bulb is used up by the initial growth of a leafy shoot. Food to be stored for a new bulb is supplied from a leafy shoot. The table onion is a good example of a commercially valuable bulb.

Tubers

Tubers are enlarged terminal portions of slender rhizomes (Figs. 4.36, 4.37E). The potato, *(Solanum tuberosum),* is a good example. The potato plant possesses three types of stems: (a) ordinary aerial stems, (b) slender underground rhizomes, and (c) their enlarged tips, tubers. In the mature potato the scar left where the tuber was broken from the rhizome is clearly visible. On the potato tuber there are nodes and internodes, lateral buds, and a terminal bud. Buds develop into stems (Fig. 4.37E). The "eyes" of the tuber are groups of buds; each group along the sides represents a lateral branch with undeveloped internodes. At the unattached "seed end" of the tuber, the "eye" is in reality a terminal branch on which only one bud is strictly terminal. In an elongated potato it is possible to make out the spiral arrangement of the eyes, because there is only one eye at a node.

Stem Tendrils

Tendrils are slender, coiling structures that are sensitive to contact stimuli and attach the plant to a support. Tendrils

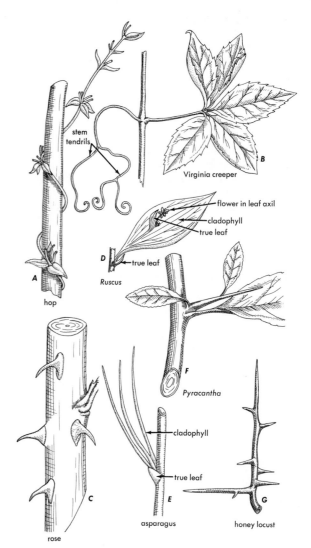

Figure 4.34   Types of stem modifications, *A* and *B,* stem tendrils; *C, F,* and *G,* thorns; *D* and *E,* leaflike stems.

are of two morphological sorts: leaf and stem. In the trumpet flower *(Bignonia),* for example, several uppermost pairs of leaflets have no blades, but instead form very slender leaf tendrils. Obviously, these are leaf tendrils. In grapes *(Vitis)* and Virginia creepers *(Parthenocissus quinquefolia),* tendrils are modified stems (Figs. 4.34A,B), as evidenced by their presence at nodes in leaf axils.

In the Virginia creeper, each tendril ends in a knob that flattens out when it comes in contact with a surface to which it adheres.

Cladodes

These are stems that are leaflike in form, are green, and perform the functions of leaves. They may bear flowers, fruit, and temporary leaves. Examples of plants with cladodes are *Ruscus* (Fig. 4.34D), *Asparagus* (Fig. 4.34E), *Smilax,* various species of cacti, and some orchids (e.g., *Epidendrum*).

Spines and Thorns

Most spines and thorns of plants are modified stems or outgrowths of stems. Spines, however, occur in certain plants, such as barberry *(Berberis)* and black locust *(Robinia pseudoacacia),* and in a few cases even roots become modified as spines. Good examples of stem thorns are those of fire-thorn *(Pyracantha)* (Fig. 4.34F) and honey locust *(Gleditsia)* (Fig. 4.34G). They are borne in the axils of leaves as ordinary branches are. Sometimes thorns bear leaves (Fig. 4.34F), which is further evidence that they are stems. Prickles, on the other hand, such as those on rose stems, are merely epidermal outgrowths, somewhat like hairs.

Stolons

Bermuda grass *(Cynodon dactylon)* has above-ground horizontal stems called stolons. These stems creep along the ground, and at each node, shoots and roots arise. In strawberry *(Fragaria)* (Fig. 4.35) stolons, roots, and leaves arise at every other node.

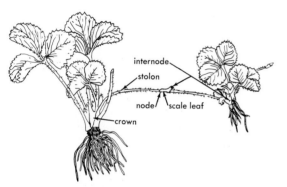

Figure 4.35   Runner of strawberry *(Fragaria)* × ⅛; roots and shoots are produced at every other node.

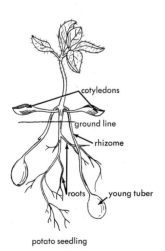

Figure 4.36   A potato seedling showing development of young tubers at the end of slender rhizomes.

Summary of Secondary Growth

1. The predominantly short life span of 3 to 5 years for cells, in general, limits the size and longevity of individual plants having only primary growth. The mechanism of secondary growth overcomes this limitation.
2. If all meristematic cells in a procambium strand become differentiated into primary vascular tissue, the vascular strand is closed to further growth.
3. If there remains an active meristematic region between primary xylem and primary phloem, continued growth is possible. These meristematic cells become the fascicular cambium.
4. An interfascicular cambium forms from ray parenchyma cells between vascular strands.
5. Union of fascicular and interfascicular cambium produces a complete cylinder of vascular cambium.
6. Divisions in cambium are longitudinal, so that the stem now increases only in girth.
7. Cork cambium is short-lived; new cork cambiums may arise each year, producing new layers of cork.
8. Cork cambium may originate in successively deeper tissues from epidermis, cortex, and phloem.
9. The production of secondary vascular tissue thus makes possible the attainment of great size and great age by individual trees, even though the functioning cells may be no older than those in a perennial herbaceous plant such as an *Iris.*
10. Most monocot stems have only primary growth that arises from a special primary thickening meristem. This meristem causes increase in both height and girth.
11. Palms supplement their lateral growth by the division of parenchyma cells throughout the stem.
12. Other monocots, such as *Agave* and *Sansevieria,* have true secondary growth, with a type of cambium that produces secondary vascular bundles and parenchyma cells.
13. In woody and herbaceous stems, vascular tissues are arranged in bundles, generally forming a circle. In monocotyledonous plants, bundles are irregularly distributed throughout the stem.
14. Rhizomes, bulbs, corms, stolons, and tubers are all modified stems.

PART TWO

Leaves

Green plants and a few species of bacteria are the only producers that use sunlight as an energy source; all other inhabitants of the earth consume what the green plants produce. In seed plants, leaves are the principal organs of production. **Chloroplasts** within leaf cells are the sites that trap light energy and use **photosynthesis** to convert it into chemical energy.

The water economy of plants calls for the absorption of much more water than can be metabolized. This excess water is returned to the atmosphere by leaves. A second role of leaves is **transpiration.** The structure of leaves is uniquely adapted to carry out these two primary roles.

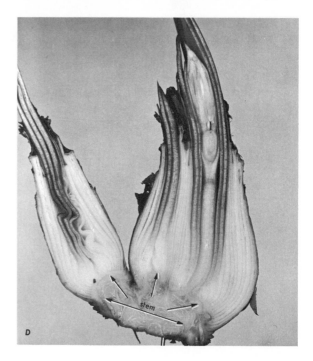

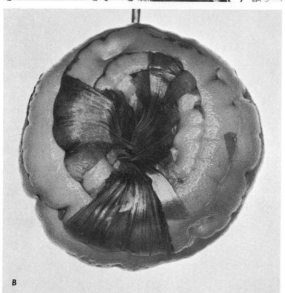

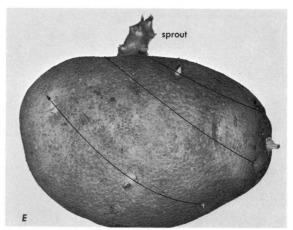

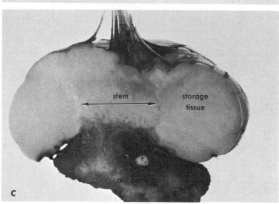

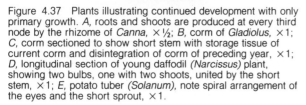

Figure 4.37 Plants illustrating continued development with only primary growth. *A*, roots and shoots are produced at every third node by the rhizome of *Canna*, ×½; *B*, corm of *Gladiolus*, ×1; *C*, corm sectioned to show short stem with storage tissue of current corm and disintegration of corm of preceding year, ×1; *D*, longitudinal section of young daffodil *(Narcissus)* plant, showing two bulbs, one with two shoots, united by the short stem, ×1; *E*, potato tuber *(Solanum)*, note spiral arrangement of the eyes and the short sprout, ×1.

Leaves are of many shapes and sizes (Fig. 4.38). Practically all leaves have veins for support and conduction, and a chlorenchyma tissue containing chloroplasts. Shape and size are such constant traits for many categories of plants that one may frequently identify an unknown plant simply by its particular type of leaf. Leaf shape, however, is not always a good diagnostic trait for use in plant identification, because it may vary within a species, and because leaf shape may sometimes be altered by its environment.

Leaf blades provide large surfaces for the absorption of light energy and carbon dioxide, both of which are required for photosynthesis. They are thin and hence no

Figure 4.38 Different kinds of leaves. *A*, poplar *(Populus deltoides); B*, castor bean *(Ricinus communis); C*, oak *(Quercus lobata); D*, rose *(Rosa odorata); E*, Virginia creeper *(Parthenocissus quinquefolia); F*, faba bean *(Vicia faba)*, × ½.

cells lie far from the surface. This form of the leaf facilitates the absorption of light energy and the exchange of carbon dioxide, oxygen, and water vapor between the intercellular spaces of the leaf and the atmosphere. The water that leaves utilize and transpire comes from roots and enters the leaf in the xylem tissue of veins. Food manufactured in leaves is transported out of the leaf in the phloem tissue of the same veins. In addition, veins may add support to the leaf.

Leaf Components

The leaves of dicotyledons are generally different from those of monocotyledons. A typical foliage leaf of a plant belonging to the dicotyledons is composed of two principal parts: (a) **blade** and (b) **petiole** (Fig. 4.38*A*). The blade is thin and expanded, the petiole slender. The thin blade is supported by a distinct network of **veins** that are composed of vascular tissues and fibers. In addition to forming a supporting framework for softer tissues of the blade, veins carry water, mineral salts, and food to and from the leaf.

The Blade

Variable external features of the leaf blade are its overall shape: **apex, margin,** and **base**. This range of variation is great, and many terms are employed by taxonomists to describe leaf shape accurately. Three of these terms seem important enough to us to mention here: leaf margin **entire** (smooth), **dentate** (toothed), or **lobed** (Figs. 4.38*F,A,C*).

The Petiole

Some leaves, such as those of peas *(Pisum)*, beans *(Phaseolus)*, roses *(Rosa)*, tulip trees *(Liriodendron)*, and many others, have two small, leaflike outgrowths at the base of the petiole, known as **stipules** (Figs. 4.38*D,F*). The petiole itself may be long or short, rounded or occasionally flat. It is usually attached to the base of the leaf blade, but in some plants, such as *Tropaeolum* and castor bean *(Ricinus)*, it is attached at the middle of the blade on the underside. This leaf is called a **peltate** leaf (Fig. 4.38*B*). The petiole is sometimes absent, as in *Zinnia*, the blade being mounted directly on the stem. Such a leaf

Figure 4.39   Leaves, showing parallel and netted venation. *A, Canna* leaf, about ×⅕; *B,* portion of similar leaf, about ×10; *C,* maple *(Acer)* leaf, about ×½; *D,* portion of similar leaf, about ×20.

is said to be **sessile** (Fig. 4.50*D,* juvenile *Eucalyptus*).

Monocotyledonous leaves such as grasses do not have distinct petioles. Instead, the leaf is divided into two parts, **sheath** and **blade**. The blade is the typical thin, expanded portion. The sheath is green, perhaps nearly as large as the blade, and it completely sheaths the stem (Fig. 4.40*A*). In many species such as corn the sheath extends over at least one complete internode. If the region of union between the blade and the sheath is examined carefully, a small flap of delicate tissue extending upward from the sheath may be seen, closely enveloping the stem. This is called the **ligule** (Fig. 4.40*B*). It may, in some cases, serve to keep water and dirt from sifting down between stem and sheath. In many species, of which barley *(Hordeum)* (Fig. 4.40*B*) is a good example, the base of the blade, at its

union with the sheath, is carried around the stem in two earlike points, the **auricles**. Ligule and auricles may both be present, or one or the other may be absent.

Simple and Compound Leaves

As to configuration of the blade, there are two kinds of leaves: (a) **simple** and (b) **compound**. In a simple leaf the blade is all in one unit (Figs. 4.38*C,B,* 4.39*A,C*). In a compound leaf the blade is composed of a number of separate leaflike parts, the **leaflets** (Figs. 4.38*D,F*). Inasmuch as leaflets have the characteristics of a simple leaf, it may sometimes be doubtful whether the structure is a simple leaf or a leaflet, especially if the leaflets are large or very numerous. A primary distinction is that buds occur

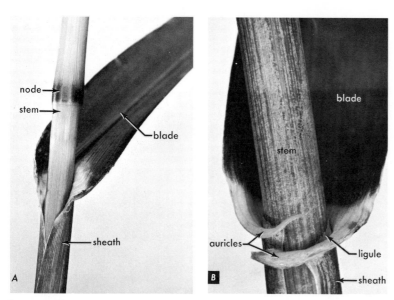

Figure 4.40 Leaves of members of the grass family. *A,* crabgrass *(Digitaria sanguinatis),* showing sheath and blade; *B,* ligule and auricles of barley *(Hordeum vulgare),* about × ¾.

in the axils of leaves, but not in the axils of leaflets.

When the leaflets of a compound leaf arise from the rachis (continuation of the petiole), as do the pinnae of a feather, the leaf is said to be **pinnately** compound, like the rose and bean (Fig. 4.38D,F); when the leaflets diverge from a common point at the tip of the petiole, the leaf is said to be **palmately** compound, like the horse chestnut *(Aesculus)* and Virginia creeper (Fig. 4.39E). Compound leaves may be once, twice, or thrice compound.

Venation

The arrangement of veins of a leaf is called venation. There are two principal types of venation: (a) **parallel venation,** usually characteristics of monocotyledonous leaves (Figs. 4.39A,B), and (b) **netted venation,** usually characteristic of dicotyledonous leaves (Figs. 4.39C,D). In netted venation there are one or more prominent veins from which smaller veins branch off to join with other small veins, thus forming a conspicuous net. In parallel veined leaves, there usually is one or a few large veins with many veins branching from them. These veins run parallel to each other and do not branch further. It is important to note that there are some dicotyledonous plants with parallel veined leaves and some monocotyledonous leaves with netted veins.

Anatomy of the Foliage Leaf

The anatomy of a leaf blade is best shown in section. In Fig. 4.41 we observe three principal tissues: (a) **epidermis,** (b) **mesophyll** (middle of leaf), and (c) **veins** or **vascular bundles.** The epidermis usually consists of a single layer of cells that covers the entire leaf surface. It protects the tissues within the leaf from drying out and

from mechanical injury. The mesophyll is composed of parenchyma cells, most or all of which contain chlorophyll, and thus are able to carry on photosynthesis. Veins possess xylem and phloem elements, which conduct water, inorganic salts, and foods. Fibers and collenchyma may be associated with conducting elements of midrib and larger lateral veins.

The petiole has its own specialized structures that enable it to support the leaf blade, conduct food, water, and inorganic salts, and disconnect itself from the stem at the close of the growing season without exposing living stem tissue to drying out or to infection.

Epidermis

The epidermis covers the entire leaf surface and is continuous with the surface of the stem to which the leaf is attached. In most leaves the epidermis is a single layer of cells. It may consist of several kinds of cells: (a) **ordinary epidermal cells,** (b) **guard cells,** (c) **hair cells** (Fig. 4.44). Ordinary epidermal cells show a variety of shapes, depending on the species. Epidermal cells are generally covered on their outer surfaces by a waxy cuticle secreted by their protoplasts. The deposition of the cuticle frequently forms a pattern, quite specific for the leaves on which it is found (Fig. 4.42).

A guard cell is a special type of epidermal cell. Guard cells occur in pairs, separated by an opening or pore (Fig. 4.43). Two guard cells plus a pore are called a stoma (plural = stomata). The size of stomata varies considerably according to the species of plants and even in any one plant. Here are some representative measurements in microns (length × breadth): bean, 7 × 3; geranium *(Pelargonium),* 19 × 12; corn *(Zea),* 19 × 5. The number of stomata per unit area varies widely depending on the species of plant and the environmental conditions under which it is growing. Usually more

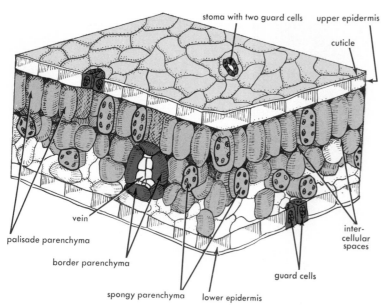

Figure 4.41   Three-dimensional diagram of a section of a foliage leaf, ×20.

Epidermis

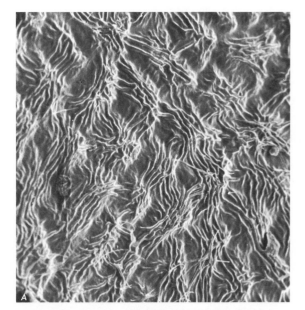

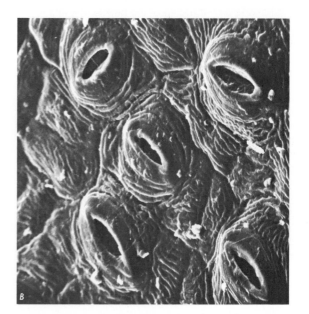

Figure 4.42   Cuticle on the epidermis of mala mujer (*Cnidoscolus*) as seen with the scanning electron microscope. *A*, upper surface, showing cuticular ridges, ×650. *B*, lower surface at higher magnification, showing cuticular ridges between stomata, ×1100.

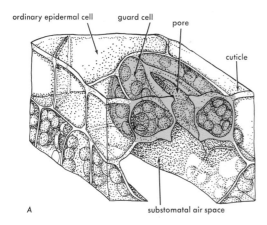

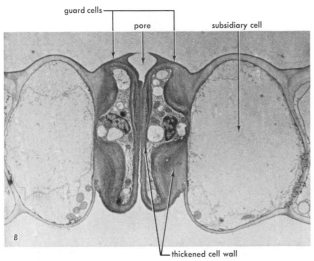

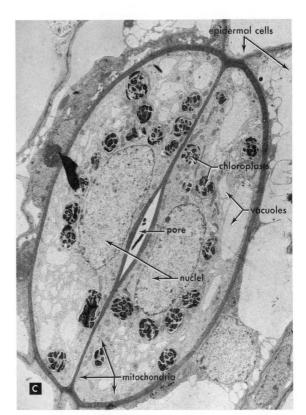

Figure 4.43   Views of stomata. *A*, diagram; *B*, corn (*Zea mays*), about ×1000; *C*, rice (*Oryza sativa*), ×900.

stomata are found on the lower than on the upper surface. Table 4.1 gives the average number of stomata per square centimeter on the upper and lower surfaces of some common plants.

Stomata are widespread in the plant kingdom. With the exception of a few submerged aquatic plants, stomata are present in all angiosperms and gymnosperms. Functional stomata have been found in liverworts, mosses, horsetails, club mosses, ferns, and cycads. In the angiosperms, they can occur on stems, petals, stamens, and pistils as well as on leaves. The major group of green plants that lack stomata is the algae. Stomata are structurally similar in the many different groups in which they are found.

A cross-sectional view of the guard cells shows that the walls are typically unevenly thickened, with the thicker, less elastic walls, adjacent to the pore. The stomata are the only openings in the leaf epidermis, and it is chiefly through them that gases pass into or out of the leaf. Although the cuticle is nearly impermeable to gases, small amounts of gases pass directly through the outer wall and the cuticle of epidermal cells. The role of stomata, including the control of their movement in transpiration, is discussed in Chapter 5.

Guard cells occur in pairs; each is crescent-shaped or semicircular in form, as seen in a surface view (Fig. 4.43*C*). Chloroplasts occur in guard cells but are lacking in ordinary epidermal cells. Both kinds of epidermal cells have long-lived protoplasts. The outer wall of epidermal

Table 4.1
Average Number of Stomata per Square Centimeter

Plant	Upper Epidermis	Lower Epidermis
Alfalfa (*Medicago*)	16,900	13,800
Apple (*Malus*)	0	29,400
Bean (*Phaseolus*)	4,000	28,100
Cabbage (*Brassica*)	14,100	22,600.
Corn (*Zea*)	5,200	6,800
English Oak (*Quercus*)	0	45,000
Nasturtium (*Tropaeolum*)	0	13,000
Oat (*Avena*)	2,500	2,300
Potato (*Solanum*)	5,100	16,100
Tomato (*Lycopersicon*)	1,200	13,000

cells, including guard cells, has a cuticle (Fig. 4.43*A*) like that of the epidermis of stems. It is effective in limiting both the inward or outward movements of water vapor and other gases. The cuticle is usually thicker on the upperside of the leaf than on the underside.

Several different types of **hairs** grow out from the epidermis of leaves and resemble hairs from the epidermis of stems. They may be unicellular or multicellular, simple or branched, scalelike or glandular. The unicellular hair,

Anatomy of the Foliage Leaf

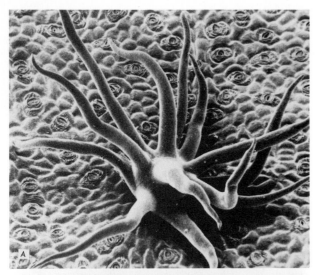

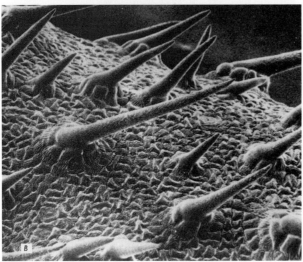

Figure 4.44 Trichomes: *A, Aleurites* leaf—branched; *B,* Croton leaf—simple, about ×350; *C, Phaseolus*—gland ×565.

the simplest kind, may be branched or unbranched. Multicellular hairs may consist of a single row of cells, but may also be branched (Figs. 4.44*A,B*). At the upper end glandular hairs bear a single large cell or a group of cells; some of these cells excrete ethereal oils, which often impart a stickiness to leaves (Fig. 4.44*C*).

Mesophyll

Mesophyll is the photosynthetic tissue between the upper and lower epidermis. It is parenchyma tissue, and is traversed by veins. Chloroplasts are present in the mesophyll cells, which, in some species, may be differentiated into two distinct layers: **palisade parenchyma** and **spongy parenchyma** (Figs. 4.41, 4.45). The palisade parenchyma is just below the upper epidermis and usually consists of from one to several layers of narrow cells with their long axes at right angles to the leaf surface. The spongy parenchyma extends from the palisade parenchyma to the lower epidermis. Cells of the spongy parenchyma are irregular in shape and loosely arranged.

Intercellular spaces are much larger in spongy parenchyma than in palisade parenchyma. Since the air spaces between mesophyll cells are interconnecting, many cells are in contact with an intercellular space. Thus, most food-making cells have free access to carbon dioxide and oxygen. To facilitate gas exchange, large air spaces, called **substomatal chambers,** are generally present beneath each stoma (Fig. 4.43*A*).

The leaves of some plants, especially those that grow in very dry habitats, may have modified leaf structure. For example, their mesophyll cells are tightly packed together and lack the differentiation between a palisade and spongy layer. Such plants may also have a thick cuticle and sunken stomata. All of these modifications are believed to reduce transpiration.

Vascular System

The veins, or vascular bundles, form a network that extends throughout the leaf. The conducting elements are xylem and phloem. Veins conduct water, mineral salts, and foods, and also mechanically support the mesophyll

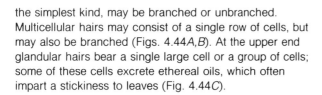

chapter **4** | **The Plant Body**

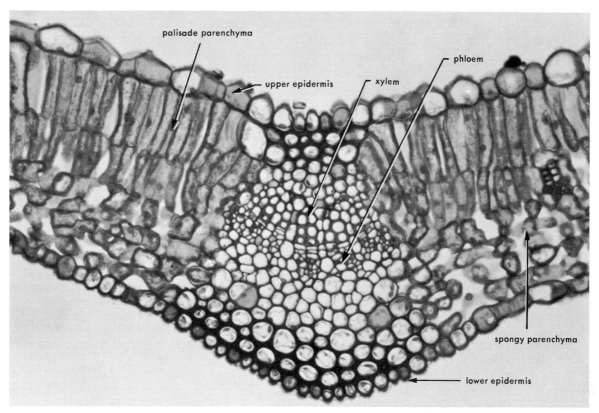

Figure 4.45 Photomicrograph of a cross section of a midrib of a lilac *(Syringa vulgaris)* leaf, ×150.

tissue. In addition to the midrib and larger lateral veins that are visible to the naked eye, innumerable minute branch veins can be seen with the aid of a microscope. The large veins contain vessels, tracheids, sieve tubes, companion cells, and also some mechanical tissue (Fig. 4.45). Such veins, in some leaves, may have both primary and secondary vascular elements.

In larger veins that have both xylem and phloem elements, the xylem is toward the upper surface of the leaf, and the phloem is toward the lower surface (Fig. 4.45). Smaller veins have few vascular elements and few or no mechanical elements. The very end of a vein is usually a single tracheid with a spiral wall (Fig. 4.46A). Veinlets are usually surrounded by one or more layers of parenchyma cells, the **bundle sheath,** which may or may not possess chloroplasts. Water and solutes must pass from conducting elements of the veinlet through the bundle sheath in order to reach the mesophyll. The smallest veinlet has an unbroken connection with vascular elements of the midrib, petiole, and stem to which the leaf is connected. Plants such as corn and other tropical grasses use a photosynthetic pathway called C_4 photosynthesis (Chapter 6). Leaves of these plants usually

are characterized by the presence of an enlarged bundle sheath (Fig. 4.46B), and lack mesophyll differentiation into palisade and spongy types.

The Petiole

In the petiole, phloem and xylem maintain their relative positions; as in the stem the phloem is on the underside of the petiole (and leaf blade) and xylem is on the upperside. One or more vascular bundles are embedded in parenchyma. Fibers may be associated with vascular tissues of bundles and, not infrequently, groups of collenchyma cells occur beneath the epidermis. Collenchyma cells are also commonly present in leaves. Usually they form part of the epidermis, but may be formed in the mesophyll near the leaf surface adjacent to the vascular bundles.

Leaf Development

Leaf development is intimately associated with the differentiation of the young shoot apex.

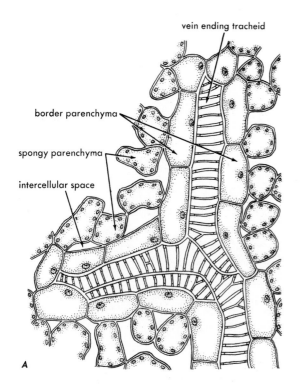

vein ending tracheid

border parenchyma

spongy parenchyma

intercellular space

A

The leaf is initiated on the flanks of the apex, as a slight bulge resulting from the enlargement and division of several cells in the outer layers of the apical meristem (Fig. 4.47). This bulge continues to enlarge until it forms a thin, elongated **leaf primordium** (Fig. 4.48A). The flanks of the primordium (the marginal meristem) will initiate the formation of the leaf blade (Figs. 4.48B,C). This region of growth remains active until the leaf has fully formed. Most cell divisions are completed early in leaf development; cell enlargement accounts for most of the increase in leaf size.

During the early stages of leaf formation, the cells immediately beneath each primordium also become meristematic. These cells, the procambium, will form the vascular tissue that connects the leaves to the stem's vascular system (Fig. 4.47).

One problem that has continually intrigued plant biologists is how the plant regulates the site of new leaf initiation. The leaves of vascular plants are distributed on the stem in four distinct patterns **(phyllotaxy)—opposite, alternate, whorled,** and **spiral** (Fig. 4.49). The position of a new leaf primordium is closely correlated with the position of the next older primordium. This has been proven by the experimental removal or isolation of the youngest visible primordium on a shoot apex, causing the

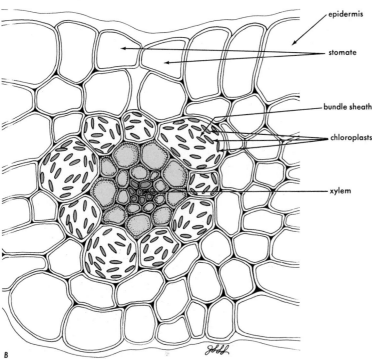

epidermis

stomate

bundle sheath

chloroplasts

xylem

B

Figure 4.46   Diagram of leaf vascular bundles. *A,* note the parenchyma tissue bordering the vein and the presence of only annular wall thickenings in the tracheids at the vein tip. *B,* cross section of a leaf with C$_4$ photosynthesis, note the prominent bundle sheath cells.

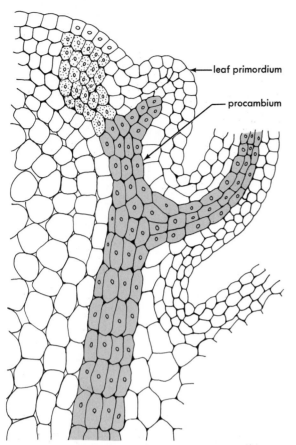

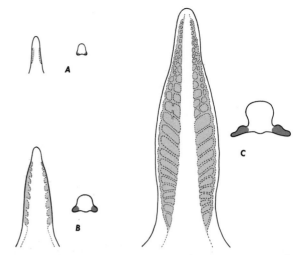

Figure 4.48 Growth of a young leaf. *A*, pencil stage, formation of marginal meristems; *B*, activity of marginal meristems initiates lateral expansion of a young leaf; *C*, continued lateral growth through the activity of marginal meristems.

next formed primordium to develop closer to it. It is also known, at least in ferns, that if the procambium beneath a suspected leaf primordium is severed, the primordium will not develop.

Leaf Shape

Internal Control

Leaf shape is under direct genetic control. Several single gene mutations of tomato *(Lycopersicon esculentum)* result in striking changes in leaf shape (Figs. 4.50*A,B*).

Figure 4.47 Diagram of a shoot apex showing the direct relationship between the initiation of a leaf and procambium strand. The second leaf shows a definite primordium and the third leaf is well-formed. The procambium strands associated with these leaves are shown.

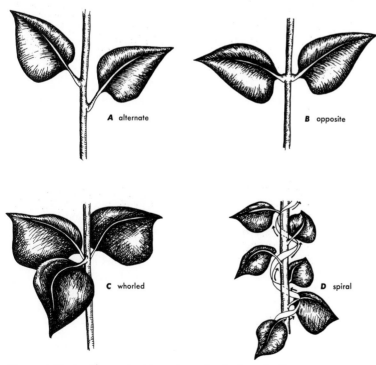

Figure 4.49 Leaf arrangement on stem axis; *A*, alternate; *B*, opposite; *C*, whorled; *D*, spiral.

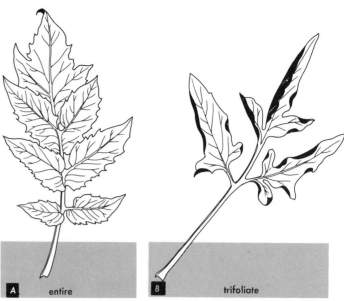

A entire   B trifoliate

Leaves are also affected by the availability of internal growth regulators. When treated with the hormone gibberellic acid, thistle plants *(Centaurea),* for example, formed abnormal, entire leaves instead of normal, lobed leaves.

Leaf shape may also vary considerably with the physiological age of plants. Thus, some plants have distinctive juvenile leaves for the first few years of growth. Subsequent adult leaves of a different form are characteristic of the older or adult plant (Figs. 4.50*C,D*).

Environmental Factors

Leaf shape is also influenced by environmental factors such as light, moisture, and temperature. The total intensity, wavelength, and daily duration of light have separate but interrelated effects. Plants growing in intense sunlight **(sun leaves)** usually have thick leaves with a thick palisade tissue and a dense, spongy parenchyma. Their intercellular spaces are small, and the epidermis is heavily cutinized and generally glossy, with the stomata confined to the lower epidermis. These plants may also have woolly epidermal hairs. Leaves of the same species growing in shade **(shade leaves)** have contrasting traits. They are thin, possessing a single palisade layer and a spongy parenchyma with many intercellular spaces. The epidermis has a thin cuticle that is usually dull, and stomata may be present on both upper and lower epidermal surfaces.

Light is required for the normal development of leaves, including the differentiation of chloroplasts to a state where they are capable of carrying on photosynthesis. In darkness, the leaves of a typical dicot remain small and pale yellow while the stems grow long and slender. This condition is called **etiolation.** Regular daily illumination that is intense enough to support photosynthesis is required for the continued healthy existence of leaves. Excessive shading usually results in the death of a leaf.

Submerged leaves of semi-aquatic plants may be vastly different in shape from aerial leaves of the same plants (Fig. 4.50*E*). Similar changes may be induced in the aerial portion of the shoot by reducing the CO_2 content of the air and lowering the temperature.

Nitrogen starvation may also induce the plant to acquire leaf traits similar to those of sun leaves; water stress may change the plant in a similar way. Thus, we see that environmental stress is an important, but complex, factor.

Leaf Modifications

Leaf shape may be considerably modified to perform functions other than photosynthesis (Fig. 4.51).

Bud scales are short, thick, sessile, often covered with dense hairs on the outer surface, and sometimes waxy or resinous. When present, they protect the delicate meristematic tissue of the shoot tip and the rudimentary leaves from drying out.

The **spines** of various species of cacti and those of *Fouquieria* represent entire transformed leaves. In the black locust (Fig. 4.51*A*), the spines are stipules, rather than leaves.

In some species of *Lathyrus,* the **tendrils** are transformed leaflets; in other species, the whole leaf is transformed into a single tendril, and leaflike stipules perform the normal functions of the leaves. In *Bignonia capreolata* (trumpet flower), the third leaflet is transformed into a tendril (Fig. 4.51*B*). Both leaf tendrils and stem tendrils serve to attach the plant to a support.

Leaves are sometimes modified as food or water storage organs. The thick, fleshy bases of leaves that comprise much of the daffodil *(Narcissus)* bulb (Fig. 4.37*D*) accumulate large quantities of food. Succulents found in deserts and saline soils have thick, fleshy leaves with special water-storage tissue.

A striking adaptation of leaves to a special function occurs in insectivorous plants. In these plants, the leaves have taken on forms and various structural features that enable them to capture insects and obtain food from their

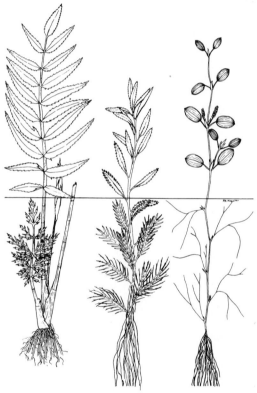

Figure 4.50 Examples of induced leaf shape changes. *A, B,* different leaf shapes induced in leaves of tomato (*Lycopersicon esculentum*) by single gene mutations. *C, D,* adult and juvenile leaves of *Eucalyptus* sp., × ¼ and × ½. *E,* examples of submerged and air forms of leaves of three aquatic plants.

bodies. Well-known insectivorous plants are sundew (*Drosera*, Fig. 4.51*E*), pitcher plants (*Darlingtonia*, Fig. 4.51*C*), and the Venus' fly trap (*Dionaea muscipula*, Fig. 4.51*D*).

Leaves sometimes function effectively in vegetative reproduction. In certain species of *Bryophyllum* (Fig. 4.51*F*), patches of tissue located in notches along the leaf margins remain meristematic. This tissue will eventually develop small, new plants while the parent leaf is still active. The little plants later drop from the leaf to the ground, where under favorable conditions they may develop into new individuals. Leaves of *Begonia* produce plantlets by the differentiation of small clusters of epidermal cells on the upper leaf surface.

Leaf Abscission

The separation of plant parts from the parent plant is a normal, continually occurring phenomenon. Leaves fall, fruits drop, flower parts wither and fall away, and even branch tips or whole branches may be separated from the parent plant as a normal part of its life. The fall of leaves from woody dicotyledons in the autumn is the most common example of this phenomenon. In practically all cases, separation or **abscission** is the result of differentiation in a specialized region known as the **abscission zone** at, or close to, the base of the petiole (Fig. 4.52). Parenchyma cells comprising the abscission layer may be smaller and lack lignin, compared to the cells of adjacent tissues. Even vascular elements may be shorter, and fibers may be absent from the bundle in the abscission zone. These anatomical features definitely make this zone an area of weakness.

Previous to the fall of leaves changes may normally occur in this zone. Cell divisions, though apparently not necessary, frequently take place, and produce a layer of brick-shaped cells across the petiole. Actual separation of the leaf may be brought about in several ways. In some

Figure 4.51 Examples of leaf modifications. *A, Robinia* stipular spines; *B, Bignonia* tendrils; *C,* pitcher plant leaf *(Darlingtonia* sp.); D, Venus fly trap *(Dionaea* sp.); *E,* sun dew *(Drosera* sp.); *F,* foliar plantlets in *Kalanchoe* sp.

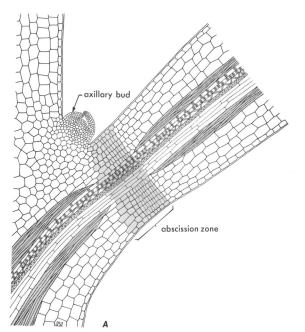

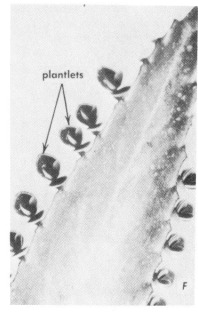

plantlets

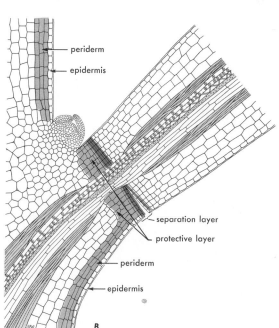

Figure 4.52 The formation of the abscission zone.

axillary bud

abscission zone

A

periderm

epidermis

separation layer

protective layer

periderm

epidermis

B

species, the middle lamella is dissolved away and cells separate. In other plants, walls and cells are dissolved. In a third small group of plants, a layer of cork forms across the petiole so that the leaf simply withers in place and is blown away. In all cases, a protective, corky layer of cells develops across the leaf scar. It is continuous with the stem cork. Furthermore, the vessels are likely to become plugged with tyloses or gums. These devices prevent the fall of a leaf from leaving an open wound or a point of entrance for organisms that might cause disease.

The abscission zone has two functions: (a) to bring about the fall of the leaf or other plant part and (b) to protect the region of the stem from which the leaf has fallen against insect damage or rot caused by bacteria or fungi.

Environmental cues are important in synchronizing abscission with the seasons. These cues are most commonly cold temperature or short days that in turn induce hormonal changes that then affect the formation of the abscission zone (Chapter 8).

Summary—Leaves

1. The chief functions of leaves are photosynthesis and transpiration.
2. Angiosperm leaves are generally supplied with flat, thin blades that are attached to the stem by petioles or sheaths. Veins strengthen the blades and transport food and water. Leaf blades may be simple or compound; leaf margins may be entire, dentate, or lobed.
3. A cross section through the blade of a typical leaf shows the following tissues: upper epidermis, mesophyll, differentiated into palisade parenchyma and spongy parenchyma, and lower epidermis. Generally, a waxy cuticle coats the epidermis. Guard cells of the epidermis form stomata that control gas exchange.
4. Mesophyll tissue is comprised of palisade and spongy parenchyma tissues and veins. Chloroplasts are present in both parenchyma tissues. Mesophyll is adapted for photosynthesis. Some plants do not differentiate mesophyll into spongy and palisade regions.
5. Leaf shape is under genetic and hormonal control, and is also influenced by light, moisture, and the physiological age of the plant. Leaves may become modified, serving as bud scales, spines, tendrils, the source of propagules, food or water storage organs, and insect traps. Such environmental factors as strong light intensity, nutrient deficiency, and water stress can affect leaf morphology and anatomy.
6. Leaf primordia develop in a definite spatial sequence on the flanks of the shoot apex.
7. The young leaf lamina is produced by marginal meristems that are also responsible for the shape of the leaf.
8. Cell divisions are completed while the leaves are enclosed in the bud; leaf expansion results mainly from cell enlargement.
9. The formation of a definite abscission zone across a petiole or fruit stem is responsible for leaf fall or fruit drop. Environmental cues and hormonal changes affect formation of the zone.

PART THREE

Roots

Functions of the Root

The functions of the root system are **anchorage**, **storage**, **conduction**, and **absorption**. You need only walk through a stream bed and observe the exposed root systems of large trees, or attempt to pull weeds, to get a first-hand

Figure 4.53 Exposed roots of white elm.

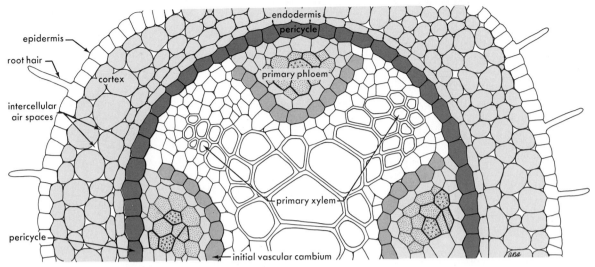

Figure 4.54 *A.* Cross-section of the root showing the primary tissues. *B.* Development of a root hair.

idea of the anchorage function of roots (Fig. 4.53). Plants with unusual roots provide an anchorage function in atypical ways. Adventitious roots on ivy, for example, develop from the stems; these short roots have a flat end-pad that anchors the vines to their growing surface. The sticky substance that is secreted for this purpose tends to erode building surfaces, often requiring the ivy to be pulled down before the building crumbles. Parasitic plants, like dodder, anchor themselves by sinking haustorial roots into their host's vascular tissue in order to tap its water and nutrient supply.

The storage function of roots is seen in carrots, which have a large tap root that undergoes secondary growth (Fig. 4.55*E*). Its mass of secondary xylem, however, is composed mostly of storage parenchyma cells stuffed with water and carbohydrates. Root storage usually assists the plant at some developmental stage, such as providing nutrients during flower production. Sugar beets, for instance, are biennial plants that begin to fill their storage root with food at the end of the first growing season. This stored material is partially used to assist the growth of early shoots during the following season, but flowering during summer uses most of the stored energy.

In a herbacious plant water is responsible for about 95% of root weight. Water is needed for all root processes and for every metabolic reaction. Soil water contains dissolved salts and minerals, such as potassium, sulfur, phosphorous, calcium, magnesium, which are needed by the plant (Chapter 2). For these reasons, plants have developed an elaborate system for the absorption and conduction of water (Chapter 5). Root hairs penetrate between small soil particles and are in intimate contact with them and the water film that surrounds them (Fig. 4.54). The cell walls of the root hairs and epidermal cells are made of cellulose and pectin. The pectic coat on the outside of the cell wall makes the root hairs gummy, facilitating their adherence to soil particles. Water is

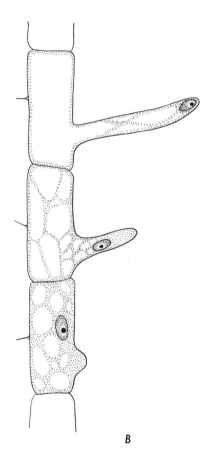

B

absorbed by the root hairs and epidermal cells and is then passed on by diffusion and active selective processes to the conducting elements (Fig. 4.54; also see Chapter 5).

Types of Root Systems

When pulled up, many grasses and small garden plants take with them a massive clump of soil. This happens because the **fibrous root system** of these plants consists

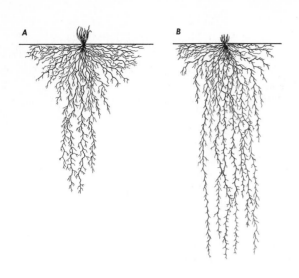
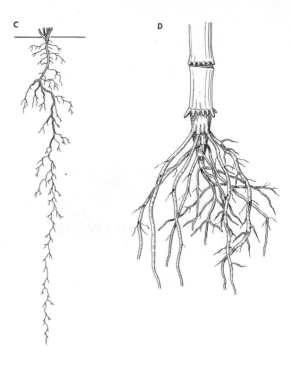

Figure 4.55 Different types of root systems. *A*, shallow, spreading, fibrous root system; *B*, fibrous root system penetrating the soil evenly from 1–1.5 m; *C*, tap root system, in which main primary root penetrates soil 2.5 m or more; *D*, fibrous root system developed from adventitious roots growing at lower nodes of stem; *E*, tap root system in carrot *(Daucus carota)*.

of several main roots that branch to form a dense mass of intermeshed lateral roots (Fig. 4.55).

Vegetables with a large storage root, like the carrot, have a **tap root system** that consists of one main root from which lateral roots radiate (Fig. 4.55). Some desert plants have a rapidly growing tap root system that enables them to penetrate the soil quickly to reach deep sources of water. The depth of root penetration of various plants is quite remarkable. A University of Nebraska botanist, John Weaver, exhumed the entire root systems of 45 species of native plants from the Midwest prairie. Only four species maintained root systems primarily in topsoil; most of them had roots that extended to 4 m deep. Even in very hard soils, the depth of root penetration well exceeded the height of the above-ground plant. A typical annual plant, like corn or rye, will build an immense fibrous root system in one growing season. In a study on rye, a single plant 50 cm tall, with 80 tillers (shoot branches), had a root system amounting to 210 m² of surface area compared to only 4 to 5 m² of the above-ground portion (Table 4.2).

In trees and shrubs, most functioning roots are localized in the upper 1 m of soil, with the majority of the feeder roots in the upper 15 cm (Fig. 4.56). However, these plants may extend lateral roots well beyond the expansion of overhead branches. In areas of closely packed trees, or on sandy soils, competition and low surface moisture may reduce the amount of area covered and encourage deeper root penetration.

Development of Root Systems

The seeds of higher plants contain a small undeveloped plant, the embryo. When the seed germinates, the embryonic root, or **radicle**, extends by the division and elongation of its cells to form the **primary root** (Fig. 4.57). Tap root systems develop from only one enlarged primary root, which then forms **lateral** or **secondary roots**. Further branching of secondary roots give rise to succeeding orders of roots, for example, tertiary, quarternary, etc. Fibrous root systems develop in a different way. The embryo of some grasses, like barley, consists of not only one radicle, but also of several additional embryonic roots called **seminal roots**. The seminal roots emerge soon after the radicle during germination. They elongate rapidly, and soon it is not possible to distinguish the primary root. Usually the seminal roots do not persist, but in some grasses they may function throughout the plant's life.

Roots that develop from organs other than roots are called **adventitious roots**. In most instances, adventitious roots develop at the nodes of the stem. There are several common examples of adventitious roots. In a young corn plant, soon after the emergence and development of a rudimentary root system, **prop roots** develop from the shoot node nearest the soil level (Fig. 4.55). These prop roots function as roots, but also assist in the support of the plant in the soil. Banyan *(Ficus bengalensis)* trees grow in saline mud in tropical lagoons and tidal marshes.

Table 4.2
Rye Root System Measurements

Root	Number	Length (m)
Main roots	143	65
Secondary roots	35,600	5,181
Tertiary roots	2,300,000	174,947
Quarternary roots	11,500,000	441,938
Total	14,000,000	609,570

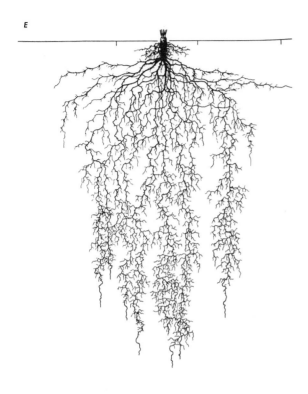

Figure 4.56 Root system of Cox's Orange Pippin on M. II, excavated when 16 years old from brick-earth soil. Note main scaffolding of roots about 25 cm deep and vertical "sinkers."

Figure 4.57   Radish seedling *(Raphanus sativus)*. *A,* photograph showing root hair growth, ×2; *B* and *C,* diagrams showing growth regions.

Branches of these trees form adventitious prop roots that extend down into the mud where they enlarge and actually prop up the large branches. The banyan is considered a sacred tree in some parts of India. Indian merchants have in the past held open-air bazaars among the prop roots and expansive branches of the banyan (Fig. 4.58).

In some instances, adventitious roots do not form at nodes. Pieces of stem, like a cane from a blackberry plant, or a branch of willow, can be made to root from their cut ends simply by placing them in moist soil. Leaves from *Begonia* and several other species also can be rooted simply by soaking them in water. Many commercially important ornamentals, in fact, are reproduced by root propagation from leaves or stems (Chapter 8).

Internal Anatomy

Root Apex

In our discussion of root structure let us first begin with the embryonic root, the radicle. The **root apical meristem** at the tip of the radicle is composed of small, regularly shaped cells that are capable of dividing. The emergence of the radicle during germination is dependent on the initiation of cell division in this region. After the radicle emerges, the cells of the root apical meristem continue to

divide; on the average, one daughter cell of each division remains as part of the meristem and divides again, while the other daughter cell differentiates into a mature, specialized cell.

The central portion of the apical meristem is composed of a pellet-shaped region of cells known as the **quiescent center** (Fig. 4.59). These cells divide, but at an extremely slow rate. They act as a regulatory center by releasing dividing cells, **initials,** just fast enough to continuously maintain the root's shape and to keep pace with its growth. The function of the quiescent center is unknown; one hypothesis, however, is that it may be the location where growth regulators are synthesized and released to control development of the root as new cells are made.

Some initials divide and produce cells in front of the apical meristem to form the **root cap.** The root cap is a thimble-shaped region of cells that surrounds the apical meristem and precedes it as the root elongates and forces its way through the soil. The root cap not only protects the apical meristem and lubricates its passage through the soil, but also is the site that perceives gravity and controls the direction of root growth.

Initial cells behind the quiescent center divide to supply the cells for the primary root tissues. These cells continue to divide, so that the region of cell divisions may extend as far as 1.5 mm from the tip of the root. Beyond this region, as far as 5 mm from the tip of the root, the cells are

Figure 4.58 Banyan tree with ancient market under aerial prop roots.

rapidly elongating in preparation to perform their mature functions.

The root can be conveniently divided into regions of cells (tissues) that perform different and specific functions in the mature root. These tissues originate from zones called **primary meristems**; they are the **protoderm, ground meristem,** and **procambium** (Fig. 4.59). These are similar to the primary meristems in the shoot. Derivative cells from these three primary meristems elongate and differentiate into the three primary tissues: **epidermis, cortex,** and **vascular tissue.**

Formation of Primary Tissues

The protoderm consists of uniformly shaped meristematic cells. At some distance from the root apex, some cells of the protoderm develop into **root hairs** by the extension of their cell wall into the surrounding soil. Root hairs form in the zone of differentiation in the root. It is most likely that root hairs have as their principal function to reach and tap for water in the soil (Figs. 4.54, 4.59). New root hairs continually form as old ones die. In one rye *(Secale)* plant, 100 million new root hairs are estimated to be formed each day.

Ground meristem cells produce the cells of the cortex. The cortex consists mostly of storage parenchyma and occasionally of sclerenchyma cells. The innermost cell layer of the cortex is the **endodermis.** Endodermal cell walls are usually thin, except for a suberized band on the radial and transverse walls, the **Casparian strip** (Fig. 4.60). Electron micrographs show that the plasmalemma of the endodermal cell is fused to the Casparian strip. This arrangement creates a barrier in these walls, making them impermeable to the passage of water and forcing water to pass through the protoplasm of the endodermal cell layer.

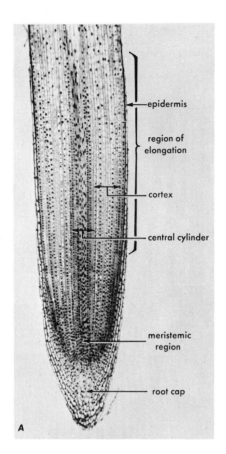

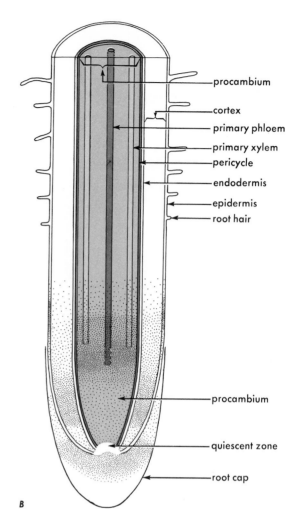

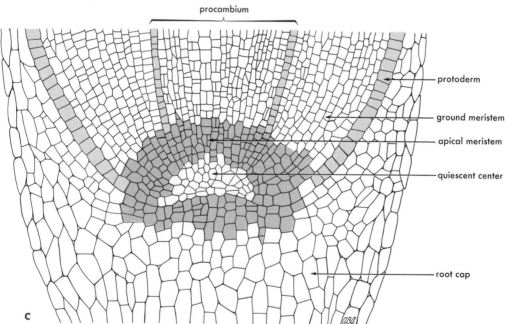

Figure 4.59   Photographic (*A*) and diagrammatic (*B*) representation of a longitudinal-section through a root from the tip to the initiation of the vascular tissue. *C.* Longitudinal-section through root tip meristem.

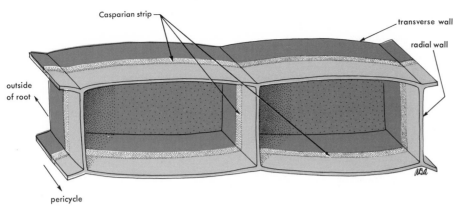

Figure 4.60 Three-dimensional view of two endodermal cells. The suberized Casparian strip is shown in gray. Water from solution outside the root wets the cellulose walls until it reaches the Casparian strip. Since water cannot wet and cross the Casparian strip, the cell walls inside the Casparian strip are wet by water that has passed through the protoplast of the endodermal cell.

This enables the root to regulate the movement of dissolved materials (Fig. 4.54).

Endodermal cells in mature regions of the root often develop partial secondary cell walls. Passage cells adjacent to xylem points often do not form secondary walls, thus ensuring the continued passage of water.

The **vascular cylinder** or **stele** develops from the procambium and forms the central core of the root. In dicotyledonous roots, the xylem occupies the middle of the stele, with the periphery consisting of alternating radii of xylem and phloem (Fig. 4.61A). In monocotyledonous roots, the central core contains parenchyma cells (Fig. 4.61B). The first matured xylem-conducting elements, the **protoxylem,** develop at the points of the radiating xylem. Dicot roots with two protoxylem points are called diarch roots, those with three points are triarch, and so on; most monocot roots have more than five protoxylem points and are therefore called polyarch. **Metaxylem** cells form within the mass of xylem to complete the central core.

The protoxylem is modified so that it is capable of transporting water while the root is elongating; that is, protoxylem cells must be able to withstand the forces that move water, and still be flexible enough to elongate. This is accomplished by the formation of a secondary cell wall in the shape of annular rings or as a spiral (Fig. 4.14A). When the root has completely elongated, the metaxylem cells mature. They no longer are required to elongate and, consequently, form thick secondary cell walls with pitted areas through which lateral exchange may take place (Fig. 4.14A). Protoxylem cells often become crushed while the metaxylem develops. Xylem of roots consists of vessel members (in flowering plants), tracheids (in most ferns and conifers), parenchyma, and fibers.

Phloem cells form in the spaces between the protoxylem arms. The protophloem is the first part of the vascular system to become functional. These cells form at the periphery of the phloem and function primarily during root elongation. Metaphloem cells develop toward the inside, and they function during the plant's adult life. Phloem of roots consists of parenchyma and fibers; sieve-tube

members and companion cells (in flowering plants); and sieve cells (in conifers and ferns).

The outer layer(s) of cells that comprise the stele is called the **pericycle.** This cell layer(s) is meristematic and plays several important roles in the life cycle of the root. In the shoot, recall that new leaves are formed at the surface and very near the apical meristem of the shoot. In roots, lateral roots are instead initiated below the surface and at a distance from the meristem.

Initiation of lateral roots at particular locations is controlled by chemical growth regulators that cause pericycle cells to begin to divide (Fig. 4.62). The **lateral root primordia,** which result, continue to form new cells that in turn elongate. Endodermal cells subtending this region often also divide for a short time, contributing cells to the tip of the new lateral root. As it expands, the lateral root pushes its way through and destroys the cortical cells and the outer epidermis. As it emerges, the new lateral root becomes organized into an apical meristem with primary meristems identical to its parent root. This system of lateral root formation may present a hazard in the plant's development, because the lateral root forms a wound in the cortex and epidermis of the primary root. If the wound doesn't heal rapidly, it is a good point of entry for plant pathogens.

Secondary Growth in Roots

Roots grow at such fast rates that the primary tissues are not capable of supplying the needs of a long-lived plant body. In these cases, secondary tissues are developed. As in the stem, two secondary or lateral meristems are required to form these tissues, the **vascular cambium** and the **cork cambium.**

In roots, secondary growth is initiated by the active division of (residual) procambium cells between the arcs of xylem and phloem. These cells form secondary xylem to the inside and phloem to the outside (Fig. 4.63). After a time, the crescent-shaped region of dividing cells joins

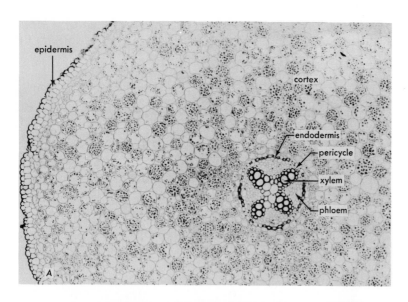

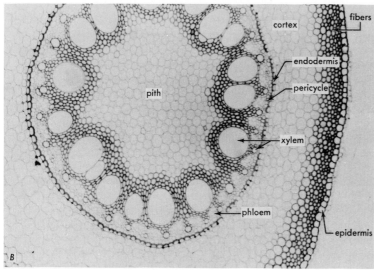

Figure 4.61   Cross section views of typical roots. *A*, dicotyledonous root of *Ranunculus* sp., X83. *B*, monocotyledonous root from corn *(Zea mays)*, X79.

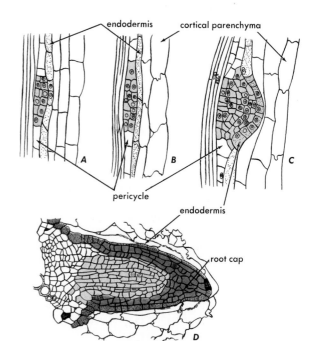

Figure 4.62   Branch roots, *A*, initiation of branch carrot *(Daucus carota)* root through formation of meristematic cells in pericycle; *B* and *C*, enlargement of meristematic region; *D*, young root pushing through cortex, X50.

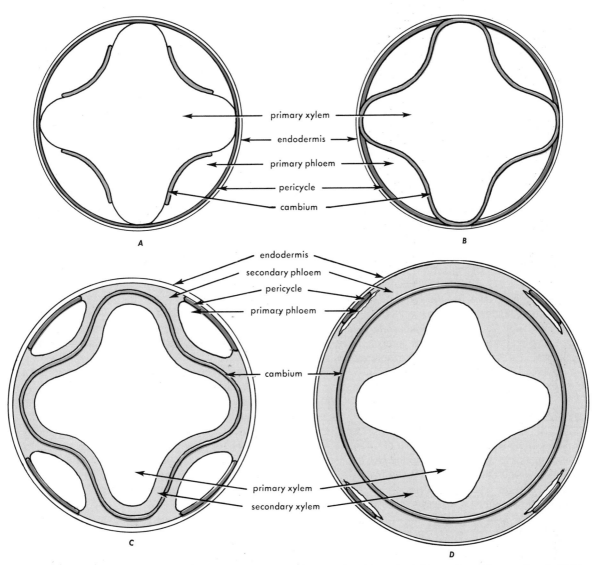

Figure 4.63 Diagrammatic representation of the development of the secondary plant body in a root. *A*, at the completion of primary growth a row of procambium cells remains (light green) and a complete circle of pericycle is present (dark green). *B*, the procambium (light green) joins with the pericyclic cells outside the xylem arms to form a continuous cyclinder of vascular cambium (light green). *C*, the vascular cambium forms secondary xylem internally and secondary phloem externally (light gray). The primary phloem is being pushed outward and a small amount of pericycle (dark green) is still associated with it. *D*, a smooth circle of vascular cambium forms, producing secondary xylem and phloem. The primary remains in the center of the stem, the primary phloem has been crushed, and only a small amount of the pericycle remains.

with the pericycle, which in turn also begins to divide. This connection forms a complete ring of vascular cambium around the entire vascular cylinder (stele). Secondary tissues are produced during the growing season to form annual growth rings, just as in the stem. Continued growth expands the root and finally causes the splitting, sloughing off, and destruction of the cortex and epidermis. However, early stresses of expansion, plus the activity of growth hormones, stimulates the pericycle between the xylem arms to resume division and form the cork cambium. This meristem forms cork cells to the outside, and parenchyma to the inside, just as in the stem.

Special Roots

Roots of a great many plants do not have the general characteristics common to most roots. We have already discussed examples of adventitious roots that arise from nonroot origins, the roots of parasitic plants, and the uniquely shaped sucker roots of clinging ivy vines. Other plants, especially those forming partnership with microorganisms, have specialized root structures.

Mycorrhizae

Mycorrhizae are short roots common to many plants; they represent an association with a soil-borne fungus. Two types of mycorrhizal roots may be found, depending on whether the fungus penetrates into the root cells or not. **Ectotrophic** types are found in roots of such trees as pines *(Pinus)*, birches *(Betula)*, willows *(Salix)*, and oaks *(Quercus)*. This type causes a drastic change in the root shape (Fig. 4.64). These mycorrhizal roots are about 0.5 cm long, have no root cap, and exhibit a simple monarch stele. In addition, the fungus penetrates between the cell walls of the cortex and forms a covering sheath, or mantle, of fungal hyphae (Chapter 12) around the entire root. Mycorrhizal roots are short and forked and sometimes are borne as tight clusters. **Endotrophic** mycorrhizae do not form a mantle over the root, and the fungus actually enters the cortex cells. Mycorrhizal roots are more efficient in mineral absorption, but they are apparently not absolutely essential for the growth of the usual host plants. This is known, because plants that are artificially fed adequate nutrients can grow without mycorrhizae. Mycorrhizae also may be beneficial to their host plants by secreting hormones or antibiotic agents that reduce the potential of plant disease. Mycorrhizae are also discussed in Chapter 12.

Bacterial Nodules

Some plants, like certain legumes, are capable of fixing nitrogen, that is, changing N_2 in the soil into the NH_4^+ (ammonia) form of nitrogen that is usable by the plant. This is done by an unusual relationship between the bacterium *Rhizobium* and the roots of legume plants. Root cells are infected by the passage of a thin infection thread of the bacteria into the root hair cells of roots. The infection thread passes through the root hair by causing the cell membrane to enfold as it penetrates the cell. The infection thread grows through the epidermal cell into the cortical cells. The bacteria then divide and also stimulate the cortical cells to divide, thereby forming the root nodule (Fig. 4.64). The bacteria are the actual agents for fixing the nitrogen.

Contractile Roots

Roots of the dandelion *(Taraxacum)*, ginseng *(Panax quinquefolium)*, and some other plants are capable of contracting. The result is that above-ground parts are kept near the soil surface. This contraction is caused by the radial expansion of cells in the root cortex. Sometimes, the collapse of cortical cells is thought to cause contraction. The vascular tissue in contracted roots forms a twisted, undulated mass (Fig. 4.64).

Summary

1. The principal functions of roots are absorption of water and nutrients, conduction of absorbed materials and food, and anchorage of the plant in the soil.
2. There are two general types of root systems: a fibrous root system and a tap root system.
3. Roots differ from all stems because they lack nodes and internodes.
4. The root tip is divided into four zones of specialization: (a) root cap, which protects the (b) meristematic region as it moves through the soil; (c) a region of elongation; and (d) a region of differentiation characterized externally by root hairs and internally by the formation of primary vascular tissues.
5. Soil water and nutrients move easily through epidermal and cortical tissues.
6. The endodermis forms the innermost cell layer of the cortex. A suberized band, the Casparian strip, in radial and transverse walls, completely encircles endodermal cells.
7. Water cannot move across the Casparian strip. Therefore, all water with dissolved nutrients inside the endodermis has passed through the protoplasts of endodermal cells.
8. The pericycle, a row of cells internal to the endodermis, represents the outermost row of cells of the vascular cylinder; they have differentiated from procambium. Cells of pericycle may eventually form part of vascular cambium or cork cambium or give rise to branch roots.
9. In cross section, the primary xylem is star-shaped and generally triarch to pentarch. Primary phloem arises between the arms of primary xylem. There is no pith in the roots of dicotyledons.
10. Water and inorganic nutrients enter primary xylem from parenchyma cells without passing through cells of phloem.
11. Water and inorganic nutrients enter root-hair cells, pass through the cortex, are filtered by protoplasts of endodermal cells, cross the pericycle, and move directly into the vessels or tracheids of primary xylem.
12. In roots having secondary growth, a vascular cambium is formed by pericyclic cells over xylem arms and procambial cells remaining between primary xylem and phloem. This vascular cambium at first forms a wavy line in cross section. It eventually becomes circular, and there is little difference in the appearance of wood from root or stem.

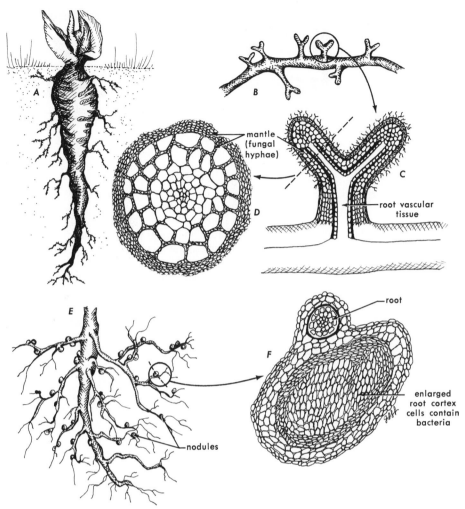

mantle
(fungal
hyphae)

root vascular
tissue

root

enlarged
root cortex
cells contain
bacteria

nodules

Figure 4.64   Views of root modifications. *A,* contractile root; *B,*
branched habit of mycorrhizal root; *C,* longitudinal section
through one of the Y-branched mycorrhizal roots; *D,* cross
section through one root showing the fungal mantle; *E,* soybean
(Glycine max) root with bacterial nodules; *F,* cross section of root
and nodule.

the absorption and transport systems

The living plant collects materials from the environment and assembles them into plant cells and organs. It therefore requires systems for absorbing materials, assembling new substances, and transporting materials within the plant. The assembly systems, metabolism and photosynthesis, are discussed in Chapters 2 and 6. The present chapter deals with the absorption and transport systems.

Water Absorption and Transport

Green plants require only a few simple materials from the environment: water, carbon dioxide, oxygen, and several minerals. Of these materials, water is the substance that plants absorb and transport in the greatest quantities. Some of the absorbed water is consumed in metabolic reactions such as photosynthesis and some of it is retained in the vacuoles as cells grow. But as much as 98% of the water that enters the plant is lost through the shoot system by evaporation. This evaporative loss is termed **transpiration**.

Plants transpire a large amount of water. A single corn plant in Kansas, between May 5 and September 8, transpired 196 liters. It has been estimated that the amount of water lost by a grove of red maple trees in a growing season may be equivalent to a sheet of water 72 cm deep over the entire grove.

An alfalfa plant may transpire 900 units of water for each unit of dry matter produced, but a millet plant may transpire 248 units of water for each unit of dry matter. Although these values vary with the environmental conditions in which the plants are grown, it is evident that plants differ in their water economy.

Plants draw their major water supplies from the soil solution, the bulk water that lies between soil particles. Water first enters the plant (Fig. 5.1) through the epidermal cells of the root tip. The flow moves through the cortex, the endodermis, and the pericycle, and finally into the xylem conduits: the vessels or tracheids. The liquid in the xylem, a solution known as **xylem sap**, moves up the plant to any region where water is being removed: to growing shoots and fruits where water is accumulated in the vacuoles of enlarging cells, and to leaves where water

is consumed by photosynthesis and lost through evaporation. In leaves, the water passes from the xylem vessels and tracheids to the mesophyll cells, from which it evaporates into the intercellular spaces and finally escapes from the leaf as water vapor moving through the stomata.

At most stages of its journey through the plant, water molecules have the option of moving either through the protoplasts or through cell walls (Fig. 5.2). Thus, for example, water first enters the walls of the root hair cells. Consisting of a porous meshwork of cellulose microfibrils, these walls readily accept water and allow it rather free movement. The epidermal walls touch cell walls in the cortex, permitting the passage of water molecules from cell to cell through the walls. In addition, there are water-filled spaces between cortical cells through which bulk water may move.

Alternatively, water molecules may pass through the plasma membranes of the cells and enter the protoplasts. The membranes offer little barrier to such movement of water in either direction. Once in the protoplast of a cell, water may move to other cells by means of the protoplasmic bridges known as plasmodesmata.

The endodermis restricts the movement of water. The endodermal cells are tightly pressed together and their Casparian strips (Chapter 4) present an impermeable layer that water cannot cross. The endodermal barrier prevents water from moving deeper into the root unless it travels through the protoplasts of the endodermal cells. This limitation is considered an important means for giving the plant control, via the selective permeability of the plasma membrane, over the movement of solutes (not water) between the soil and the vascular system. This could also be important in limiting the entry of toxic materials from the soil and the loss of important materials to the soil.

In the xylem, the principal movement of water occurs as a bulk flow in the vessels and tracheids. Flow is probably not appreciable within the secondarily thickened walls of these dead cells.

Water Potential and Water Movement

We have considered the path that water follows through the plant. Now we ask what is the force that drives the flow of water? The answer requires the concept of **water potential**.

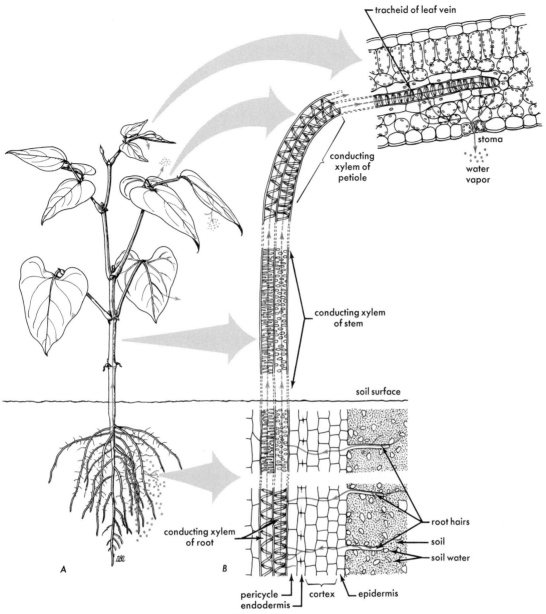

Figure 5.1 *A*, diagram showing water movement in a whole bean plant. *B*, details of the xylem pathway.

Labels in figure:
- tracheid of leaf vein
- conducting xylem of petiole
- stoma
- water vapor
- conducting xylem of stem
- soil surface
- root hairs
- soil
- soil water
- conducting xylem of root
- pericycle
- endodermis
- cortex
- epidermis
- *A*
- *B*

Whenever a wet and a dry place are joined by a water mass, water flows toward the dry region. This is the principle behind the movement of water in plants. "Wetness" and "dryness" are relative terms. Pure water is perfectly wet, and a region that contains no water is perfectly dry. However, between these extremes there are many systems that contain water mixed with other materials. Physical chemists, who measure the degree of wetness in a system, call it the water potential. Pure water is assigned the highest potential; if anything is added to the water, such as cellulose microfibrils or sugar molecules, the water potential decreases. In moist air, where water is present as a gas, the water potential increases if more water is added to the air. The presence of other gases does not influence the water potential of humid air to any significant degree.

If two ends of a water column are at different water potentials, water will spontaneously move toward the end with the lower water potential. This is a formal way of saying that water tends to move from wet places to dry places.

The concept of water potential is useful in explaining **osmosis** and **turgor pressure**, two phenomena that are important in the mechanisms of water movement. Osmosis is the diffusion of water across a differentially permeable membrane (Fig. 5.3). It occurs when the solutions on the two sides of the membrane differ in water potential. If the two solutions are at the same pressure and temperature, then the one with the higher solute concentration will have a lower water potential; it is less like pure water. If the membrane will not permit solutes to pass, water will then move along the water potential gradient toward the

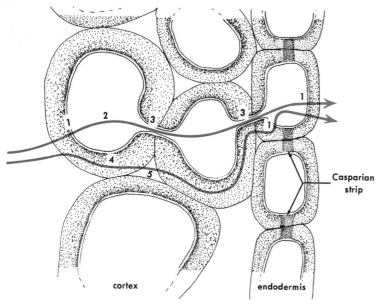

Figure 5.2 Paths of water and ion movement in outer root tissues. Two alternative pathways are shown. Water and ions move similarly but ions may be aided by transport mechanisms in crossing plasma membranes. 1 = plasma membranes; 2 = protoplasm; 3 = plasmodesmata; 4 = cell walls; 5 = intercellular space.

solution that contains the most solute particles. This water flow across a membrane is osmosis; it brings the two solutions closer to the same water potential. Unless an outside force intervenes, osmosis will continue until the two solutions achieve equal water potentials. Then equilibrium prevails.

The living plant cell is in contact with a wet cell wall (Fig. 5.4). The water between the wall's carbohydrate polymers is a dilute salt solution. On the other side of the plasma membrane, the contents of the protoplast form a second more concentrated solution. Thus, water will tend to move by osmosis from the wall into the protoplast. This transfer of water raises the volume of the protoplast. Against the swelling, the wall stretches a little but soon comes to the limit of its elasticity and no more volume

expansion is possible. Every incoming water molecule adds to the hydrostatic pressure within the cell. This can be appreciated by comparing the molecules to people trying to crowd into a full room. The added pressure is termed turgor pressure. It is an outward pressure that is exerted on the cell wall by the crowded molecules in the cell. Against this turgor pressure the wall exerts an opposite, inward-directed force called wall pressure. When the cell is swelling, turgor pressure exceeds wall pressure; the wall yields and stretches. At equilibrium the turgor and wall pressures are equal and opposite.

Turgor pressure together with the structure of the wall give the cell a rigid shape. Cells go limp when they lose their turgor pressure through water loss. A plant with greatly reduced turgor pressure is said to be wilted.

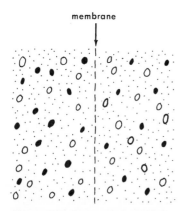

Figure 5.3 System that would show osmosis. A membrane separates two solutions that contain water (o) and solute (●) molecules. The pores in the membrane permit water but not solutes to pass. Water potential will be lower in the left-hand solution; water will move into that solution.

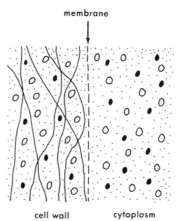

Figure 5.4 Conditions at the surface of the cell, affecting water movement. The solution in the cell wall contains fewer solutes than cytoplasm, but polymers such as cellulose are present.

chapter 5 | The Absorption and Transport Systems

If a cell or group of cells is immersed in a solution that has a higher solute concentration than that of the cell sap, water diffuses outward, and the turgor pressure in the cell is reduced. Although the volume of the cell decreases somewhat, more striking is the withdrawal of the protoplast from the cell wall and the decrease in the size of the vacuole. This phenomenon is called **plasmolysis**.

The space between the cytoplasm and the cell wall in a plasmolyzed cell is filled with the solution in which the cells are immersed. If plasmolyzed cells are immersed in water or in a solution whose concentration is less than that of the cell sap, the cells regain their turgor; water molecules diffuse inward.

A cell may die if plasmolysis is pronounced and prolonged. For example, if a heavy application of ordinary salt is placed on the soil where weeds are growing, so that the root cells are surrounded by a solution of high concentration, water diffuses from the cells, and they become severely plasmolyzed. If this state is prolonged, the roots die. The salt killed the roots, not because it was toxic to the root cells, but rather because severe plasmolysis was brought about by the high concentration of the soil solution. In parts of the western United States where rainfall is low and the evaporation rate high, salts of the soil may accumulate on the surface and form what are known as "alkali flats." In such soils, the salt concentration of the soil solution may be so high that many crop plants cannot grow; only species that are especially adapted to tolerate high salt concentration are able to survive. These plants are known as **halophytes**.

For reasons that cannot be simply stated, turgor pressure (any kind of pressure, for that matter) raises the water potential of the solution. Water stops entering the cell when the turgor pressure has raised the protoplast's water potential to equal that of the water in the cell wall. Then equilibrium prevails. Thus we see that an interplay between pressure and solute concentrations determines the direction of water flow. *The effect of increased pressure is to make a concentrated solution behave more like a dilute solution.*

In these matters, the carbohydrate polymers in the wall do more than restrain the protoplast. They also influence the potential of the water in the cell wall. The effect is to decrease the water potential and the effect depends very strongly on the amount of water in the wall. In a completely saturated wall, dripping wet, the effect of the wall polymers is very slight. But when the wall begins to dry out, the water potential quickly becomes very sensitive to further drying. The drying wall develops an immensely powerful tendency to keep its remaining water and to draw water from neighboring regions. This is the same principle behind the absorptive powers of blotting paper. The water is said to move in response to **capillary action**. The driving force occurs because the wall polymers form bonds with the adjacent water molecules. Such binding between water molecules and polymers is known as **adhesion**. The polymers may be viewed as having a tightly held blanket of bound water.

Equally important, water molecules tend to bond to one another. These intermolecular bonds, termed hydrogen bonds (Appendix), are weaker than the covalent bonds within a molecule. However, each water molecule can

Table 5.1
Factors That Affect Water Potential in a Solution

Water Potential Is Increased by:	Water Potential Is Decreased by:
Pressure Heat	Solutes Polymers (adhesion)

participate in four hydrogen bonds with as many as four other water molecules. The result is that the entire water mass shows **cohesion**, a tendency to hang together. The forces of cohesion not only join molecules of the bulk solution into a loose unit, but also they join the water mass to the layer of bound water adhering to the cell wall polymers. The effect is as if the polymers exert an anchoring influence on water. This effect weakens in proportion to the distance water is from the wall polymers. The water in the saturated wall that is far from the polymers acts like free water. But when this free water is withdrawn, the remaining water is closer to the polymers and is more strongly bound by the forces of adhesion. In brief, *the adhesion forces exerted by wall polymers tend to make a dilute solution behave osmotically as if it were more concentrated in solutes.*

If we take capillary action and adhesion into account, it appears now that water movement between the wall and the protoplast is controlled not only by the concentrations of solutes and by turgor pressure but also by the influence of polymers in the cell wall. These factors are summarized for convenience in Table 5.1. Equilibrium prevails (no net water flow) when the sum of these effects on water potential is the same on both sides of the membrane. In any other condition the direction of water movement (and its pace) varies with the balance of factors.

The Mechanism of Water Movement in the Plant

Water moves within the plant toward any location where the water potential is low; that is, toward any region where the solute concentration is rising or where water is being removed. Photosynthesis both consumes water and produces solutes, making the locations of photosynthesis low in water potential and the targets of water flow. The water potential is also relatively low in the cells of growing regions; here the cells have walls that continuously stretch under turgor pressure. This plasticity of the walls keeps the turgor pressure from rising high enough to stop the osmotic movement of water into the cells. The incoming water provides the bulk needed for growth.

In transpiration, water is moved by a third means. Here, the primary event is the diffusion of water vapor from the humid air in the leaf to the dry air outside the leaf. This movement leaves the air channels within the leaf with a reduced water potential. Water evaporates from the moist walls of the leaf cells, obeying the rule of movement toward regions of low water potential. The walls are left more dry and the remaining water drops in potential because of the capillary effect. Thus in transpiration, water movement is driven by withdrawing water from the wall and exploiting the effect of capillary action or adhesion.

Once a point of low water potential has been established, the remaining question is how the motive force is transmitted throughout the plant and into the soil solution from which the water is obtained. Cohesion is a major part of the answer. The water mass within the plant–soil system is a continuous whole that is bound together by the forces of cohesion. One may view the cell walls that stand between the soil solution and the top of the plant as a series of porous obstructions that exert a drag on the flow of water while leaving the forces of cohesion in operation. Thus, when the water potential is reduced by drying the walls of the leaf cells, the adhesion forces that attract nearby water molecules are effectively pulling on a chain of water molecules that extends all the way down to the soil solution and into each cell of the plant.

The forces that move water up the plant must work against gravity as well as the drag (resistance) exerted by cell walls. The taller the tree, the greater the weight of the water columns that must be drawn up the xylem and the greater the force that must be sustained to keep up a given flow rate. The driving force in transpiration is generated by the drying leaf cell walls. Therefore, to do its work against gravity and resistance, a tall tree must be able to survive with drier leaf cell walls than a short plant.

Plant physiologists refer to this picture of water movement as the **cohesion and transpiration-pull theory** (among other names). In this theory, note that water is being pulled, not pushed, and that the water column may be under considerable tension. It has often been asked whether water columns really can sustain the tensions needed for lifting water as high as the tops of tall trees, which may exceed 100 meters in height. Laboratory experiments have shown that thin columns of water do indeed have the required tensile strength and can withstand pulling forces great enough to raise water up to a height of 300 meters.

But the presence of a single, tiny gas bubble destroys the tensile strength of a water column. With a bubble in the column, tension can raise water no higher than 9.75 meters (at sea level). Beyond that point, further increases in tension merely expand the bubble with no further rise in the water column.

Under sufficient tension, a water column may spontaneously produce bubbles. This is called **cavitation**. With sustained tension, bubbles will expand until they meet a barrier such as a cell wall. Bubbles cannot cross cell walls because the wall polymers exert too great an attraction for water. In the plant, the conducting area of the xylem is divided into a large number of microscopically narrow vessels and tracheids. Cavitation may occur in many of these during transpiration, but the bubble will be confined to the vessel or tracheid of its origin and the rest of the xylem can continue to function. When transpiration stops (e.g., at night when stomata close), bubbles formed by cavitation may be resorbed so that the blocked conducting cell can function again.

Root Pressure and Guttation

It is an old observation that the rooted stump of a freshly cut plant often "bleeds" xylem sap from the cut end. Such "bleeding" represents the movement of water up the xylem without transpiration as a driving force. The xylem sap is pushed rather than pulled upward in this case. The pushing force is termed **root pressure**. It is an osmotic phenomenon: the xylem sap has a higher concentration of ions than the soil solution because of the action of ion-transport systems in the plasma membranes of root cells. Thus water moves osmotically from the soil into the xylem. This raises the volume of liquid in the xylem and forces it up the stem. In this case the xylem contents are moved by positive pressure rather than by tension.

Root pressure is unimportant as a source of water movement when transpiration is occurring, but it may be a significant water-moving force in small plants when transpiration is not taking place. For instance, when a well-watered, vigorously growing tomato plant is placed under a bell jar, transpiration ceases as the atmosphere in the jar becomes saturated. Continued absorption of water by osmosis in the roots then results in a slow exudation of water from the tips of the leaves (Fig. 5.5). This loss of liquid from leaves is termed **guttation**. Many plants have specialized openings called **hydathodes** at the tips of their leaves, through which guttation liquid passes outward.

Figure 5.5 Guttation from tips of barley (Hordeum vulgare) leaves.

How Roots "Locate" Water

Without extensive irrigation or frequent rainfall, plants usually remove water from the soil faster than it can be replaced by movement from neighboring regions. Hence, for continued water absorption the roots must continually grow into new areas that have not yet been dried.

Geotropism (Chapter 8) gives the growth of roots a general downward orientation, increasing the probability that the root will contact untapped reserves of water. In addition, pressures exerted by contact with soil particles often force roots to grow laterally and branch roots tend to invade any moist soil they meet.

It is evident that roots do not "go in search of water." However, they grow only in moist soil. In practice, this means that if the top 30 cm of soil are moist, and dry soil lies below it, the roots of plants will be confined to the upper, moist soil. In irrigation practice, this relationship of soil moisture and root growth is important. In irrigating a garden or lawn with a hose, one is easily misled into believing, because the soil surface is wet, that sufficient water has been applied. Examination may reveal that at depths of 12 cm or more the soil is very dry—too dry for the growth of roots. In growing plants, it is nearly always desirable to stimulate maximum root growth and a root system that attains its normal depth and spread. This may be accomplished in part by keeping the soil moist to the proper depth. The loss of water from soils at levels below the first few centimeters is essentially due to plants. Thus one of the principal reasons for removing weeds from a growing crop is that they deplete water from the soil.

If the soil solution increases in solute content, the rate of water absorption by the roots declines. When the concentrations of the two solutions are the same, water intake ceases. And, if the soil solution becomes more highly concentrated than the cell sap, owing to excessive applications of fertilizers or of saline water, water will be withdrawn from the root cells. Moreover, root growth is inhibited at high salt concentrations and the roots are not able to extend into new soil areas.

Carbon Dioxide Absorption

Carbon dioxide is often the factor that most limits plant growth: in most plants it is the sole source of carbon for building organic compounds, and it is only a minor component of the air from which it is absorbed (about 0.03% of air is carbon dioxide). Correspondingly, there is an advantage in being able to trap the largest possible fraction of the carbon dioxide that passes the plant in currents of air.

When air that contains carbon dioxide meets the plant surface, the only reason it enters the plant is that photosynthesis keeps the carbon dioxide concentration in the tissues very low. This causes CO_2 to enter by diffusion along the concentration gradient. The steeper the gradient, the faster the flow of CO_2 into the plant and the larger a fraction of the available CO_2 the plant will be able to trap from a moving air stream. The concentration gradient is steepest when the photosynthetic cells are directly in contact with the air.

In line with these principles, the plant body tends to be structured in a way that brings as many photosynthetic cells as possible near to the outside air. The broad, flat shape of the leaf greatly increases the area of plant tissue that is close to the air. Within the leaf, air spaces form an extensive system of channels, giving almost all the photosynthetic mesophyll cells direct contact with an air space that leads to the outside of the leaf.

Control of Carbon Dioxide Absorption and Water Loss

Cells that are directly exposed to dry air quickly lose too much water by evaporation. Thus, the same air exposure that promotes CO_2 capture also raises dangerous problems of water loss. The balance between these two factors has doubtless played a great part in the evolution of leaf structure.

In land plants, fatal water loss is prevented by impermeable cork on older stems and by an impermeable waxy cuticle that covers the young shoots and leaves. These coatings limit gas exchange to special apertures, the stomata, which can be opened and closed to get the best possible balance between CO_2 capture and water loss. A typical leaf may have as many as 6 million stomata (Chapter 4).

The stomata function in the following way (Fig. 5.6). Each pore is lined by a pair of special epidermal cells known as guard cells. These cells are sausage-shaped; they also differ from ordinary platelike epidermal cells in having thicker walls and an ability to perform

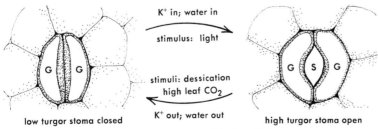

K⁺ in; water in
stimulus: light

stimuli: dessication high leaf CO_2

K⁺ out; water out

low turgor stoma closed   high turgor stoma open

Figure 5.6 Diagram summarizing the mechanism by which guard cells open and close stomata. The gain or loss of K⁺ involves active transport; water follows by osmosis. G = guard cell; S = stoma.

photosynthesis. The walls of the guard cells are thickest on the side that faces the pore. This feature allows the stoma to be opened and closed by changing the turgor of the guard cells. When turgor pressure is low, the guard cells press together and the stomatal pore is closed. An increase in turgor pressure stretches the walls of the guard cell slightly, the thinner portions stretching the most. This causes the guard cells to become curved, opening the stomatal pore.

The turgor in the guard cells is controlled by adding and removing solutes. If solutes are added, the water potential decreases; water enters the guard cells from adjacent cells, and the turgor pressure rises. The reverse happens if the guard cells lose solutes.

Guard cells take up potassium (K^+) ions from adjacent cells as the stomata open. They release the K^+ ions as the stomata close. This ion transfer seems to depend on an active transport "pump" in the plasma membrane. The movements of K^+ probably play a major part in controlling the stomata.

Organic solutes may also be important. Thus in some cases starch grains in the guard cells are partially converted to free sugar as the stomata are induced to open. The reverse occurs when the stomata are caused to close. These facts show that the solute relations that control the stomata are likely to be complex.

Control of stomatal opening by turgor pressure guards the plant against excessive water loss: if the leaves lose water too fast, they wilt—that is, they lose turgor. This closes the stomata, limiting further water loss. (It has been discovered recently that the hormone abscisic acid is rapidly produced by leaves when they experience water stress. This hormone may signal the stomata to close before the loss is severe.)

In addition, two relevant environmental factors act as control cues for the stomata. Carbon dioxide inside the leaf causes the stomata to close, and light causes stomata to open. The result is that the stomata open only when there is light for photosynthesis, and when the carbon dioxide has been reduced to a point where more is needed.

However, some plants have their light signals reversed, so that the stomata open and take in CO_2 at night while closing in the daytime. As CO_2 enters these plants, it is stored by attachment to an organic carrier molecule. In daylight, with the stomata closed, the stored CO_2 is gradually released by the carrier and used in normal photosynthesis. This reversed control system is especially common in succulent plants of dry regions. Its value is that less water is lost with the stomata open at night because the lower night temperatures yield slower evaporation.

The rate of transpiration and its effect on the plant depend on the humidity and temperature of the air. Low humidity and high temperature promote evaporation, increasing the rate of transpiration and the extent of drying in the walls of leaf cells. At moderate transpiration rates, the mesophyll cells of leaves have a solute concentration that is high enough to enable them to maintain turgor pressure even though the walls are partially dried. But with increased rates of evaporation, as in the midday heat of a dry summer day, the leaf cell walls may develop such a low water potential that they extract water from the protoplasts as well as from the xylem. This removes the turgor pressure needed to maintain leaf shape and the leaves wilt. Midday wilting is common in big-leafed herbs such as squash plants. It is a temporary condition; normal turgor returns when the rate of transpiration decreases in the evening and the walls within the leaf can regain moisture. Plants that are native to hot, dry areas often have reinforcing sclerenchyma in their leaves, a feature that makes the leaf less sensitive to drying.

Certain plants called xerophytes (Chapter 9) are able to survive in extremely dry habitats, whereas certain other plants succumb. Such surviving plants undoubtedly maintain a balance between water outgo and water intake, whereas in plants that fail to survive the dryness, the loss of water, at least for a period, exceeds the absorption of water. Plants adapted to dry habitats often have roots that penetrate deeply into the soil; or, like many desert cacti, they have an extensive surface root system that can rapidly absorb limited rainfall from spring or summer showers. Many desert plants have special water storage tissue, as in the plants with fleshy stems and leaves that are commonly known as succulents (Fig. 9.30).

Anatomical features that are advantageous from the point of view of preventing water loss are (a) the cuticle of leaves, young stems, and fruits, (b) sunken stomata, (c) distribution of stomata, and (d) reduction of the transpiring surface. In addition, stomatal behavior is an important factor in controlling water loss.

Most of the water lost from a plant passes out through the stomata. Some water, however, is lost through the cuticle. Various modifications in leaf structure reduce cuticular transpiration; for example, thickening of the outer wall of the epidermal cells and the presence of a waxlike material, cutin, in this wall. Most plants of arid and semi-arid climates have a thicker cuticle than do those of humid climates.

In some plants (Fig. 5.7), the stomata are below the general level of the leaf surface. When this condition occurs, the water vapor must diffuse through a relatively long and sometimes tortuous passageway, with the result that the diffusion rate is lessened.

The leaves of many plants have stomata only on the undersurface or, if on both surfaces, mostly on the lower. Loss of water from leaves that have stomata only on the lower surface is usually less than loss from leaves that have an equal number of them on both the upper and lower sides.

In most plants, the principal organs of transpiration are leaves. Therefore, any decrease in leaf surface will reduce transpiration and conserve the water absorbed by roots. In corn and some other monocots, the leaves roll up during drought, thus exposing less surface to the air than do fully expanded leaves. Cacti and some euphorbias have no foliage leaves; in these plants the transpiring surface is restricted to the stem surface.

The leaves of many plants are clothed with hairs. It's possible that hairs reflect light, reducing leaf temperature. In some plants, such as common mullein, the hairs reduce transpiration. But experiments with some plants have shown that water loss from the leaves is greater when the hairs are present than when they are lacking.

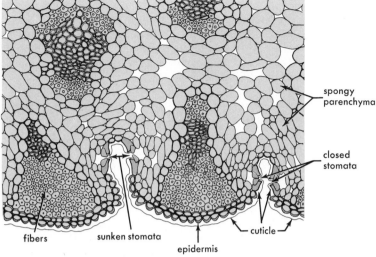

Figure 5.7 Sunken stomata in crypts of a yucca leaf. Note also the thick epidermal cuticle and the large bundles of fibers.

Mineral Ion Absorption

Principal Mineral Elements

A variety of mineral elements are needed by the plant. These and the typical quantities withdrawn from the soil by wheat plants are shown in Table 5.2. The mineral elements, discussed below, are largely absorbed in the form of ions dissolved in the soil water.

Nitrogen is an essential component of proteins, nucleic acids, and many other molecules that play important metabolic roles. The atmosphere is rich in N_2 gas; it comprises 78% of the air. Some 8000 kilograms of N_2 lie above every square meter of land. However, ordinary green plants get their nitrogen mainly as nitrate ions (NO_3^-) from the soil. To a lesser degree, ammonium ions (NH_4^+) also serves as a nitrogen source, and some plants apparently can derive nitrogen from certain organic compounds in the soil.

Nitrogenous fertilizers, both natural and commercial, are usually the most important fertilizers applied to growing plants. The chief commercial nitrogenous fertilizers contain nitrogen (a) in the nitrate form; (b) in the form of ammonia or its compounds; (c) in organic compounds, such as cottonseed meal, and (d) in the amide form, such as urea and calcium cyanamide. Organic nitrogen in complex molecules cannot be used directly by plants but must first be converted into available forms through the action of

micro-organisms in the soil.

The rate of growth of plants is influenced greatly by available nitrogen. Nitrogen is very mobile in the plant and can be translocated from mature to immature regions of a plant. An early symptom of nitrogen deficiency is a yellowing of leaves, particularly the older leaves; then follows a stunting in the growth of all parts of the plant (Fig. 5.8E).

An excess of available nitrogen results in vigorous vegetative growth and a suppression of food storage and of fruit and seed development.

Sulfur comprises a small but vital fraction of the atoms in many protein molecules. As disulfide bridges (SS), the sulfur atoms aid in stabilizing the folded protein, while sulfhydryl (SH) groups participate in the active sites of some enzymes. Some enzymes require the aid of certain small molecules that may contain sulfur. Finally, the machinery of photosynthesis includes some sulfur-containing compounds (e.g., ferredoxin).

Ordinary green plants obtain sulfur in the form of sulfate ions. When plants deficient in sulfur are given sulfur fertilizers, the most noticeable response is an increased root development and deeper green color of the leaves. Sulfur deficiency symptoms are shown in Fig. 5.8F.

Phosphorus is a component of nucleic acids, energy-carrying molecules such as ATP, and the important phospholipids of cellular membranes. Some proteins also contain phosphate groups. The highest concentrations of phosphorus occur in rapidly growing plant parts, such as

Table 5–2

Amounts (in kg. per hk.) of Macro- and Micronutrients Removed from the Soil in One Growing Season by a Wheat Crop

	Macronutrients						Micronutrients				
Element	N	K	P	Ca	S	Mg	Fe	Mn	B	Zn	Cu
Amount	85	47	17	13	12	9	0.8	0.6	0.3	0.2	0.03

maturing fruits, seeds, and shoot tips.

Applications of phosphorus to deficient soils promote root growth and hasten maturation, particularly in cereals. Phosphates (potash) are the principle source of phosphorus for plants. A tomato plant grown in culture solution without phosphorus is shown in Fig. 5.8C.

Potassium serves as an enzyme activator. Over 40 enzymes have been found to require potassium for maximum activity. In addition, potassium ions are important in controlling the stomata. This is a mobile element, which will migrate from older tissues to meristematic regions. For example, during the maturing of a fruit crop potassium moves from leaves into fruits.

Any water-soluble inorganic compound of potassium, such as potassium sulfate, phosphate, or nitrate salts, can be used by plants as a potassium source. Potassium deficiency symptoms in the tomato are shown in Fig. 5.8D.

Calcium is required by all ordinary green plants. It is one of the constituents of the middle lamella of the cell wall, where it occurs in the form of calcium pectate. Calcium affects the permeability of cytoplasmic membranes and the hydration of colloids. Calcium may be found in combination with organic acids in the plant. Oxalic acid, for example, is a by-product of metabolism. It is a soluble substance and is toxic to the protoplasm if it reaches a high concentration in the cell. When united with calcium, however, the soluble oxalic acid is converted into the highly insoluble calcium oxalate, which does not injure the protoplasm. There is also evidence that calcium favors the translocation of carbohydrates and amino acids and encourages root development. Calcium deficiency is frequently characterized by a death of the growing points (Fig. 5.8G) because it is not readily translocated in the plant from mature to immature regions. General disorganization of cells and tissues also results from calcium deficiency. This effect is consistent with one of the key roles of calcium—maintaining the normal structure of cell membranes.

Magnesium is a constituent of chlorophyll; it occupies a central position in the molecule (Fig. 6.4). Many enzyme reactions, particularly those involving a transfer of phosphate (energy metabolism) are activated by Mg^{+2} ions. Magnesium is also vital to the function of ribosomes in protein synthesis.

A deficiency in magnesium results in **chlorosis**—a lack of chlorophyll and, therefore, foliage that is pale yellowish rather than green (Fig. 5.8I). This disease is common in cultivated plants and in many cases applications of magnesium effect a cure. However, chlorosis can also be caused by deficiencies in other elements such as iron.

The materials listed above provide elements that are known as **macronutrients** (macro = large) because plants require them in large quantities. The macronutrient elements are C, H, O, N, P, S, K, Ca, and Mg.

In addition, plants require tiny amounts of several other elements. These, known as **micronutrients**, include at least those listed below. It is difficult to study the plant's need for an element if that element is needed only in trace quantities, because the element may be hard to eliminate from the nutrient medium or the air around the plant. Thus physiologists anticipate that in the future more essential elements will be discovered. (There is already evidence

that some plants may require small amounts of Na and Ni.)

Iron ions are components of several electron-carrying compounds (cytochromes and ferredoxin) essential in respiration and photosynthesis. In picking up and giving off electrons, the iron ion is alternately reduced and oxidized by other compounds.

Although iron is not a component of chlorophyll, it is essential for chlorophyll synthesis. Iron deficiency may be responsible for chlorosis (Fig. 5.8H). The quantity of iron required by plants is very small. For example, chlorosis of pineapples in Hawaii, because of the unavailability of iron in the soil, is cured by spraying the leaves with iron salt solutions. Orchard trees suffering from iron chlorosis may be cured by injecting iron compounds into the trunk or applying iron salts to the soil.

Boron deficiency causes varied and complex symptoms such as decreased root and shoot elongation, inhibition of flowering, and darkening of tissues. The metabolic basis of boron action is not clear, though boron is known to play a part in regulating carbohydrate breakdown. Deficiency diseases may be cured by applying very small quantities (10–25 kg) of sodium tetraborate (borax) per acre. However, excessive amounts of boron in the soil or water may be toxic; in fact, borates are used as weed killers.

Zinc is an activator or a part of several enzymes. One of the earliest effects of zinc deficiency is a drop in production of the growth hormone auxin (Chapter 8), resulting in inhibition of growth. Deficiency diseases such as the well-known "little-leaf" of deciduous fruit trees (Fig. 5.9) can be cured by spraying the trees with zinc salts, by injecting dilute solutions of zinc salts into the trunks, or by driving zinc brads into the trunks. These correctives show how small are the quantities of zinc needed by the plant.

Manganese activates several enzymes and also plays an essential role in the oxygen-liberating steps of photosynthesis. The most striking deficiency symptom is chlorosis (Fig. 5.10), but this chlorosis is somewhat different from that caused by iron deficiency. In iron

Figure 5.9 Disease of peach *(Prunus persica)* known as "little leaf," caused by a deficiency of zinc. Branch at left untreated; branch at right cured by driving in zinc-coated nails.

chapter **5** | **The Absorption and Transport Systems**

Figure 5.8   Tomato plants grown in the nutrient culture solutions to show visual symptoms of mineral deficiencies. *A,* complete solution; *B,* deficiency symptoms in leaves; *C,* minus phosphorus; *D,* minus potassium; *E,* minus nitrogen; *F,* minus sulfur; *G,* minus calcium; *H,* minus iron; *I,* minus magnesium; *J,* minus all micronutrients. All plants subjected to deficient solutions were first grown in a complete nutrient solution for two weeks.

Figure 8.1 *Above:* Pea seedlings grown in light and in darkness. Note minimum stem elongation and maximum leaf development in light. The center two plants have had only one day in light, which caused straightening of the stem (hook-opening) and start of leaf expansion and greening, as compared to the fully etiolated seedlings on the left.

Figure 8.2 *Above:* Pea leaf development as a function of light. Fully developed green leaves from plants in continuous light; early stage of development after one day in light (center); etiolated leaf from dark-grown plant protected by plummular hook. All eight days old.

Figure 8.3 *Right:* Corn seedlings grown in light or in darkness showing the long mesocotyl growth that pushes the coleoptile and its enclosed first leaf and shoot apex to the soil surface. In the light-grown seedlings the mesocotyls are only a few millimeters long.

Figure 8.4 *Right:* Corn (*Zea mays*) leaf development. A series showing the leaf enclosed in the coleoptile, emergence from the coleoptile in darkness where unrolling is prevented, an artificially unrolled leaf, and a similar leaf grown in light, fully unrolled with greening completed.

8.1

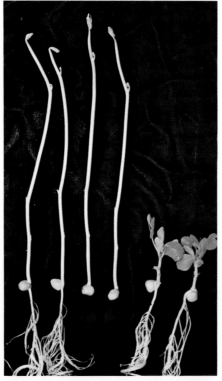

8.3

8.2

8.4

Figure 5.10 Chlorosis of tomato *(Lycopersicon esculentum)* leaf caused by a deficiency of manganese in the nutrient solution.

chlorosis the young leaves become white or yellow with prominent green veins, while manganese chlorosis gives the leaf a mottled appearance. Spraying or dusting crops with as little as 22 kg of manganese sulfate per hectare often cures deficiencies.

Copper is a constituent of certain enzyme systems such as ascorbic acid oxidase and cytochrome oxidase, and of the compound plastocyanin, which is part of the electron transport chain in photosynthesis. Deficiency causes abnormalities of growth (Fig. 5.11), especially in plants growing in marsh and peat soils.

Molybdenum is important in enzyme systems involved in nitrogen fixation and nitrate reduction. Plants suffering from molybdenum deficiency can absorb nitrate ions but cannot then metabolize this form of nitrogen. If nitrogen in the form of ammonium is supplied to these plants, the effects of molybdenum deficiency are less severe. For

Figure 5.11 Tomatoes *(Lycopersicon esculentum)* growing in solution with all essential chemical elements except copper. Leaves at left were sprayed with solution containing copper.

some Australian soils that are low in molybdenum only 140 gm of MoO_3 per hectare, applied once every 10 years, increased pasture yield by 6 to 7 times.

Chlorine participates in the oxygen-evolving steps of photosynthesis. This element is present in the soil solution as the very soluble Cl^- ion. Because of the very small quantities required by plants and its almost universal occurrence in soils and air (salt spray travels long distances), chlorine is never intentionally added as a component of fertilizer. In fact rainfall may contain enough chlorine to satisfy the requirements of plants.

The Absorption of Minerals

As mentioned above, the minerals required by plants are normally absorbed by roots from the soil solution. There are exceptions: the pitcher plant and the Venus' flytrap, for example, capture insects that, on decomposition, release minerals that are absorbed by the specialized leaves that serve as traps. **Epiphytes** (plants that use the aerial parts of other plants as supports for growth and are not themselves rooted in the soil) may obtain minerals by trapping airborne dust and debris. Highly specialized parasitic plants such as dodder and mistletoe sink modified stems known as **haustoria** into the vascular tissues of host plants and obtain their minerals in that fashion.

Finally, many plants form cooperative associations with soil microorganisms. The microorganisms bring in minerals from the soil. In this respect two classes of symbiotic associations can be distinguished. **Mycorrhizae** (Fig. 5.12) are associations in which a fungus invades the tissues of the roots and, through its own absorptive capabilities, extracts minerals from the soil and brings them into the host plant. The fungus obtains vitamins or other organic materials in exchange. In this association the fungus takes over the role of root hairs and the formation of root hairs may be strongly suppressed. Mycorrhizal associations are especially common in forest trees on nutritionally poor soils.

Another kind of association is seen in **nitrogen-fixing nodules** (Fig. 4.64E). Here, bacteria that can convert N_2 to NH_4^+ dwell within thickened regions (nodules) of roots. The most common instance is that of legumes (e.g., clover and alfalfa) where the bacteria are species of the genus *Rhizobium*. Without the bacteria, the plants would be unable to use the nitrogen that is present as N_2 in the soil air. In return, the bacteria obtain photosynthetic products that serve their own growth as well as providing the energy needed for nitrogen fixation.

But aside from these special cases, it is absorption of dissolved soil minerals by roots that provides the plant with its mineral supply. Since the mineral ions are passively carried along with the absorbed soil water, the path of their movement through the plant is similar to that of water. The dissolved soil minerals can pass through the endodermal layer of the root only by going through the plasma membrane and into the endodermal protoplasts. This gives the plasma membranes an opportunity to screen the passage of minerals in and out of the vascular system, between the soil and the main plant body. The

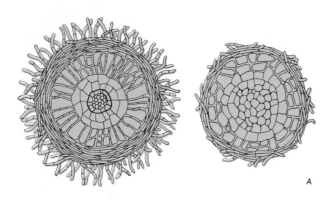

Figure 5.12 Mycorrhizal roots of *Pinus virginiana. A,* diagram showing cross section of root with attached fungal hyphae; *B,* roots with fungal sheath.

extent to which this screening actually discriminates between desirable and undesirable mineral elements is undetermined, because it is clear from examination of plants grown in soils with toxic minerals that poisonous elements may enter the vascular system.

The plasma membranes of most cells, including those of root cells, contain active transport systems that discriminate between various ions. Some ions may accumulate (e.g., K^+) to high concentrations inside the protoplasts; others may be pushed back into the wall space (e.g., Na^+). These active transport systems may raise the concentration of a given ion within the protoplasts as much as 1000 times above the concentration in the soil solution. This may be a great aid to the plant in acquiring rare elements. However, the active transport systems have a cost: their operation

consumes ATP, which must be produced by respiration with a consequent demand for O_2 and glucose.

Oxygen Absorption

Oxygen is required for respiration by all the cells in the plant. Photosynthesis produces more than enough oxygen to meet the needs of nearby cells during the daylight hours, but much of this oxygen leaks away to the atmosphere and little of it is transported within the plant to regions as far away as the root tips. Consequently, leaves may absorb oxygen from the air at night, and roots usually require a supply of oxygen from air trapped in the soil.

The soil contains a reservoir of atmospheric gases in

Figure 5.13 Effect of aeration on root growth in tomato (*Lycopersicon esculentum*). *Left,* plants growing in complete nutrient solution through which air was bubbled; *right,* plants growing in same solution without aeration.

the spaces between solid particles. This trapped air may exchange with the open air above ground if the soil has a suitably loose texture. Heavy or compacted soils are deficient in aeration, and cultivation may improve the air supply of roots. Roots may absorb oxygen directly from the soil gases (the oxygen dissolves in the water of cell walls); alternatively, gaseous oxygen may dissolve in soil water, from which it is taken up along with the water and minerals. There seems to be no special transport mechanism for trapping soil oxygen; entry is by diffusion along the concentration gradient that results from the consumption of oxygen in respiration.

The amount of air in the soil depends not only on the pore space but also on the water content of the soil. If water occupies the pore space, air is forced out. Water-soaked (saturated) soil contains practically no air except the amount dissolved in water. If the soil about the roots is continuously water-soaked, plants die because of insufficient oxygen and possibly as a result of the accumulation of carbon dioxide. Most species of land plants will grow normally with their roots in a water solution, if it is well-aerated by bubbling air through it (Fig. 5.13). Evidence indicates that inadequate soil aeration results in a reduction in the rate of water and mineral absorption.

Most land plants, including agricultural plants, will not long survive with the root system submerged in unaerated water or surrounded by a soil that is water-soaked. But some plants flourish under such conditions. Among them are rice, various swamp and marsh plants, and the bald cypress (*Taxodium distichum*, Fig. 5.14). Almost all such

Figure 5.14 Aerial "stump roots" of bald cypress (*Taxodium distichum*).

plants contain in the stem and roots large communicating air spaces. Thus, air absorbed into the leaves and stems may reach the living cells of the root in sufficient quantity. Bald cypress and certain species in tropical mangrove swamps develop special root branches that grow upward until their ends are above the water level. These special root branches have a central core of loose tissue through which air moves downward to the submerged organs.

Transport of Organic Materials

Within a living plant, substances are constantly moving from one place to another. Although they are interrelated, we can detect four types of movement based on the rates at which movement occurs: (a) slow diffusion of molecules and ions; (b) moderate movement of materials carried by cytoplasmic streaming in living cells; (c) more rapid flow of material in the sieve tubes; and (d) very rapid conduction of water and mineral solutes in the xylem.

Diffusion and Protoplasmic Streaming

The movement of water or of solute molecules and ions into a cell across the cell wall is a relatively slow process. Diffusion occurs along activity gradients and may take place through plasmodesmatal connections between cells as well as directly through the permeable cellulose walls of most cells. Once a solute molecule or ion diffuses into a cell, its rate of movement may be increased many times as it is picked up by the protoplasmic stream. Cytoplasm may stream at rates of a few to several hundred millimeters per hour. The highest rate of protoplasmic streaming that has been measured was observed in the protoplasm of a slime mold in which the cytoplasm was streaming at a rate of 486 millimeters per hour.

In terms of the dimensions of a single cell, cytoplasmic streaming is frequently very rapid, since a substance may be transported across the length of a cell in a matter of seconds. In terms of the whole plant, however, this is a very slow process. It would take days, for instance, for a molecule in the leaf to be carried upward to the growing shoot tip or downward to the roots. Figure 5.15 shows the way in which one solute may be moving by diffusion and protoplasmic streaming in one direction through a series of cells while another solute may be moving in another direction. Diffusion and protoplasmic streaming are the chief methods by which solutes move through living plant tissues, except in the sieve tubes.

The Mechanism of Phloem Conduction

The phloem sieve tubes move organic materials rapidly throughout the plant. That the phloem is the major path of food transport is shown by the results of ringing experiments. When a ring of bark is removed from a stem down to the cambium (thus removing the phloem), the carbohydrate content and especially the sugar concentration of the sap in the leaf, bark, and wood above

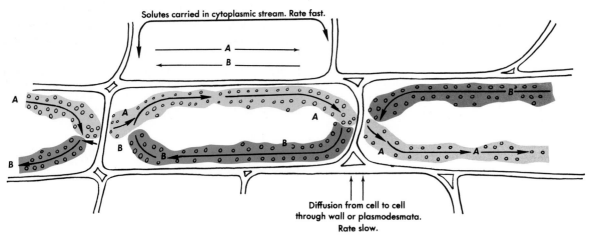

Solutes carried in cytoplasmic stream. Rate fast.

Diffusion from cell to cell through wall or plasmodesmata. Rate slow.

Figure 5.15 Diagram illustrating how two solutes may be moving in opposite directions by diffusion and protoplasmic streaming in the same tissue.

the ring increases after a few hours. Below the ring, the concentrations of these solutes in the bark and wood decreases. These conditions would not prevail if foods were moving downward in the xylem tissues.

Translocation of material in the phloem may reach a rate of 100 centimeters per hour. When this is compared to the much lower rates of protoplasmic streaming, it is evident that normal protoplasmic streaming is insufficient to cause this rapid transport. In fact, the mature functional sieve-tube members have highly modified cellular contents (Chapter 3) in which protoplasmic streaming does not seem to occur.

The contents of the sieve-tube members of one sieve tube make up a continuous liquid system. The pores in the sieve plates connect the sieve-tube members into a continuous tube. According to one widely accepted theory (the mass flow theory), water and solutes move together as one mass in the sieve tubes along their length.

The mechanism of mass flow can be illustrated by the flow of photosynthetic products from the leaves to the roots or to other sites of storage and consumption (Fig. 5.16). The process begins with the production of glucose molecules in the mesophyll cells of the leaves. From this site of high concentration, glucose diffuses to the bundle sheath cells that line the leaf veins. These cells contain enzymes that assemble glucose molecules into the double sugar, sucrose. An active transport system in the sheath cells or the adjacent sieve tubes then transfers the sucrose molecules into the protoplast of the sieve tubes. As in all active transport, this process consumes ATP. The accumulation of sucrose molecules in the sieve tubes decreases the water potential of the phloem sap, causing water to move into the sieve tubes from neighboring cells such as the xylem elements. This increases the hydrostatic pressure in the sieve tubes. This pressure pushes the phloem sap *en masse* away from the leaf blade, down the petiole, and into the stem of the plant. At any site where solutes (chiefly sucrose) are removed, the water potential increases and water diffuses out of the sieve tubes and into adjacent cells and the xylem. This results in a lowered pressure and movement of solution

along the sieve tube from regions of high pressure to regions of low pressure.

When we consider this mechanism, it appears that the direction of movement in the phloem must always be from sites where solutes are being produced to sites where solutes are being consumed. Given this general rule, the direction of flow in the phloem may change from time to time depending on the activities of the plant. For example, during its first growing season the sugar beet transfers a great deal of sucrose from the leaves to the roots via the phloem. Next spring the process is reversed; sucrose moves from the roots to the young, growing shoot system.

Although sucrose is usually the principal organic substance carried in the sieve tubes, other organic materials such as hormones and amino acids, and even inorganic ions, may also be carried.

The rate of phloem transport may be controlled by the rates at which solutes are added and removed. But there are at least two other mechanisms that may affect the rates of movement. These are discussed below.

Callose, a special polysaccharide with unusual linkages between the glucose subunits, is deposited on the sieve plates by the action of enzymes. Other enzymes remove callose from the same locations. The thickness of the callose layer on the sieve plates is therefore dynamically maintained. The thicker the callose, the smaller the pores in the sieve plates. To the extent that pore size influences the rate of flow in sieve tubes, regulation of the callose system may affect the flow rates. Recent experiments have shown that the enzymes of the callose system are highly sensitive to mechanical stimuli such as vibrations or crushing. Such stimuli cause an extensive buildup of callose in a matter of seconds, a response that is slowly reversed when the stimuli cease.

Also a network of protein strands occurs in the conducting space of each sieve-tube element. Under normal conditions of smooth flow in the sieve tubes, the protein network seems to be firmly anchored to the ends of the cells and does not affect the transport of materials. But abrupt changes in the flow rate, such as occur when a stem is cut, cause the protein network to rip loose and

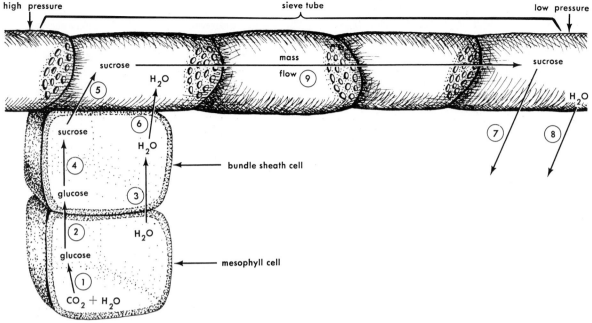

Figure 5.16 Mass flow hypothesis of phloem transport. In mesophyll cells, photosynthesis (1) produces glucose, which moves by diffusion (2) into bundle sheath cells. Water follows by osmosis (3). Enzymes (4) in the bundle sheath convert glucose to sucrose, which moves by active transport (5) into the sieve tube. Water follows by osmosis (6), raising pressure in the sieve tube. Elsewhere, sucrose diffuses (7) from the sieve tube; water follows (8) by osmosis, reducing the local pressure. The sieve-tube contents flow en masse (9) from high to low pressure regions.

pile up against the nearest sieve plate, partially blocking the sieve pores. Both the callose and the protein network may be significant in limiting the loss of phloem sap from stems and petioles that have been injured.

Summary

1. Water is absorbed chiefly from the soil by roots and moves throughout the plant in the xylem.
2. Some water is consumed in metabolism or retained in growth but most of it is lost by evaporation (transpiration).
3. The force that moves water is the tendency of water to move from regions of high water potential to regions of low water potential.
4. The water potential of a solution is raised by adding pressure and is lowered by adding solutes or materials such as wall fibrils that absorb water.
5. The water potential of air increases as its humidity increases.
6. If two solutions are separated by a membrane that will permit water but not solutes to pass, water will move through the membrane from a high to a low water potential. This water movement is called osmosis.
7. Because plant cells are enclosed in walls, water uptake creates turgor pressure, which gives the cell a rigid shape. Wilting results from the loss of turgor.
8. The transpiration-pull theory states that evaporation of water from the leaf pulls water upward through the xylem.
9. Cohesive and adhesive forces between molecules prevent the xylem water columns from breaking under tension.
10. When there is no transpiration, water may enter the xylem by osmosis, creating positive root pressure and pushing water up the plant.
11. Roots do not seek water, though growth is generally directed downward. Roots grow only in moist soil.
12. Carbon dioxide is absorbed from the air, reaching leaf cells through stomata and intercellular spaces.
13. Gas exchange is limited by cork and cuticles and by stomata.
14. Stomata are opened and closed by raising and lowering the turgor pressure of guard cells. The changes in pressure are achieved by pumping potassium ions into and out of the guard cells. Water follows by osmosis. Light exposure and CO_2 depletion stimulate the stomata to open.
15. Minerals enter the plant through the roots, moving passively with the water flow except when they meet membranes, which may present a barrier. Carriers control ion movements through membranes.
16. The impermeable Casparian strip, in the walls of the endodermis, prevents water and solutes from using the cell walls as an avenue for moving between the vascular system and outer root tissues. This permits membranes to control traffic.
17. Leaves may absorb oxygen from the air at night. Roots usually absorb O_2 from air in the soil.

18. Organic solutes move through the phloem between regions of production, storage, and use.

19. The mass flow hypothesis states that the solution in the sieve tubes moves *en masse* from regions of high turgor pressure to regions of low pressure. The high source pressures are thought to result from active pumping of sucrose into sieve tubes, followed by osmotic entry of water. Opposite movements occur at the low pressure regions.

20. The rate of movement in the phloem may also be affected by a protein network that can plug the sieve pores when sieve tubes are injured; and it is also affected by enzymes that control the size of the sieve pores by depositing and removing callose.

6 | CHAPTER

photosynthesis

The living body is a metabolic system that maintains itself by consuming energy-rich fuel molecules. Many organisms (animals, fungi, and most bacteria) depend on supplies of fuel molecules that they find ready-made in the environment. The green plants and a few forms of bacteria differ from them in this respect: they can trap the energy of sunlight to produce their own fuel molecules, using common low-energy molecules such as carbon dioxide and water as raw materials. The capture of light to produce fuels is known as photosynthesis.

Ultimately, the whole living world depends on photosynthetic organisms for its energy. Without them the supply of chemical fuels would soon be exhausted. Therefore plants have a primary importance in the economy of nature; this gives photosynthesis a unique value among metabolic pathways. It seems appropriate, therefore, to discuss photosynthesis in a separate chapter, and in more detail than was done for other phases of metabolism.

The Discovery of Photosynthesis

Prior to the early seventeenth century, the general belief was that plants derived the bulk of their substance from soil humus. This idea was overthrown by a simple experiment performed by the Flemish physician and chemist Jan van Helmont, who planted a willow branch in a box and supplied rainwater to the plant as needed. In five years it grew to a weight of 169 pounds (77 kilograms). The soil had lost only 2 ounces (57 grams); consequently, van Helmont reasoned that the plant substance must have come from water. This was a logical deduction, though we now know it was not entirely correct.

Our knowledge of photosynthesis was further improved by a religious reformer, philosopher, and spare-time naturalist, Joseph Priestley. In 1772, Priestley reported that a sprig of mint could restore confined air that had been made impure by burning a candle. The plant changed the air so that a mouse was able to live in it. The experiment was not always successful, probably because Priestley, who did not know the role of light, did not always have adequate light. Seven years later, Jan Ingen-Housz noticed that air was revitalized only when green parts of plants were in the light.

It was three years later, in 1782, that a Geneva pastor, Jean Senebier, discovered another important part of the process—that "fixed air," carbon dioxide, was required.

Thus, it could be said that green plants in the light use carbon dioxide and produce oxygen.

But what was the fate of the carbon in the CO_2? Ingen-Housz answered this question in 1796 when he said that the carbon went into the nutrition of the plant; that is, it became part of the organic matter. Overall, these discoveries can be summarized by writing the equation for photosynthesis as:

$$\text{carbon dioxide} + \text{water} \xrightarrow[\text{green plants}]{\text{light}} \text{oxygen} + \text{organic matter}$$

The nature of the organic material formed in photosynthesis could be guessed by observing that starch accumulates in leaves during photosynthesis and is used up when the leaves are kept in darkness. Starch is built from glucose molecules. Assuming that glucose is the direct product of photosynthesis, we can write and balance the overall equation for the process as:

$$6CO_2 + 6H_2O \xrightarrow[\text{green plant cell}]{\text{light energy}} C_6H_{12}O_6 + 6O_2$$

This formulation tells us that water is required in the same proportion as CO_2. More advanced methods using radioactive tracers have confirmed this relationship, but have also shown that other sugars besides glucose are formed early in photosynthesis. Hence it is probably more realistic to write photosynthesis as in the formula below.

$$CO_2 + H_2O \xrightarrow{\text{light}} \underset{\text{carbohydrate}}{(CH_2O)} + O_2$$

If we look at the overall equation for photosynthesis, we might logically propose that the oxygen liberated is released from the carbon dioxide molecule. However, van Neil, working at Stanford University, made a comparative study of the photosynthesis that occurred in several organisms belonging to different groups of plants. The green and purple sulfur bacteria are able to use hydrogen sulfide instead of water. Sulfur instead of oxygen is liberated.

$$CO_2 + 2H_2S \xrightarrow{\text{light}} (CH_2O) + H_2O + 2S$$

Some other bacteria and some algae can use hydrogen instead of water to reduce carbon dioxide:

$$CO_2 + 2H_2 \xrightarrow{\text{light}} (CH_2O) + H_2O$$

Comparing these cases, van Neil concluded that a general equation for photosynthesis should be written as follows:

$$CO_2 + 2H_2A \xrightarrow{\text{light}} (CH_2O) + H_2O + 2A$$

carbon dioxide   hydrogen donor   carbohydrate   water

The hydrogen donor H_2A can be H_2O, H_2S, H_2, or any other substance capable of donating hydrogen to CO_2 in the process of photosynthesis.

To test van Neil's hypothesis, a group of biochemists supplied plants with water that contained the unusual oxygen isotope, ^{18}O instead of the common isotope ^{16}O. The oxygen liberated in photosynthesis contained ^{18}O. If plants were supplied instead with carbon dioxide that contained ^{18}O, along with water that contained ^{16}O, then the oxygen liberated did not contain ^{18}O. This demonstrated that van Neil was correct: the oxygen liberated in photosynthesis comes from water.

The Chloroplast—Site of Photosynthesis

Although the overall equation of photosynthesis seems simple, the discovery that O_2 comes from H_2O rather than from CO_2 makes it clear that there are hidden complexities. A good understanding of photosynthesis requires that the process be taken apart and studied piece by piece. Plant biochemists are still working on this task.

One successful approach has been based on the examination of chloroplasts, the organelles that perform photosynthesis. Chloroplasts are abundant in the cells of leaves and other green parts of plants; the green color is due to light-absorbing compounds or **pigments** within the chloroplast. It has been mentioned (Chapter 3) that each chloroplast has three principal parts: (1) a surrounding envelope consisting of a pair of membranes; (2) a liquid called **stroma** that is enclosed by the envelope; and (3) a set of flattened membrane-lined sacs called **thylakoids** (grana and frets) that are embedded in the stroma (Fig. 6.1).

Chloroplasts are very fragile, but if properly treated they can be removed in good working condition from the cell. In the test tube, their operation can be explored without the confusion that goes with having other kinds of organelles in action at the same time.

The procedure for isolating chloroplasts or other organelles (fractionation) begins with a method for breaking open cells. Fragments of tissue are placed in a dilute solution of salts and organic compounds. The precise composition of the medium is important, and must be determined by trial and error. The cells are broken by grinding with sand in a mortar, or with a kitchen blender.

This releases some of the chloroplasts intact, though others are broken. The ground material is filtered to remove large solid fragments. Then the suspension of organelles is spun in a centrifuge to separate the

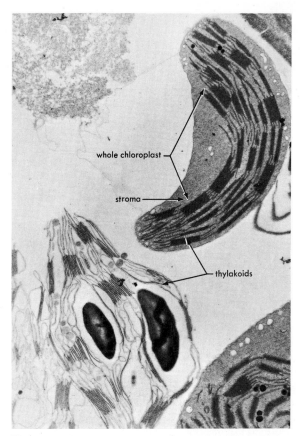

Figure 6.1   Electron micrograph of isolated chloroplasts of *Vicia faba*, showing one plastid with stroma and one plastid that has lost its outer envelope and stroma, ×2500.

organelles according to their density or mass. Again, trial and error are needed to determine the most effective procedure. The final result is a solution in which chloroplasts are suspended almost free of other organelles. The isolated chloroplasts can produce carbohydrate and oxygen if they are given light, carbon dioxide, and water.

The electron microscope has shown (Fig. 6.1) that chloroplast suspensions contain both intact chloroplasts that have all their membranes, and damaged chloroplasts that have lost their envelope and stroma and retain only the thylakoids. This important fact makes it possible to isolate thylakoids and to study their activity apart from the stroma.

The thylakoids prove to be the part of the chloroplast that contains the green pigment. They cannot convert CO_2 into carbohydrates. But if they are given light, water, and an organic compound that can accept electrons, the isolated thylakoids can produce O_2. It appears therefore that the thylakoids are the light-capturing and O_2-producing part of the chloroplast (Fig. 6.2). Because of their dependence on light, the reactions of the thylakoids are commonly called the **light reactions**. The reactions of the stroma, which do not use light, are known as the **dark reactions**.

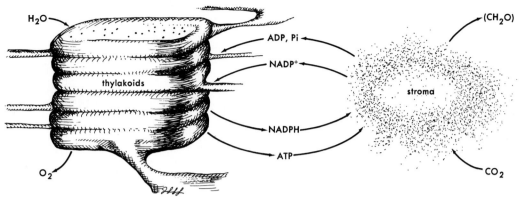

Figure 6.2 The raw materials and products of the thylakoids and the stroma.

The Thylakoid (Light) Reactions

The basic process in the thylakoids is a light-driven flow of electrons (Fig. 6.3). Light energy displaces certain electrons from their equilibrium locations, and their movements back toward equilibrium are channeled so as to build useful, energy-rich molecules. The ability of moving electrons to do work should not seem unusual, because we see in everyday life that electrical currents can be used to do many kinds of construction work.

The two primary working components of the thylakoid are (1) a light-capturing system, and (2) a system of electron transport pathways or chains.

The Light-Capturing System

The light trap consists of pigment molecules that are firmly attached to the membranes of the thylakoids. The most important pigment is chlorophyll (Fig. 6.4), a complex molecule that contains a light-absorbing "head" and a long inert tail. The head is a flat array of four connected rings, with a magnesium ion held in the center. The tail, a hydrocarbon chain, may help to hold the chlorophyll molecule in position. Within the thylakoid membrane, chlorophylls bind to proteins of several kinds (Fig. 6.5). These proteins differ in the number and types of

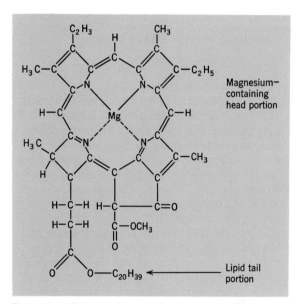

Figure 6.4 Chlorophyll a molecule.

chlorophyll molecules they will bind. In turn, the proteins join into larger working groups in the membrane.

There are several different varieties of chlorophyll. Each differs according to the small groups that are attached to the light-absorbing ring. One of these, called **chlorophyll a** (Fig. 6.4), is present in all chloroplasts and seems to be

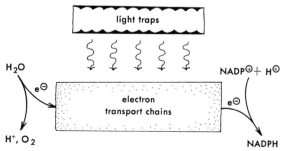

Figure 6.3 Light traps and electron transport chains are the two major functional systems of the thylakoid. (diagrammatic).

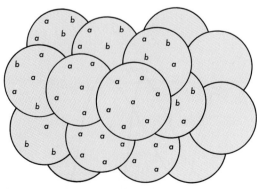

Figure 6.5 Chlorophyll molecules (a,b) are bound to protein molecules (circles) that in turn form groups (chlorophyll-protein complexes) within the thylakoid membrane. This illustration is highly schematic; the detailed arrangements are unknown.

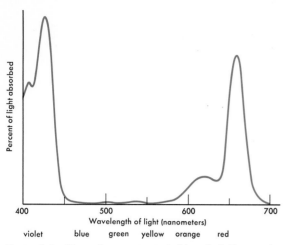

Figure 6.6 Absorption spectrum of chlorophyll. The graph shows the fraction of received light that is absorbed when the pigment is exposed to various wavelengths of light. The relation between wavelength and color of light is also shown.

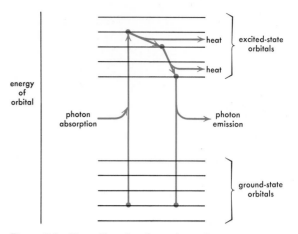

Figure 6.8 Absorption of a photon by a pigment causes an electron to move from a low-energy, ground-state orbital in the molecule to a high-energy, excited-state orbital. The energy difference is the excitation energy, equal to the photon energy. The excited electron may give up energy to other molecules as heat or light (fluorescence).

seems to be essential for photosynthesis.

In addition to chlorophyll *a*, the chloroplast contains other pigments, such as carotenoids, phycocyanin, phycoerythrin, and other chlorophylls. These are called **accessory pigments** because, although they may contribute to light capture in photosynthesis, they are not universal in occurrence and the light they absorb must be passed on to chlorophyll *a* before it can be used.

The color of a pigment is an indicator of its light-absorbing ability. For instance, chlorophyll absorbs red and blue light very effectively but is a poor absorber of green and yellow light. The absorption behavior is best shown by means of an absorption spectrum (Fig. 6.6). In this figure the subjective quality of color is supplemented with a more exactly measurable property of light, the wavelength. The accessory pigments differ from chlorophyll *a* in absorption behavior (Fig. 6.7) and may improve the photosynthetic efficiency by capturing some of the light that the chlorophyll would have allowed to escape.

The absorption properties of chlorophyll account for the green color of leaves. Sunlight is a mixture of all colors (wavelengths) of light. From this mixture chlorophyll subtracts the red and blue portions, leaving green light to be reflected or to pass through the leaf so that it may be seen.

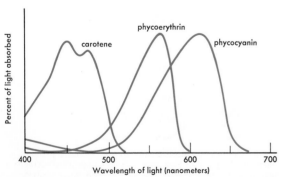

Figure 6.7 Absorption spectra of three accessory pigments.

When light interacts with pigment molecules, the beam of light acts as if it were composed of light particles, called **photons.** Each photon carries a definite amount (quantum) of energy, and there is a definite relationship between the quantum size and the wavelength of the light: longer wavelengths mean less energy per photon. In absorbing light, a pigment molecule must accept a whole photon or nothing at all. Whether absorption can occur depends on the energy carried by the photon as well as the internal structure of the molecule. The electrons in the molecule move about constantly in a set of pathways known as **orbitals.** Each orbital is associated with a definite amount of energy (Fig. 6.8); to be in a certain orbital an electron must have exactly the right amount of energy. At normal temperatures and in darkness the electrons within a molecule tend to occupy the orbitals with the least energy, those closest to the atomic nuclei, toward which the electrons are drawn by electrical forces. In this low-energy condition the molecule is said to be in the **ground state.** Absorption of a photon causes one of the electrons to shift to an orbital of higher energy. The exact amount of additional orbital energy needed to make this shift must be supplied by the photon: a photon with too much or too little energy will not be absorbed.

Here, then, is the primary event of light capture and the primary conversion of light energy into chemical energy. The pigment molecule after absorption is said to be in an **excited state;** the pattern of electron movement within the molecule has been thrown out of equilibrium. We say that the light energy has been converted to **excitation energy.**

Pigment molecules remain in the excited state only for a small fraction of a second before giving up their extra energy and returning to the ground state (Fig. 6.8). In photosynthetic pigments, there are many closely spaced orbitals, and the excited electron can dispose of its extra energy by stepping through these orbitals, donating the difference in energy to other molecules that collide with the excited pigment. The effect is to increase the motion of the surrounding molecules. This extra motion appears as heat, which is a waste to the system. Also, at any moment the excited molecule may dispose of any remaining excitation energy by giving off a new photon. The emitted light is known as **fluorescence** and it, too, is a waste to the system.

Reaction Centers

Light absorption is useful only if there is an organized system available to exploit the excitation energy of pigments before it can be converted to heat or fluorescent light. This depends on the action of specially placed chlorophyll *a* molecules that can give up an electron to an acceptor compound when they receive excitation energy. By contrast, ordinary pigments—including most chlorophyll *a* molecules—do not ionize on excitation.

The ionizable chlorophyll *a* molecules occur at sites in the membrane called **reaction centers**. The detailed arrangement is not known, but each reaction center seems to be bound to an antenna system, which is built of many chlorophylls bound to proteins. The reaction center and its antenna form a **photosystem**. Two kinds of photosystems (I and II) are known. Each thylakoid has several hundreds of them (Fig. 6.9). **Photosystem I** seems to contain six protein molecules with about seven chlorophyll *a* molecules per protein. One or two of the chlorophylls are at the reaction center. Other compounds may also be present in the photosystem, such as the unknown compound (Z, in Fig. 6.10) that accepts electrons from chlorophyll. **Photosystem II** is not as well known, but is thought to be broadly similar to Photosystem I.

As shown in Fig. 6.9, the thylakoid also contains units called **light-harvesting complexes**. These groups of pigment-bearing proteins do not have reaction centers, but collect light energy to feed the photosystems (mainly photosystem II). The light-harvesting complexes contain all the accessory pigments and some chlorophyll *a*.

The Transfer of Excitation Energy

Excitation energy finds its way from the antenna and light-harvesting pigments to the reaction centers by a process called **resonance transfer**. Here the full quantum of excitation energy (*) is passed from one pigment molecule to another without any loss.

$$pigment_1^* + pigment_2 \xrightarrow[transfer]{resonance} pigment_1 + pigment_2^*$$

In this event the pigment that originally had the excitation energy drops to its ground state and the recipient pigment becomes excited. There is no transfer of electrons, no emission of photons or chemical interaction between the pigments. The exact mechanism of transfer is not known, but we do know that efficient transfer depends on the pigments being bound close together.

Electron Transport

The events that follow the arrival of excitation energy at the reaction centers are diagrammed in Fig. 6.10. Overall, light energy drives a flow of electrons along a system of **carriers** from water to $NADP^+$. The carriers are bound to the membrane between the reaction centers. Though their precise arrangement is not known, the carriers seem to be organized so that the electron flow causes H^+ to move from the stroma to the space within the thylakoid. This is analogous to charging a battery.

The resulting difference in H^+ concentration across the membrane represents a store of energy that is thought to drive the formation of ATP. According to current ideas, the enzymes that form ATP are bound to the thylakoid membrane and are arranged so that the formation of ATP releases H^+ to the stroma and OH^- to the space within the thylakoid. The OH^- combines with H^+ to form H_2O. These events would decrease the H^+ difference across the membrane, "discharging the battery" to form ATP.

Electrons move spontaneously along the electron transport chain because each carrier in the chain has a greater tendency to capture and hold electrons (a greater electron affinity) than the carrier before it. This is shown in

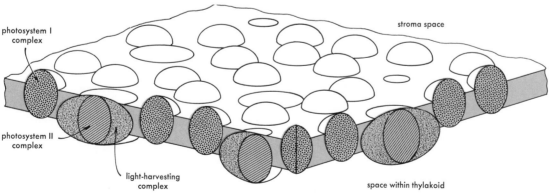

Figure 6.9 Proposed model for the arrangement of photosystems in the thylakoid membrane. Each complex is a group of proteins to which chlorophyll molecules are bound. The Photosystem I and Photosystem II complexes include reaction centers. Adapted from P. A. Armond, L. A. Staehelin, and C. J. Arntzen, *J. Cell Biol.* 73:400–418 (1977).

photosystem I complex

photosystem II complex

light-harvesting complex

stroma space

space within thylakoid

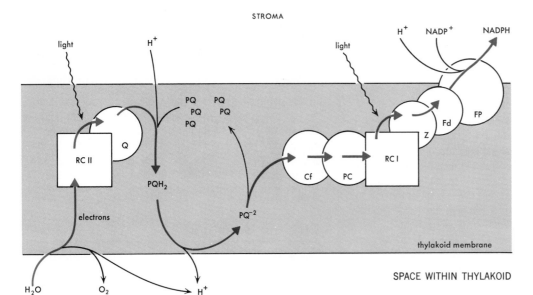

Figure 6.10 Proposed flow of electrons (green arrows) and H⁺ in the thylakoid membrane. Overall, light is thought to drive a flow of H⁺ across the membrane in one direction and a flow of electrons in the opposite direction. RC I and RC II are reaction centers; circles and PQ represent compounds that take part in electron transport chains. Sizes of the parts are not drawn to scale and locations are schematic.

Fig. 6.11. Here the carriers with the greatest electron affinity are lowest on the vertical scale. The **oxidation-reduction potential** shown on this scale is an exactly measurable property of the carrier that is directly related to the electron affinity: the higher the potential, the lower the electron affinity of the carrier.

Chlorophyll supplies electrons for the transport chains. But ground-state chlorophyll, and the chlorophyll ion, have a great tendency to hold their electrons. Only the excited, un-ionized chlorophyll molecule can give up an electron to the chain.

As Fig. 6.11 shows, the chain of carriers that starts with

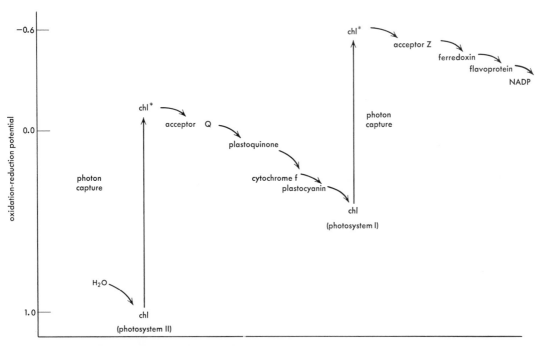

Figure 6.11 Oxidation-reduction potential diagram for the electron carriers and Photosystems I and II. The arrows show the path of electrons in photosynthesis. Carriers high on the scale tend to give up electrons to lower carriers; electrons lose energy as they move to carriers that are lower on the potential scale. Symbols: chl = chlorophyll *a* in the ground state; chl* = chlorophyll *a* in the excited state.

Photosystem I ends by giving its electrons to the mobile compound **NADP⁺**, which picks up H⁺ from the surrounding solution at the same time:

$$NADP^+ + H^+ + 2\text{ electrons} \longrightarrow NADPH$$

The chlorophyll of Photosystem I is left as a positive ion, which cannot accept photons or donate more electrons to the chain. These chlorophyll ions take electrons from the chain fed by Photosystem II. This restores neutrality and allows the chlorophyll of system I to absorb another photon.

Meanwhile, the chlorophyll of Photosystem II has been left as an ion with a high electron affinity. The system II chlorophyll ions get electrons from water molecules (Fig. 6.11). Hydrogen ions and O_2 are released from water when the electrons are removed. The water molecule has a great electron affinity, but the chlorophyll ion has a still greater affinity. The precise mechanism that transfers electrons from water to chlorophyll ions is unknown. However, we do know that manganese ions and at least one enzyme are involved.

Energy Storage by the Thylakoid

There is a close relation between the oxidation–reduction potential of a carrier and the energy of its electrons: an electron has more energy if it is attached to a carrier that is high on the potential scale. One can view the carriers in Fig. 6.11 as a series of stations that a moving electron must pass; the vertical position of the carrier represents the energy the electron has at each point in its travels. With this view, we see that excitation gives the electron an initial dose of energy; with each transfer along the chain the electron loses some of its energy. The final product, NADPH, carries electrons that still retain a large fraction of the original energy. NADPH is a high-energy product of the thylakoids that can carry high-energy electrons and hydrogen to the stroma for use in building sugar molecules. Having given up its load, the "empty" NADP⁺ molecule can return to the thylakoid for more electrons.

O_2 is also a high-energy product of the thylakoids. The oxygen atoms in O_2 have a high affinity for electrons and will take them from other molecules if the opportunity arises. The role of O_2 in driving respiration has already been discussed in Chapter 2.

The thylakoids also condense ADP and inorganic phosphate (Pi) to form the energy-rich molecules of ATP:

$$ADP + Pi + \text{energy} \longrightarrow ATP + H_2O$$

This ATP formation in the thylakoids is called photophosphorylation. The ATP is released to the stroma, where it supplies part of the energy needed for sugar production.

Since ATP carries more energy than the raw materials from which it was made, clearly some process in the thylakoids must have supplied the needed energy. At present the details are not known. We do know that under normal conditions the energy is provided by the movement of electrons between Photosystems I and II (Fig. 6.12). One view is that electron flow causes a gradient of H⁺ across the membranes, and ATP formation draws energy by discharging the H⁺ gradient (p. 115). In any case, we say that ATP formation is somehow **coupled** to the flow of electrons. This process is known as **noncyclic photophosphorylation**.

If the supply of NADP⁺ runs low, the electrons moved by Photosystem I spill into the chain from system II and find their way back to the original chlorophyll of system I, forming a closed cycle of electron movement that is driven by excitation energy (Fig. 6.12). This cyclic electron flow may also cause coupled ATP formation. ATP formation in this cycle is termed **cyclic photophosphorylation**.

Summary of the Thylakoid Reactions

The events in the thylakoids are summarized in Fig. 6.12. The useful high-energy products are NADPH, O_2, and ATP. Light acts by driving a flow of electrons from water to NADP⁺. The removal of electrons splits water to release O_2 and H⁺. Finally, some of the energy flow during electron transport is captured in the coupled formation of the ATP.

CO₂ Reduction and the Formation of Sugar (Dark Reactions)

Two sensitive analytical methods have been especially useful in showing the sequence of reactions by which sugar is built from CO_2. These new methods were the use of carbon dioxide in which the carbon was radioactive and the use of a special process, chromatography, with which the investigator could easily and accurately separate minute amounts of different organic compounds from one another.

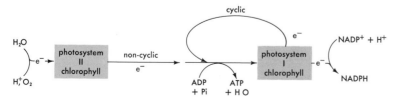

Figure 6.12 Noncyclic photophosphorylation (ATP formation) draws energy from the flow of electrons between Photosystems I and II. In cyclic photophosphorylation the electrons start and end in Photosystem I.

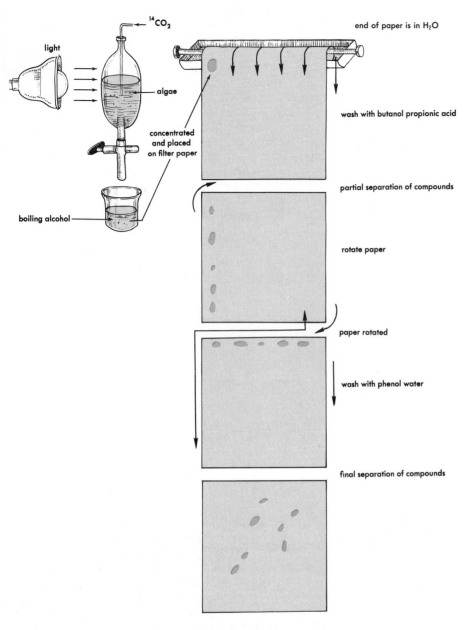

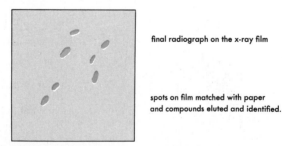

Figure 6.13 Method for labeling and isolating photosynthetic products. Illuminated green cells are exposed to $^{14}CO_2$, then dropped into boiling alcohol to extract the products that have incorporated ^{14}C. Chromatography follows as shown in the diagram and described in the text. An autoradiograph completes the process; identity of materials in the radioactive spots is determined by comparison with the positions of spots made when known compounds are subjected to similar chromatography.

The problem confronting physiologists was to determine the sequence of compounds through which carbon atoms taken into the leaf pass during the synthesis of sugar. When the radioactive isotope of carbon (^{14}C) became available, scientists treated leaves with carbon dioxide that contained radioactive carbon ($^{14}CO_2$) and determined the carbon compounds formed. This may easily be done by the method outlined in Fig. 6.13. The mixture of substances extracted from the leaf is first separated by chromatography: A spot of the extract is placed on one corner of a sheet of filter paper. One edge of the paper is then dipped in a carefully chosen solvent solution. The solvent flows across the paper, carrying various materials from the leaf extract along at different and characteristic rates. This converts the original spot of material into a row of spots, each spot containing a different collection of molecules. The paper is now dried and rotated 90 degrees; then another edge is dipped into a second solvent solution. The second solvent differs from the first and causes solutes to migrate differently on the paper. Again as the solvent moves, it draws each spot out into a series of spots, each containing only one or a few substances.

By comparing the positions of spots on the finished chromatograph with the positions reached by known substances on similar runs, we can identify the substances in each spot.

The method is completed by placing the chromatograph in contact with a sheet of photographic film for several days. Spots that contain radioactive carbon expose the film. The developed film therefore shows which of the materials in the extract were produced from $^{14}CO_2$ in photosynthesis.

A group of scientists at the University of California under the direction of Melvin Calvin were particularly successful in using this technique for working out the early carbon pathways of photosynthesis. Calvin eventually received the Nobel Prize for this work.

To introduce radioactive carbon into the photo-synthesizing cells, these investigators placed an algal culture in a glass flask in the light, exposed the cells to radioactive CO_2, and killed the cells in boiling alcohol. By varying the time after the plant was exposed to $^{14}CO_2$, they could determine the first compounds synthesized during exposure to ^{14}C.

If photosynthesis proceeds for an hour or so in an atmosphere containing radioactive carbon dioxide, most of the labeled carbon will be found in carbohydrates (sugar or starch). If photosynthesis is stopped after only a few seconds, most of the labeled carbon is found in a three-carbon acid containing phosphorus, phosphoglyceric acid (PGA).

PGA is thus an intermediate between carbon dioxide and sugar. By stopping photosynthesis at increasingly longer intervals (from a few seconds to several minutes), investigators have found that many carbon compounds become labeled with radioactive carbon.

It has been possible by means of such experiments to draw up an outline representing the various reactions that occur (Fig. 6.14).

Here are the key points of this carbon cycle:

1. Catalyzed by the enzyme carboxydismutase, CO_2

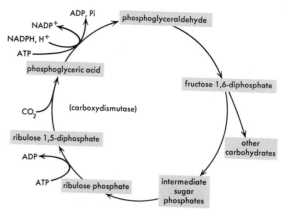

Figure 6.14   Diagram showing some steps in the carbon cycle of photosynthesis, the Calvin, or C_3, cycle.

combines with ribulose diphosphate, a five-carbon-atom sugar phosphate continually being produced in the cell. Two molecules of phosphoglyceric acid (PGA), a three-carbon-atom compound, are produced.

2. Two molecules of PGA are reduced to form two molecules of a three-carbon-atom sugar phosphate, phosphoglyceraldehyde. The energy that drives this reaction comes from NADPH and ATP, and energy from these molecules is stored in the newly formed triose sugar. ADP and NADP are regenerated.

3. Two triose phosphates may be combined to form a six-carbon-atom sugar phosphate, fructose diphosphate. By this process the plant has essentially added one CO_2 molecule to a five-carbon-atom sugar to produce one molecule of a six-carbon-atom sugar.

4. As this process continues, some of the fructose phosphate may be transformed through other reactions into other carbohydrates, including sucrose and starch.

5. Some of the fructose phosphate molecules are used to form new molecules of ribulose diphosphate, the compound that accepts CO_2 in step 1. Thus the whole process forms a cycle of reactions, with CO_2 from the air and hydrogen from water entering the cycle, and various sugars being produced.

This carbon cycle of photosynthesis is often called the Calvin cycle or the C_3 path. It is almost universally present in photosynthetic plants.

There is a special added path of carbon fixation that is common in plants that have evolved in regions of high light intensity, high temperature, and drought. This is the C_4 pathway, also called the Hatch-Slack pathway after the two Australian workers who studied it most extensively.

In this pathway, CO_2 is first taken up by an enzyme called PEP carboxylase. This enzyme has a stronger affinity for CO_2 than does carboxydismutase. It adds CO_2 to the three-carbon compound phosphoenolpyruvate (PEP), to form oxaloacetic acid, a four-carbon-atom compound. Later, oxaloacetic acid may be converted to two other C_4 acids, malic and aspartic. These acids may accumulate to high levels. They seem to serve chiefly as a means of capturing CO_2 and storing it, because they later release their CO_2, which is taken up by carboxydismutase and used in the ordinary C_3 cycle to produce sugars. Thus the C_4 path is an addition, rather than an alternative, to the Calvin cycle.

CO$_2$ Reduction and the Formation of Sugar (Dark Reactions)

Under the hot and dry conditions of the tropics, plants that have the C_4 path are much more efficient at extracting CO_2 from the air and building sugars than plants that can use only the C_3 path. This efficiency is reflected in the economic value of sugar cane, which has the C_4 path. Maize (corn) and sorghum are also C_4 plants, among other examples from over 100 genera. These include both dicots and monocots, which suggests that the C_4 pathway has arisen independently many times in evolution.

Efficiency of Photosynthesis

Although the green leaf is the major organ used in the production of man's food, it is not particularly efficient in utilizing the sun's energy. We speak of the efficiency of a machine, such as a diesel engine, a gasoline motor, or an electric motor. We calculate the energy value of fuel used by the machine and compare it with the machine's energy output. Much of the fuel's potential energy is lost. The ratio of energy outgo to energy intake represents efficiency. Of the total radiant energy that falls upon green leaves, about 80% is absorbed. Of the remaining 20%, a part is reflected from the leaf surface and a part passes through the leaf. Part of the radiant energy absorbed is changed to heat and raises the temperature of the leaf; a large part of the energy absorbed is used in transpiration; the remainder of it is utilized in photosynthesis and stored in carbohydrate molecules. Only about 0.5 to 3.5% of all the light energy that falls on a leaf is used in photosynthesis. Although this is a very low percentage compared to machine efficiency, the supply of solar energy is continuous and abundant.

However, there is reason to believe that the leaves of higher plants, under the most favorable conditions, should be about 10 times more efficient in converting solar energy into chemical energy by photosynthesis. Research will be needed if we are to raise the photosynthetic efficiency of plants growing out of doors to the efficiency that is theoretically possible. The large research effort that has been devoted to photosynthesis also aims for a basic understanding of how nature converts light energy into chemical energy. With this information we might someday devise artificial alternatives to photosynthesis as a way to meet some of the world's food problems. And, too, some of the lessons we learn about the plant's use of solar energy could prove useful in our search for alternatives to nuclear and fossil fuels.

The productivity of plants is determined partly by the environment and partly by the hereditary characteristics of the plant (e.g., its structure and metabolic capabilities). Thus one way to solve the problem of producing enough food for the world population is to breed highly productive varieties of plants. This approach already has been so successful in some areas (e.g., the development of highly productive types of rice and wheat) that we speak of the "green revolution." The importance of this approach was recognized by the recent award of the Nobel Prize to Norman Borlaug, who succeeded in breeding new strains of high-producing grains. Unfortunately, these strains require high levels of fertilizer application—an expensive and pollution-creating practice.

One aspect of photosynthesis that seems a likely point of attack for improving productivity is the process known as **photorespiration.** In some plants, photorespiration may reoxidize and release as much as 30% of the CO_2 that has been incorporated in photosynthesis. Photorespiration differs from ordinary respiration in that it does not generate ATP. At the present time it seems to be the result of a flaw in the photosynthetic system. It involves the oxidation of glycolic acid, one of the early products of the carbon reduction system. Investigations are currently underway to determine whether new strains of crop plants, or new growing conditions, can be found that will have lower rates of photorespiration and therefore better retention of the products of photosynthesis.

The Environment and Photosynthesis

The principal external conditions that affect the rate of photosynthesis are (a) temperature; (b) light intensity, quality, and duration; (c) carbon dioxide content of the air; and (d) mineral elements in the soil.

Temperature

Plants of cold climates carry on photosynthesis at much lower temperatures than do those of warm climates. The process is known to occur in certain evergreen species of cold regions, even at temperatures below 0°C. Algae in the water of hot springs may carry on photosynthesis at a temperature as high as 75°C. Most ordinary temperate-climate plants, however, function best between temperatures of 10 and 35°C. If there is adequate light intensity and a normal supply of carbon dioxide, the rate of photosynthesis of most ordinary land plants increases with an increase in temperature up to about 25°C; above

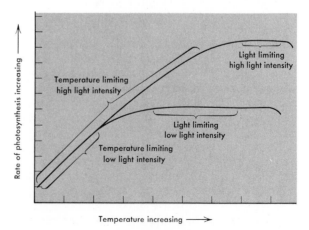

Figure 6.15 Graph showing interaction of light and temperature on the rate of photosynthesis. When light is limiting, the rate of photosynthesis is independent of temperature. When temperature is limiting, the rate of photosynthesis is independent of light intensity.

this range there is a continuous fall in the photosynthetic rate as the temperature is raised. At these higher temperatures, the length of exposure is of importance. At a given constant high temperature (for example, 40°C), the rate of photosynthesis decreases with time.

Under conditions of low light intensity, an increase in temperature beyond a certain minimum will not produce an increase in photosynthesis (Fig. 6.15). These conditions may occur in the winter in greenhouses. If the temperature is raised too high, the plants will suffer because the rate of photosynthesis has not been changed but respiration has been increased by the higher temperature.

Light

In discussing the effect of light upon the rate of photosynthesis, we must consider three elements: (a) intensity, (b) quality (wavelengths), and (c) duration. Given the proper temperature and sufficient carbon dioxide, carbohydrates produced by a given area of leaf surface increase with increasing light intensity up to a certain point (optimum light intensity), after which their production decreases.

Intense light appears to retard the rate of photosynthesis. The usual light intensity in arid and semi-arid regions is well above the optimum for photosynthesis in many plants, especially introduced crop plants. In these regions on days when the sky is overcast, the light intensity is probably closer to the optimum for photosynthesis than on clear, sunny days. Leaves on the surface of plants receive light of greater intensity than those beneath that are shaded. Therefore, some of the leaves receive light of optimum intensity, whereas others may receive light either above or below the optimum.

Carbon Dioxide

The carbon dioxide utilized by land plants is absorbed by the leaves from the atmosphere.

Let us consider the problem in this way: If all the chlorophyll-bearing cells of the plants of the world are constantly taking carbon dioxide from the atmosphere during daylight, the quantities of this gas used must be enormous, and there must necessarily be processes in nature that are continually replenishing this supply. It is known that the amount of carbon dioxide in the air is low (0.03%) and that it remains fairly constant. It has been estimated that a hectare of corn (25,000 plants) during a growing season of 100 days will accumulate 6350 kg of carbon; all this carbon is derived from the carbon dioxide of the atmosphere. It would require 23,300 kg of carbon dioxide to furnish this quantity of carbon. These estimates serve to emphasize the fact that enormous quantities of carbon dioxide are used in the photosynthetic process of green plants.

Obviously, the amount of carbon dioxide is limited. The present atmospheric supply would be used up in about 22 years were it not constantly being renewed. Through several natural processes carbon dioxide is continually released to the atmosphere:

1. The living cells of all plants (both green and nongreen) and of all animals release carbon dioxide in the respiratory process (Chapter 2).
2. The dead bodies of plants and animals, and the excretions of animals, contain large quantities of carbon and other elements in the form of organic compounds; in the decay of these compounds, resulting from the activities of bacteria and fungi, large quantities of carbon dioxide are released to the atmosphere.
3. Carbon dioxide is also added to the atmosphere when wood, coal, oil, gas, or any other carbon compound burns.
4. Carbon dioxide is released to the atmosphere from mineral springs and volcanoes.
5. The oceans are important reservoirs of dissolved carbon dioxide, and from them carbon dioxide probably escapes whenever its concentration in the atmosphere decreases.

If light intensity and temperature are favorable, the carbon dioxide of the atmosphere frequently limits the rate of photosynthesis. This may be particularly true in greenhouses that are kept closed in the winter. Under these conditions, the carbon dioxide in the air may be reduced far below the 0.03% average.

It has been determined experimentally that at usual temperatures and light intensities, an artificial increase of carbon dioxide up to a concentration of 0.5% may result in an increased rate of photosynthesis, but only for a limited period of time. It appears that this high level of carbon dioxide is harmful to plants; after 10 to 15 days' exposure, the plants show injury.

Minerals

The chemical formula of chlorophyll a is $C_{55}H_{72}O_5N_4Mg$ and of chlorophyll b is $C_{55}H_{70}O_6N_4Mg$. The synthesis of chlorophyll, therefore, depends on a supply of nitrogen and magnesium, both derived from salts in the soil. Moreover, chlorophyll is not formed unless iron is available, although this element is not a component of the chlorophyll molecule. Leaves of plants deficient in nitrogen, magnesium, or iron are pale and yellow, a condition termed chlorosis. This abnormal condition may also be caused by other factors, but whenever it does occur, the rate of photosynthesis is lowered.

Summary

1. Photosynthesis is the primary energy-storing process of life. In it, light energy is stored as chemical energy in organic compounds.
2. The raw materials of photosynthesis are carbon dioxide and water; the products are sugar and oxygen (O_2).
3. Chloroplasts are organelles that perform photosynthesis in green plants.
4. The energy of photons (light) removes electrons from chlorophyll. Some of the energetic electrons, plus hydrogen ions from water, are taken up by NADP to form NADPH. The rest of the photoelectrons return to chlorophyll ions after passing along chains of electron

carriers. During this transport, some of the electron energy is transferred to produce ATP.

5. The electrons lost by chlorophyll in making NADPH are replaced by electrons from water. This splits water molecules to release H^+ and O_2.

6. The reactions that require light to form ATP, NADPH, and O_2 are called the light reactions. They occur in thylakoids of the chloroplasts.

7. In the stroma of the chloroplast, the H and electrons of NADPH are transferred to organic compounds. These compounds are used in a series of reactions that incorporate CO_2 and produce molecules of sugar. ATP is used as an energy source. These steps are called the dark reactions. Two major pathways of carbon fixation are the C_3 and C_4 pathways.

8. The photosynthetic process captures only a small fraction of the available light energy.

9. Photosynthesis may be limited by the CO_2 supply, light, temperature, minerals, and the hereditary efficiency of the plant.

flowers, fruits and seeds

PART ONE

The Flower

The flower initiates the sexual reproductive cycle in all Anthophyta (the flowering plants) and, in so doing, it terminates the growth of the shoot bearing the flower. With some plants—annuals, biennials, and some perennials—death follows flowering and seed set. With most perennials the capacity for vegetative growth is regained after sexual reproduction and the development of the next generation. In woody perennials, provision is always made for continued vegetative growth, through mixed buds or associated shoot buds following flower production. The function of the flower is to facilitate the important events of gamete (haploid reproductive cell) formation and fusion.

Of all the characteristics of flowering plants, the flower and fruit are the least affected by changes in the environment. For instance, we have seen that leaf shape is influenced by age, light, water, and nutrition. The basic morphology and anatomy of flowers and fruits are not affected in these ways and thus are important parts for angiosperm classification.

In addition, flowers have a high esthetic value, and the resulting fruits and seeds are of major importance in food production.

The essential steps of sexual reproduction, meiosis and fertilization, take place in the flower. The complete sexual cycle involves (a) the production of special reproductive cells following meiosis, (b) pollination, (c) fertilization, (d) fruit and seed development, (e) seed and fruit dissemination, and (f) seed germination. The seed completes the process of sexual reproduction in the angiosperms, and the embryo in the seed is the first stage in the life cycle of new individuals.

Flower Structure

A typical flower is composed of four whorls of modified leaves: (a) **sepals,** (b) **petals,** (c) **stamens,** and (d) a **carpel** or **carpels,** all attached to the modified stem end that supports these structures, the **receptacle** (Fig. 7.1).

The sepals enclose the other flower parts in the bud and are generally green. All the sepals taken collectively

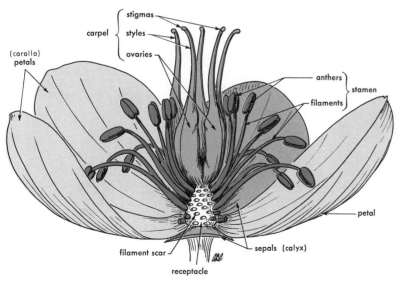

Figure 7.1 A diagram of a longitudinal section of a flower of Christmas rose *(Helleborus)*. The perianth consists of two similar whorls; there are numerous stamens arranged in a spiral on a cone-shaped receptacle. Five separate carpels form the central whorl of floral parts.

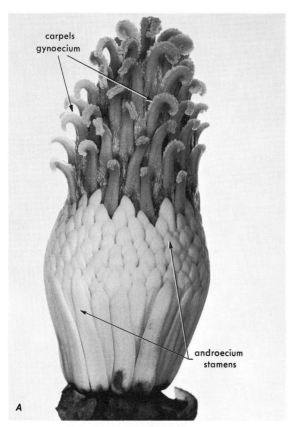

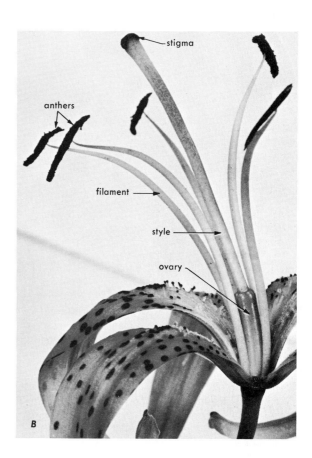

Figure 7.2 The essential floral organs. *A, Magnolia,* showing many separate stamens and separate carpels arranged in a spiral on the receptacle; *B,* regal lily *(Lilium regale),* showing six stamens with anthers and filaments, and three carpels in a single pistil with stigma, style, and ovary; *C,* carpels suggestive of foliage leaves. *Sterculia plantanifolia:* dehiscence of matured ovary showing seed attached to the margins of the five leaflike carpels, $\times \frac{5}{8}$.

constitute the **calyx.** The petals are usually the conspicuous, colored, attractive flower parts. Taken together, the petals constitute the **corolla.**

The stamens form a whorl, lying inside the corolla. Each stamen has a slender stalk or **filament** at the top of which is an **anther,** the pollen-bearing organ. The whorl or grouping of stamens is called the **androecium** (Fig. 7.2).

The carpel or carpels comprise the central whorl of modified floral leaves (Fig. 7.2). Collectively, the carpels are spoken of as the **gynoecium.** Each individual structure in the gynoecium is referred to as a **pistil.** A pistil may be composed of a single carpel, or of several united carpels in the center of the flower.

There are generally three distinct parts to each pistil: (a) an expanded basal portion, the **ovary,** in which are borne the ovules; and (b) the **style,** a slender stalk supporting (c) the **stigma** (Fig. 7.1).

The term **perianth** is applied to the calyx and corolla collectively. It is frequently used to describe flowers, such as the tulip *(Tulipa),* in which the two outer whorls are similar in appearance. The individual parts of such a perianth are called **tepals.**

Sometimes individual flowers or compact clusters of flowers will have a whorl of small leaves or **bracts** subtending them. A collection of bracts subtending flowers is called an **involucre.**

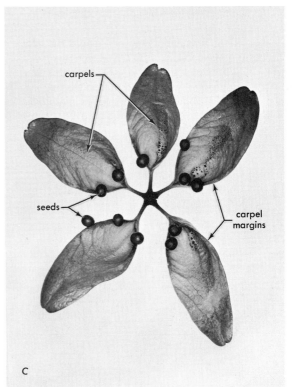

Carpels and Stamens as Modified Leaves

The question arises as to why carpels and stamens are considered modified leaves. One reason is because the

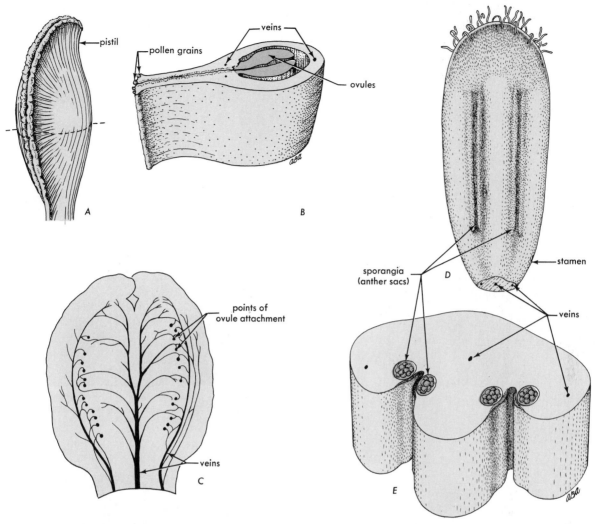

Figure 7.3 Primitive pistil and stamen from living plants in southeast Asia. *A,* pistil of *Drimys piperita; B,* cross section of pistil (dotted line); *C,* pistil laid open; *D,* stamen of *Degeneria vitiensis; E,* cross section of stamen.

early developmental stages of floral parts closely resemble those of leaves. The second reason, also circumstantial and speculative, concerns the shape of spore-bearing leaves in primitive angiosperms. If we were to examine the stamens of primitive flowers such as *Magnolia* or *Degeneria* (Fig. 7.3) (see Chapter 16), we would see that the shape of the stamen is very leaflike. In addition, the folded open carpels of primitive angiosperms are leaflike in appearance. If a young pea pod, which is a single carpel, is opened carefully along its ventral suture (toward the axis), the margins may be folded back, showing the seeds, which have developed from ovules, along the margin of a leaflike carpel (Fig. 7.17*D*).

The relationship between carpels and stamens is even more striking in *Sterculia platanifolia,* because in this plant there are five simple pistils united only by their stigmas (Fig. 7.2*C*). When these stigmas mature, they open to show five very leaflike carpels that bear seeds along their margins. Carpels may thus also be compared with leaves. The evolutionary steps still remain speculative.

Variations in Floral Structure

General Description

Complete and Incomplete Flowers

The parts of a typical flower—sepals, petals, stamens, and carpels—are all attached to the receptacle. A flower with all four sets of floral leaves is said to be a **complete flower.** An **incomplete flower** lacks one or more of these four sets. For example, there are (a) flowers that partially or completely lack a perianth (Figs. 7.4*A,B*), (b) flowers with carpels but no stamens, and (c) flowers with stamens but no carpels (Fig. 7.5).

Perfect and Imperfect Flowers

Unisexual flowers are either **staminate** (stamen-bearing) or **pistillate** (pistil-bearing). Unisexual flowers are said to be **imperfect,** whereas bisexual flowers are **perfect.** When staminate and pistillate flowers occur on the same

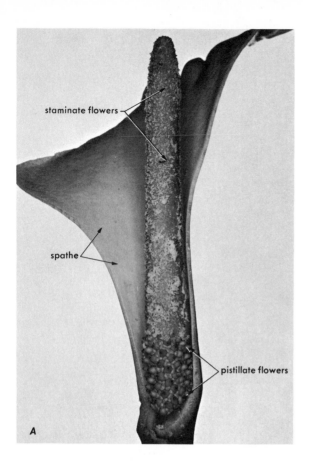

Figure 7.4 Flowers with perianth parts lacking. A, spathe of calla lily *(Zantedeschia aethiopica)*, sepals and calyx absent; B, flower of *Clematis*, petals absent and sepals prominent.

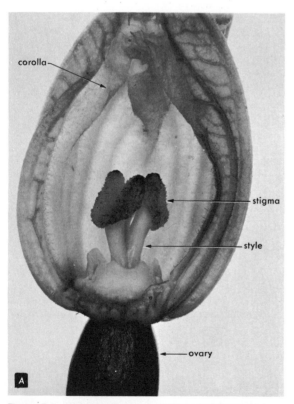

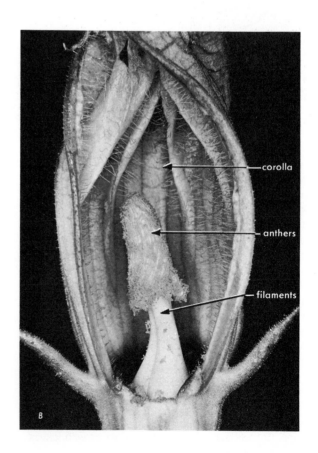

Figure 7.5 Imperfect flower. A, pistillate and B, staminate flowers of squash *(Cucurbita)*, × ½.

chapter **7** | **Flowers, Fruits and Seeds**

individual plant, as they do in corn (Zea), squash (Cucurbita), walnut (Juglans) (Figs. 7.5A,B, 7.6, 7.7), and many other species, the plant is said to be **monoecious**. In corn (Fig. 7.5), for example, the tassel (borne at the top of the stalk) consists of a group of staminate flowers, and the young ear is a group of pistillate flowers. When staminate and pistillate flowers are borne on separate individual plants, as in Asparagus, willow (Salix), and many other species, the species (or the plant) is said to be **dioecious**.

Floral Symmetry

In many flowers such as those of columbine (Aquilegia) (Fig. 7.8A) or cherry (Prunus) (Fig. 7.9B) the corolla is made up of petals of similar shape that radiate from the center of the flower and are equidistant from each other. Such flowers are said to be **regular**. In these cases, even though there may be an uneven number of parts in the perianth, any line drawn through the center of the flower will divide the flower into two similar halves. They may be exact duplicates or mirror images of each other.

Irregular flowers, such as pea (Pisum) (Fig. 7.8C) and mints (Mentha) (Fig. 7.8B), have whorls either (a) with dissimilar flower parts, (b) with parts that do not radiate from the center, or (c) are not equidistant from one another. In most of these flowers, only one line will divide the flower in equal halves; the halves are usually mirror images of each other. Some flowers such as, bleeding

Figure 7.6 Flowers of corn (Zea mays). A, the tassel, ×⅕; B, exserted anthers of staminate flowers, ×1; C, pistillate flowers forming a very young ear, ×½; D, several pistillate flowers with attached styles, ×1.

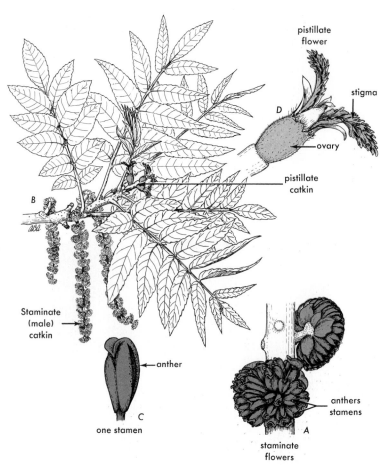

Figure 7.7 English walnut *(Juglans regia)* flowers. *A,* staminate flowers; *B,* staminate inflorescence (catkin); *C,* dehiscing anther; *D,* pistillate flower.

heart *(Dicentra),* though irregular, may be bisected by any number of lines into similar mirror images.

The irregular bean or pea flower has a corolla composed of the following (Fig. 7.8*C*): one broad conspicuous petal, the **standard** or **banner;** two narrower petals **(wings),** one on each side; and, opposite the banner, two smaller petals that are united along their edges to form the **keel.**

Union of Flower Parts

In the flower of columbine illustrated in Fig. 7.8*A,* all parts of the flower are separate and distinct; that is, each sepal, petal, stamen, and carpel is attached at its base to the receptacle. In many flowers, however, members of one or more whorls are to some degree united with one another, or are attached to members of other whorls. Union with other members of a given whorl is termed **connation.** Union of flower parts from two different whorls is known as **adnation** (Figs. 7.8*A,C;* 7.5*A,B;* 7.9*B,C*). A precise terminology has been developed to indicate these fusions. For example, syncarpy refers to the situation where carpels of the gynoecium are connate.

Elevation of Flower Parts

In flowers of magnolia (Fig. 7.2*A,* 16.3) and tulip (Fig. 7.9*A*), the receptacle is convex or conical and the different flower parts are arranged one on top of another. The gynoecium is thus situated on the receptacle above the points of origin of the perianth parts and androecium. An ovary in this position is said to be **superior.** In the daffodil (Fig. 7.9*C*) the ovary appears to be below the apparent points of attachment of the perianth parts and the stamens. This is an **inferior** ovary. With an inferior ovary, the anatomy of the flower parts indicates that the lower portions of the three outer whorls, calyx, corolla, and androecium have fused to form a tube or **hypanthium.** The ovary in the daffodil is completely adnate with the hypanthium.

In a flower with a superior ovary, sepals, petals, and stamens arise from the outer, lower portion of the concave receptacle, below the point of origin of the carpels (Fig. 7.1). The perianth and stamens are **hypogynous,** or with reference to the three outer whorls, a condition of **hypogyny** exists. In a flower with an inferior ovary, the perianth and stamens appear to arise from the top of the ovary, and they are **epigynous.**

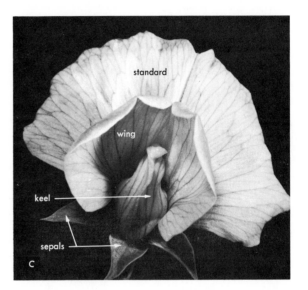

Figure 7.8 Floral symmetry. *A,* regular flower of columbine *(Aquilegia); B,* irregular flower of a mint *(Salvia),* × ½; *C,* flowers and essential organs—flower or garden pea *(Pisum sativum).*

In some flowers such as the plum and apricot *(Prunus* sp.), the hypanthium does not become adnate to the ovary. The perianth parts and stamens arising from the rim of cuplike hypanthium around the ovary are **perigynous** (Fig. 7.9*B*).

Inflorescences

In most flowering plants, flowers are borne in clusters or groups. Morphologically, an **inflorescence** is a flower-bearing branch or system of branches. In manuals for the identification of flowering plants, there are many different terms descriptive of various kinds of inflorescence. Only the most common ones are discussed and illustrated here (Fig. 7.10).

A very simple type of inflorescence may be found in such plants as currant *(Ribes)* and radish *(Raphanus).* It is called a **raceme.** In this type, the main axis has short branches, each of which terminates in a flower. Each flower is on a short branch stem called the **pedicel.** The main axis of a raceme continues to grow in length more or less indefinitely; the apical meristem persists. Primordia of leaves arise in the usual manner along the margin of this

apical meristem, and in the axil of each leaf a flower is borne. The oldest flowers are at the base of the inflorescence and the youngest are at the apex.

In a simple raceme, the flowers are on pedicels that are about equal in length. In a **spike** (Fig. 7.10), the main axis of the inflorescence is elongated, but the flowers, each in the axil of a bract, are sessile (without a pedicel). The **catkin** is a spike that usually bears only pistillate or staminate flowers. The inflorescence as a whole is shed later. Examples are willow, cottonwood *(Populus),* and walnut (Figs. 7.7, 16.7).

In all the inflorescences just mentioned, the flowering axis is elongated. If it is short, flowers appear to be arising umbrellalike from approximately the same level. An inflorescence of this kind, in which pedicels are of nearly equal length, is called an **umbel** (Fig. 7.10). The onion *(Allium)* is a good example. The **head** is an inflorescence in which the flowers are sessile and crowded together on a very short axis. Members of the family Asteraceae (Compositae), including thistle *(Cirsium)* and sunflower, have this type of inflorescence (Fig. 7.10).

Inflorescences of the raceme type may be compound, that is, branched. A branched raceme is called a **panicle** (Fig. 7.10). Compound spikes occur in wheat, rye, and

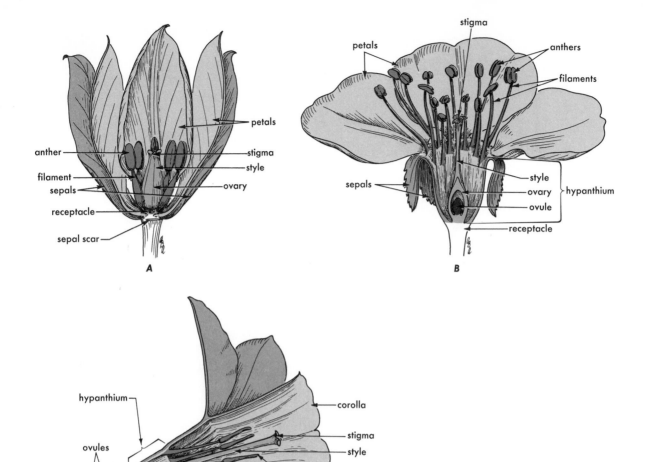

Figure 7.9 Diagram of elevation of floral parts. *A*, hypogyny *(Tulipa); B*, perigyny in cherry flower *(Prunus avium); C*, epigyny in daffodil flower *(Narcissus pseudonarcissus)*.

certain other grasses.

In contrast to the raceme types of inflorescences, just described, is the **cyme** (Fig. 7.10). In the cyme, the apex of the main axis produces a flower that involves the entire apical meristem; hence, that particular axis ceases to elongate. Other flowers arise on lateral branches farther down the axis of the inflorescence and, thus, usually the youngest of the flowers in any cluster occurs farthest from the tip of the main stalk. The flower cluster of chickweed *(Cerastium)* is an example of the cyme.

Angiosperm Life Cycle

The essential floral organs necessary for sexual reproduction are the stamens (androecium) and carpels

(gynoecium) (Fig. 7.1). The perianth, composed of calyx and corolla, is a protective covering of the stamens and pistils and when it is present it also serves to attract and guide the movements of pollinators.

The Androecium

Each stamen consists of an anther supported by a filament (Fig. 7.11*A*). The anther usually consists of four elongated and connected lobes called pollen sacs. Early in the development of the anther, the pollen sacs each contain a mass of dividing cells called **microsporocytes** (microspore mother cells) (Fig. 7.11*B*). Each microsporocyte divides by meiosis to form four haploid (1*n*) microspores. The nucleus of each microspore then divides by mitosis to form a two-celled **pollen grain** that

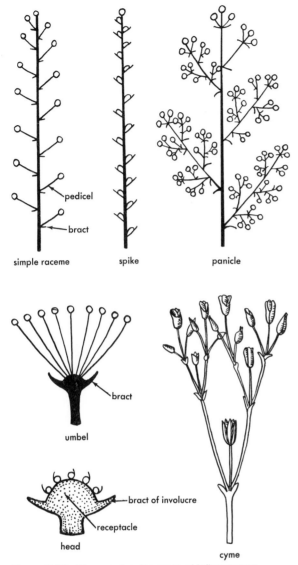

pedicel

bract

simple raceme   spike   panicle

bract

umbel

bract of involucre

receptacle

head

cyme

Figure 7.10   Diagram showing types of inflorescence.

contains a **tube cell** and a smaller **generative cell** (Fig. 7.11*E*). The role of this two-celled, haploid, **male gametophyte** (gamete-producing plant) is to produce sperm nuclei necessary later for fertilization.

The pollen grain is surrounded by an elaborate, ornate cell wall. The pattern of the pollen wall can be intricate and beautiful; at the same time it is useful both to the plant and to plant taxonomists. The wall is composed of an extremely hard material called sporopollenin (Fig. 7.13). This material is so hard that pollen grains many thousands of years old still retain their same pollen wall texture and pattern. (Fig. 7.12).

Remember that floral structures are relatively unaffected by environmental changes. Consequently, the form of such structures as pollen grains can be used to make genetic comparisons between plant species. Plant taxonomists study pollen wall patterns and are able to

make intelligent guesses about relationships in their evolution.

After the pollen grains are mature, the anther wall splits open and the pollen are shed. In some manner (which will be discussed later), the pollen are transported to the stigma of adjacent or distant flowers. This process is called **pollination**. Once on the stigma, if the pollen and the stigma are genetically compatible, it will germinate to form an elongate **pollen tube** (Fig. 7.11*F*). The ornate pattern of the pollen grain wall serves as the means for receptive stigmas to recognize compatible pollen. The troughs and ridges of the pollen wall (Fig. 7.13) apparently contain proteins in specific patterns and concentrations. If the patterns correlate with those of the stigma, metabolic events are triggered within the pollen grain to stimulate the pollen tube to grow into the pistil's tissue.

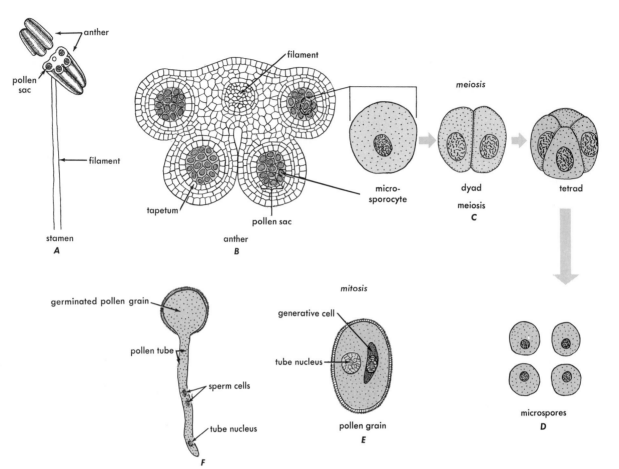

Figure 7.11 Development of pollen from a microsporocyte to the pollen grain. *A*, stamen; *B*, cross section of anther; *C*, development of tetrad of cells from the pollen mother cell by meiosis; *D*, four microspores; *E*, pollen grain; *F*, germination of pollen grain into pollen tube.

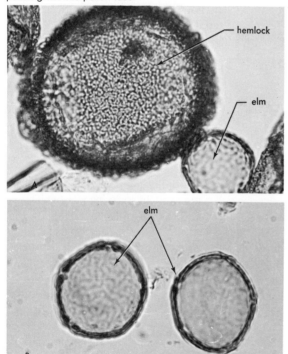

Figure 7.12 Fossil *(A)* and modern *(B)* pollen of the same genus.

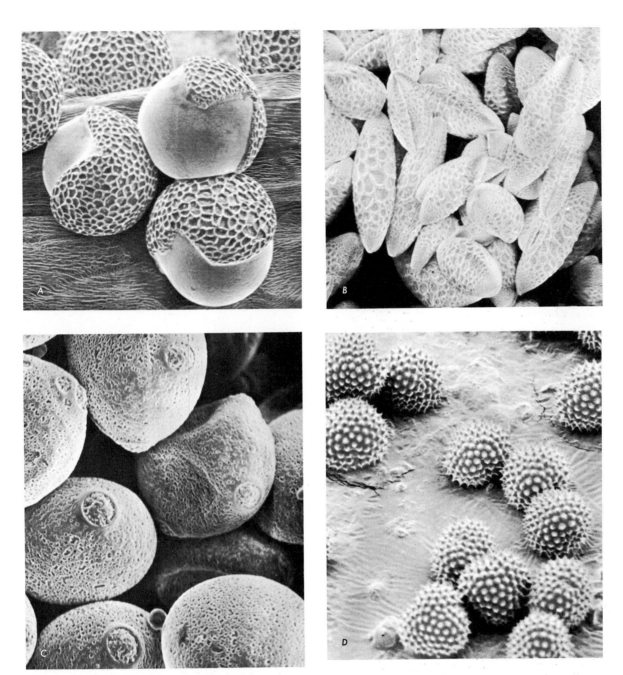

Figure 7.13 Pollen grains as seen with the scanning electron microscope. *A, Iris*, ×210; *B*, day lily *(Hemerocallis fulva)*, ×275; *C*, cucumber *(Cucumis sativus)*, ×870; *D*, ragweed *(Ambrosia)*, ×1000.

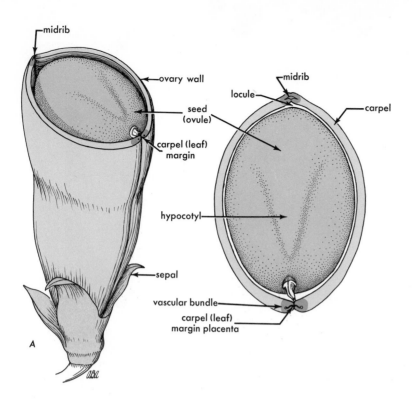

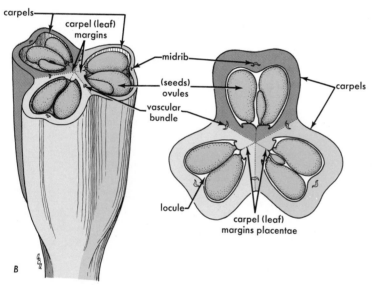

Figure 7.14 Diagram comparing gynoecia consisting of a single carpel and of three connate carpels. *A*, section of ovary of pea consisting of a single carpel; *B*, section of ovary of *Tulipa* showing three connate carpels.

The Gynoecium

The structure of the gynoecium depends on the number and arrangement of carpels comprising it (Fig. 7.8*C*). In the pea flower for example, there is a single carpel forming the gynoecium (Figs. 7.14, 7.18). It consists of three parts: (a) the **ovary**, an expanded basal portion, (b) the **style**, a slender stalk, and (c) a hairy irregular tip portion, the **stigma**. When the **pistil** is composed of one carpel, such as pea, it is referred to as a simple pistil. In many other instances a compound pistil, consisting of two or more fused carpels, may occur (Fig. 7.14*B*).

The ovary is a hollow structure having from one to several chambers, or **locules** (Figs. 7.14, 7.15). The number of carpels in a compound pistil is generally related to the number of stigmas (Fig. 7.9*A*), the number of locules and, sometimes, the number of faces of the ovary (Fig. 7.14*A*). The pea has a single stigma and the ovary has one locule (Fig. 7.14*A*). There are three stigmas in the tulip gynoecium, the ovary is three-sided (Fig. 7.9*A*), contains three locules, and is composed of three carpels.

Placentation

The tissues within the ovary to which the **ovules** are attached are called **placentae** (singular, **placenta**, Fig. 7.16). The manner in which the placentae are distributed in the ovary is termed **placentation**. When the placentae are on the ovary wall, as in the pea *(Pisum)* and bleeding heart *(Dicentra,* Fig. 7.17*D*), the placentation is **parietal**. When they arise on the axis of the ovary, which has several locules as in lilies, *Fuchsia,* and tulip, the placentation is **axile** (Fig. 7.17*C*). Less frequently, the ovules are on the axis of a one-loculed ovary, in which event the placentation is **central**, as in the primrose family (Figs. 7.17*A,B*).

Style and Stigma

The style is a slender stalk that terminates in the stigma. It is through stylar tissue that the pollen tube grows. In some flowers, the style is very short or entirely lacking; in others, it is long. In *Zea mays,* the corn silks are the styles (Figs. 7.6*C,D*). In general, the style withers after pollination, but in some plants (e.g., *Clematis*) it persists and becomes a structure that aids in the dispersal of the fruit. The stigmatic surface often has short cellular outgrowths that aid in holding the pollen grains; and sometimes it secretes a sugary and sticky solution, the **stigmatic fluid**. In many wind-pollinated plants, such as the grasses, the stigma is much branched, or plumelike.

The Ovule

The ovule, the structure that will eventually become the seed, arises as a dome-shaped mass of cells on the surface of the placenta. The outer cells of the dome develop to form one or two protective layers, the **integuments** (Figs. 7.15*C,* 16.1). The integuments do not fuse at the apex of the ovule, thus leaving a small opening, the **micropyle** (Fig. 7.15*C*). While this development is taking place, one of the internal dividing cells of the ovule, the **megasporocyte**, is enlarging in preparation for meiosis. The tissue within which the megasporocyte has differentiated is known as the **nucellus**. The ovule then is composed of one or two outer protecting integuments, along with the micropyle, nucellus, and megasporocyte (Fig. 7.15*E*).

Embryo Sac Development

Meiotic division of the **megasporocyte** (megaspore mother cell) is considered the first step in the development of a seed. As a result of these divisions, a row of four cells called **megaspores** are produced in the nucellus. As a rule, three of the cells (the ones nearest the micropyle) disintegrate and disappear, whereas the one farthest from the micropyle enlarges greatly (Figs. 7.15*D,E*). This megaspore now develops into the mature **embryo sac**. The usual stages are as follows: (a) a series of three mitotic divisions, forming an eight-nucleate embryo sac (Fig. 7.15*H*); (b) migration of nuclei, (Fig. 7.15*I*); (c) cell wall formation around nuclei (Fig. 7.15*I*).

At the micropylar end of the embryo sac there is one egg cell associated with two **synergid cells**. Since it is frequently difficult to differentiate the egg cell from the other two cells, all three cells are sometimes referred to as the **egg apparatus**. The two nuclei that migrated toward the center approach each other; these nuclei, known as **polar nuclei** (Figs. 7.15*I,J*), form a binucleate cell called the **endosperm mother cell**. The three nuclei remaining at the end of the embryo sac opposite the micropyle form the **antipodal cells**. The embryo sac, which may be considered to be a seven-celled haploid plant, (female gametophyte), is now mature and ready for fertilization.

Not every species of flowering plant produces a seven-celled embryo sac, but there is always an egg cell and an endosperm mother cell.

Fertilization

Prior to fertilization, the pollen tube has grown down through the stigma and style and has entered the ovary (Figs. 7.11*F,* 7.15*J,* 16.1). Many pollen grains may germinate, and their pollen tubes may grow through the pistil, but only one usually enters into the embryo sac for fertilization. While the pollen tube is growing, the generative cell within it divides by mitosis to form two sperm cells. In some plants, like sunflower, the sperm cells form even before the pollen are shed from the anther.

Reaching the ovary, the pollen tube grows toward one of the ovules, usually enters the micropyle, penetrates one or more layers of nucellar cells, enters the embryo sac (Figs. 7.15*J,* 16.1), and approaches the egg cell and the endosperm mother cell. These stages are not easy to study, but it appears that the tip of the tube ruptures, one sperm cell enters the egg, and the other sperm enters the endosperm mother cell. Within the egg, sperm and egg nuclei fuse. Within the endosperm mother cell, the other sperm nucleus and two polar nuclei fuse (Figs. 7.15*I,J*). This double fusion of egg with sperm and polar nuclei with sperm is called **double fertilization**. The *zygote* and *primary endosperm cell* result from these fusions, the antipodal cells and synergid cells will degenerate, and conditions are set for further development of the seed and fruit.

One interesting question concerns how the pollen tube is directed into the embryo sac. The answer has been worked out partially for cotton *(Gossypium)*. In the embryo sac of cotton, one or both of the synergid cells begin to shrivel and die before the pollen tube enters. It has been suggested that this activity results in the flow of some chemical substance from the synergid that possibly influences the direction in which the pollen tube grows. Two observations favor this hypothesis. First, at the base of both synergids is found a highly convoluted cell wall, known as the filiform apparatus (Fig. 7.15*I*). Cell walls in this configuration are commonly believed to be associated with active chemical secretion. The second bit of evidence is that the pollen tube actually penetrates one of the synergids and empties its contents into it. The two sperm then pass through the incomplete upper cell wall of the synergid; subsequently, one of them moves through the incomplete side wall of the egg to fertilize the egg cell, and

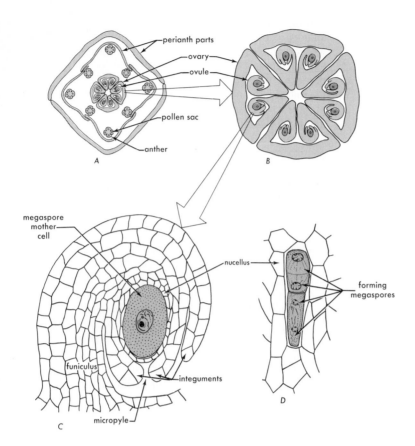

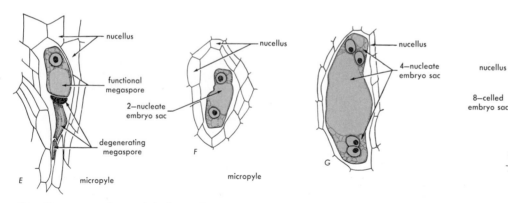

the other moves deeper into the embryo sac to fuse with the endosperm mother cell.

The fusions that occur in the embryo sac initiate the following changes in the entire ovary:

1. Development of the zygote to form the embryo plant.
2. Development of the primary endosperm cell to form the endosperm (the reserve food supply of the seed).
3. Development of integuments to form the seed coat.
4. Absorption or disintegration of nucellar tissue. In some plants, however, a portion of the nucellus, rather than the endosperm, may become the storage tissue of the seed. Such tissue of nucellar origin is called **perisperm**.
5. Development of ovary tissue forms the fruit.
6. Possible stimulation of accessory flower parts, such as the receptacle, sepals, or petals, to increased growth and incorporation into the fruit.

Thus, the importance of fertilization is to stimulate the growth of certain floral parts and to bring about a withering of others. The net result is the development of the ovary wall into the fruit and the ovule into the seed.

Pollination

Pollination may be brought about by either wind or insects, and occasionally by water, birds, bats, and other small mammals. Flower structure and pollen type are generally adapted to one or the other of these pollinating

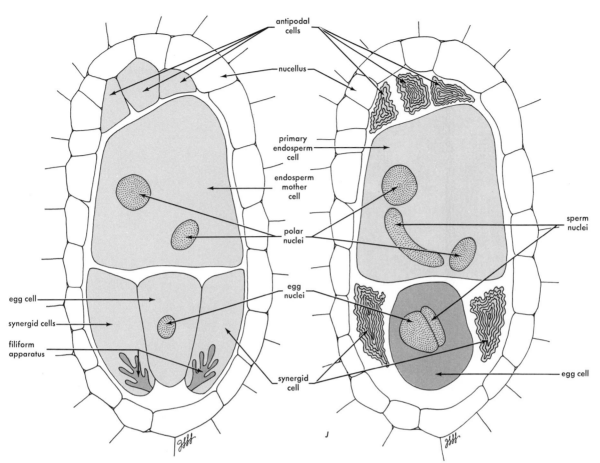

Figure 7.15 Diagram of embryo sac development. *A,* cross section of a flower bud showing the four whorls, four-carpelate ovary in center; *B,* enlarged view of ovary showing four carpels and ovules; *C,* an ovule, megasporocyte in prophase of meiosis; *D,* late telophase of second meiotic division; *E,* four megaspores, three degenerating; *F,* two-nucleate embryo sac; *G,* four-nucleate embryo sac; *H,* eight-celled embryo sac; *I,* mature embryo sac; *J,* fertilization, synergids degenerating.

vectors. In insect pollination, adaptation may be very complex, indicating a long association between the insect vector and plants.

Pollinating Vectors

Wind

Pollen is carried chiefly by wind and insects. **Wind pollination** is common in plants with inconspicuous flowers, such as in grasses (Fig. 7.6), walnuts (Fig. 7.7), oaks, and ragweeds (*Ambrosia*). Such plants usually produce pollen in enormous quantities. The flowers usually lack odor and/or nectar and hence are unattractive to insects. Furthermore, their pollen is light and dry and

easily wind-borne. Their stigmas, in many cases, are feathery and expose a large surface to catch flying pollen. Hayfever victims suffer from windborne pollen.

Living Vectors

Plants that are pollinated by living vectors usually possess a colorful perianth that, together with other floral parts, may be arranged into a complex architecture (Figs. 7.8, 7.9). They also produce both a sweet-tasting fluid, called nectar, and volatile compounds having distinctive odors. Many investigators have shown that these characteristics are highly adapted for the attraction of pollinators.

Bees, for example, are able to detect and distinguish between many odors and colors and degrees of

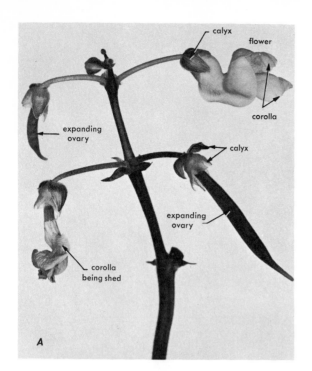

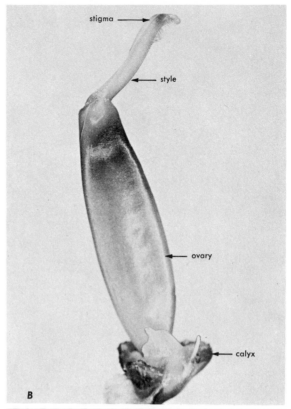

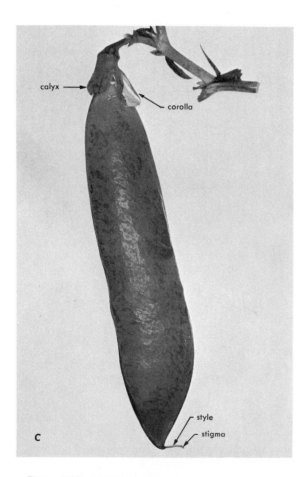

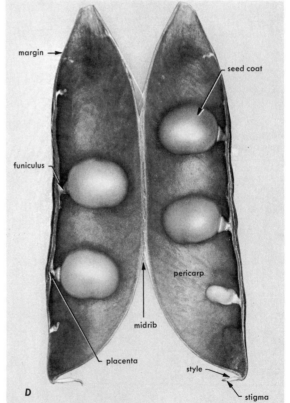

Figure 7.16 Development of legume fruits. A, bean (Phaseolus vulgaris) from flowers to young pods, ×1; B, pistil from a pea (Pisum sativum) flower, ×5; C, pea pod, unopened; D, opened pea pod showing developing seeds attached to carpel margins, ×1.

chapter **7** | **Flowers, Fruits and Seeds**

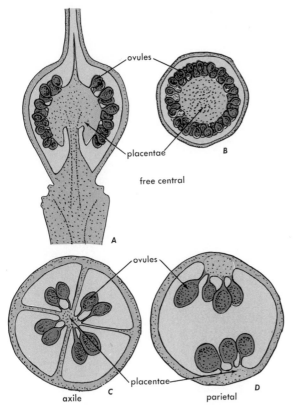

Figure 7.17   Three types of placentation. *A* and *B*, free central *(Primula)*; *C*, axile *(Fuchsia)*; *D*, parietal *(Dicentra)*.

sweetness. Bees can distinguish at least three colors in addition to ultraviolet, including the most common perianth shades (yellow and blue). Bees are directed to nectar-producing glands by splashes or lines of contrasting color. When the insect attempts to reach nectar or collect pollen, floral architecture ensures that the insect transfers pollen to the stigma. In nasturtium, for example, nectar is held in a long narrow tube or spur. A foraging bee directs its proboscis down the tube and usually rubs it against the stigma along the way. In this way, pollen, which may have been picked up in earlier visits to other flowers, is deposited on the stigma.

Orchid flowers have many very specialized mechanisms to ensure pollination. The petal shape and color of species of *Ophrys* closely resemble the appearance of a female wasp or fly. Male wasps or flies, emerging from the pupal stage before the females, mistake the *Ophrys* flower for females. The male lands on the flower and attempts copulation; repeated "pseudocopulation" results in pollination.

Types of Pollination

There are essentially two different kinds of pollination, determined by the genetic similarity of the plants involved. If anthers and stigmas, essential organs in pollination, have the same genetic constitution, the transfer of pollen from the anther to the stigma in the same flower or plant is called **self-pollination**. If two parent plants with different genetic constitutions are involved, transfer of pollen from the anther of the flower of one plant to the stigma of

another is called **cross-pollination**. Many species are capable of both self- and cross-pollination; others are obligate cross-pollinators.

Summary

1. The flower, the distinguishing structure of the Anthophyta, is formed of four whorls of parts specialized to carry out sexual reproduction, including pollination, fertilization, and seed production.
2. A floral apex has lost its ability to elongate and its potentiality for vegetative growth. The ability for vegetative growth is not regained until the completion of the two nuclear phenomena, meiosis and fertilization.
3. The four whorls of floral leaves are (a) the calyx, composed of sepals; (b) the corolla, composed of petals; (c) the androecium, composed of stamens; and (d) the gynoecium, composed of carpels.
4. There are two parts to a stamen, the pollen-producing anther and the filament.
5. A single unit of the gynoecium is frequently called a pistil. If it consists of one carpel, it is a simple pistil; if it consists of two or more fused carpels, it is a compound pistil.
6. A pistil consists of three parts, the stigma, the style, and the ovary. The stigma is receptive to pollen, and the ovary encloses the ovules. At the time of fertilization, the ovules consist of integuments, nucellus, and embryo sac.
7. Meiosis takes place both in the anthers and in the ovule.
8. Pollen may be considered a two-celled haploid plant protected by a cell wall, which is frequently elaborately sculptured. Two sperms are produced by the generative cell.
9. The embryo sac may be considered a seven-celled haploid plant. Two of its cells, the egg cell and the primary endosperm cell, are stimulated to develop by fusion with the sperm cells.
10. Pollination is the transfer of pollen from anthers to stigmas. A pollen tube growing down the style to the embryo sac serves as a pathway by which the two sperms reach the embryo sac.
11. Floral architecture facilitates pollination, which is most frequently carried out by wind or insect vectors.
12. Double fertilization, the union of one sperm with the egg cell and of a second sperm with the primary endosperm cell, occurs only in the Anthophyta.
13. The embryo plant develops from the zygote. The endosperm, developing from the fusion of sperm and primary endosperm nucleus, supplies some nourishment for embryo and sometimes for seed development.
14. Variation in floral architecture arises (a) from variation in number of parts in a given whorl; (b) from symmetry in floral parts; (c) by connation of floral parts; (d) by adnation of floral parts; (e) by elevation of floral parts; and (f) by the presence or absence of certain whorls.
15. Flowers are either solitary on flower stalks or grouped in various inflorescences such as heads, spikes,

catkins, umbels, panicles, racemes, or cymes.

16. There are essentially two types of pollination: cross-pollination, which results in much variation in the progeny, and self-pollination, which results in progeny having great similarities.

PART TWO

The Fruit

A **fruit** is a ripened ovary of a flower; it may include other floral parts. In many cases, fruit development depends on pollination and the activity of indoleacetic acid, or other growth substances.

Fruits are auxiliary structures in the sexual life cycle; they occur only in the Anthophyta. Their development has unquestionably been an important factor in the successful evolution of land plants. Fruits protect seeds, aid in their dissemination, and may be a factor in timing their germination. Fruits are highly constant in structure and are thus very important plant parts in the classification of the Anthophyta. In everyday usage, the term "fruit" usually refers to a juicy and edible structure, such as an apple, plum, peach, cherry, orange *(Citrus)*, or grape. Such structures as string beans, eggplant *(Solanum melongena),* okra *(Hibiscus esculentus)*, squash, and cucumbers, which are commonly called "vegetables," and "grains" of corn, oats, wheat *(Triticum)*, and other cereals are not popularly thought of as fruits. However, all the above are **fruits** in a botanical sense.

Development of the Fruit

Development of the fruit (pod) of pea and bean is described as an example of the changes that may occur during transformation of the ovary into a fruit. Pistils of bean and pea flowers are each composed of one carpel (Fig. 7.17); that is, one ovule-bearing leaf. Recall that the ovary of bean is formed by fusion of margins of the carpel and that ovules are attached to these margins (Fig. 7.14*A*). A cross section of the bean ovary (Fig. 7.18) shows ovary wall, ovule, locule. The ovary wall may be divided into three distinct layers, (a) an outer epidermis, (b) an inner epidermis, and (c) a middle zone consisting of several layers of cells. There are three carpellary bundles, one for each margin and one for the midrib opposite them.

Fertilization initiates a series of changes in the embryo sac and other tissues of the ovule that lead to the development of the seed. Tissues of the ovary wall undergo marked changes into three layers. The fruit wall (developed from the ovary wall) is called the **pericarp,** and the three more or less distinct parts, in order beginning with the outermost, are **exocarp, mesocarp, and endocarp** (Fig. 7.23*B,* shown in almond). When the pod is mature, floral structures such as pedicel, calyx, withered stamens, style and stigma, and even remnants of the corolla may also be present.

Mature pods usually show the presence of small, underdeveloped (abortive) ovules. It is probable that these

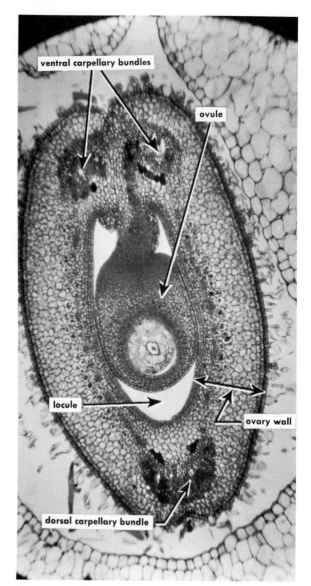

Figure 7.18 Cross section of bean ovary after fertilization, ×20.

ovules were not fertilized. If none within the bean ovary is fertilized, the ovary does not enlarge. In most plants, normal fruit development takes place only if pollination is followed by fertilization and if fertilization is followed by seed development. Some fruits, such as banana and navel oranges, develop without fertilization. These fruits, called **parthenocarpic fruits,** are seedless and frequently different in shape and taste from normal seeded fruits.

Kinds of Fruits

In classifying the different kinds of fruits, the following criteria are taken into account: (a) the structure of the flower from which the fruit develops, (b) the number of ovaries involved in fruit formation, (c) the number of carpels in each ovary, (d) the nature of the mature pericarp (dry or fleshy), (e) whether or not the pericarp splits (**dehisces**) at maturity, (f) if dehiscent, the manner of splitting, and (g) the possible role that sepals or receptacle may play in formation of the mature fruit.

Simple fruits are derived from a single ovary. They may be dry or fleshy; the ovary may be composed of one carpel or of two or more carpels; and the mature fruit may be dehiscent or indehiscent. On the other hand, **aggregate fruits** and **multiple fruits** are formed by clusters of simple fruits. The difference between these types of fruits depends on the number of flowers involved in their formation. In strawberry *(Fragaria)* (Fig. 7.19*A*) and blackberry *(Rubus)* (Fig. 7.19*B*), the aggregate fruit is derived from the many ovaries of a single flower. These matured ovaries or fruits are all attached to a common receptacle. Such groupings constitute aggregate fruits, and the simple fruits comprising them may be classified according to the scheme of classification of simple fruits given below. Mulberry, *(Morus),* fig *(Ficus),* and pineapple *(Ananas comosus)* are multiple fruits. Multiple fruits consist of the enlarged ovaries of several flowers, more or less grown together into one mass. In some plants—mulberry, for example—associated floral structures form a part of the fruit, and in the fig the receptacle enlarges to become the sweet edible portion (Figs. 7.20*A,C*).

Simple Fruits

Pericarp Dry and Dehiscent

Legume or Pod. This type of fruit, characteristic of nearly all members of the pea family, Fabaceae (Leguminosae), arises from a single carpel, which at maturity generally dehisces along both sutures (Fig. 7.17). Pods may be spirally twisted or curved as in alfalfa.

Follicle. An example of follicle is the fruit of magnolia (Figs. 7.21*A*, 16.3). The follicle develops from a single carpel, and opens along one suture, thus differing from the pod, which opens along both sutures.

Capsule. Capsules are derived from compound ovaries, that is, an ovary composed of two or more united carpels. Each carpel produces a few to many seeds. Capsules dehisce in various ways (Figs. 7.21*B,C,D*).

Silique. This is the type of fruit (Fig. 7.21*E*) characteristic of members of the mustard family, Brassicaceae (Cruciferae). The silique is a dry fruit derived from a superior ovary consisting of two carpels. At maturity, the dry pericarp separates into three portions; the seeds are attached to the central, persistent portion.

Pericarp Dry and Indehiscent

Achene. Sunflower is an example of this type of fruit. These fruits are commonly called "seeds," but as in sunflower (Fig. 7.21*F*), a carefully broken pericarp reveals that the seed within is attached to the placenta by its stalk. This pericarp may be separated easily from the seed coat, that is, from the layer of cells just beneath it. Achenes are indehiscent.

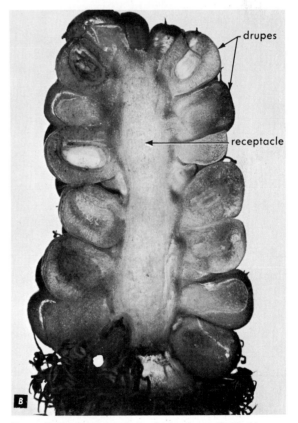

Figure 7.19   Aggregate fruits. *A,* strawberry; *B,* blackberry.

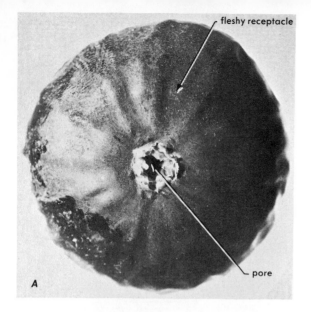

fleshy receptacle

pore

A

drupes

B

fleshy receptacle

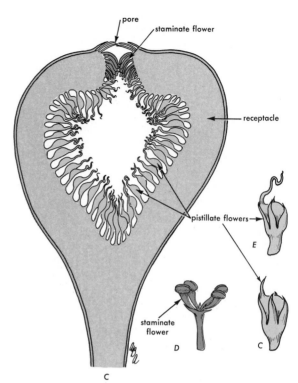

pore

staminate flower

receptacle

pistillate flowers

E

staminate flower

D

C

C

Figure 7.20 The fruit and flowers of fig *(Ficus carica)*. *A*, end of fleshy receptacle, showing pore, which is lined with staminate flowers; *B*, receptacle turned inside out, showing pistillate flowers that have matured into small drupes, ×1. Diagram of a fig. *C, D,* and *E,* inedible capri fig; *C*, section of receptacle showing location of staminate and pistillate flowers, ×1; *D*, staminate flower, ×3; *E*, top: fertile long-styled pistillate flower of the edible Calimyrna fig, ×3; *E*, bottom: short-styled abortive pistillate flower, ×3.

Figure 7.21 Simple fruits. *A, Magnolia,* follicle; *B, Datura,* capsule; *C, D, Tulipa,* capsule; *E, Lunaria,* silique; *F, Helianthus,* achene, *G, Acer,* samara.

Grain or Caryopsis. This is the fruit of the grass family, Poaceae (Gramineae), which includes such important plants as wheat, corn, and rice *(Oryza)*. Like the achene, the grain is a dry, one-seeded, indehiscent fruit (Fig. 7.26). It differs from the achene, however, in that pericarp and seed coat are firmly united all the way around.

Samara. This is a dry, indehiscent fruit, which may be one-seeded, as in the elm *(Ulmus)*, or two-seeded, as in maple *(Acer)* (Fig. 7.21*G*). These fruits are typified by an outgrowth of the ovary wall, which forms a winglike structure.

Schizocarp. This is the fruit characteristic of the carrot family, Apiaceae (Umbelliferae), which includes such common plants as carrot *(Daucus)*, celery *(Apium)*, and parsley *(Petroselinum)*. The schizocarp is a dry fruit

consisting of two carpels that split, when mature, along the midline into two one-seeded indehiscent halves.

Nut. The term nut is popularly applied to a number of hard-shelled fruits and seeds. A typical nut, botanically speaking, is a one-seeded, indehiscent dry fruit with a hard or stony pericarp (the shell). Examples are the chestnut *(Castanea)*, walnut, and acorn (Fig. 7.23*A*). An acorn, the fruit of the oak, is partially enclosed by a hardened involucral cup. The husk or shuck of the walnut, which has been removed before the product reaches the market, is composed of involucral bracts, perianth, and the outer layer of the pericarp. The hard shell is the remainder of the pericarp.

Pericarp Fleshy

Drupe. Examples of this type of fleshy fruit are cherry,

B

A

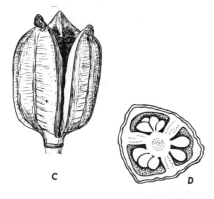

C

D

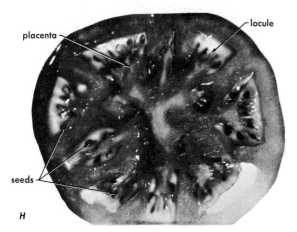

placenta

locule

seeds

H

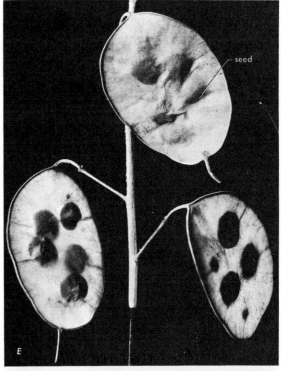

seed

E

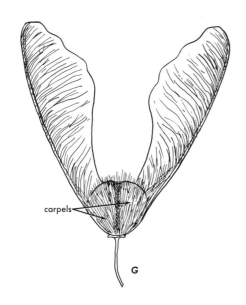

carpels

G

attachment

pericarp

seed

F

almond, peach, and apricot, all of which are members of a genus in the Rosaceae (rose family). The olive *(Olea)* fruit is also a drupe. The drupe is derived from a single carpel, and is usually one-seeded. However, if one examines young ovaries of flowers of almonds, cherries, plums, or other members of the group to which they belong, two ovules will be found, one of which usually aborts (fails to develop into a seed, Fig. 7.23*B*). The drupe has a hard endocarp consisting of thick-walled stone cells. The exocarp is thin, forming the skin, and the mesocarp forms the edible flesh.

Berry. This fleshy type of fruit is derived from a compound ovary. Usually, many seeds are embedded in a flesh, which is both endocarp and mesocarp, although the line of demarcation may be difficult to see (Fig. 7.21*H*). The tomato is a common example of a berry.

The citrus fruit (lemon, orange, lime, and grapefruit) is a type of berry called a **hesperidium**. It has a thick, leathery rind (peel), with numerous oil glands, and a thick juicy portion composed of several wedge-shaped locules (Fig. 7.22). The peel of a citrus fruit is exocarp and mesocarp; the pulp segments are endocarp. The juice is in pulp sacs or vesicles; they are outgrowths from the endocarp walls (Fig. 7.22*C*), and each mature vesicle is composed of many living cells filled with juice.

Another berrylike fruit is that of members of the family Cucurbitaceae, which includes watermelon, squash, pumpkin, and cucumber. It is called a **pepo** (Fig. 16.10). The outer wall (rind) of the fruit consists of receptacle tissue that surrounds and is fused with the exocarp. The flesh of the fruit is principally mesocarp and endocarp.

Pome. This type of fruit is characteristic of a subfamily of Rosaceae, to which belong apple (Fig. 7.24) and pear. The pome is derived from an inferior ovary (Fig. 7.24*A*). The flesh is enlarged hypanthium, and the core has come from the ovary (Fig. 7.24*B*).

Compound Fruits—Formed From Several Ovaries

Aggregate Fruits

An aggregate fruit is one formed from numerous carpels of one individual flower. These fruits, considered individually, may be classified as types of simple fruits. The strawberry flower has numerous separate carpels on a single

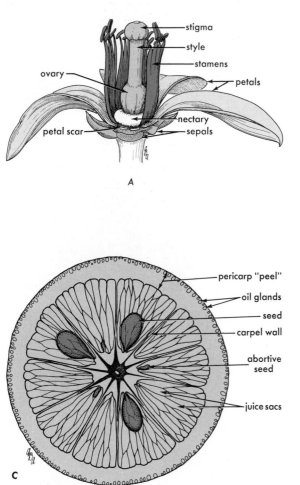

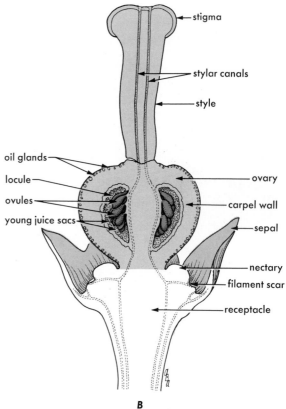

Figure 7.22 Flower and fruit (hesperidium) of orange *(Citrus sinensis)*. A, flower, dissected to show pistil and stamens; B, lengthwise section of maturing ovary; C, cross section of mature fruit.

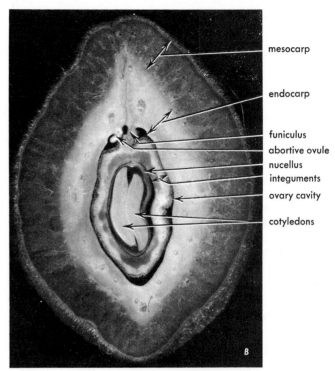

mesocarp

endocarp

funiculus
abortive ovule
nucellus
integuments
ovary cavity

cotyledons

Figure 7.23 *A,* acorns of oak *(Quercus).* The fruits are a type of nut and the cups are fused involucrol bracts, ×¾. *B,* drupe, almond *(Prunus amygdalus).*

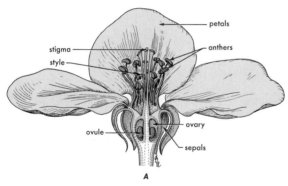

petals

stigma
style

anthers

ovule
ovary
sepals

A

Figure 7.24 Flower and fruit (pome) of apple *(Malus sylvestris).* *A,* median longitudinal section of flower showing inferior ovary and fused (adnate) petals and sepals; *B,* median section of mature fruit.

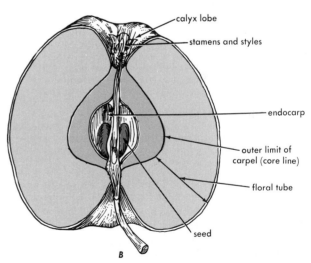

calyx lobe

stamens and styles

endocarp

outer limit of carpel (core line)

floral tube

seed

B

receptacle. The ovary of each carpel has one ovule, and the ovary develops into a one-seeded dry fruit (achene). The receptacle to which these fruits are attached becomes fleshy; the whole structure, which we call a strawberry, is an aggregate of simple fruits, each an achene. The receptacle is stem tissue and consists of a fleshy pith and cortex with vascular bundles between them. The hull of the strawberry fruit is composed of persistent calyx and withered stamens. The achenes are usually spoken of as seeds.

Flowers of raspberry and blackberry, and other species of *Rubus* have essentially the same structure as those of strawberry. In these flowers, many separate carpels attached to one receptacle develop into small drupes (Figs. 7.19*A,B*; 16.9).

Multiple Fruits

A multiple fruit is one formed from individual ovaries of several flowers. These fruits, considered individually, may be classified as types of simple fruits. The fig and

Kinds of Fruits

pineapple are examples of multiple fruits; the individual fruits composing them are nutlets in fig and parthenocarpic berries in pineapple.

The fig fruit we eat is an enlarged fleshy receptacle (Fig. 7.20). Its flowers are very small and are attached to the inner wall of this receptacle. Both staminate and pistillate flowers occur and may be borne in the same or in different receptacles.

Each ovary develops into a nutlet that is embedded in the wall of the receptacle. Thus, the fig is derived from many flowers, all attached to the same receptacle.

Accessory Parts of Fruits

In the pineapple, the edible portion is a thickened pulpy central stem in which berries are embedded. Tissues other than the ovary wall which form part of a fruit, are referred to as **accessory**. Thus, much of the fleshy fruit of pineapple, apple, and strawberry is accessory.

PART THREE

The Seed

Seed Development

The seed completes the process of reproduction initiated in the flower. The embryo developed from the zygote, and the seed coat from the integuments of the ovule. Recall that at the time of fertilization, egg cell, primary endosperm cell, synergids, and antipodals constituted the embryo sac. The embryo sac is surrounded by nucellus, and all these cells and tissues are enveloped by one or two integuments. This complete structure is the ovule (Figs. 7.25A, 7.15). Double fertilization involves (a) fusion of egg and sperm nuclei to form a zygote nucleus and (b) fusion of polar nuclei with a second sperm nucleus to form a primary endosperm nucleus (Fig. 7.25B).

After fertilization, the zygote nucleus remains quiescent for a time. The primary endosperm nucleus, however, divides rapidly and soon builds up an **endosperm tissue** which, at first, may lack cell walls (Fig. 7.25C). The zygote nucleus will divide only after the endosperm has developed. The first cell divisions of the new generation result in a filament of from four to eight cells (Fig. 7.25D). Now the cell closest to the micropyle elongates, pushing the other cells of the filament into the endosperm. At about the same time, the cell furthest from the micropyle divides at right angles to the axis of the filament (Fig. 7.25E). This is the first in a series of divisions that will finally result in the embryo of the seed. Further divisions result in a globular stage, and finally in a heart-shaped stage after the two cotyledons have developed (Fig. 7.25G). The major distinction between embryos of dicotyledonous and monocotyledonous plants is the presence of cotyledons, two or one, respectively. The embryo always consists, in addition, of an axis with a root apex at one end and a shoot apex at the other. (Fig. 7.25H). The course of events just described is very

Key to Fruits

I. Fruit formed from a single ovary of one flower. *Simple fruits.*
  A. Pericarp fleshy.
    1. The ovary wall fleshy and containing one or more carpels and seeds. *Berry.*
      a. Ovary wall a hard rind. *Pepo.*
      b. Ovary wall with a leathery rind. *Hesperidium.*
    2. Only a portion of the pericarp fleshy.
      a. Exocarp thin; mesocarp fleshy; endocarp stony; single seed and carpel. *Drupe.*
      b. Outer portion of pericarp fleshy, inner portion papery, floral tube fleshy; several seeds and carpels. *Pome.*
  B. Pericarp dry.
    1. Dehiscent fruits.
      a. Composed of one carpel.
        1. Splitting along two sutures. *Legume.*
        2. Splitting along one suture. *Follicle.*
      b. Composed of two or more carpels.
        1. Two or more carpels dehiscing in one of four different ways. *Capsule.*
        2. Two carpels; separating at maturity, leaving a persistent partition wall. *Silique.*
    2. Indehiscent fruits.
      a. Pericarp bearing a winglike growth. *Samara.*
      b. Pericarp not bearing a winglike growth.
        1. Carpels, two to many, united when immature, splitting apart at maturity. *Schizocarp.*
        2. Carpels one, if more, not splitting apart at maturity; one-seeded fruits.
          a. Seed united to the pericarp all around. *Caryopsis* or *grain.*
          b. Seed not united to the pericarp all around. Fruit large, with thick, stoney wall. Nut. Fruit small, with thin wall. Achene.
II. Fruits formed from several ovaries.
  A. Fruits developing from one flower. *Aggregate fruits* (classify the individual fruits in key for simple fruits).
  B. Fruits developing from several flowers. *Multiple fruits* (classify individual fruits in key for simple fruits).

general, for there are many variations in details of embryo development.

While development of the embryo is taking place, the nucellus, endosperm, and integuments are also undergoing changes that are characteristic of the group of plants to which the seeds belong. In the great majority of plants, nucellus and endosperm are required only for the initial stages of development. This is particularly true of the nucellus, which is generally used up as a nutritive source in early stages. It persists as a food storage tissue, the **perisperm**, in seeds of sugar beet *(Beta),* and a few other species. The persistance of endosperm as a food reserve

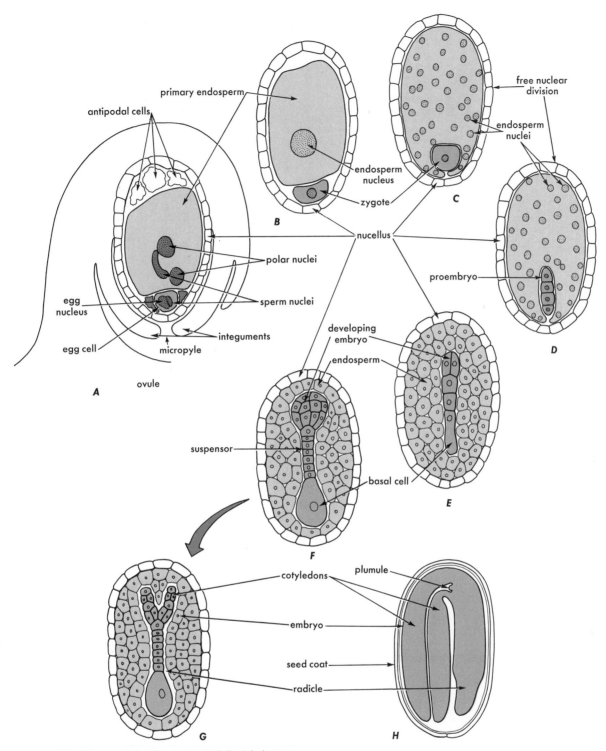

Figure 7.25 Diagram of the development of dicotyledonous angiosperm seed. *A*, ovule after fertilization; *B*, after fusion of gamete nuclei to form zygote and endosperm; *C*, free nuclear divisions in the endosperm; *D*, filamentous stage of the proembryo; *E*, elongation of suspensor cell and division of proembryo cell; *F*, globular stage of the embryo; *G*, heart-shaped stage of embryo development.

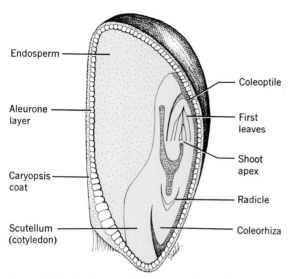

Figure 7.26 Median section of caryopsis of yellow foxtail (*Setaria lutescens*).

Labels on Figure 7.26: Endosperm, Aleurone layer, Caryopsis coat, Scutellum (cotyledon), Coleoptile, First leaves, Shoot apex, Radicle, Coleorhiza

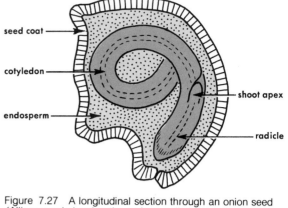

Figure 7.27 A longitudinal section through an onion seed (*Allium cepa*) showing the embryo coiled within the endosperm.

Labels on Figure 7.27: seed coat, cotyledon, endosperm, shoot apex, radicle

occurs in many monocotyledons (Figs. 7.26, 7.27). It is highly developed in grasses, some of which like rice and corn are of major economic importance. The endosperm persists as a food storage tissue in relatively few dicotyledonous seeds; the castor bean (*Ricinus communis*) is a good example.

In most seeds, integuments of the ovule become hard and horny seed coats in the mature seed. The scanning electron microscope shows the seed coats to be variously and sometimes beautifully sculptured (Fig. 7.28).

Kinds of Seeds

Seeds of angiosperms differ in two ways: (a) they have one or two cotyledons and (b) they store food either in the embryo, in the endosperm, or more rarely, in nucellar tissue, the perisperm, which lies between embryo and seed coat. In most dicotyledonous seeds food is stored in cotyledons (Fig. 7.25) while in monocotyledonous seeds the endosperm is usually the principal food storage tissue.

When food storage occurs within the embryo, the normal vascular tissues of the embryo convey the solubilized food to meristematic regions, where it is required for growth. Food stored in the endosperm, outside the embryo, is in many cases absorbed through normal protodermal or epidermal cells. In other cases, the cotyledons become specialized to act as absorbing organs (Fig. 7.35). In most grasses, the entire cotyledon has become a highly specialized absorbing organ known as the scutellum (Fig. 7.26).

Common Bean

The fruit of the bean plant is a pod; within it are seeds. The external characters of the bean seed are more clearly

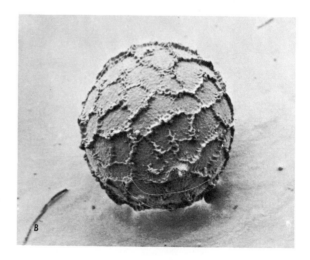

Figure 7.28 The coats of many seeds are characteristically sculptured. *A*, morning glory (*Convolvulus arvensis*), ×23; *B*, California poppy (*Eschscholtzia californica*), ×62.

seen after soaking it in water. Points of interest are the **hilum, micropyle,** and **raphe** (Fig. 7.29). The hilum is a large oval scar, near the middle of one edge, left where the seed broke from the stalk or **funiculus** when the beans were harvested. The micropyle is a small opening in the seed coat (integument) at one side of the hilum, which was observed in the ovule as the opening through which the pollen tube entered. The raphe is a ridge at the side of the hilum opposite the micropyle. It represents the base of the funiculus, which is fused with the integuments. Conducting tissue present in the funiculus will be continued in the raphe. An elevation or bulge of the seed coat adjacent to the micropyle marks the position of the radicle (embryo root) within the seed.

When the seed coat of a soaked bean is removed, the entire structure remaining is embryo; no endosperm is present. The following parts of the embryo can be observed: (a) a shoot consisting of two fleshy cotyledons, a short axis (the hypocotyl) below the cotyledons, and a short axis (the epicotyl) above the cotyledons, bearing several minute foliage leaves and terminating in a shoot tip; and (b) the root or radicle (Fig. 7.29).

Grasses

The so-called "seed" of corn, or of any other grass, is in reality a fruit, the caryopsis. It is a one-seeded, dry, indehiscent fruit with the pericarp (ovary wall) firmly attached to the seed. Pericarp and seed coats are so firmly attached to each other and to other tissues of the grain that it is impossible to peel them away.

A longitudinal section of a caryopsis of yellow foxtail (*Setaria lutescens*) is shown in Fig. 7.26. The endosperm constitutes the bulk of the grain and is composed of (a) an outermost layer (single row of cells) known as the **aleurone layer** and (b) a starchy endosperm. Cells of the aleurone layer contain proteins and fats but little or no starch.

Like all embryos of other seed plants, the grass embryo has an axis with a **shoot apex** and a **root apex** (Fig. 7.26). The shoot apex, together with several rudimentary leaves, are surrounded by a sheath, the **coleoptile**. The rudimentary root (radicle) is also surrounded by a sheath, the **coleorhiza**. At the juncture of shoot and root is a very short stemlike region. A relatively large part of the grass embryo is a single cotyledon, which has for a long time been called the **scutellum**. It is a shield-shaped structure that lies in contact with the endosperm. The outer cells of the scutellum secrete enzymes that digest the stored foods of the endosperm. These digested foods move from endosperm cells through the scutellum to the growing parts of the embryo. Unlike the bean, the cotyledon of grasses remains within the seed during germination and never develops into a green leaflike structure above ground.

It is noteworthy that the process of milling to make polished white rice removes the outside caryopsis coat, plus the entire protein-rich aleurone layer, and the outer layers of the endosperm. This means that most of the nutritious protein is removed before the rice is eaten! The removed material, composed of the outside layers and the embryo axis, is sold as bran.

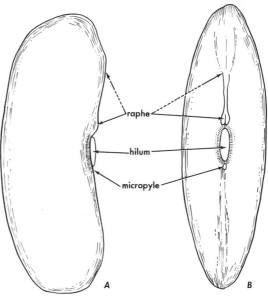

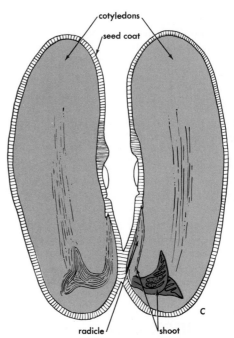

Figure 7.29   Bean (*Phaseolus vulgaris*) seed. *A,* external side view; *B,* external face or edge view; *C,* embryo opened.

Onion

The onion seed consists of seed coats enclosing a small amount of endosperm. The embryo is a simple axis, the radicle and single cotyledon being quite prominent. The shoot apex is located at about the midpoint of the axis and appears as a simple notch. The embryo is coiled within the seed coats, the radicle usually pointing toward the micropyle in the seed coats (Fig. 7.27).

Dissemination of Seeds and Fruits

Agents in Seed and Fruit Dispersal

The chief agents in seed and fruit dispersal are wind, water, and animals.

Wind

The structural modifications of seeds and fruits that aid in dissemination by wind are of several types. The most common of these are winged types (Fig. 7.30C) and plumed types (Figs. 7.30A,B).

Water

Fruits with a membranous envelope containing air like those of sedges, or with a coarse, loose, fibrous outer coat, as in the coconut (Cocos), are well adapted for dispersal by water. A great variety of fruits and seeds float in water, even though they lack special adaptations to ensure buoyancy, and are readily transported long distances by moving water. In irrigated districts, irrigation water is a very important means by which weed seeds are distributed.

Animals

Many seeds and fruits are carried by animals, both wild and domesticated. Seeds with beards, spines, hooks, or barbs adhere to hair of animals (Fig. 730E). An example is bur clover (Medicago denticulata).

Seeds of many plants pass through the digestive tract of animals without having their viability impaired. Fleshy, edible fruits may be eaten by birds (Fig. 7.30F) and carried by them long distances, and then the seeds regurgitated or discharged with the excrement. Squirrels carry nuts, such as those of walnut and hickory (Carya), and the seeds of pines. Seeds of some aquatic and marsh plants and of mistletoe (Phoradendron), which are covered with a sticky material, are carried on the feet of birds.

Seed Dormancy and Germination

Seeds can remain viable for remarkably long periods. In one study, begun in 1878, Dr. W. J. Beal of Michigan State University buried jars containing seeds from several plant species. At 5- and 10-year intervals a jar was opened and the seeds tested for germination. Most species remained viable for at least 10 years, and one species, the moth mullein (Verbascum blattaria), still germinated after more than 90 years. This, however, is no record for seed longevity. Seeds from the Oriental lotus (Nelumbo nucifera) have been removed, still viable, from archeological diggings known to be more than 1000 years old! Seeds of a few other species have been recovered from cold and anaerobic deposits, dated at over 1000 years, and have also proven viable. It is truly remarkable that seeds can remain living over such long periods.

What induces germination? Many seeds will germinate when provided with moisture, oxygen, and a favorable temperature. The water content of seeds is low, between 5 and 10%. The cytoplasm with contained organelles is scarcely recognizable because it is crowded between the large amount of reserved food materials (Fig. 7.31). Water is, at first, imbibed very rapidly, which results in the swelling of the protoplasm with a reappearance of organelles (Fig. 7.32). Increase in water content makes possible an increased metabolic activity. From this point in time the seed must have an adequate water supply for survival.

Many seeds will not germinate even when supplied with water, oxygen, and a favorable temperature. There are various factors in different plants that are associated with the breaking of dormancy. For instance, light is necessary for the germination of some grass seed and for some strains of lettuce (Lactuca) and perhaps other plants. Water and gases must pass through the seed coats. If seed coats restrict the movement of water and gases or are impermeable to one or the other or to both, the seed coats must decay, be broken, or be scratched, allowing water and oxygen to reach the embryo before germination can start. In some seeds, the embryo itself fails to germinate when provided with favorable conditions. In orchids, the seeds are disseminated while embryos are rudimentary and the embryos must develop further before germination.

Many kinds of seeds are known that will not germinate unless the seeds, while on a moist substrate, are subjected for a time to temperatures close to freezing. At the other extreme, some seeds will not germinate unless they have been subjected to the rather high heat of a fire. Another type of dormancy is produced by the presence of natural chemical inhibitors. These occur in many fruits and serve to keep the seeds dormant while they are enclosed by the fruit. In other cases, inhibitors are present in the seeds themselves or may be produced by decaying leaves, or forest litter. The influence of the plant growth substances on dormancy and seed germination is still unclear.

The first indication that the processes of germination have begun is generally the swelling of the radicle. In all cases, the radicle imbibes water rapidly and, bursting the seed coats and other coverings that may be present, starts to grow downward into the soil. This helps to assure that the young seedling has an adequate supply of water and nutrients when the shoot breaks through the surface of the soil. Although the succeeding steps of germination are essentially similar, there are variations. For instance, in germination of beans, peas, and onions, a structure with a sharp bend or **hook** is first forced through the soil (Figs. 7.33, 7.34, 7.35), but the structure forming the hook is different in each case. Once above the ground, the hook straightens. In the case of bean, a straightening of the hypocotyl raises cotyledons and shoot apex above ground (Fig. 7.33); a lengthening and straightening of the epicotyl will pull the shoot apex away from the cotyledons (Fig. 7.33).

Upon straightening of the pea epicotyl, however, the cotyledons remain in the ground, and only the apex and first leaves will be raised upward (Fig. 7.34). Cotyledons

Figure 7.30 Fruits and seeds showing various devices aiding in dissemination. *Wind: A, Clematis,* ×4; *B,* dandelion *(Taraxacum vulgare),* ×10; *C,* seed of Coulter's big-cone pine *(Pinus coulteri),* ×2.
Attachment: D, cranesbill *(Geranium),* ×3; *E,* bur clover *(Medicago denticulata),* ×5; *F,* fleshy edible fruit of Cottoneaster, ×4; Violent *dehiscence* of pericarp; *G,* vetch *(Vicia sativa),* ×2.

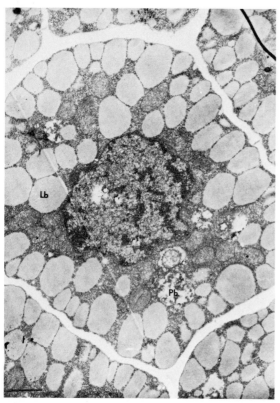

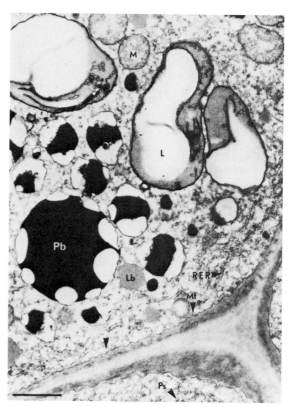

Figure 7.31 Yellow foxtail *(Setaria lutescens)* scutellum in dry seed (caryopsis or grain), × 0000.

Figure 7.32 Yellow foxtail *(Setaria lutescens)* radicle after 65 hours of germination, × 0000.

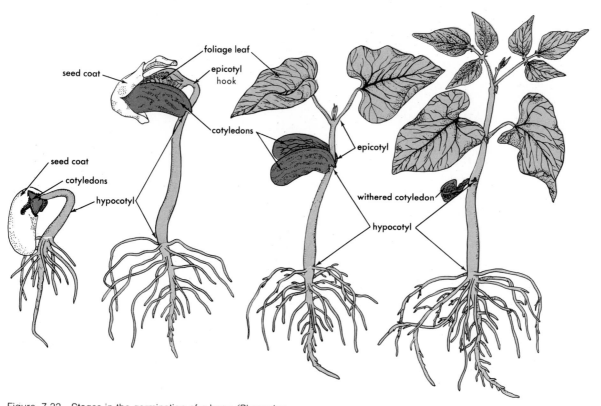

Figure 7.33 Stages in the germination of a bean *(Phaseolus vulgaris)* seed.

chapter **7** | **Flowers, Fruits and Seeds**

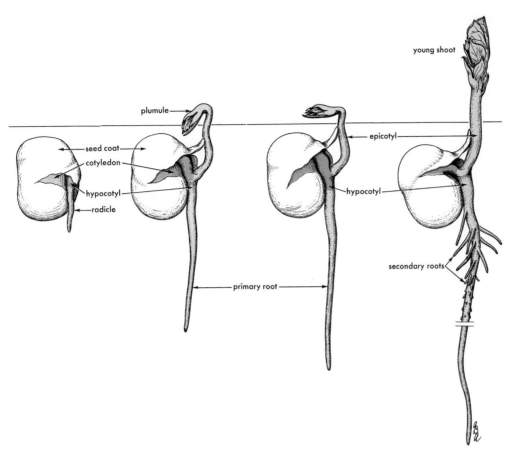

Figure 7.34 Stages in the germination of a pea *(Pisum sativum)* seed.

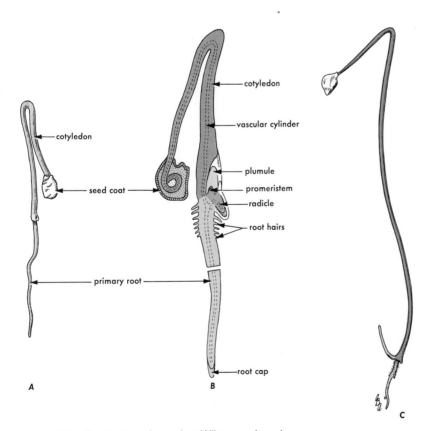

Figure 7.35 Germination of an onion *(Allium cepa)* seed.

that are raised above ground may carry on photosynthesis.

In onion, the primary root first penetrates the soil. Then a sharply bent cotyledon breaks the soil surface, and slowly straightens out. The cotyledon of the onion is tubular and its base encloses the shoot apex (Fig. 7.35). There is a small opening at the base of the cotyledon through which the first leaf finally emerges. In grasses, the situation is more complex. The shoot and root apices are enveloped by tubular sheaths known as the coleoptile and coleorhiza, respectively (Fig. 7.36). The primary root rapidly pushes through the coleorhiza. The root continues to grow, but is eventually replaced by adventitious roots that arise from the lower nodes of the stem (Fig. 7.36). The coleoptile elongates and emerges above ground, becoming 2–4 cm long. At this time, the uppermost leaf pushes its way through the coleoptile and, growing rapidly, becomes part of the photosynthesizing shoot.

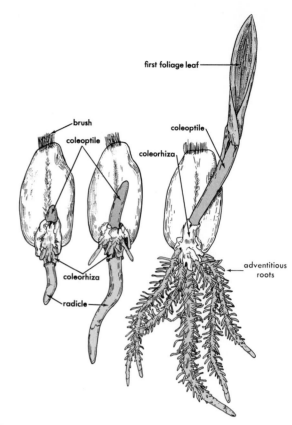

Figure 7.36   Germination of a grain of wheat (*Triticum*).

Summary

1. Seeds may store food within or outside the embryo. In most dicotyledons such as cucumber and bean, food is stored in the two cotyledons. Cotyledons may also serve as absorbing and, later, as photosynthesizing organs. Food may be stored in an endosperm rather than in the cotyledons. In monocotyledonous seeds, food is stored in an endosperm.

2. In grasses, such as corn, a single cotyledon (scutellum) has become a highly specialized absorption organ that does not emerge from the seed. In seeds like those of lilies and onion, the cotyledon emerges from the seed coats and becomes green, but its tip continues to absorb food from the endosperm.

3. Mature seeds are dormant and, depending on the species and the immediate environment of the seeds, they may remain viable and dormant from a few months to many years.

4. Dormancy is usually broken by the provision of moisture, oxygen, and a favorable temperature. Other factors, such as light, the removal of chemical inhibitors, or the destruction of the seed coats, may be required in some instances.

5. In germination, the cotyledons may be elevated above the ground, sometimes to become photosynthetically active, or in the case of some monocotyledons, to continue absorption of the endosperm. In other cases, the cotyledons remain below the ground.

6. In the grasses, the single cotyledon functions as a specialized food-absorbing organ.

8 CHAPTER

the control of growth and development

Which are the more alike: cattle in a herd or oaks in a woodland? The cattle are decidedly more uniform than the trees in terms of the number of limbs and appendages they have. The systems that control growth and development in the higher animal body are tightly programmed and conservative; they resist influences from the environment far more than they adjust to them.

Plants are the opposite. They have stimulus-response systems that allow the environment to modify their path of development. This is the principal way in which plants adjust to the environment; this is how they compensate for the lack of muscles. If in a poor location, they detect the situation by means of environmental signals and they modify their morphology in a way that permits them to persist. And whereas animals meet the threat of predators by running away, the stationary plant suffers the attack and later replaces the lost parts.

Environmental Adaptation by the Young Plant

The plant's use of environmental signals can be illustrated by tracing the behavior of a seedling as it starts the life of an independent plant.

Once the seed germinates, the root senses gravity and directs its growth downward, in the direction that usually leads to water, minerals, and anchorage. The shoot grows upward in response to gravity, toward the most likely location of light.

The newly emerging shoot is a delicate structure that must push its way through soil and past obstacles. Dicotyledonous plants do not expand their leaves until they reach the soil surface (see Color Plate 2, Figs. 8.1, 8.2). The cluster of leaf primordia and the apical meristem are usually inverted and pulled up through the soil behind the recurved portion of the hypocotyl or epicotyl (the hook). The delicate leaves and apex are thus protected. In some monocot plants, the young leaves are rolled up and completely ensheathed in a protective tube, the **coleoptile.** When the coleoptile reaches the surface, the leaves push out of the splitting coleoptile and unroll. Light, received by a pigment called **phytochrome,** signals that the shoot has reached the surface. Plants that are grown in darkness do not receive this signal and grow in the

pattern adapted for movement through the soil (see Figs. 8.3, 8.4, Color Plate 2). The dark-grown seedling is said to be **etiolated.**

In addition to leaf expansion, other changes are initiated by light. The stem usually elongates very rapidly in darkness (or in the soil) because light inhibits elongation. Supporting tissues of the vascular system are weakly developed in darkness when the soil would normally give all the support necessary. Light stimulates increased development of xylem, by triggering the formation of enzymes that in turn catalyze lignin synthesis two or three hours after the start of illumination. This adds the strength that is needed to support the stem above ground.

Whereas light slows the elongation of much of the stem, cells on the lower surface of the hook increase their elongation. This causes the stem to straighten so that the leaves are oriented upward (Fig. 8.1). These effects of light are all initiated by phytochrome.

The first light received by the plant may be coming from only one side. The usual response is for the stem or coleoptile to initiate a bending response that orients the tip toward the light.

The mesophyll cells of a leaf and the cortex of a stem are programmed to delay the formation of the photosynthetic apparatus until they are exposed to light (Figs. 8.1, 8.4). In darkness, the immature plastids contain a small amount of **protochlorophyll,** an incomplete form of chlorophyll, which is attached to a special catalytic protein. On illumination, the protochlorophyll absorbs light and is converted to chlorophyll. More protochlorophyll is formed in its place. As chlorophyll accumulates, membranes in the plastids are organized into thylakoids and the full battery of chloroplast enzymes is assembled. The leaf can be fully green within 36 to 48 hours (Figs. 8.2, 8.3). Phytochrome is involved in these events, too. The delay in developing mature chloroplasts prevents the plant from needlessly spending stored food on plant parts that may never receive light for photosynthesis.

As this introduction has shown, the systems that control development in the plant can respond to signals from the environment. Physiologists have long felt that a study of these signals and the responses might lead us to valuable insights about the plant's developmental systems. Charles Darwin, for example, held such a belief; and the book that he and his son Francis published in 1881, "The Power of Movement in Plants," helped to launch a revolution of knowledge about developmental control. But to understand this revolution, we must first shift our focus to the level of the cell.

Hormones and the Control of Growth and Development

The typical plant begins its life as a single cell, the zygote. Many cell divisions occur, and the organism becomes a colony of cells that cooperate to form and maintain an integrated plant body. The cells in one region form a leaf; those in another region, a root or a flower. Some cells become green and perform photosynthesis; others lose their nuclei, form sieve plates, and join the phloem. What causes the cells to develop so differently? What tells each cell the proper path for its development? And how do the cells achieve cooperation?

Acorns always produce oak trees; corn kernels always produce corn plants. This shows that development depends on hereditary information. At the outset, all the information is contained in the zygote. Thus one might suppose that cells develop differently because each cell gets only a fraction of the zygote's original store of hereditary information.

But experiments have shown that mature parenchyma cells can give rise to whole plants, if the cells are removed from their original location and are given suitable stimuli and raw materials. This demonstrates that mature cells may contain all the hereditary information that the zygote contained. Evidently, then, each cell uses only a part of its hereditary information. A common idea is that each cell follows a particular reading program as it refers to the hereditary information. This selective pattern of reading guides the cell through a sequence of developmental changes. Cells come to differ because they follow different reading programs, but if so, how is the program determined?

The simplest suggestion is that cells follow a path of development in response to signals from the surrounding environment. This agrees with the fact that seedlings can modify their development according to cues such as light and gravity. But, in addition, internal signals may pass between cells within the plant to coordinate their development. Such internal signals were postulated by Charles and Francis Darwin, and much of modern plant physiology has been concerned with identifying these signals and tracing their action. Today we know of five classes of molecules that act as developmental signals in plants. These compounds, called **hormones,** act at very low concentrations and function in the plant only as signals.

We do not yet know in detail how the hormones initiate their effects. But even small changes in their molecular structures tend to change their activity. This suggests that hormones are precisely shaped to fit receptor sites, much as modulators attach to regulatory enzymes (Chapter 2). Proteins are likely candidates for receptors since other types of molecules generally lack the structure-recognition capabilities that proteins have. One popular proposal is that the hypothetical hormone receptors inhibit or

stimulate the reading of particular genes, as in the operon mechanism (Chapter 2). According to another current idea the hormones bind to membranes, affecting their permeability to other molecules that in turn affect regulatory enzymes and the reading of genes.

Although we cannot define exactly how hormones work, we can describe the kinds of developmental processes that they help to control, and we can illustrate the hormones in action. Development combines the results of three major processes: **cell division,** which produces new cells; **growth,** which increases the size of cells and organs; and **differentiation,** the changes by which cells become specialized in structure and function. Each of these processes can vary in speed and direction. And all of these variables must be controlled in each part of the plant body if development is to be normal. The following passages outline what we know about the part that hormones play.

Auxins

Our knowledge of hormones began with the discovery of the **auxins,** a class of hormones that were first detected because they stimulate elongation growth in grass coleoptiles. The kind of experiment that led to their discovery is shown in Fig. 8.6.

The only naturally occurring auxin to be identified so far is **indoleacetic acid (IAA),** which is shown in Fig. 8.5A. But many man-made compounds have been developed that affect plants in the same way as IAA, and these are also called auxins.

Control of Cell Enlargement by Auxin

Cell enlargement is a process critical to the life of almost every cell in a plant, and the ability of the plant to control this process precisely is central to growth and morphogenesis. A plant cell may be thought of as an inflatable bag (the protoplast) surrounded by a cover (the cell wall) that may be rigid or may be expanded under pressure, depending on its makeup. By using respiratory energy and carrier systems, a variety of solutes are pumped into the vacuole, and water follows osmotically, creating turgor pressure. The turgor pressure in a typical cell might be maintained at five times the pressure of the surrounding air. For comparison, automobile tires are usually inflated to more than three times atmospheric pressure. Whether or not growth occurs will depend on both the pressure and the extensibility of the wall.

As the cell grows, the wall is forced to stretch. Stretching hardens the wall and makes it less extensible. The same thing happens with a stretched rubber band, as one may easily discover. To continue growth, the wall must therefore be continually softened. It is in this softening process that auxin seems to exert its most immediate effect. The wall is built of carbohydrate polymers, such as cellulose and pectin as well as protein. These framework polymers are joined in a network by cross-links that are believed to limit the extent to which the wall can be stretched. For continued growth,

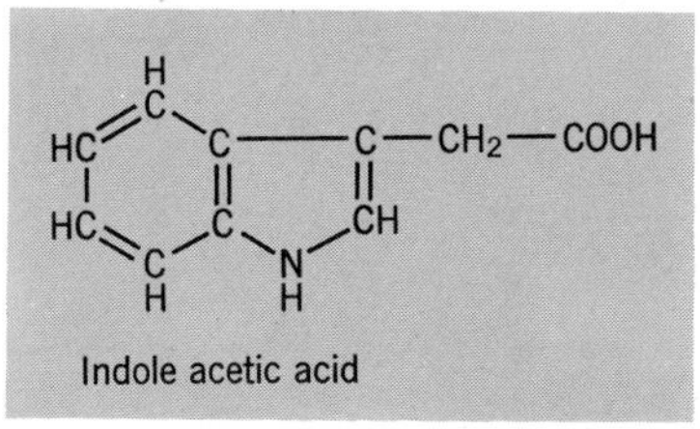

A: Molecule of the auxin, indole acetic acid (IAA), made up of two rings and a side chain

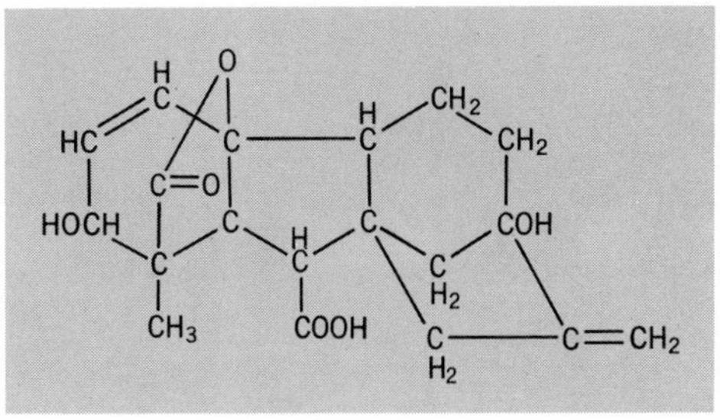

B: Molecule of gibberellic acid

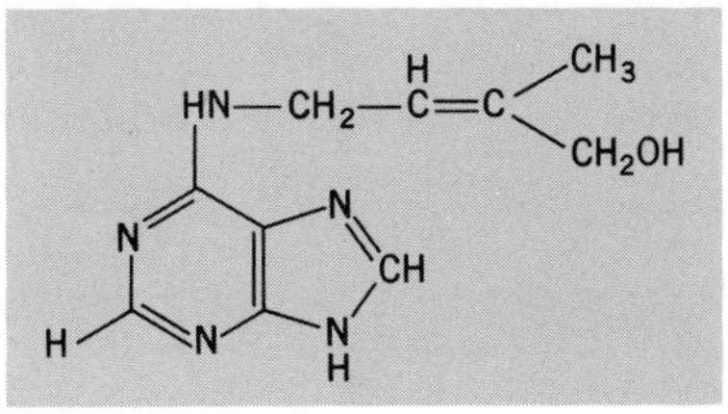

C: Zeatin

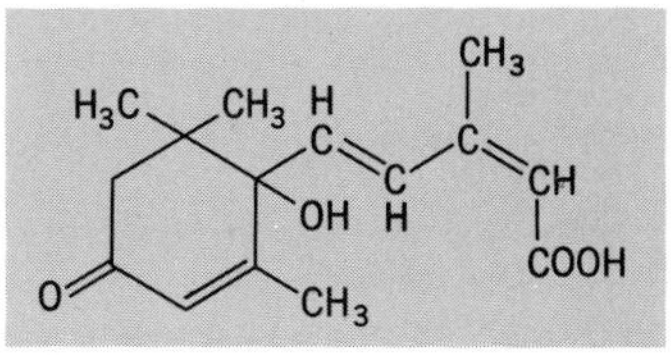

D: The structure of abscisic acid (also called dormin or abscisin II)

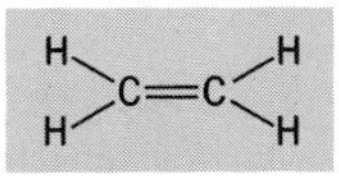

E: Ethylene

Figure 8.5 Plant hormones. A, indole acetic acid, IAA, the most common natural auxin; B, gibberellic acid, GA_3, one of more than 40 very similar natural gibberellins; C, zeatin, one of several natural cytokinins, D, abscisic acid, ABA; E, ethylene, a gaseous growth regulator.

physiologists believe that the cross-links between wall polymers must continually be broken and re-formed in new positions. By this theory the turnover of cross-links is the basic softening process. Recent evidence suggests that in some plant cells, hydrogen ions stimulate the breaking of cross-links in the wall. The required hydrogen ions originate in the protoplast and are pumped through the plasma membrane and into the wall by a carrier system that is embedded in the membrane. In dark-grown plants of several species, auxin has been found to stimulate the action of this ion pump. Hence the higher the auxin concentration, the faster hydrogen ions are supplied to the wall, and the faster the cell grows. It has long been known that auxin can act only if respiration is also taking place. This makes sense if the ion pump in the membrane uses ATP as an energy source.

There is evidence that in some cells, auxin may soften walls and promote growth by a mechanism other than causing hydrogen ion secretion. This is true, for example, in light-grown pea stems. It is not yet known how auxin promotes growth in such cases.

If growth is to continue for a long period of time without the wall thinning to the point where it breaks under turgor pressure, new framework polymers must continually be added. Materials are secreted through the membrane and deposited in the wall. Respiratory activity is needed to provide energy for both the synthesis and the secretion process.

The rate of wall secretion must be coordinated with growth if the wall is to remain at a constant thickness. In addition to its wall-softening effect, auxin also appears to play a role in stimulating and coordinating wall formation. The mechanism for this effect of auxin is unknown.

Controlling Auxin Concentration

Elongation rates respond within minutes if auxin is added or removed. This emphasizes the importance of factors that control the concentration of auxin.

Auxin is formed in regions of the plant separate from the zones of rapid elongation. The tip of the coleoptile and the stem tip with its cluster of young leaves make IAA from the amino acid, tryptophan. The auxin is then moved down the stem through the elongating region (Fig. 8.6). The movement is called polar transport because it is a one-way, energy-requiring movement away from the tip. Typical rates of movement are 10 to 15 mm per hour, much slower than movement of materials in the xylem and phloem, which may exceed 1 meter per hour. Furthermore, polar transport in the youngest stem regions is opposite to the direction of flow in the phloem and xylem and has been shown to occur even in stem tissue from which all vascular elements were removed.

Thus, auxin moves down from the tip through the growing zone, and the concentration in this stream establishes the growth rate of the tissue. Remove the tip of a coleoptile, the auxin source, and the remaining auxin is rapidly drained out of the growing zone. Within a few minutes the growth rate drops. Put back the tip or replace the auxin supply with a block of gel containing IAA and growth is soon restored (Fig. 8.6). If, however, the auxin

Hormones and the Control of Growth and Development

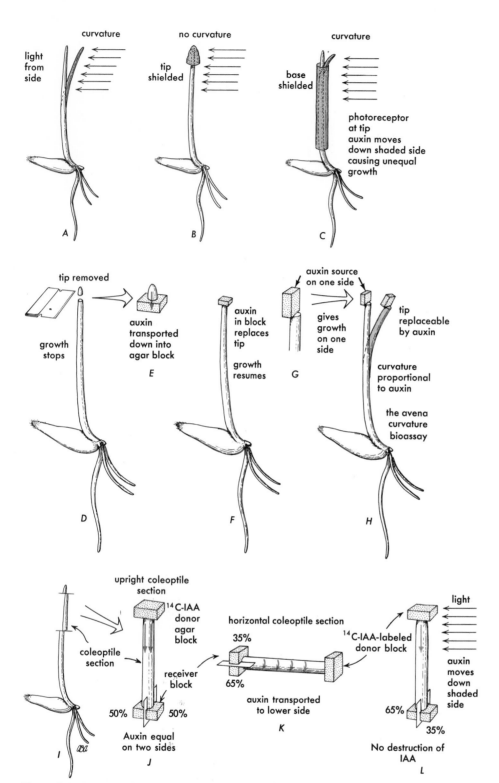

light from side

curvature

tip shielded

no curvature

base shielded

curvature

photoreceptor at tip auxin moves down shaded side causing unequal growth

A

B

C

tip removed

growth stops

auxin transported down into agar block

D

E

auxin in block replaces tip

growth resumes

F

auxin source on one side

gives growth on one side

G

tip replaceable by auxin

curvature proportional to auxin

the avena curvature bioassay

H

upright coleoptile section

¹⁴C-IAA donor agar block

coleoptile section

receiver block

50%   50%

Auxin equal on two sides

I   J

horizontal coleoptile section

35%

65%

auxin transported to lower side

K

light

¹⁴C-IAA-labeled donor block

auxin moves down shaded side

65%

35%

No destruction of IAA

L

Figure 8.6 Diagram of experiments showing how regulation of auxin transport can result in phototropic curvature (*A, B, C, L*) or in geotropic curvature (*I, J, K*), and the *Avena* curvature bioassay used in estimating minute amounts of auxin (*D, E, F, G, H*). Experiments in which IAA labeled with radioactive carbon was used (*I, J, K, L*) show how geotropic and phototropic stimuli cause auxin to be transported laterally. Unequal amounts of IAA accumulate in the receiver blocks.

source is placed on only one side of the cutoff stump, the cells that line up below the auxin supply grow more than those on the other side. This leads to a bending of the coleoptile or stem.

Tropic Curvatures

Plants can orient the direction of growth of organs in response to environmental cues. These growth responses are called **tropisms**. When the shoot and root emerge from the seed, it is of obvious advantage to direct shoot growth upward and root growth downward. The direction of gravitational force is a reliable cue. We do not yet know how the plant senses gravity, but many physiologists think the answer may be **statoliths**—particles in the cell that are heavy enough to sink to the bottom, comparable to the stones in the inner ear that give human beings a sense of the vertical.

Colorless plastids, heavy with starch grains, may act as statoliths. When a plant is laid on its side, these **amyloplasts** tumble to the lowermost side of the cell.

Once the shoot has sensed its orientation in space, we know that the direction of auxin transport changes, so that by the time the auxin has traveled from the tip of the stem to the main growing zone there may be twice as much auxin moving through the lower half of the stem compared with the upper side (Fig. 8.6). The greater growth on the lower side pushes the tip until it is pointing up. The amyloplasts settle back to their original position, and auxin transport becomes uniform on both sides again. The plant requires a few minutes to sense its original change in position, and by 15 minutes there is enough auxin concentration difference to cause a curvature. In some rapidly growing seedlings, the stem may turn upright within an hour from the time that it is placed horizontally. This tropic curvature in response to gravity is called **geotropism**.

The geotropic curvature of roots downward has many aspects in common with the shoot response, except that there is considerable doubt that auxin is the hormone involved. It is clear that the root cap produces some controlling substance that moves away from the tip to control curvature. However, auxin is transported toward the root tip. The identity of the growth substance involved in root geotropism is being actively investigated. At present it seems likely that the controlling substance is a growth inhibitor.

Phototropism is a change in the direction of growth that results when the stem or coleoptile is exposed to light from one side (Fig. 8.6). The response depends on a pigment system that absorbs blue and violet light. The nature of the pigment is unknown, but its absorption suggests that it is not phytochrome or chlorophyll. In an unevenly illuminated shoot, there is more pigment activation on the lighted side than on the shaded side. Careful experiments have shown that IAA labeled with radioactive carbon and applied to the tips of coleoptiles moves symmetrically down the coleoptiles in the dark. Light treatments that cause curvature result in twice as much auxin moving down the shaded side as down the lighted side, but the total amounts transported are the same as in dark controls. Thus somehow the light-sensing pigment causes a lateral diversion of the auxin flow, as in geotropism. The result is a curvature toward the light.

Apical Dominance

In the typical pattern of growth, the growing tip of the stem with its young leaves inhibits the sprouting of lateral buds below the apex. This phenomenon is called **apical dominance**. In a related effect, called **apical control**, the main apex slows the growth of lateral branches that do sprout. This dominance of apical growth over lateral growth decreases with the distance from the tip of the stem, and varies with the age of the plant, the genotype, nutrition, and other environmental factors. A plant with strong apical dominance (e.g., the sunflower) has little or no branching. Weak apical dominance leads to a bushy appearance, as in tomato plants. In grasses, branching occurs at the very basal nodes of the stem. These branches are called **tillers** and the process is **tillering**.

The suppressed lateral buds act as a reservoir of shoot tips that can rapidly take over growth to replace a damaged main shoot tip. In garden pea seedlings, removal of the growing tip is followed within four hours by increased metabolism in the first lateral bud below the old tip. Another early event is the differentiation of xylem to form a vascular connection between the bud and the main stem. Completion of the vascular connection is followed by growth of the new shoot.

The main shoot tip apparently gives off an inhibiting influence that passes down the stem. If the tip is replaced by a supply of auxin in a lanolin paste or agar block, the lateral buds remain inhibited. This suggests that lateral buds are inhibited in the intact plant by auxin that the shoot tip produces. But paradoxically, after the branch has started to grow, auxin applications increase its growth and the active branch forms its own auxin supply. Thus auxin acts in opposite ways before and after the bud sprouts. Auxin inhibits the formation of vascular connections to the lateral bud, a fact that may help to resolve the paradox. This is an effect of auxin on differentiation, which is apart from its effect on growth. Here we see an example of a common observation: each plant hormone exerts a diversity of effects in different situations.

There are many other "growth correlations" where the growth and development of one plant part is related to that of another (Fig. 8.7). Thus apical control may cause lateral organs such as branches, leaves, rhizomes, and stolons to grow horizontally or at some angle with respect to gravity that is different from the vertical habit of the main shoot. Removal of the main shoot may be followed by a bending of a nearby branch into an upright position as it takes over the dominant position in the plant. The direction of growth of some rhizomes and stolons has been shown to depend on hormones coming from the shoot tip.

One might suppose that continued production of auxin at the tip and its transport downward should lead to an accumulation somewhere. But no region of accumulation has been found. Auxin is destroyed or inactivated along the way. There are enzymes in plants that can destroy much more auxin than the plant can ever produce. And there are enzymes that inactivate auxin by tying it to another molecule, forming an inactive compound:

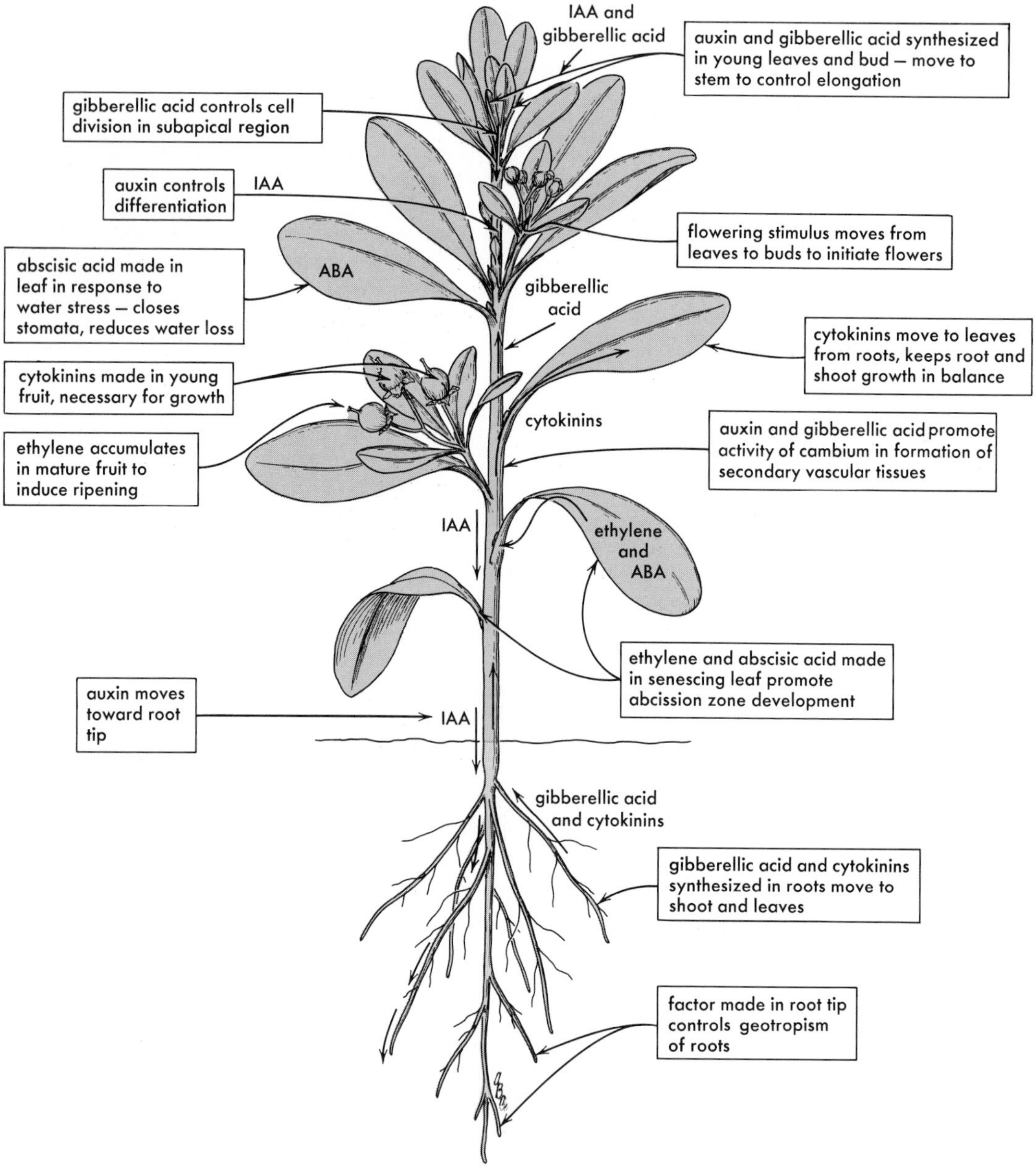
IAA and
gibberellic acid
auxin and gibberellic acid synthesized in young leaves and bud — move to stem to control elongation
gibberellic acid controls cell division in subapical region
auxin controls differentiation
IAA
flowering stimulus moves from leaves to buds to initiate flowers
abscisic acid made in leaf in response to water stress — closes stomata, reduces water loss
ABA
gibberellic
acid
cytokinins move to leaves from roots, keeps root and shoot growth in balance
cytokinins made in young fruit, necessary for growth
cytokinins
auxin and gibberellic acid promote activity of cambium in formation of secondary vascular tissues
ethylene accumulates in mature fruit to induce ripening
IAA
ethylene
and
ABA
ethylene and abscisic acid made in senescing leaf promote abcission zone development
auxin moves toward root tip
IAA
gibberellic acid
and cytokinins
gibberellic acid and cytokinins synthesized in roots move to shoot and leaves
factor made in root tip controls geotropism of roots

Figure 8.7 A diagram illustrating some typical hormonal interrelationships among various portions of the plant.

$$\begin{array}{ccc} \text{indoleacetic acid} & \longrightarrow & \text{indoleacetylaspartic} \\ + \text{ aspartic acid} & & \text{acid (inactive)} \end{array}$$

Supplying the plant with high concentrations of auxin frequently stimulates the synthesis of the enzymes responsible for inactivation.

Cell Differentiation

When xylem elements differentiate, the process is easily seen because the changes in the cell are extreme. The wall develops heavy thickenings with characteristic patterns, and eventually the protoplast dies. These visible signs of change have made the xylem a favorite tissue for studying cell differentiation.

One experimental approach has been to perform delicate operations on the shoot tip to study how xylem forms in the young leaf primordia and the nearby stem. These studies show that a leaf primordium stimulates differentiation in the procambial strand leading to it (Fig. 8.8*A*). If the leaf primordium is sliced away, the same differentiation of vascular tissue in the stem can be induced by applying auxin (Fig. 8.8*C*).

In *Coleus* stems, a wound that severs a vascular bundle is followed by cell divisions and differentiation of xylem elements from parenchyma cells around the wound. The new xylem elements reconnect the injured bundle. Observed more closely, it has been found that new phloem cells are formed even earlier than the xylem cells in this regeneration process. These events require a supply of auxin that is normally transported out of the

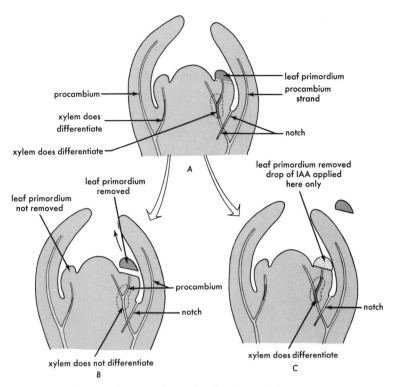

Figure 8.8   Diagram of an experiment showing how a leaf primordium provides a stimulus for xylem differentiation in the procambium. The notch severing the procambium isolates the tissue of interest above it. *A*, control with intact leaf primordium; *B*, leaf primordium removed; *C*, leaf primordium removed and a drop of auxin substituted. Auxin is an effective replacement for the stimulus from the primordium.

leaves above the wound and then down the stem. If the leaves are removed at the time of wounding, regeneration does not occur. The effect of the missing leaves can be replaced by applying auxin to the petiole stumps.

Further down the stem where elongation has ceased, even an herbaceous plant like the tomato may have some secondary growth taking place from a vascular cambium. This cambial activity depends on a supply of auxin that is received from the shoot tip. In woody plants that have been dormant over winter, the buds that sprout in the spring give off auxin that stimulates the cambium. A wave of cambial divisions can be traced down the stem, following the flow of auxin.

Further advances concerning the differentiation process have come with **tissue cultures.** Under suitable conditions, blocks of parenchyma tissue from stems can be induced to continue cell divisions and growth without differentiation. This produces a uniform mass of unspecialized cells, a **callus.** Blocks of callus tissue made from lilac stems show a remarkable response to auxin and sucrose: if these two substances are applied to the top of the block, a discontinuous ring of vascular tissue develops within the tissue block, at a distance below the surface. The diameter of the ring and its distance from the droplet can be increased by raising the concentration of auxin. If auxin spreads by diffusion, its concentration must decrease with distance from the source. If we put these facts together, it seems that the signal for cells to become vascular elements must involve their exposure to a particular auxin concentration.

Whether the differentiating cells become xylem or phloem elements in lilac callus depends on the concentration of sucrose that is applied. High concentrations induce nothing but phloem; low concentrations induce only xylem; and intermediate levels stimulate both xylem and phloem to differentiate. Significantly, when both xylem and phloem are formed, their arrangement is concentric; the xylem is toward the center, as in the cross section of an intact stem.

Gibberellins

While European scientists of the 1920s were occupied with the auxins, Japanese workers discovered another class of hormones. These hormones were first identified in studies of a disease of rice plants, the *bakanae* (foolish seedling) disease. Afflicted seedlings grow very tall and eventually fall over. The disease is caused by a fungus, *Gibberella fujikuroi*. The abnormal growth could be duplicated by treating normal seedlings with the liquid in which the fungus has grown. The active agent in the liquid proved to be a group of substances that were given the name **gibberellins.** They form an extensive group of related compounds, variations on the basic structure shown in Fig. 8.5*B*. Although these compounds were first discovered in a disease, they have since been found as natural hormones in a wide array of flowering plants. The *bakanae* disease was a result of oversupply.

Hormones and the Control of Growth and Development

Gibberellins and Stem Growth

The effect of gibberellin on a monocot (rice) was described above. Some dicotyledonous plants have very few cell divisions in the subapical region of the stem tip during the early months of growth, but leaves are formed normally at the apical meristem. (Fig. 8.14). The result is a shoot that has many leaves separated by very few short internodes. A plant with this form is called a **rosette**, whereas plants with elongate internodes are called **caulescent**. Rosette plants typically respond to gibberellin applications by rapid stem elongation, or **bolting**. The bolting response can be caused naturally by an environmental factor such as winter cold or long days. These factors trigger an increase in gibberellin, which promotes cell divisions as outlined above.

Auxins and gibberellins often act simultaneously to control stem elongation. In some cases, gibberellin may stimulate cell division, thus providing more cells on which auxin can act. In the coleoptile, gibberellins act at an early stage of development and the auxin-sensitive stage shown in Fig. 8.6 comes later in life. Woody stems need both auxin and gibberellin to maintain cambial activity.

In Europe, commercial use is made of the growth retardant CCC or Cycocel (2-chloroethyltrimethyl ammonium chloride) to inhibit stem elongation in wheat plants. Cycocel inhibits the production of gibberellin in the plant. Shorter, stronger stems result, and the plants are much more resistant to lodging, that is, to being knocked down by wind and rain. Lodging makes harvesting difficult. (Short stems in grains are also achieved genetically in breeding programs and are important factors in Mexican wheats and Philippine IR8 rice, plants of the "green revolution," that have raised crop productivity in Asia and Central America.)

"Dwarf" forms in a variety of plants are often due to a diminished gibberellin synthesis or to an enhanced production of compounds that oppose the action of gibberellin. Dwarf forms of corn and peas are used in estimating the concentrations of gibberellins in extracts of plants or plant parts. Such methods are called **bioassays**. Figure 8.9 shows a bioassay where quantities of gibberellin are estimated by comparing their effect on growth to that of known hormone concentrations. Bioassays have been important in studying all the plant hormones, because plant materials are often sensitive to quantities of hormone that would be far too small to detect chemically.

Gibberellins and Enzyme Synthesis

As a grain such as barley or corn starts to germinate, the embryo begins to grow but has limited food reserved in itself. The main reserves are in the starchy endosperm, a tissue with cells loaded with starch, reserve proteins, and some nucleic acids. A special layer of living cells, the aleurone layer, surrounds the main endosperm tissue and is instrumental in digesting the stored materials to soluble forms that can diffuse to the embryo. Early in germination, the embryo synthesizes gibberellin (GA), which diffuses to the aleurone layer and triggers the synthesis of enzymes to digest stored materials in the endosperm. If the embryo is removed from the seed prior to germination, very little digestion of the remaining endosperm takes place. Adding gibberellin in minute quantities to the embryoless endosperm induces the synthesis and secretion of enzymes just as if the embryo had provided the stimulus (Fig. 8.10).

The barley aleurone layer can be isolated and studied independently of the embryo and storage endosperm. It is a collection of cells with no growth activities and no division; it does have, however, a very active ability to synthesize a few proteins and to secrete them. As such it has been used to study the mechanism of action of gibberellin.

About 6 hours after adding gibberellin, the aleurone cells start to make enzymes that break down starch, proteins, and nucleic acids. Prior to the appearance of the enzymes, new ribosomes and endoplasmic reticulum membranes are formed, as well as enzymes involved in membrane lipid synthesis. We have much to learn about the way in which gibberellin unleashes all these changes, though it does appear that an activation of genes (new enzyme synthesis) is involved.

| 0.0 | 0.0001 | 0.0005 | 0.001 | 0.0025 | 0.005 | 0.01 | 0.025 | 0.05 | 0.1 |

μg GA$_3$/PLANT

Figure 8.9   Dwarf peas *(Pisum sativum)* showing the promotion of stem elongation by gibberellic acid applied seven days prior to photograph. This response is used to bioassay the gibberellin contents of plant extracts.

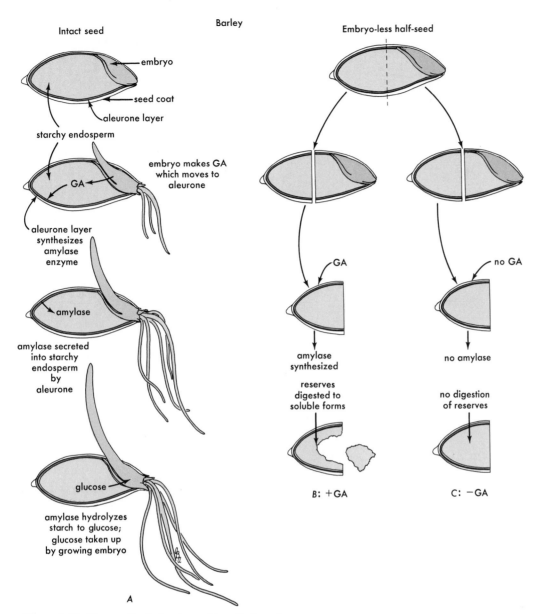

Figure 8.10 Diagram illustrating how gibberellin from the embryo induces the synthesis of the starch-degrading enzyme, α-amylase, in the aleurone layer.

In the manufacture of beer, the starch stored in the grain endosperm must be hydrolyzed to a soluble form before its conversion into alcohol by the glycolytic enzymes of yeast. The natural production of amylase occurs in the malting barley. The addition of extra GA speeds up amylase synthesis by the aleurone enough to be used commercially.

Cytokinins

Hormones of a third major class, the **cytokinins**, were discovered through work with tissue cultures. Physiologists have found that keeping cells alive and active after they have been removed from the plant depends on the exact constitution of the medium to which the cells are transferred. Finding the best medium is a

matter of trial and error. Such trials led to the discovery, in the mid-1950s, that cell division could be stimulated by adding adenine, one of the nucleic acid bases. Since then a variety of related compounds have been found to work much better than adenine. These are the cytokinins, one of which is shown in Fig. 8.5C. Some cytokinins are synthetic, while others occur naturally, especially in young developing fruits.

Besides affecting cell divisions, the cytokinins play a part in controlling the initiation of plant organs. In tissue cultures, for example, cells of tobacco pith will enlarge if supplied with nutrients and auxin, but will divide only if small amounts of a cytokinin are added. Moreover, by varying the balance between auxin and cytokinin, it is possible selectively to initiate the development of roots and shoots (Fig. 8.11). High auxin-to-cytokinin ratios cause root initials to differentiate. A low ratio of auxin to cytokinin

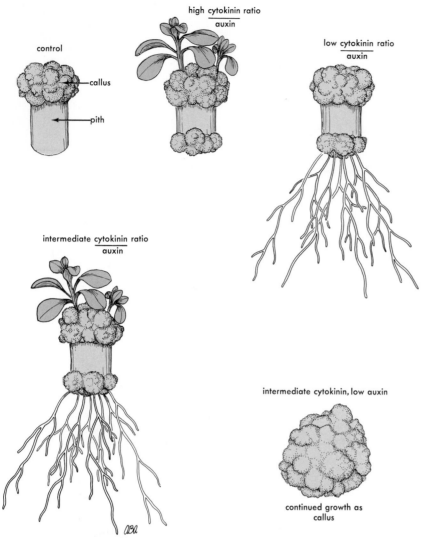

control

callus

pith

high cytokinin ratio / auxin

low cytokinin ratio / auxin

intermediate cytokinin ratio / auxin

intermediate cytokinin, low auxin

continued growth as callus

Figure 8.11 Diagram of the control of differentiation exerted by interaction of auxin and cytokinin. Pieces of tobacco pith tissues were aseptically grown on nutrient medium (tissue culture) supplemented with various levels of the two hormones.

causes clumps of cells to become apical meristems that grow into shoots. Intermediate concentration ratios cause callus to form. Shoots can be removed, rooted, and grown to mature plants.

Cytokinins also seem to contribute to apical dominance. If a cytokinin is applied directly to a lateral bud that is being suppressed by an active shoot tip, the bud may grow out. The applied cytokinin promotes the formation of vascular connections between bud and stem, a process that is inhibited by auxin. Cytokinins appear to be made in the tips of roots, and they travel upward in the xylem. They may accumulate at the cut end of a rooted stump, where adventitious buds often form. The opposed movements of auxin and cytokinin may give each region of the plant a unique balance between quantities of these two hormones, and their opposing effects on bud growth may help to determine the pattern of branching as the plant grows.

Plants normally maintain a close balance between shoot

and root growth. The upward movement of cytokinins from the root may help to maintain this balance. Increases in root growth cause more abundant cytokinin supplies, which cause a corresponding increase in shoot growth. Leaves also contribute to the shoot/root balance by producing carbohydrates and vitamins (especially vitamins of the B group, needed as cofactors in respiration) that roots cannot build for themselves.

Cytokinins also prevent leaf deterioration, or **senescence.** If a leaf is removed from a stem, senescence is initiated. The changes include the breakdown of storage carbohydrates, proteins, nucleic acids, and chlorophyll; and a cessation of protein synthesis. If the leaf is put under conditions where it forms adventitious roots, the senescence is halted. Applications of cytokinin have a similar effect in halting the deterioration of detached leaves. The maintenance of active RNA and protein synthesis seems to be an important part of cytokinin action in delaying senescence.

Ethylene

Some years ago it was the practice among lemon growers to pick lemons green and to allow them to ripen in heated boxcars as they were shipped across the United States to distant markets. This system worked nicely, until the leaky kerosene heaters in the freight cars were replaced by more efficient steam heaters. The shippers were dismayed to find that the lemons no longer ripened in time for marketing. Investigators found that the kerosene stoves had been leaking small quantities of a simple gas, **ethylene,** which acted as a ripening stimulus. This is one of many early observations that ethylene as a pollutant affects plant development. More recently, ethylene has been found to be a potent natural hormone that contributes to normal development in plants. Its structural formula is shown in Fig. 8.5E.

Ethylene is built in the plant from the amino acid **methionine,** a constituent of all cells. Since ethylene has only limited solubility in the aqueous phase of a cell, it diffuses into the atmosphere. Natural ethylene action in a tissue depends on its continued manufacture. Its concentration depends on the balance between synthesis and diffusion. Enclosing plants—especially fruits—in a tight container can cause the concentration of ethylene made by the plant to increase to levels that have drastic effects on its growth and development. Effective levels are often in the range of 0.1 to 1 part per million of air. Even the levels of ethylene present in urban air pollution are sometimes enough to cause biochemical changes in plants.

The action of ethylene in controlling vegetative growth can be illustrated by its effect on the pea seedling. As the plumule emerges from the seed, the tip of the shoot has a hooked curvature that protects the apex. The young plumule synthesizes ethylene rapidly in darkness. The ethylene maintains the hook and prevents the leaves from enlarging. When the shoot breaks through the soil, light is absorbed by phytochrome, which in turn triggers a decrease in ethylene synthesis. The leaves are released from ethylene inhibition and expand. The hook "opens" to present the leaves to sunlight, which further stimulates leaf development. (Fig. 8.1).

While growing underground, the shoot encounters obstacles such as clods. The pressure of the obstacle against the shoot causes a dramatic increase in ethylene production. The results can be simulated by gassing exposed seedlings with ethylene (Fig. 8.12). Ethylene induces the stem to swell and inhibits elongation. The thickened stem can exert more upward force against the obstacle. If the stimulus is prolonged, the stem's geotropic response is modified; it starts to grow almost horizontally. These responses increase the likelihood that the shoot will penetrate or grow around the obstacle to reach the open air.

Some of these effects involve changes in the direction of cell growth, and they raise one of the most fundamental unsolved problems in plant development: the question of what controls the direction of cell growth; what causes some cells to grow equally in all directions while others become cylindrical or spindle-shaped. Studies of growing cells have suggested that the direction of growth is determined when new cellulose microfibrils are added to the wall. If the microfibrils are laid down at random, the cell expands uniformly. If they are deposited around the cell like the hoops around a barrel, the cell will grow

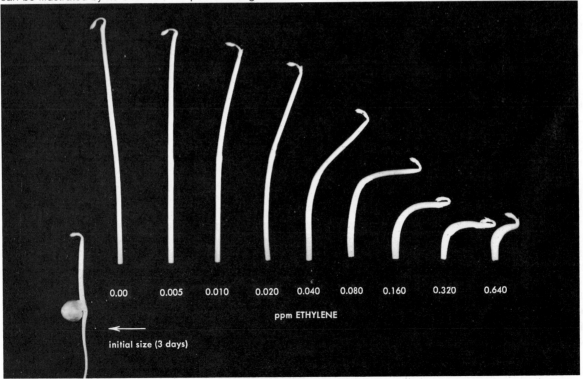

Figure 8.12 The response of dark-grown pea seedlings to various levels of ethylene during four days (ppm = parts per million ethylene in air). Internally generated ethylene under stress can induce similar responses.

chiefly in length, because the hoop arrangement prevents a large increase in cell diameter.

What does this have to do with ethylene? In some of the responses in which ethylene induces a thickening, as in the shoot that is trapped beneath an obstacle, the effect seems to depend on a change in the direction of cell growth. Ethylene causes some of the cells to deposit their new cellulose microfibrils more randomly, resulting in the consequences mentioned above.

We have no idea at present how the protoplast, *inside* the plasma membrane, controls the orientation of wall polymers that are laid down *outside* the plasma membrane, or how an information store such as DNA can specify such a directional process. But the effect of ethylene seems to offer a step toward understanding these problems.

Another effect of ethylene is seen in the growth of young trees, where the relationship between height and diameter of the trunk is very much a function of the motion it undergoes as a result of wind. If a young tree stem is tied to a rigid stake, as is frequently the case in container-grown trees, or is supported by other plants in a crowded nursery, the usual response is a tall, slender, weak stem. If the stake is flexible enough to allow some movement in response to gentle winds, the plant has a sturdier form resembling one grown without support. The plant that grows without support from a young stage typically has a shorter stem, larger in diameter, and with a tapered form coming from a wide base at the soil level. Routine exposure to high winds can greatly increase this response, as can be seen in trees on windy sea coasts or mountain ridges. Controlled gassing of tree trunks with ethylene stimulates enlargement, and it is thought that stress-induced ethylene synthesis in the trunk contributes to this growth response.

Ethylene and Fruit Ripening

During the early development of a fruit, the seeds are immature and delicate. Dispersal at this time would be premature. Many fruits have a programmed pattern of changes that convert the fruit from a seed-manufacturing to a seed-dispersal organ. These changes are called **ripening.**

Often the fruit changes its color from green to yellow, orange, red, or blue. This increases visibility to potential animal dispersal agents. To make this change chlorophylls are broken down, and other pigments such as the red and blue anthocyanins are synthesized. Another change is the conversion of starch and organic acids to sugar so that the fruit becomes sweet. Some of the materials in the cell walls are broken down; the cells now become more loosely bound to each other, resulting in a softer fruit. Volatile flavor components are synthesized, and these contribute to much of the flavor we value. One important link in this grand plan is the rise in ethylene to a critical concentration within the fruit. When this critical concentration is reached, ripening is initiated. Usually this is signaled by, or coordinated with, the maturation of the seeds.

Ethylene may cause ripening when it is not wanted. In fruit held in cold storage, elaborate precautions are used to prevent ethylene from accumulating in the atmosphere in quantities that would trigger ripening. When a fruit goes through the phase of rapid ripening, the production of ethylene may increase many times beyond the rate that was initially necessary for it to reach the critical triggering level. Thus, ''one bad apple in a barrel'' producing ethylene may trigger the ripening of all the apples in the barrel. Or the ethylene produced by the mold on an orange may be enough to trigger undesirable changes in the rest of the oranges.

Abscisic Acid

Many plants of the temperate zone become inactive or *dormant* during the cold season. In addition, seeds are usually dormant at the time of release. Even in active, photosynthesizing trees in midsummer, the terminal buds on the young twigs may go dormant after producing a sufficient number of leaves to clothe the tree. These phenomena illustrate the importance of dormancy and its release in the plant's repertoire of control and capabilities.

Although the mechanisms that occur during dormancy are not fully understood, one factor that is often involved is the hormone **abscisic acid**. Abscisic acid (Fig. 8.5*D*) is so named because it is plentiful in leaves that have recently been dropped or **abscised**. Applications of abscisic acid (often abbreviated ABA) to some plants will cause leaf and fruit abscission.

Some dormant buds and seeds are also rich in ABA, and their release from dormancy is correlated with a loss of ABA. In some seeds the ABA may be lost by leaching as seeds are exposed to rains. In other seeds and dormant buds, ABA may be gradually broken down by enzymes during the winter.

Just how ABA causes dormancy is not entirely clear. In some instances it seems that dormancy is actually due to a lack of gibberellin. Where this is true, dormancy may be overcome by adding gibberellin. If ABA is removed (e.g., by leaching), the level of gibberellin in the tissue increases as a prelude to the release from dormancy.

When transpiration exceeds water absorption, the water content of leaves decreases, causing a loss of turgor and wilting. It has been shown that under water stress in less than half an hour, the leaf may begin to make and accumulate abscisic acid. Within a few hours, several species display a manifold increase in ABA concentration. It has also been shown that ABA fed to leaves with open stomata can start stomatal closure in just five minutes. Abscisic acid appears to act by interfering with the uptake or retention of potassium (or sodium) in guard cells. Since potassium ions are required to provide the high turgor that guard cells need to maintain their stomata open, a lack of these ions will close the stomata. When water loss by transpiration is thus retarded, turgor in the leaf as a whole is regained, and the leaf recovers from wilting. The high level of ABA is metabolized away in one or two days and the stomata begin to open normally.

Interactions Between Hormones

For each class of hormone, there are examples where one hormone alone seems to control a particular process: for instance, the control of growth by auxin in the oat coleoptile. But it is more usual for the various hormones to interact in complex ways to control development. This has already been seen in the control of root and shoot formation in tissue cultures by auxin and cytokinin. Also, the effects of ethylene and auxin are often interrelated. An experimental treatment of tissues with high auxin levels often induces the synthesis of ethylene. Ethylene may then inhibit the transport of auxin. These results suggest that auxin and ethylene levels may mutually control each other in many tissues.

The control of flower morphogenesis in cucurbits (cucumbers, squash, melons) is a complex case of hormonal interaction. A typical cucurbit plant first forms male (staminate) flowers; at later nodes it forms perfect flowers. Some strains produce plants with only female (pistillate) flowers. It has been shown that applied auxin raises the internal ethylene level, and that the ethylene stimulates the initiation of female flower parts on flowers that would normally lack them, or it tends to make pistillate flowers from normally perfect flowers. Gibberellin, either in naturally high concentrations or artificially applied, promotes the initiation of anthers. There appears to be an ethylene–gibberellin balance that acts at the time of flower initiation to determine which organs the flower will have.

Ethylene and auxin may interact in controlling the abscission of leaves and fruits. Ethylene causes an abscission layer to form and mature at the base of the petiole. Auxin both promotes ethylene formation and blocks ethylene action. If the auxin supply drops, as it does in an aging leaf, ethylene can act, the abscission layer forms, and the leaf soon abscises.

The Practical Use of Hormones

Natural hormones and synthetic compounds that act like them are used in growing and marketing plant products. Several examples are mentioned below. Bear in mind that these are only a sampling of the compounds and methods used.

Fruit Ripening

Ripe bananas store poorly. Today it is common practice to pick, ship, and store bananas while they are still green. Gassing with ethylene triggers ripening, and the bananas will be ready to eat in five days.

Bud Sprouting

Stored potato tubers are prevented from sprouting by treatment with auxins. This is a case of bud inhibition, comparable to natural apical dominance.

Fruit Drop

Fruits such as apples, pears, and citrus may drop from the trees before harvest time. The fruits can be induced to cling to the trees for several days longer by spraying with synthetic auxinlike compounds, chiefly naphthyleneacetic acid and 2,4-dichlorophenoxyacetic acid (2,4-D). These compounds apparently retard the formation of the abscission layer at the base of the petiole.

Flowering

Pineapple plants can be induced to flower and form fruits on a precise schedule by applying auxin, ethylene, or a compound that breaks down to release ethylene. The auxin acts by inducing ethylene production in the plant. This aids the pineapple industry in coordinating harvesting, processing, and shipping. The same principle can be used at home; an ornamental bromeliad house plant can be induced to flower by putting it in an airtight container with a ripe apple as an ethylene source.

Rooting of Cuttings

In many commercially valuable plants the preferred method of propagation is by means of cuttings—either because the seeds are hard to germinate, or because valuable hybrid properties are lost in seed formation, or because cuttings produce a large plant faster than a seedling. A "cutting" is a detached shoot. To form a new plant, it must form roots, usually at the cut end (Fig. 8.13). In some species cuttings spontaneously form roots only if they are kept moist. Other species are harder to root, and for them various methods have been devised as rooting aids. Treatment of the cut end with auxins such as naphthaleneacetic acid or indolebutyric acid is often effective. The bases of the cuttings are immersed in the growth regulator solution for 12 to 24 hours or dusted with a dry powdered preparation; they are then placed in a bed of planting compound or moist sand. The conditions required for rooting (e.g., hormone concentration and

Figure 8.13 The promotion of root initiation by the synthetic auxin, indolebutyric acid, on American holly *(Ilex opaca)* cuttings. Row *A*, cuttings that stood in an aqueous solution of 0.01% indolebutyric acid for 17 hours before being placed in sandy rooting medium. Photographed after 21 days. Row *B*, untreated controls.

treatment time) vary with the species. Interested amateur gardeners can buy these chemicals at retail nurseries.

Growth Regulators as Herbicides

Though auxin is essential in small quantities, an excess can cause serious physiological disturbances. Some synthetic auxinlike compounds have proven to be very potent weed-killers (herbicides) for this reason. The so-called "phenoxy" compounds are especially effective: for instance, 2,4-D and 2,4,5-trichlorophenoxyacetic acid (2,4,5-T). These compounds are selective, killing broad-leafed plants and leaving the grasses relatively unharmed. This makes them useful for destroying dicot weeds in grain fields, dandelions in lawns, and brush in rangeland.

Light and the Control of Development

Plants have at least three pigment systems that initiate developmental changes in response to light. **Phototropism**, the directional control of growth by light, involves an unknown pigment and has been described previously. The **greening** of plants when they break out of the ground is initiated by a pigment called **proto-chlorophyll**, which was discussed at the beginning of the chapter.

The third and by far the most versatile pigment system that regulates development is the **phytochrome** system. Phytochrome is a pigment that exists in two stable forms, active and inactive, each having a different color. The inactive form absorbs red light most strongly and is called P_{red} or P_r, and the active form is called $P_{far-red}$ or P_{fr} because it most strongly absorbs the far-red light that lies just at the extreme of human visual sensitivity. When either form absorbs light, it is converted to the other form.

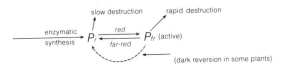

P_r is continuously synthesized from precursors, and is slowly destroyed by enzymes. P_{fr} may exist and exert its effect for many hours, but is unstable and is more rapidly destroyed by enzymes than is P_r. In some plants, P_{fr} is spontaneously converted back to P_r in darkness.

Continuous sunlight with its broad spectrum of wavelengths constantly converts the available phytochrome molecules back and forth between the P_r and P_{fr} forms, so that a steady-state mixture results with the P_r/P_{fr} ratio determined by the spectral composition of the light.

Phytochrome has a wide variety of effects on plant development, of which the following are important examples.

Seed Germination

Some seeds do not germinate in darkness. After a brief soaking, a short exposure to red light will induce them to germinate. Red light acts by converting P_r to P_{fr}, which in turn relieves some unknown block in metabolism and permits germination. If far-red light is given immediately after the red treatment, P_{fr} is converted back to the inactive P_r form and the seeds will not germinate. For this latter cancellation effect, the far-red treatment must occur before P_{fr} has had time to act.

Responses to light that are cancelled by a second dose are said to be **photoreversible**. Red/far-red photoreversibility is a unique phenomenon that is seen only in events governed by phytochrome.

Shoot Growth

The main signal to a seedling that it has emerged from the soil is the formation of P_{fr} by light. The complex response of the seedling has already been described. In these de-etiolation responses the action of P_{fr} may involve such factors as the release of active gibberellin from an inactive form and the activation of genes to form new enzymes.

When a plant finds itself almost continuously in the shade of other plants, the quality of light falling on the plant is changed in a way that can markedly affect the phytochrome status of the tissue. As light passes through successive layers of leaves, the red wavelengths are filtered out by chlorophyll, while the far-red wavelengths pass through. In the shade created by other plants, light rich in far-red tends to markedly lower the P_{fr} concentration. The result is more rapid stem elongation and an increased probability of growing out from under the shade that is cast by other plants. Even in plants grown in light, phytochrome is present, and experiments have shown that it exerts subtle controls on stem elongation. In this example, a change in light quality can trigger a response that has obvious survival advantages.

Although the mechanism of action of phytochrome has not yet been established, some work suggests that it may be attached to membranes and may exert some control over membrane function.

Photoperiodism and Flowering

Over most of earth's surface, there are pronounced seasonal differences in temperature, water supply, and illumination. The developmental system of the flowering plant is equipped to prepare for these changes. Well before the weather is harsh, broad-leafed trees begin to shut down the anabolic systems in the leaves. There is an orderly withdrawal; the leaves are stripped of some nutrients, abscission layers are formed, and the leaves are abandoned to the elements. If the plant can make underground storage organs, they are filled up before there is a great danger of the loss of food. In addition the metabolic system gradually shifts its composition to protect against frost damage before ice appears. The broad-leafed tree buttons down for the winter and goes dormant. After the cold season, the plant still waits until it is well into the fair season before stirring to active life, disdaining the occasional day of false spring that occurs in late winter before the weather is reliable.

Not all plants are so good at predicting seasons. But on the other hand, the preceding comments were scarcely a sampling of the seasonal adjustments that plants can show. There are desert plants, for example, that prepare themselves for a dangerously hot spell rather than a dangerously cold one. And more subtly, plants may time their activities during the growing season in a way that minimizes competition for light, moisture, and pollinators. This also may be based on a sensing of the seasons.

Of all the earth's seasonal variables, the most reliable is the annual cycling of day and night lengths. Plants have been found to contain a system that measures the lengths of the nights. The control of development by this timing system is called **photoperiodism.**

All forms of photoperiodism are based on the system that monitors the day/night cycle. That system is located in the leaves. Given suitable light/dark cycles, photoperiodically sensitive leaves send out stimuli through the phloem to other parts of the plant: to the shoot apex, where flowers or winter buds may be induced; or to underground stems, which may be induced to form tubers.

To explore photoperiodism, then, the principal focus of attention is on the leaves and their timing system. But to illustrate its operation in the passages below, let us take the control of flowering as our example.

In regard to flowering, some plants achieve their timing without reference to light stimuli, relying instead on factors such as maturation. These plants are said to be **day-neutral.** Other plants, however, use photoperiodism to time the flowering response. These plants fall into two major groups that are traditionally designated as **long-day** and **short-day** plants. Actually, current research has

shown that the plants are really measuring the lengths of nights rather than days. But the names are too firmly embedded in the language of physiologists to be changed.

Long-day plants usually flower in the spring when nights are growing shorter (Fig. 8.14). In these plants the signal for flowering is sent from leaves to shoot tips when the nights are less than a particular critical length. The critical night length is characteristic of the species and in general is set so that the plant has a suitable period of vegetative growth in the spring before flowering. Thus plants in the more northerly regions, where winter ends later, tend to have shorter critical night lengths; they therefore flower later in spring.

Short-day plants, such as many chrysanthemums, typically time their flowering for fall or winter when the nights are progressively lengthening (Fig. 8.15). Here again there is a critical night length, but in short-day plants the signal for flowering occurs when nights become *longer* than the critical length. These plants fail to flower in the shorter days of early spring because at that time they are immature.

Presumably the advantage of flowering in the fall is that the plants have been able to accumulate photosynthetic products all summer for better seed production. A well-set, accurate timing system should choose the optimum time to initiate flowering so that the plant has the longest possible period of active growth while still allowing time for seeds to mature before the cold sets in.

It is not entirely understood how the leaf measures lengths of night. We do know that phytochrome signals the end of the day. During the daylight hours there is always some P_{fr} present, but at night the P_{fr} gradually disappears. It seems that the disappearance of P_{fr} somehow "tells" the plant that night has begun. The next light exposure results in new P_{fr} being formed; this acts as a signal that the night is over. Since P_{fr} formation requires only a little light, even a brief flash of red light during the night can seriously interfere with the timing of night length. As would be expected with a phytochrome-

Figure 8.14 Effect of day length on behavior of the long-day plant henbane (Hyoscyamus). Plants were grown with 8 hour photoperiod until they were about a month old and were then subjected to 24 photoperiods of 10, 11, 12, 13, 14, or 16 hours. Initiation of flowers and accompanying stem elongation (bolting) occurs on days longer than 12 hours.

Figure 8.15 Effect of day length on behavior of *Chrysanthemum*, a short-day plant. *A*, plant that received light of natural short days of autumn and blossomed at the usual time. *B*, plant that received an hour of light near the middle of each night for several weeks beginning just before flower buds would normally have been initiated; thus, each long dark period was divided into two short ones. This interruption was sufficient to delay flowering.

mediated response, the effect of a red light flash at night can be cancelled when it is immediately followed by a flash of far-red light.

Though phytochrome signals the beginning and end of night, it appears that in darkness some other component in many plants actually measures the time. This timing system has been called the **biological clock**. Its nature is unknown, but it is highly accurate and insensitive to such disturbing factors as differences in temperature and nutrition. Some plants can discriminate between night lengths that differ by as little as 10 minutes, a degree of precision that allows flowering to occur within a week of the same date every year.

The biological clock is a metabolic system that completes a cycle approximately once each 24 hours. It synchronizes many physiological functions on a daily cycle and is present in organisms ranging from humans to algae.

In regard to the timing of flowering, the disappearance of P_{fr} after nightfall somehow provides a time marker that the clock uses to begin to measure night length. At dawn P_{fr} is re-formed and the timing is stopped. For the control of flowering, the important point is whether the clock has

run long enough to exceed the critical night length. The answer determines whether or not the leaf will send a flowering stimulus to the shoot apex.

In considering the flowering stimulus we have only mentioned that it is made in the leaves and moved to the apices. We do not yet know its chemical nature. Some experiments strongly suggest that the same flowering stimulus serves long-day, short-day, and day-neutral plants. They apparently differ only in the conditions required to bring about its synthesis.

Whatever the nature of the flowering stimulus, it must be produced by a metabolic system that we might call a *stimulus generator*. In short-day plants the stimulus generator is activated if the clock passes its critical point during one or more successive nights (Fig. 8.16). To produce a short-day flowering habit, by contrast, the stimulus generator must be set to produce its flowering stimulus automatically if the clock does *not* pass the critical point for one or more nights in succession. That is, long-day and short-day plants differ according to whether the clock activates or inhibits the stimulus generator when the critical night length is exceeded.

Temperature and Development

Photoperiodism is not the only system that times flowering. Some plants require a prolonged exposure to winter cold as a prerequisite to flowering.

Plants having this type of control include both winter annuals and biennials. Winter annuals typically germinate just prior to or during the winter so that the young plant passes through a period of weeks of exposure to low temperatures prior to the warmer weather of spring and summer when they flower.

Biennials generally have a full season of vegetative growth the first year, exposure to cold weather during the next winter, and then they flower in the second spring and summer. Without exposure to the low temperatures, flowering will be delayed or prevented (Fig. 8.17). Gibberellins can sometimes substitute for the cold requirement. This promotion of flowering in response to cold treatment is called **vernalization**. The cold treatment generally requires temperatures of 10°C or less for several weeks. Often, the cold requirement is followed by a requirement for long days. This prevents the plant from flowering too early in spring. The growing tip itself, rather than the leaves, may perceive the cold stimulus in vernalization.

"Winter" cereals, wheat and rye, are normally planted in the fall so that they germinate before the onset of winter and flower promptly in the following spring and summer after making a minimum number of leaves. If they are planted in the spring and are not vernalized, flowering is delayed until many more leaves are formed. "Spring" cereal strains are normally planted in the spring and flower promptly in response to long days. Winter cereal strains can be artificially vernalized prior to a spring planting date by moistening the seeds and holding them at about 1°C for a period of weeks.

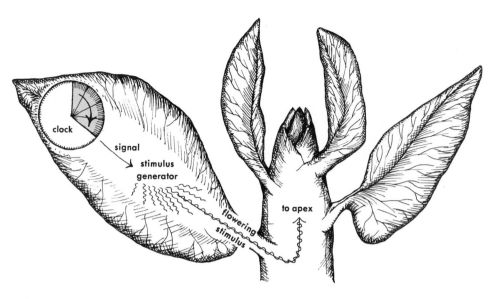

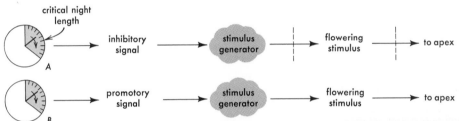

Figure 8.16 Control of flowering in long-day (A) and short-day (B) plants. Details are hypothetical. In both cases the timing of night length is started by the disappearance of P_{fr} after nightfall. In both, the same kind of flowering stimulus originates from a metabolic "generator" in the leaves and moves to the shoot apex. Whether a stimulus is sent depends on whether the clock reaches the critical night length measurement before being interrupted by light. The plants act as if the clock sends a signal on reaching the critical length; short-day and long-day plants differ as to whether the signal promotes or inhibits the stimulus generator.

In the discussion of photoperiodism, we indicated that many woody perennial plants use the photoperiod to induce a dormant condition in the bud prior to the onset of unfavorable weather. The bud typically goes through a period of increasing dormancy until it can no longer be rapidly reactivated by favorable environmental conditions or a shock treatment such as defoliation. By fall it may be in a state of rest in which no treatment will activate it other than the passage of time at low temperature, a process called **chilling**. The chilling requirement generally requires temperatures a little above freezing but below 10°C. Normally, the requirement is met by midwinter and at that time a branch brought into the greenhouse will have buds bursting in a few weeks. The buds are no longer in a condition of rest; they can respond to the warmer days of spring with renewed growth. The photoperiod induced the dormant condition, and the chilling process signaled that it was safe to respond to springlike weather.

Fruit trees typically need 250 to 1000 hours of chilling to produce good growth and fruit production in spring and summer. The lack of sufficient cold weather sets the southern limit for peaches, apricots, cherries, plums, and apples, and for effective flowering of ornamentals such as lilac (Syringa vulgaris).

Figure 8.17 Substitution of gibberellic acid for the cold requirement (winter) in the flowering of the biennial carrot (Daucus carota). A, control plant under long days only; B, long days plus gibberellic acid, no cold treatment; C, long days plus cold treatment, no gibberellic acid.

Temperature and Development

Another temperature-related phenomenon is the development of **frost hardiness**. In the summertime, temperatures that are a few degrees below freezing will kill leaves and buds. During the fall, many plants gradually develop resistance to temperatures as low as 50°C below freezing. Short photoperiods and exposure to temperatures just above freezing are the stimuli that induce frost resistance.

The seeds of some species will not germinate until they have been exposed to several weeks of temperatures near freezing. Such a cold, wet treatment of seeds to promote germination is called **stratification**.

Summary

1. Plants are able to respond to a wide variety of environmental situations because they have systems for monitoring important environmental factors and they have systems that modify the course of development in response to these stimuli.

2. The control of seed germination and the control of growth and development in the seedling are instances in which developmental responses to light, gravity, temperature, and other environmental cues have survival value.

3. The coordination of activities between various parts of the plant, and the course of development within each cell and organ, are controlled in part by the production and distribution of hormones.

4. Plant hormones are natural compounds that act in small quantities as signals to regulate development.

5. Auxin, the gibberellins, the cytokinins, abscisic acid, and ethylene are the major plant growth hormones.

6. Auxins promote cell enlargement by altering the cell walls, making them more extensible. Growth is controlled by factors that affect the synthesis, transport, and inactivation of auxin. Phototropic and geotropic curvatures in the shoot are regulated by changing the transport of auxin to the growing cells.

7. Auxins also help to control other developmental processes such as apical dominance, cell division, and differentiation of specific cell types in the vascular system.

8. Gibberellins control stem elongation through effects on cell elongation and division. They also affect seed germination and cambial activity. In certain instances they may induce the synthesis of enzymes.

9. Cytokinins control cell division in specific tissues, they interact with auxin to control organ initiation and bud growth, they help to prevent aging in leaves, and they help to coordinate root and shoot growth.

10. Ethylene interacts with auxin to control stem elongation. It also helps to maintain the etiolated growth habit in darkness and to regulate ripening in many fruits.

11. Abscisic acid helps to regulate the leaf's response to water stress through stomatal closure, contributes to seed dormancy, and promotes abscission in leaves and fruits.

12. Light acts as a developmental signal when it is absorbed by specific photoreceptor pigments. The chief receptor pigments are phytochrome, the phototropic pigment, and protochlorophyll.

13. Phytochrome exists in active and inactive forms that are interconvertible by red and far-red light. Phytochrome senses when light is present or absent. When stimulated by light, it starts seedlings on the development of the above-ground mode of growth. It may inhibit stem elongation, promote leaf expansion, promote the germination of seeds, and participate in many other developmental events.

14. Photoperiodism, the control of development in response to the lengths of days and nights, is the basis for many seasonal adjustments in plants.

15. The photoperiodic timing system occurs in the leaves and seems to involve a "biological clock" that is governed by phytochrome.

16. There are two main types of photoperiodically controlled plants, as regards flowering: long-day and short-day plants, which respond respectively to night lengths shorter than critical or longer than critical. Many plants are day-neutral; in them, flowering is not photoperiodically controlled.

17. Buds may be induced to go dormant by a photoperiodic system and can be activated again by favorable conditions only after passing through a prolonged chilling period, which is normally provided by the winter season.

18. A prolonged period of low temperature may be needed to vernalize certain plants, that is, to remove a block that prevents the formation of flowers by the shoot apex. Low temperature may also be a requirement for seed germination.

9 CHAPTER

plant ecology

Through evolution, existing plant species have achieved a balance with their environment. In each type of habitat, certain species group together. Fossil records indicate that similar groups of species have lived together for millions of years. The species of these groups share incoming solar radiation, soil, water, and nutrients to produce a relatively constant amount of living matter. They recycle nutrients from the soil to living tissue and back to the soil again; and they divide the environment's resources. Plant ecology attempts to explain this balance between plants and their environment. For example, what stresses does the environment put on plants, and how do plants respond to them? Plant ecology also attempts to determine how this natural balance may be artificially imitated or manipulated to improve human life.

Plant ecology deals with levels of biological organization beyond the organism level: the population, the community, the ecosystem. A **population** is a group of closely related organisms that normally interbreed. A **community** is composed of all the populations in a given habitat. An **ecosystem** consists of the living community plus the nonliving factors of its environment.

Components of the Environment

The environment of an organism includes all of the living and nonliving things around it. Some important environmental components are moisture, temperature, light, soil, and living organisms, and all of them work separately and jointly to influence plant distribution and behavior. Species differ in their range of tolerance to these environmental factors. For example (Fig. 9.1), coast redwood *(Sequoia sempervirens)* is restricted to a narrow strip of California and adjacent southwestern Oregon, about 9500 square km, that experiences heavy summer fog. Coast redwood is not tolerant of variations in moisture

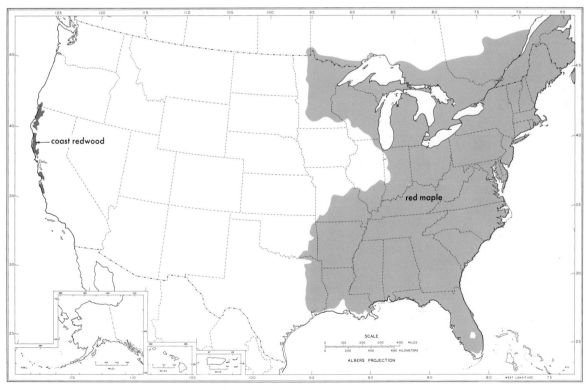

Figure 9.1 Distribution of coast redwood *(Sequoia sempervirens)* and red maple *(Acer rubrum)* in the United States.

or temperature. Red maple *(Acer rubrum),* however, occurs over about 4,000,000 square km of eastern North America in wet, dry, cold, or warm environments. It is tolerant of wide variations in moisture and temperature.

There are two types of environment. The **macro-environment** is influenced by the general climate, elevation, and latitude of the region. Weather Bureau data on rainfall, wind speed, and temperature are measures of the macroenvironment. Measurements are taken at a standard height of 1.5 m above ground in a clear area away from buildings or trees.

The **microenvironment** is the environment that is close enough to the surface of an organism or object to be influenced by it. For example, bare soil tends to absorb heat; consequently, the temperature just above or below the soil surface is much higher than air temperature (Fig. 9.2). Light quality and quantity are much different for herbs beneath a forest canopy than for the leaves of the canopy itself (Fig. 9.3). Air as far as 10 mm from the surface of a leaf on a still day is less turbulent and higher in humidity than free air further away from the leaf. In marshy areas, where the water table is close to the surface, minor dips or rises in the topography greatly influence the root microenvironment. Plants growing in shallow depressions are often subject to more frequent frost than those growing on higher ground, because cold air will settle in the depressions. Ecologists are convinced that the microenvironment is as important for plant growth as the macroenvironment.

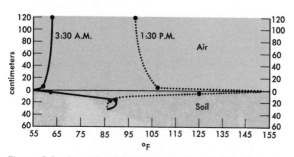

Figure 9.2 Air and soil temperatures at several positions just above and just below the soil surface. Temperatures were measured during the warmest and coolest parts of the day.

Moisture

Water is a most important factor in determining both the distribution of plants over the earth's surface and the character of an individual plant. Probably no single factor is so largely responsible for the abundance of plants in a habitat as is the supply of water. So important are the water relationships of plants that various attempts have been made to classify plants on the basis of these relationships. One such classification divides plants into (a) **xerophytes,** which are able to live in very dry places, (b) **hydrophytes,** which live in water or in very wet soil, and (c) **mesophytes,** which thrive best with a moderate water supply.

Plants with xerophytic characteristics, which limit transpiration or in other ways closely control water balance, occur in different climatic zones, especially deserts. Xerophytic plants do not necessarily have a lower transpiration rate than do mesophytes when water is ample, but they do possess one or more characteristics that enable them to survive periods of drought. The most effective of them are a thick cuticle, early or daytime stomatal closure, reduction of the transpiring surface, and water storage tissue.

The time of precipitation is as important to plant growth as the total annual amount. For example, tropical rain forests and tropical savannahs may both receive the same total yearly rainfall—let's say 200 cm—but the forest receives an equal share of the total each month, while the savannah has pronounced dry and wet seasons. The result is that the two vegetation types are quite different: tropical rain forests support tall, evergreen trees and vines in profusion, but tropical savannahs support grasses, shrubs, or short trees, which are often deciduous.

Plants adapt to dry climates in several ways. Cactus plants (Fig. 9.30) develop water storage tissue in the stem and reduce transpiration by the loss of leaves; all their photosynthetic activity is performed in stem tissue. Mesquite shrubs *(Prosopis)* and salt-cedar trees *(Tamarix)* possess long roots that tap ground water, sometimes at depths below 53 m. **Epiphytes,** growing on tree trunks above the soil (Fig. 9.4), trap falling rain in leaf axils or in

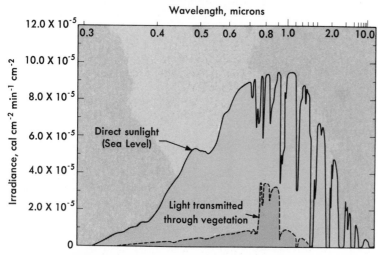

Figure 9.3 Distribution of solar radiation energy by wavelength. Energy was measured in direct sunlight and beneath a forest canopy.

Figure 9.4 *Tillandsia*, a genus with many epiphytic species common to tropical and subtropical regions of the western hemisphere.

dead cells on the surface of roots. These epiphytes are not parasites; they depend on trapped water and debris for all growth requirements. Annual herbs adapt to drought by completing a brief life cycle only during periods of sufficient soil moisture. *Boerhaavia repens* of the Sahara Desert is a small annual that can go from seed to seed in 10 to 14 days, but most annuals have a life span of 3 to 8 months.

There is some evidence to show that plants can absorb dew through their leaves, but the amount of water obtained in this way is not great. Usually, dew, fog, or high humidity are of more value in reducing the rate of transpiration, than by physically entering the plant. (Exceptions include small epiphytes, certain annual plants, mosses, and lichens.)

Plants may also utilize fog by condensing the mist into large drops that trickle down the stem or drip from the branches, increasing soil moisture considerably. The amount of water added to soil in this way by trees on the San Francisco Peninsula was measured as 5 to 74 cm in one summer, the exact amount within that range depending on the type of exposure. Normal precipitation (rain) for the area is only 64 cm a year, with almost no rain in summer.

Temperature

Temperature influences moisture availability and the rate at which chemical reactions occur. At freezing temperatures, water changes to ice and is unavailable for plant growth; consequently, in cold climates one of the principal factors limiting growth may be drought.

Absorption of water by roots is slower in cold soil than in warm soil. At high temperatures, evaporation may remove much of the soil moisture before the new growth of shallow roots can reach it, and transpiration will also be stimulated, resulting in further rapid depletion of soil water.

Brief extremes in temperature may be more important in determining plant distribution than are long-term moderate temperatures. Palms and many cactus plants seem excluded from areas that experience a specific frequency of frost. California lilac *(Ceanothus megacarpus)* seeds germinate poorly unless they have been exposed briefly to high temperatures. Many temperate zone plants are not frost-resistant until they have been exposed to moderately low temperatures for a short time and thus "hardened." Other plants require an alternation of day and night temperatures (a **thermoperiod**) for best growth.

Topography greatly influences soil and air temperature. In mountains latitude and elevation combine to restrict species to certain belts (as shown for mountain hemlock in Fig. 9.5). Even at the same latitude and elevation, minor differences in the direction of slope **(aspect)** create different microenvironments. A gorge running east-west through an Indiana forest was studied to illustrate the effect of aspect on plant growth. The gorge was 65 m wide at the top and 45 m deep; its sides supported scattered trees, shrubs, and herbs. Meteorological instruments were placed 15 cm above and below the soil surface midway down each side. During spring, the south-facing slope was found to exhibit a greater daily range of topsoil temperature, a higher mean air temperature, a higher rate of soil water evaporation, and lower relative air humidity than the north-facing slope. Of nine spring-flowering species present on both banks, the average flowering time was six days earlier on the south-facing bank. A similar difference in flowering time could be achieved on level ground only over a distance in latitude of 180 km.

Temperature is only an estimate of the heat energy available from solar radiation. Solar radiation at the limits of our atmosphere is equivalent to about 2 cal per cm^2 per min. Much of this radiation is absorbed, scattered, or reflected within the atmosphere, and only half may reach the ground and heat it. In turn, the warm earth reradiates energy back to space—terrestrial radiation. The difference between solar (incoming) radiation and terrestrial (outgoing) radiation is termed **net radiation**. During a clear day at the equator, solar radiation is greater than terrestrial and, thus, net radiation is positive; during the night, the reverse is true and net radiation is negative. But, over a 24 hr period, a month, or a year, net radiation may be positive or negative, depending on the location. For example (Fig. 9.6), at Aswan, Egypt—a warm desert climate—net radiation is positive each month, with a peak in June. Net radiation is positive only five months of the year at Turukhansk, Siberia—a cold subartic climate—and the annual net radiation is very close to zero.

Light

Figure 9.3 shows how solar radiation is changed in quality and quantity by its passage through green leaves. Light

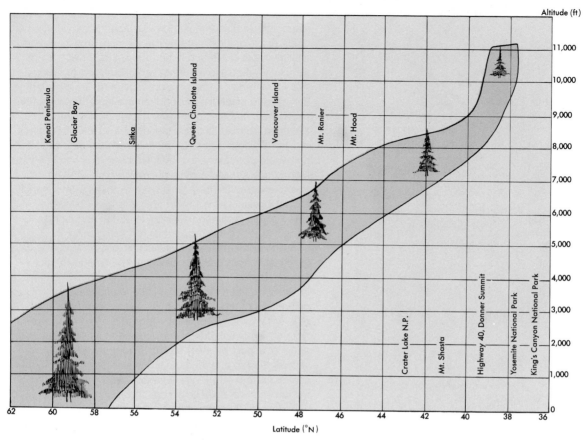

Figure 9.5   Distribution of mountain hemlock *(Tsuga mertensiana)*, showing the relationship between altitude and latitude.

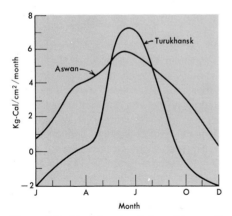

Figure 9.6   Monthly net radiation of Aswan, Egypt and Turukhansk, Siberia.

intensity on the forest floor may be only 1 to 5% that of full sunlight. Some herbs of the deciduous forest complete their life cycle in early spring, before the trees above them develop their complete foliage and reduce light intensity. Leaves of plants that develop in shade exhibit different morphology, anatomy, and physiology than those that develop in sunlight, even when both are attached to the same plant. Shade leaves are larger, thinner, contain less chlorophyll per gram of tissue, have less well-defined palisade and spongy mesophyll layers, and achieve maximum rates of photosynthesis at much lower light intensities than do sun leaves. Recent biochemical evidence has shown that the difference in photosynthetic rates is due to a difference in the activity of the enzyme responsible for CO_2 fixation.

When a large tree dies in a mature forest, the canopy is opened and increased light intensity reaches the ground beneath. This increased light radiation induces seedlings there to begin rapid growth. Seedlings must be able to survive many years of slow growth in shade before such an opening "releases" them. Eastern hemlock *(Tsuga canadensis)* is remarkably shade-tolerant. Mature trees of this species live for 1000 years. They can retain the capacity for rapid growth when released from shade to an age of 400 years. Saplings only 2 m tall and 2 to 3 cm in diameter may have a ring count indicating an age of 60 years.

Trees that are not shade-tolerant may become established and grow to maturity if some disturbance such as fire or logging first removes the shade. Seedlings of shade-tolerant species grow beneath them, however, and replace them as they die, while their own seedlings do not remain alive. For this reason, it is difficult to maintain pure stands of valuable shade-intolerant species such as teak, mahogany, Douglas fir, or the southeastern yellow pines. These species only invade disturbed sites and do not long maintain themselves.

Light is also reduced in intensity and quality by passage through water. Algae are not commonly found below a depth of 60 m except in very clear water. A record depth for algae may have been found in Lake Tahoe, California. A sample of water and bottom mud from a 140 m depth contained the alga *Chara*. Light intensity at that depth, despite the water clarity, is only 0.1% of full sunlight.

Brief exposure to light may be as important for some plants as long-term exposure. Recall from Chapter 8 that seeds of some species can be stimulated to germinate in the dark if they are first exposed briefly to full sunlight; this is done by simply shaking them for a few seconds in soil held in a glass container. This treatment could be duplicated in nature by a plow turning the soil and seeds in it, briefly exposing the seeds to sunlight, then reburying them. Plowing aerates the soil, makes it more permeable for root growth, and removes plants at the surface—all conditions of advantage to an emerging seedling.

Soil

Soil is the part of the earth's crust that has been changed by contact with the living and nonliving parts of the environment. It is a weathered, superficial layer typically 1 to 3 m thick, made up of decomposed and partly decomposed parent rock material with associated organic matter in various stages of decomposition. There are many kinds of soils and many different soil conditions. The character of natural plant covering and the behavior of crops depend on soil conditions as well as climatic conditions. Environmental influences that operate through soil are called **edaphic factors**; those that act on the plant through the atmosphere are called **climatic factors**.

Each environment creates a unique soil type. These soil types have their own history of development, morphology, and chemical attributes. In the United States alone, there are 10,000 soil types.

Soil Formation

Most soils consist primarily of mineral particles that are formed by a slow, continual process of weathering of the parent rock. Mineral matter is derived from fragmented rock. The kind of parent rock (e.g., granite, lava, sandstone, limestone, and shale) and the degree of weathering determine the nature of the mineral or inorganic components of the soil.

Weathering may involve simply **mechanical breaking** of the parent rock. For instance, the action of strong winds or wave action may hurl sand against rock outcroppings and wear them down, or water collecting in pores of the rocks may expand at freezing temperatures to create fissures that ultimately fragment the rock. **Chemical weathering** results in more profound changes in the mineral matter itself. Atmospheric gases, such as CO_2 or SO_2, become dissolved in rainwater and produce acidic solutions that dissolve the parent rock material. Plant roots secrete weak acids. Certain algae, bacteria, and lichens hasten the decomposition of the least resistant mineral materials.

Weathering of the parent material, under the influence of climate and organisms, proceeds along a definite series of

stages from a young soil in which the processes of soil development are continuing to a mature soil that has approached a steady state.

Soil formation may proceed rapidly if the type of parent material and type of climate are favorable. Soils form rapidly where there is a warm, humid climate, forest vegetation, flat topography, and a parent material that is easily broken down. For example, Fort Kamenetz, built of limestone slabs in the Ukrainian part of the Soviet Union, was abandoned in 1699. Today, a mature soil 10 to 40 cm thick has developed from the limestone. Soils form slowly in regions with a cold, dry climate, grass vegetation, steeply sloping topography, and parent rock material that is not easily broken down. It has taken from 1000 to 10,000 years for mature soils to develop in northern areas scoured during the Wisconsin glaciation.

Plants influence soil development in several ways. They move ions through the soil by absorbing them through their roots. These ions accumulate in the leaves and are returned to top soil after leaf abscission. Plant roots and shoots decay and add nitrogen and sometimes acids to the soil. In addition, plant canopies shelter the soil surface, reducing erosion and water loss.

The **texture** of soil is determined by the mineral fractions of a soil; we customarily speak of coarse sand, fine **sand, silt,** and **clay**. The size of soil particles decreases in the order given (Table 9.1). Clay is not only the smallest mineral soil particle, but because of chemical weathering it has also undergone the most change from the parent rock.

Although most soils contain both mineral and organic matter, the organic matter they contain constitutes only 2 to 5% of their weight. These predominantly **mineral soils** may be classified according to the relative amounts of the sand, silt, and clay they contain. Soils that contain roughly equal amounts of sand, silt, and clay are called **loams**. Depending on the relative amounts of each component in the soil, we speak of clay, clay loam, silt loam, loam, sandy loam, loamy sand, and sand textures (Table 9.2). The suitability of a soil for plant growth is greatly influenced by its sand, silt, and clay content. **Organic soils** are defined as soils having a layer, at least 30 cm thick, that is more than 30% organic matter.

Soil Water and Its Dissolved Substances

An important difference among the various soils is their ability to hold water. It is greatest in clay and organic soils, and least in coarse sand.

Table 9.1

Classification of Soil Mineral Matter According to Size of Particles (International System)

Type of Particles	Range in Diameter of Soil Particles, mm
Coarse sand	2.0–0.2
Fine sand	0.2–0.02
Silt	0.02–0.002
Clay	0.002 and smaller

Table 9.2
Composition of Three Soils According to the Size of the Mineral Particles

Soil Type (texture)	Coarse Sand %	Fine Sand %	Silt %	Clay %
Sandy loam	67	18	6	9
Loam	27	32	21	20
Clay	1	9	22	68

If water is applied to a soil in the field, the spaces between the soil particles (**pore spaces**) become filled only for a short time to the depth wetted. With drainage, the water begins to move downward under the influence of gravity. After a while, this movement downward stops; the soil particles hold a certain amount of water against the pull of gravity. The amount of water held by the soil after drainage is called the **field capacity** of that soil. When a soil is at field capacity, or even at a moisture content well below this amount, a film of water completely surrounds each soil particle, and water also exists in the form of wedges between soil particles. Clay particles and organic matter in the soil are **colloidal** particles, and such particles hold water by **imbibition**. Water imbibed by soil particles is much more difficult to remove from the soil than water that exists as a film or as wedges.

Soil particles themselves, particularly finely divided clay and organic matter, act as giant negatively charged ions that attract a cloud of positively charged ions such as Ca^{2+} and K^+. Thus, colloidal soil particles, while not in solution themselves, serve as reservoirs to which many of the various ions essential for plant growth are loosely attached and from which these ions may be released into solution by a process known as **cation exchange** (Fig. 9.7).

Acidity influences the physical properties of the soil, the availability of certain minerals to plants, the biological activity in the soil, and consequently strongly influences plant growth. Certain plants such as azaleas, camellias, and cranberries grow best in acid soils; most plants do best in soils near neutrality, while a few plants grow satisfactorily in slightly alkaline soils.

Acid soils, such as those beneath northern coniferous trees that shed acidic foliage, range from pH 3.5 to 5.0; agricultural soils in the humid areas range from pH 5.0 to 7.0; and soils in arid or saline regions may reach pH's as high as 11.0, but a range of 8.0 to 9.0 is more common.

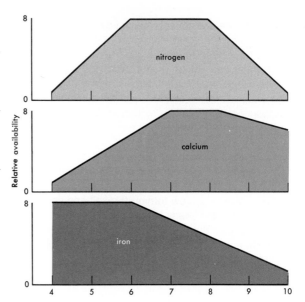

Figure 9.8 Relationship of relative nutrient availability to soil pH.

Acidity *per se*—that is, H^+ or OH^- ions—is not directly responsible for limiting plant growth. Rather, soil pH affects the availability of plant nutrients in two ways. In an acid soil, hydrogen ions (H^+) replace other **cations** (like K^+, Ca^{2+}, and Mg^{2+}) on the negatively charged clay particles, allowing the nutrient ions to float free in the soil water and to be easily leached from the soil as that water moves downward. Soil pH also affects the solubility of plant nutrients. As shown in Fig. 9.8, some nutrients (e.g., iron, manganese, and aluminum, increase in solubility as pH decreases. Aluminum may even reach toxic levels in some acid soils. Other nutrients (e.g., calcium and magnesium), increase in solubility as pH increases. Extremes of soil pH may also affect plant growth indirectly by suppressing bacterial growth.

Soil Profile Development

In wet temperate regions, such as those that support dense coniferous forests, soil development is principally caused by percolating rainwater that continuously dissolves nutrient salts in the upper portion of the soil and carries them, along with particles of finely divided, chemically weathered mineral matter and bits or organic matter, downward into the lower levels of the soil. This process is called **leaching**. In time, the soil consists of a series of superimposed layers or **horizons** that differ in color, texture, and chemical attributes (Fig. 9.9).

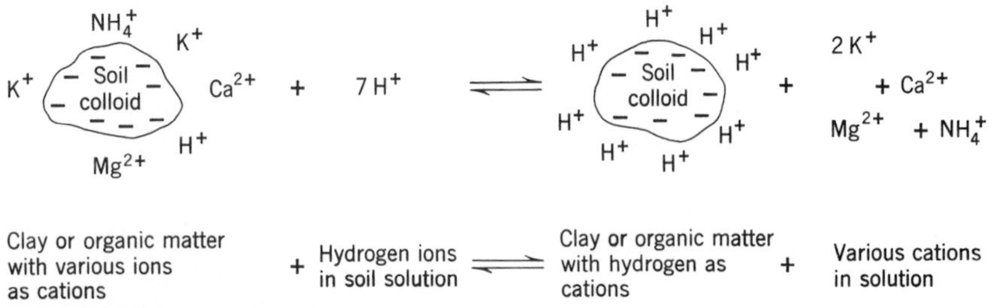

Figure 9.7 Cation exchange on soil colloids.

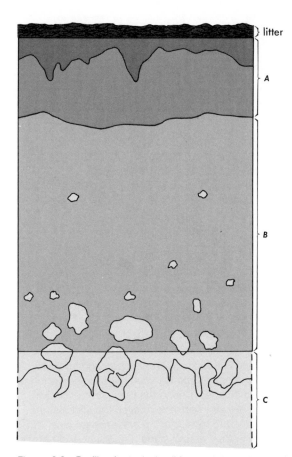

Figure 9.9 Profile of a typical soil in a wet, temperate region. The A horizon, except for a thin region just beneath the litter, is leached of organic matter, clay, and salts. These leached substances accumulate in the B horizon. The C horizon, removed from weathering processes, represents parent material from which the A and B horizons have been formed.

Table 9.3

The Derivation and General Significance of the Names of the 10 Soil Orders

Order Name	Derivation of Name	Characteristics
Entisol	Artificially created	Soils of very limited development. Profile properties largely inherited from the parent material.
Inceptisol	L. *inceptum,* beginning	Soils exhibiting moderate development. Identifying horizons are of types that form relatively quickly.
Aridisol	L. *aridus,* dry	Soils of arid regions. Limited change in parent material because of low climatic intensity.
Mollisol	L. *mollis,* soft	Soils with friable surface horizons noticeably darkened by organic matter. The degree of base saturation is high.
Spodosol	L. *spodus,* wood ash	Soils with illuvial *B2* horizons having significant accumulations of free Fe and Al oxides, noncrystalline clays, and humus; usually without structure and often partially cemented.
Alfisol	Artificially created	Soils containing a *B* horizon with significant amounts of crystalline clays and generally a moderate to high degree of base saturation.
Ultisol	L. *ultimus,* last	Soils containing a *B* horizon with significant amounts of crystalline clays but generally with a low degree of base saturation.
Oxisol	Gr. *oxid,* acid, sharp	Highly weathered soils containing a *B* horizon consisting primarily of sesquioxides or 1:1 clays.
Vertisol	L. *verto,* to turn	Soils that form large cracks on drying; self-plowing.
Histosol	Gr. *histos,* tissue	Organic soils.

The upper, or A, horizon of soil beneath a conifer forest is rich in organic matter only in the few uppermost centimeters, where freshly deposited organic litter is decomposing. The lower portion of the A horizon is sandy and light-colored, has a pH of 3 to 5, and may be deficient in nutrients, organic matter, and clay. Gradually, about 30 cm below the surface, the B horizon begins. The B horizon is often reddish-brown from the oxidized iron that has accumulated in it; it is relatively rich in mineral nutrients and clay and usually has a more neutral pH than the A horizon. The B horizon, then, is a layer of accumulation. Finally, 100 to 130 cm below the surface, the C horizon, made up of slightly weathered parent material, appears. The relative thickness of the A and B horizons depends on the climate and amount of plant cover. Soils like these are called **spodosols** (old name = podzols). There are nine other soil orders (Table 9.3), four of which are discussed below.

Alfisols and **ultisols** lie beneath the deciduous forests of the eastern United States. They are related to spodosols in that they are leached and acidic, but not as extremely as spodosals.

Mollisols (old name = chernozem) develop beneath grasslands that experience moderate rainfall and high summer temperatures. Moisture is no longer sufficient to leach the soil severely; the pH is 7 to 8. Fibrous root systems of grasses thoroughly permeate this soil, and their decay evenly enriches it with organic matter. An A horizon, containing about 10% organic matter by weight, occupies the top 60 cm. It may be almost black in the upper part, but becomes dark brown below. Just beneath the A horizon is a calcium carbonate accumulation, leached from above. The layer is sometimes called the A_c horizon. Below it is the parent material; there may be no B horizon.

Oxisols (old name = laterites) form in regions of high rainfall with continuously warm temperatures, such as those that support tropical rain forests. Mineral cycling is rapid because of fast plant growth and fast litter decomposition. Soil is acidic in pH. Removal of the forest canopy quickly leads to soil deterioration because litter decomposition no longer compensates for leaching. Excessive amounts of fertilizer must be added if the land is farmed. Silicates (sand) are leached, but clay and iron are not; the iron in clay is oxidized to Fe_2O_3, giving the topsoil a brilliant reddish color.

Fire

In the past 40 years, fire has been rediscovered as a major ecological factor. We say "rediscovered" because primitive human beings were well aware of its effect and learned to use it for their own purposes. It may well be that their most important food crops, their domestic animals, their routes of migration, and some of their cultural attributes have been molded by natural and man-made fires. Certainly, many animals are today attracted to fire and exhibit behavioral patterns in relation to fire that imply thousands of years of evolution in response to it.

Most natural fires are started by lightning, and in North America many vegetation types, including grassland, chaparral, and at least some conifer forests, owe their distribution and community structure in large measure to lightning fires. On U.S. National Forest land, during the 22-year period 1945 to 1966, lightning fires accounted for 64% of all fires, an average of 5000 fires each year. In addition, early explorers and settlers invariably commented on the frequency of fires in forests and grasslands. In the words of E. V. Komerek, a recent investigator: "Lightning fires are an integral part of our environment and though they may vary in both time and space, they are rhythmically in tune with global weather patterns. Our environment can be called a fire environment."*

One of the first vegetation types shown to be dependent on frequent fire for its maintenance in this country is the pine savannah along the southeastern coastal plain (Fig. 9.10). Tall, scattered loblolly, slash, shortleaf, and longleaf pines dominate the area, and a thick growth of grasses (mainly *Andropogon* and *Aristida*), with some herbs, cover the ground beneath. During the nineteenth century when this vegetation was protected from fire, it was noticed that the pines gave way to oak; in time the result was a thick oak forest with little grass. As the pines were valuable for lumber and turpentine, and the grass for forage, landowners were concerned about this trend. But the importance of fire was not conclusively demonstrated until 1930, with longleaf pine *(Pinus palustris)*.

---

* E. V. Komarek, Sr. 1968, The nature of lightning fires, pp. 5–41. In *Proceedings of the 7th Tall Timbers Fire Ecology Conference,* Tall Timbers Research Station, Tallahassee, Florida.

Figure 9.10 The pine savannah of the southeastern coastal plain. The pine shown is longleaf *(Pinus palustris);* the area is in North Carolina.

Longleaf pine is tolerant to fire, and dependent on it. The seeds germinate in the fall soon after they drop to the ground from the cones. During their first year of growth, the seedlings are very sensitive to even the slightest fire. However, during the next 2 to 4 years, longleaf pine seedlings are in the "grass" stage (Fig. 9.11). Most of the growing is done by the roots, and the stem apex remains close to the ground. The terminal bud is covered with a dense mat of hairs that protects it from surface fires that may sweep the area during this time. Fire is even beneficial at this stage, because it kills a particular fungal disease that parasitizes the long needles.

By the time the seedling is 3 to 5 years old, a surface fire is desirable to remove grasses that may have covered

Figure 9.11 Close-up of longleaf pine seedling in the "grass" stage.

the seedlings and shaded it from the sun. The seedling will then make a spurt of stem growth so rapid, that by the age of 8 to 9 years the young tree's canopy will be high enough above the ground to be out of reach of surface fires. In addition, its thick bark can endure the heat of a fire without permitting damage to the cambium. If fire is kept from such an area for 15 years, grasses and young oak and pine saplings of other species become so dense that reproduction of longleaf pine is suppressed. If fire continues to be kept out beyond that point, oak trees gradually replace the pines and form a closed forest.

The standard practice today is to burn the grass and pine debris about every four years. The grass is quickly able to regenerate itself, and pine reproduction is favored, while oak seedlings are killed. If longer periods go by without fire, too much dry litter collects on the ground, increasing the hazard of setting a fire so hot that mature trees are damaged. Of course, no matter how "cool" the fire, great care is taken to control it. The cost for control burning amounts to considerably less than $1 per acre.

On the west coast recognition of the importance of controlled burning lagged behind its perception on the east coast. In California, the absence of fire in some areas has produced potentially catastrophic conditions, especially in the mixed-conifer forest of the Sierra Nevada mountains. Before settlement by Europeans, natural fires swept these forests about once every eight years (as revealed when fire scars in tree trunks are dated by ring count). However, those lightning fires did not create raging, rampaging, extremely hot fires, because there was only a relatively small amount of litter on the ground. The appearance of those fire-adapted forests was stately and open, according to early reports by travelers. As shown in Fig. 9.12, however, much of the mixed-conifer belt of the Sierra Nevada (about 1500 m elevation) today is no longer stately and open. We now have evidence that in the absence of fire, these forests are transformed into crowded fir (Abies) and incense-cedar (Calocedrus) forests, and that the longer fire is excluded the more catastrophic is the eventual, inevitable fire.

California experiences a high frequency of lightning strikes during dry, late summer, and it is impossible to prevent fires from starting. A considerable amount of litter—needles, twigs, and bark—is shed each year. So long as fires move through an area frequently, say every 8 to 25 years, the fire does not become hot enough to reach the upper canopy and significantly damage the trees. The native trees are adapted to such frequent, light burns. The seedlings of many of them require contact with mineral soil in order to become established because litter does not retain moisture. Fire, of course, clears the ground of litter. In controlled experiments, seedlings of ponderosa pine, Jeffery pine, and big tree are much more abundant on burned plots than on unburned plots. If the seedlings are protected from fire for 10 to 15 years, they become tall enough and develop a bark that is thick enough to withstand a light surface fire. However, should fire be excluded for much longer periods, so much brush and litter collects that even mature trees cannot withstand the flames that eventually come. Many magnificent stands of big tree (Sequoiadendron)—the most massive tree in the world, and found nowhere else in the world—are now, for this reason, in great danger of being destroyed by fire (Fig. 9.13).

Biological Factors

It is typical for several plant and animal species to coexist in a given habitat. Very rarely does a single species, to the

Figure 9.12   Thick understory of shade-tolerant trees that has developed in a Sierran mixed-conifer forest as a result of fire exclusion.

Figure 9.13 A grove of giant sequoia, or big trees
(Sequoiadendron gigantea) in Yosemite National Park, California.

exclusion of all others, control a habitat. If all the species in a group are individually examined, they will be seen to utilize different parts of the environment or alternate with each other in time. Some plants (green plants) are producers of carbohydrates, but others (parasitic and saprophytic fungi) are consumers or decomposers of it; some animals feed on plants, others feed on insects, and others feed on small mammals; some plants are trees that utilize full sunlight, others like ferns utilize weak light; some animals are nocturnal, others are diurnal; some plants are evergreen, others deciduous; and so on. The portion of the environment utilized by each species is called its **niche**. The niche has often been called the "occupational address" of a species. In a particular area at a given time it is thought that one and only one species can occupy or fill each niche. Two species that occupy similar niches compete strongly with each other.

Competition is one form of biological interaction. It may be defined as the decreased growth of two species or plants because of an insufficient supply of some necessary factor(s) (Table 9.4). Sometimes only a single factor, such as a mineral element, is lacking, but usually two or more factors are limiting and it is difficult to separate them and determine which one creates the greater competition stress. Competition is very important in determining plant distribution. Many species restricted to saline, dry, or nutritionally poor soils would actually grow better on "normal" soil if other plants were removed. These species are restricted to their niches because they are poor competitors in comparison with the numerous plants that populate more moderate niches. They will die because the other plants have faster root and shoot growth rates, and remove water and sunlight. Sometimes, introduced plants become widespread pests and actually replace native species because they are better competitors than the native species. This seems to be the reason for the enormous increase of cheatgrass *(Bromus tectorum)* in the intermountain region of the United States.

Amensalism, another form of biological interaction, may be defined as the inhibition of one species by another. Whereas competition results from the removal of a resource, amensalism results from the addition of something to the environment. In the coastal hills of southern California, sage shrubs *(Salvia)* cover the slopes, and grasses carpet the valleys. Occasional pockets of shrubs occur in the grassland. Figure 9.14 is an aerial view of these pockets of shrubs. The ground beneath and about the shrubs is bare of grass, and the grass is stunted as far as 9 m from the shrubs. Since the zone of stunting is well beyond the limits of shrub root growth, competition between shrubs and grass for soil moisture is not the cause of stunting. If clumps of grass are transplanted to the bare zone, their growth is severely retarded. Analysis of the soil beneath the shrubs, the bare zone, or the grasses, shows no major differences in physical or chemical attributes. The reason for this distribution is that many volatile oils are emitted from sage leaves, and two of the oils in particular—cineole and camphor—are very toxic to grass seedlings. When both fresh sage leaves and grass seeds are placed in a closed chamber, germination is often reduced to zero, and radicle growth is considerably reduced. These same toxins can be isolated from the air around sage shrubs, from the natural soil, and from seed coats of grasses in the soil. The conclusion of some ecologists is that the bare areas are caused by toxins emitted by *Salvia* that inhibit the germination and growth of grasses. More recent studies indicate that a high intensity of foraging by birds and small mammals, which nest in the shrubs, also contribute to maintaining the bare zones.

Amensalism by chemical means may be very common in nature. This happens because plants are leaky systems, passively contributing all sorts of substances to their environment. One investigator grew seedlings of 150 species in a nutrient culture that included several radioactive elements as markers. The plants took up the isotopes through their roots and transported them to all parts of the plant, including the leaves. The plants were then exposed to a mist, and the water that condensed and

Table 9.4
Some categories of Biological Interaction in Terms of the Effect the Relationship Has on Each Partner. "On" Indicates that the Two Are Not in Contact. No Effect = 0, Stimulation = +, Depression = −

| Name of Interaction | Effect | | | |
| | On | | Off | |
	#1	#2	#3	#4
Neutralism	0	0	0	0
Competition	−	−	0	0
Mutualism	+	+	−	−
Protocooperation	+	+	0	0
Commensalism	+	0	−	0
Amensalism	0	−	0	0
Parasitism/herbivory	+	−	−	0

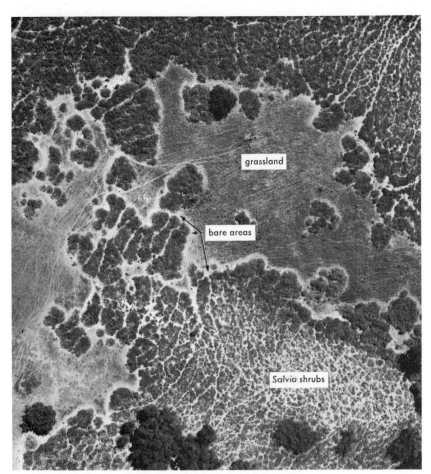

Figure 9.14 Aerial photo of *Salvia* shrubs adjoining grassland. The light bands between and around the shrubs are bare of most plant cover.

ran off the leaves was collected for analysis. He found that 14 elements, 7 sugars, 23 amino acids, and 15 organic acids—all radioactive—had been leached from the plants. In nature, these substances would have been leached by rain from the leaves and would have accumulated in the soil. Other studies have shown that roots are similarly leaky.

Herbivory, another form of interaction, plays an obvious part in plant distribution. A glance along pasture fences reveals that normally abundant but palatable plants are absent from the pasture, yet are common outside it; weedy species, which are not palatable, are common in the pasture, yet are rare outside it. Grazing animals are responsible for this difference. Plant defenses against herbivores include: (1) a spatial distribution that makes it difficult for herbivores to locate and damage host plants; (2) a life cycle timing, such as seasonal leaf drop, that makes the host plant unavailable to herbivores; and (3) the manufacture of substances that are distasteful or otherwise inhibit herbivores from feeding on host plants. Parasitism is herbivory by a plant rather than by an animal, but the result and type of interaction are the same.

Animal pollinators form a **mutualistic** relationship with plants. The bee, moth, and beetle, which cross-pollinate flowers while feeding on nectar or pollen, are familiar examples. Lichens are mutualistic associations of fungi and lichens and mycorrhizae are mutualistic associations of fungi and higher plants. Epiphytes, mentioned earlier, form a **commensal** relationship with the host tree. **Protocooperation** is exhibited by adjacent members of some species that form natural root grafts, thus sharing soil resources.

Ecology and Plant Populations

Ecologists would like to use species as indicators of particular environments. They would like to know, for example, that a certain species only grows in a certain range of temperature or moisture. Then the temperature and moisture patterns of an area could be accurately predicted by just noting whether that plant species grows there. Unfortunately, most species are tolerant of a wide range of environments and they are not genetically uniform. Individuals of species X that grow in one environment are slightly different from others of species X that grow in another environment. Their genetic variation makes it possible for them to compete successfully in several different environments.

In the early twentieth century, the Swedish botanist Göte Turesson collected seeds of species that had a wide

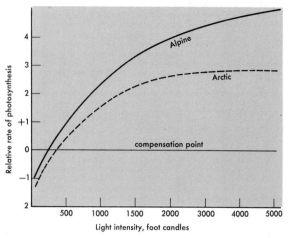

Figure 9.15 Photosynthetic response to different light intensities for alpine and arctic ecotypes of alpine sorrel (Oxyria digyna).

range of habitats—from lowland, southern, and central Europe to northeastern Russia in the Ural Mountains. When he germinated the seed and grew the plants in a uniform garden in Akarp, Sweden, he found that members of widespread species were not uniform. Plants that grew from seeds collected in warm, lowland sites often were taller and flowered later in the year than those that grew from seeds collected in cold, northern sites. Yet these variants had very similar flower morphology and were interfertile, traits indicating that the variants were all part of one species. Turesson called these variants within species **ecotypes**. He hypothesized that they were ecologically and genetically adapted to particular environments.

Evidence that has since accumulated indicates that most wide-ranging species are composed of a continuum of ecotypes, each differing slightly in morphology and/or physiology. For example, alpine sorrel (Oxyria digyna) is a small perennial herb that grows in rocky places above timberline in mountains and north of timberline in the arctic. Both the alpine and arctic environments experience long periods of freezing weather, but alpine areas receive

Figure 9.16 A community composed of essentially a single species: spinifex, or porcupine grass (Triodia basedowii) in central Australia.

much higher light intensities during the growing season than do arctic areas. Physiological ecotypes have resulted. For example, the optimum light intensity for photosynthesis in alpine plants is much higher than that for arctic plants (Fig. 9.15). The two ecotypes also differ morphologically in leaf color, leaf shape, and the frequency of rhizome production.

Applications of the ecotype concept have been especially useful in forestry. The U.S. Forest Service conducts tests to determine the best ecotype of a species to be used for reforestation in a given area. Seed is collected throughout a species' range and is germinated in a uniform garden; seedling growth is carefully watched, and the ecotype best suited for the garden climate is selected for nearby reforestation programs.

Ecology and Plant Communities

Some environments support only one species. Dense clusters of cattail (Typha) grow in water-filled ditches, and all other vascular species are excluded; widely spaced spinifex (Triodia) covers hundreds of square km of arid Australia, and few other perennial species are present (Fig. 9.16). Most environments, however, support groups of species associated together in **communities**.

Provided that distances are not too great, a given community repeats itself wherever the environment is suitable. For example, high peaks in the Smoky Mountains of North Carolina all exhibit red spruce (Picea rubens), mountain ash (Sorbus americana), Fraser fir (Abies fraseri) and a ground carpeted with several moss and fern species (Fig. 9.17). The spruce, fir, ash, mosses, and ferns are important members (but not the only ones) of this high-elevation forest. High elevations at different latitudes or on different continents support other species, so there are other types of high-elevation forest communities. There are hundreds of nonforest communities as well. Along slopes of the California Coast Range at moderate elevation, a shrubby community of scrub oak (Quercus dumosa), manzanita (Arctostaphylos), and Ceanothus repeats itself. Black spruce (Picea mariana), Sphagnum moss, and the shrub Labrador tea (Ledum palustre) are repeatedly associated together in virtually every bog across Canada.

Communities are named after their most common large species—the plants that receive the full force of the macroenvironment. The spruce and fir of the high-mountain community in the Smoky Mountains occur in greater numbers than does the mountain ash, and their leaf canopies influence the microenvironment in which the carpet of ferns and mosses grow. The spruce and fir are said to **dominate** the community, and the community name is **spruce-fir**. For similar reasons, we have **black spruce** communities and **scrub oak** communities.

Communities exhibit a unique architecture or structure, produced by the leaf canopies of their dominant and subordinate species. In the tropical rain forest of British Guiana, one group of species forms a canopy at 30 m, a second group at 18 m, and a third at 9 m (Fig. 9.18). The

Figure 9.17 Subalpine forest in the mountains of North Carolina. The trees are red spruce *(Picea rubens)* and fraser fir *(Abies fraseri).*

ground floor is almost bare of shrubs or herbs. In deciduous forest communities of the United States, there is one tree canopy layer at about 18 m, a layer of short trees (like dogwood) at 3 m, a rich shrub layer at 1 m, and a seasonally present layer of herbs on the ground. Some desert communities exhibit only a single (shrub) layer.

Fossil impressions millions of years old sometimes reveal communities very similar to those that exist today.

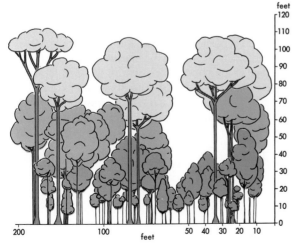

Figure 9.18 Profile diagram of the architecture of a tropical rain forest in British Guiana. There are three tree strata, at approximately 30, 18, and 9 m. Shrubs and ground herbs are uncommon, but tree saplings are included in the 9 m stratum.

These fossil communities are so similar to modern communities that we are able to reconstruct past climates from them. A fossil community near the town of Goshen in central Oregon, dated to the Eocene epoch (50 million years ago), included the following genera: *Drimys, Magnolia,* holly *(Ilex),* oak *(Quercus), Ocotea,* and fig *(Ficus).* The closest living approximation of this fossil community today occurs 2400 km south of it in the temperate uplands of Central America. Apparently, the climate in North America is now much colder than it was in the Eocene. The reconstruction of past vegetation types and climates by the use of fossil evidence is called **paleoecology.**

Another form of fossil evidence is pollen that has been trapped in lake sediments. As pollen is shed near a lake, some sinks to the bottom and is incorporated with silt and organic matter that forms sediment. A chronological sequence of pollen is thus preserved; the deeper the pollen occurs in the sediment, the older the pollen. Pollen of many species is resistant to decay under such anaerobic conditions, and may remain intact for thousands of years. Cores of sediments can be examined under the microscope and the pollen identified as to genus, and sometimes even to species. Within limits, paleoecologists assume that the abundance of pollen in the core is proportional to the abundance of that species in the past.

How May Communities Be Measured?

If communities are to be useful as tools with which to estimate the nature of the environment, they must be described more completely than just by their name, because minor variations in species composition indicate different environments. For example, in areas where spruce and fir are both dominant and very abundant, the environment is probably different from where they are still dominant but less abundant. How can the abundance of a species or the composition of a community be measured?

1. One way to characterize a community is to measure the amount of ground covered by each species. The amount of ground cover is important because it affects the microenvironment of the community. To sample the ground cover, one could stretch a tape measure across the ground; the length of tape covered by each plant canopy is proportional to the total ground cover of each species. This method of determining ground cover is called a **line transect.** Numerical information from line transects often contradicts first impressions. In desert shrub communities, for example, a person looking across the landscape would think that the shrubs covered much of the ground (Fig. 9.19), but such a estimate is in error because of the low perspective. If seen from above, when the canopy edges can be projected perpendicularly to the ground, it is apparent that actually only 10 to 20% of the ground is covered. Forest communities, on the other hand, exhibit more than 100% cover because of overlap in canopy layers at different levels.

2. Another method of estimating cover, which is especially useful when plants are low to the ground, is by the use of quadrats. **Quadrats** are variously shaped frames that

Figure 9.19 Desert scrub dominated by creosote bush *(Larrea tridentata)*, an abundant warm desert plant of North America.

Figure 9.21 A stand of longleaf pine *(Pinus palustris)*, about 50 years old, showing a well-developed understory of broad-leaved species of oak and hickory. In time, the pines will die and be replaced by maturing members of this understory.

can be placed on the ground directly over the plants. The amount of ground covered by each species can be estimated as a percentage of the area enclosed by the quadrat frame. A community can be sampled by placing quadrats at random or at regularly spaced locations throughout it. Usually, not more than 10% of the total community area is included in the quadrat samples.

3. Other methods of sampling, based on mathematical assumptions, can be used to determine whether individuals are distributed at random, regularly (like trees in an orchard), or in clumps (Fig. 9.20). Nonrandom distribution (regular or clumped) may result from propagation by asexual means, the method of seed disperal, competition, or some pattern (like terracing) in the microenvironment. Most temperate zone species are distributed in clumps.

Succession

As we have already mentioned, the microenvironment beneath a plant community is very different from that in the open. Temperature, humidity, soil moisture, and light are all affected by the canopy. A stable community consists of species whose seedlings can survive in its unique microenvironment; seedlings of other species do not survive. If the stable community is removed by some catastrophe, leaving the soil exposed to full sunlight, the first species that colonize the site are not those of the old community. They are seedlings of species adapted to growth in full light intensity.

In parts of the southeastern United States where oak–hickory communities are stable, the first species to

invade following disturbance are annual and perennial herbs, which grow best in high light intensity, such as horseweed *(Conyza canadensis), Aster,* and broomsedge *(Andropogon virginicus)*. These make up the **pioneer** community. Pine seeds blow in from neighboring areas during the first several years after the disturbance, and within five years there are many pine seedlings. As the seedlings grow, they compete with the herbs for moisture, and their expanding canopies begin to change the microenvironment. Within 30 years of the disturbance, a tall stand of pine results. Examination of the forest floor, however, shows many oak and hickory seedlings and very few pine seedlings. Pine seedlings grow poorly in shade and under conditions of root competition, but those of oak and hickory do well. Within 50 years of the disturbance, the hardwood seedlings form a well-defined understory beneath the pines (Fig. 9.21). Whenever a pine tree dies, it is replaced in the upper canopy by an oak or a hickory. Within 200 years of the disturbance, the forest again consists only of mature oak and hickory trees, and the forest floor is covered with many seedlings of the same

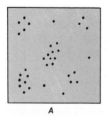
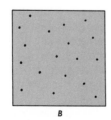
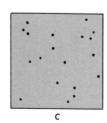

Figure 9.20 Diagrammatic representation of three types of plant distribution: *A,* clumped; *B,* regular; *C* random.

trees. The shrubs and herbs associated with an oak-hickory community are also present. Such a sequence of change in plant communities is called a **succession** or **sere**, and the stable community that is capable of indefinitely maintaining itself is called a **climax community**. In this case, oak-hickory is the climax community. All the intermediate, temporary communities are called **seral stages**.

Sometimes, climax communities are prevented from becoming established because of a recurring disturbance. There is some evidence that part of the prairie of the central United States that greeted pioneers could have been a forest were it not for recurring prairie fires. The fires destroyed slow growing tree seedlings, but grasses quickly regenerated or germinated. When fire is controlled, prairie near the forest edge soon supports many tree seedlings.

The first ecologists to elaborate successional pathways—men like Josef Paczoski in Europe and Frederick Clements in the United States—had no experimental evidence for documentation. They could not record changes as time passed, for most successions take hundreds of years to complete. They determined the existence of successional pathways by comparing a climax community with adjacent small areas that had been disturbed at known times in the past. They examined freshly disturbed or denuded areas (as a result of, for example, ice, landslides, fire, logging, and wind-throw), which revealed the likely pioneer community, or the first stage of succession. Sites that had been disturbed at some time earlier (verified from records of fires, grazing, or logging) exhibited later successional communities because the passage of time was greater. Complete sequences of succession were sometimes represented as bands of different communities about a gradually filling bog or behind a retreating glacier (Fig. 9.22). The community closest to the lake or glacier was the most recently established and represented the pioneer community. The next closest community occupied soil that had been exposed for a longer time; at some earlier time it had supported the pioneer community, but enough time had passed for the next successional stage to appear. Beyond the second community was a third, representing the third successional stage, and so on until the climax community was reached. Ring counts of trees closest to the pond or glacier, coupled with estimates of the rate of silting or of glacial movement, gave clues as to the time necessary for each successional stage to be reached.

The conclusions of these early workers have yet to be experimentally or empirically verified, and only recently have some of the causes of succession (such as competition, amensalism, and growth requirements of seedlings) been experimentally dealt with.

Vegetation Types of the World

The term **vegetation** refers to the life form of plants dominating a community: trees, shrubs, or herbs; deciduous or evergreen; coniferous or broadleaved. A community dominated by evergreen angiosperms belongs

Figure 9.22 Succession of communities about a pond near Juneau, Alaska. Standing water in the pond supports leaves of cow lily *(Nuphar)*; mats of *Sphagnum* moss and sedge *(Carex)* encroach on the pond along its edge; beyond is a shrub-dominated community of *Empetrum, Cassiope,* and *Kalmia,* with some mountain hemlock trees *(Tsuga mertensiana)*; finally, sitka spruce *(Picea sitchensis)*, a tall, dark-green tree, dominates the rest of the area.

to a different vegetation type than one dominated by shrubs or one dominated by conifers. On the other hand, many different communities dominated by different conifers belong to the same vegetation type. We will briefly examine six major vegetation types: tundra, taiga, deciduous forest, tropical rain forest, savannah–prairie, and scrub.

Tundra

If one were to travel north or south toward the poles, it would be evident that the trees gradually become stunted and less common, and finally disappear completely. At the same time, low shrubs, perennial herbs, and grasses become dominant. The resulting meadowlike vegetation is called **tundra** (Fig. 9.23). Annual plants are very rare, and the reproduction of perennials is chiefly by vegetative means such as rhizomes. Shrubs are dwarfed, gaining normal height only in the protection of the lee of boulders or small hills. The winter wind, carrying ice, acts like a sand blast and prunes back stems wherever they project beyond the boulder or hill.

The warmest month has an average daily mean

Figure 9.23 Alpine tundra, Beartooth Plateau, Wyoming, 3300 m elevation.

temperature of 10°C or less and the growing season is short. Only 2 to 4 months of the year have average daily temperatures above freezing. Approximately the top 30 cm of soil thaws during the growing season and roots may freely penetrate it, but below this level soil water may be permanently frozen; this soil is called **permafrost**. Annual precipitation is about 25 cm, very little falling as snow. The terrain is generally flat, but small depressions are common and water tends to stand in them during the growing season. Day length fluctuates from 24 hr in June (northern hemisphere) to none at all in December.

A similar tundra vegetation occurs in mountains at elevations above treeline. The climate, however, is slightly different. Solar radiation is more intense, day length does not fluctuate so extremely, and frosts are more common during the growing season. Many plants are tinged with red from abundant anthocyanin. This substance absorbs some of the high light intensities that may damage the plant's photosynthetic apparatus.

The lower limit of these alpine areas depends on latitude. At 60°N in Alaska, treeline is at 900 m; at 45°N in the Rocky Mountains, it is at 3000 m; at 20°N in central Mexico, it is at 4000 m; at 5°S in the Andes of South America, it is back to 3600 m; and at 50°S in Chile, it is back to 900 m. Because summer temperatures are cooler near the coast, timberline also depends on distance from the ocean. Thus, timberline is lower in the Cascades than in the Rockies.

Taiga

Just south of the tundra is the **taiga**, a broad belt of communities dominated by conifers. A similar belt occurs in mountains, just below the alpine zone (Fig. 9.24). Trees are slender, short (15–20 m), and relatively short-lived (less than 300 years), but the forests are dense. The taiga is made up of a patchwork of forests on upland sites that alternate with bogs formed in depressions. Soils are acidic and relatively infertile spodosols; in northernmost or

Figure 9.24 Taiga, northern Canada.

Figure 9.25 Coniferous forest, Olympic Peninsula, Washington. Note the moss and lichen-covered trunks and branches and the dense carpet of ground vegetation.

uppermost parts of the taiga the soil has permafrost below a surface layer.

The growing season is 3 to 5 months long, and temperatures above 30°C are occasionally reached. Winter temperatures are very severe; the mean daily temperature is below freezing for six months of the year, and at Yakutsk, Siberia, the average January daily temperature is −43°C. Annual precipitation is less than 50 cm, and the air is exceptionally dry. The taiga is monotonous and silent; animal life is not abundant.

At lower elevations in some mountains, and along the coast (where temperature extremes are not so severe), coniferous forests are much more luxurious. Trees are taller and the undergrowth denser. Many favorite recreation areas of the western United States are situated in these richer forests. The Olympic Peninsula of Washington receives over 200 cm of annual precipitation and supports the most spectacular coniferous forest in the world (Fig. 9.25). Douglas fir, the most heavily cut lumber species of the United States, reaches its best growth in this area.

Deciduous Forest

In contrast to the taiga, the **deciduous forest** supports a diversity of plant life and is vibrant with animal activity. Deciduous angiosperm trees dominate communities of this forest. When day length becomes shorter and nights cooler in late summer, hormone concentrations in such trees change, and the destruction of chlorophyll follows, allowing other pigments to show through in brilliant fall colors of red, yellow, and orange. In spring, before new leaves have fully expanded from buds, beautiful and distinctive herbaceous perennials come into flower. Shrubs are common, but not dense.

These forests occur in areas with a cold (though not severe) winter, and a warm, humid summer. Snowfall may be heavy, but most of the annual precipitation falls as

Figure 9.26 Tropical rain forest, New Guinea. The dense, lower vegetation is absent farther into the forest. Note the hanging lianas.

summer rain. Annual precipitation ranges from 75 to 125 cm. Since the growing season may be as long as 200 days, it is not surprising that much of the deciduous forest in Europe, Asia, and North America has been cut and replaced with cropland.

Soils are alfisols in the north and west, ultisols in the south. (The latter have striking yellow or red colors in the B horizon, which is often exposed if plowed.)

Tropical Rain Forest

No other vegetation can quite equal the **rain forest** in sheer diversity of species and complexity of community structure (Fig. 9.26). The deciduous forest may exhibit 5 to 25 different tree species per acre, but the tropical rain forest supports 20 to 50. The tree canopies typically form three layers, and beneath this dense overstory there are very few shrubs and herbs. Light intensity on the ground is less than 1% of full sun. Most of the herbaceous plants grow as epiphytes on trunks and branches of trees. Vines cling to tree trunks and wind their way up to the canopies. Seeds of a few species, such as the well-named strangler fig, germinate in the canopy and grow down to the soil. Many of the tree trunks flare out at the base in peculiar buttresses.

The species are evergreen. A given leaf may be shed or a given bud may break at any time of the year. The leaves tend to be very large, with long, tapering tips that may serve to drain the surface of moisture. Rainfall is high, often over 250 cm a year, but it is evenly distributed throughout the year. Temperature is also uniform throughout the year, averaging 27°C. Warm temperatures and high humidity couple to produce a climate oppressive to humans. Afternoon showers, which reduce the temperature, offer only short relief.

Farming peoples of the rain forest have for centuries practiced agriculture on a cut-burn-cultivate-abandon plan. The vegetation is removed in a small area by cutting and burning, cultivation follows for a few years, and the plot is abandoned. Abandonment follows because crop growth declines each year. The rain leaches nutrients from the exposed topsoil and the nutrients are not replaced by litter as they are beneath the normal forest canopy, as mentioned earlier in this chapter. Much of the rain forest of Asia and Africa has been disturbed in this way. More recent disturbance involves bulldozing and clear-cutting, especially in South America; the rain forest may be incapable of reinvading these areas. The vegetation that invades these abandoned plots is shrubby, dense, and often includes bamboo. It is this successional stage, which eventually will give way to rain forest, that serves as the "jungle" of television and films. Soils are typically in the oxisol order.

Savannah and Prairie

Forests give way to grassland as rainfall decreases and becomes more seasonally distributed. At first, trees remain close enough to retain a closed canopy, but gradually they become more widely spaced, and eventually they drop out altogether except along river banks and canyons. **Savannah** is the term applied to grassland with widely spaced trees whose canopies cover less than 30% of the ground (Fig. 9.27); **steppe** or **prairie** refers to grassland without trees (Fig. 9.28).

Grasslands of the United States are heavily used for agriculture and grazing purposes, and little of the original

Figure 9.28   The original steppe, or prairie vegetation of central California, dominated by the perennial bunch grass *Stipa pulchra*.

vegetation or animal life remains today. Only from the records of explorers and early settlers are ecologists able to reconstruct in their imagination what these thousands of square miles once looked like.

The eastern part, with relatively high rainfall, supported a dense "sea" of grass some 1 to 2 m tall. Here is a description from a journal dated 1837:

The view from this mound . . . beggars all description. An ocean of prairie surrounds the spectator whose vision is not limited to less than 30 or 40 miles. This great sea of verdure is interspersed with delightfully varying undulations, like the vast swells of ocean, and every here and there, sinking in the hollows or cresting the swells, appears spots of trees, as if planted by the hand of Art for the purpose of ornamenting this naturally splendid scene.

This is now the corn belt and is probably one of the most agriculturally fertile areas of the world. Productivity is so high that it permits us to export grain to many countries around the world. Its mollisol soils are dark brown in color from the rich amount of organic matter left by decaying fibrous grass roots.

The drier, western part supported shorter and sparser stands of grass, but there was still enough growth to allow graziers to cut and bale it as hay. Today, the western

Figure 9.27   African savannah in Kenya. The flat-topped trees are *Commiphora*, which is the source of the biblical myrrh.

prairie supports a shorter, sparser cover of grass. Reasons for such a decline include too much grazing pressure by herds of cattle and a slight drying trend in the climate of the area over the past 75 years. Shrubs from the desert and junipers from the foothills have invaded (possibly because of fire suppression), and they have further reduced grass cover.

Grasslands are the home of the largest terrestrial herbivores and carnivores. Vast herds of bison that roamed the North American prairie are gone, but animal preserves in African savannahs still excite visitors by a spectacular animal cast that includes eland, wildebeest, giraffe, elephant, and lion.

Scrub

Scrub is vegetation dominated by shrubs, and it occurs in areas with hot, dry summers and moderate winters. In areas with high, but seasonal, precipitation, the shrubs are tall and thorny, but if rainfall is only 40 to 50 cm a year, scrub can be a nearly impenetrable thicket of shrubs with evergreen, hard, spiny, small leaves (Fig. 9.29). Scrub of this sort is found around the Mediterranean Sea, on coastal hills of California and Chile, in southwestern Australia, and at the tip of South Africa. Even though the species of plants in these widely separate areas are entirely different, the vegetation has come to look strikingly similar. In the western U.S., this type of scrub is called **chaparral**.

Desert scrub (Figs. 9.19 and 9.30) grows in areas of rainfall below 25 cm a year. Shrubs are widely spaced and have far-reaching root systems (much of which runs just beneath the soil surface). The shrubs may be associated with **succulents,** plants that store water in stems or leaves, like the common cactus of the southwestern United States. Some shrubs have deep roots that tap permanent supplies of ground water, while others reduce transpiration by dropping their leaves every dry period and developing new ones every rainy period. Seeds of desert annuals may remain viable for long periods in the soil until a favorable year, when rainfall is sufficient to trigger

Figure 9.30 Desert scrub in southern Arizona with succulents (cacti).

germination and a colorful show of flowers a few weeks or months later. Although desert scrub seems to be made up mainly of perennial species, an actual count of all species appearing throughout the year reveals nearly as many annuals as perennials.

A Final Note

We have described major vegetation types as if they have sharp boundaries with, and differ greatly from, all other types. Actually, vegetation types typically grade into one another or form complex mosaics that may extend for

Figure 9.29 Chaparral in the foothills of California.

thousands of hectares. Meeting zones of two or more vegetation types are called **ecotones**.

Ecology and Ecosystem

Energy Flow

We have examined plant ecology at the population, community, and vegetation levels. A final level of organization is the ecosystem. The ecosystem of a particular habitat includes the plant and animal communities, the physical (nonliving) environment, and all the interactions between them.

The organisms of an ecosystem can be divided into three categories: producers, consumers, and decomposers. **Producers** are green plants and certain bacteria that perform photosynthesis. **Consumers** are parasitic, herbivorous, or carnivorous plants and animals that feed on other plants or animals. **Decomposers** are saprophytic bacteria, fungi, and certain animals that obtain nutrition by breaking down dead organic matter. The three groups form a complex web of interdependence: consumers feed on producers, and decomposers feed on the organic remains of both and convert them to lower-energy, smaller molecules that are taken up again by producers.

Caloric energy is also passed through the ecosystem, and in this transfer organisms are linked to each other by a **food chain**. Food chains are short in the arctic tundra: meadow plants trap light energy and convert it to chemical energy (e.g., proteins or carbohydrates); caribou transfer some of this energy to themselves, retaining some of it, excreting some, and respiring some; and human beings transfer some of the caribou caloric energy by eating a fraction of the total caribou. Aquatic ecosystems exhibit longer food chains: algae (plankton especially) produce the caloric energy; some is passed to the small animals that feed on plankton, then to the small fish that eat the plankton feeders, then to larger fish, and then to shore birds or land animals like bear or man.

There is tremendous variation in net productivity from one ecosystem to another. **Net productivity** equals calories produced in photosynthesis minus calories lost in respiration. Desert ecosystems, for example, exhibit less than 0.5 gram of dry matter produced per square meter of land per day; grasslands, 0.5 to 3.0; deciduous and coniferous forest, 3 to 10; agricultural areas utilized all year (as in sugar cane production), 10 to 25.

Energy transfer through the ecosystem involves losses each time energy passes from one type of organism to another. Figure 9.31 summarizes a very simple food chain consisting of California rangeland on which steers are grazed. The range receives 700,000 kilocalories (kcal) of sunlight energy per square meter per year. After deducting the energy lost by reflection and heating, the fraction spent by plants on their own respiration, and other energy channeled into roots and other unconsumable parts of the plants, only 800 kilocalories are taken in by the grazing animals. The animals in turn spend nearly all their energy unproductively (from the point of view of human beings) in

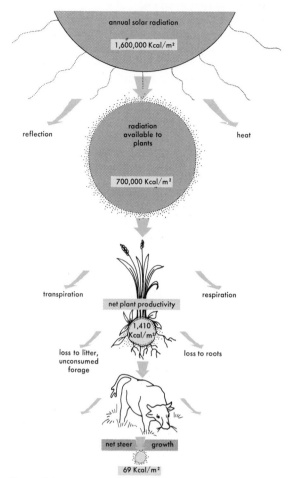

Figure 9.31   Loss of energy in western rangeland. Of 700,000 kcal of radiant energy that reach each square meter of surface, only 0.1% becomes available as plant food for the cattle grazing the rangeland. Only 0.01% of the plant food consumed during an eight month period is converted to meat available for human consumption, that is, only 0.004% of the original incident solar energy.

respiration, waste matter, and inedible bones and hide, leaving a net of only 69 kilocalories per square meter, of the original 700,000, that goes into meat. This represents about 139 kg of edible beef on the hoof per ha per year.

Pollution

Human beings add to the environment many substances whose effect on plants and animals is poorly understood (Fig. 9.41). If organisms in food chains or ecosystems shared by humans are damaged by these substances, then humans may be indirectly, but nevertheless severely, damaged.

These substances are called **pollutants**. Some are liquids or solids that have been introduced to waterways and may affect the metabolism of aquatic plants, such as algae (see Chapter 11). Others are gases released to the atmosphere; they may affect the growth of wild or cultivated terrestrial plants. Among the most important of these air-borne pollutants are sulfur dioxide (SO_2), ozone (O_3), and peroxyacetyl nitrate ($C_2H_3O_5N$, also called PAN).

Sulfur is present in iron, copper, lead, and zinc ores;

during refining it is released as SO_2. It is also released from the burning of coal. The SO_2 enters leaves through stomata and goes into solution on the wet mesophyll cell walls as sulfurous acid. Inside the cells, such vital metabolic processes as protein synthesis are disrupted because of the reducing nature of this acid. Ultimately, cells shrink and collapse, producing visible chlorotic lesions on the leaves. Chronic exposure of such sensitive crops as alfalfa, barley, and lettuce to only 0.1 to 0.5 ppm (parts per million) of SO_2 causes visible damage and reduced yield. The effect of SO_2 on natural vegetation is dramatically revealed in Copper Basin, Tennessee, where early in this century a refinery freely released SO_2 before its toxic nature was understood. Over an area of perhaps 40 km^2 the original forest was destroyed (Fig. 9.32). The barren land was so severely eroded that subsequent attempts to revegetate the area have been unsuccessful.

Ozone and PAN are important components of urban smog. In large part they are a result of the incomplete combustion of automobile fuel (Fig. 9.33). Our chemical understanding of smog and its affect on plants started less than 30 years ago. Ozone is naturally present in air, but smog contains 10 to 100 times the natural concentration, and this concentration is sufficient to cause millions of dollars of crop damage a year. Ozone enters through the stomata, and its oxidizing nature causes many metabolic malfunctions, including chloroplast breakdown. In the

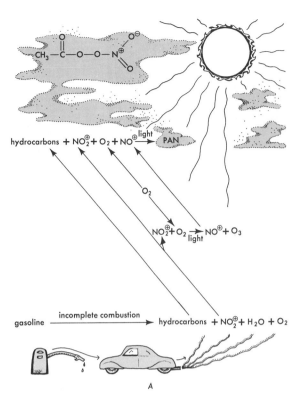

Figure 9.33 The formation of ozone (O_3) and peroxyacetal nitrate (PAN) in smog. The formula for PAN is given at the top of the diagram.

mountains near Los Angeles, air circulation takes sufficient quantities of ozone to a height of 1700 m; this causes needle drop, reduced growth, and the ultimate death of trees in ponderosa pine forests (Fig. 9.34).

Other experiments in the Los Angeles Basin have shown the effect of PAN on citrus. Established lemon and orange trees were encased in miniature greenhouses. Some greenhouses circulated normal, smoggy, outside air; others circulated air that had first been filtered of PAN. Over the period of six years, the average yearly fruit yield of lemon trees was reduced 39% by PAN, and the orange tree yield was reduced 64%. Evidently, PAN interferes with the light reactions of photosynthesis and also with respiration.

It is hoped that ecologists will eventually show us how ecosystems can be rationally manipulated to simultaneously satisfy all our needs: industry, urbanization, agriculture, conservation, and recreation. Knowledge must be gained and compromises made between our needs and what the environment can support.

Summary

1. Plant ecology is the study of how the environment influences plant distribution and behavior. It is also the study of the structure and function of nature. Ecology deals with the population, community, and ecosystem levels of organization.

2. The components of the environment include moisture, temperature, light, soil, fire, and living organisms. Each species utilizes a different part of the environment: its niche.

Figure 9.32 Copper Basin, Tennessee. The normal eastern deciduous forest was destroyed by sulfur fumes from a smelter in the early twentieth century and revegetation attempts have been unsuccessful.

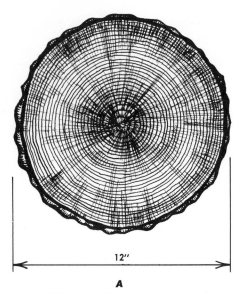
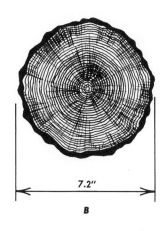

12"

A

7.2"

B

Figure 9.34 The effect of smog on ponderosa pine growth in a mixed-conifer forest near the Los Angeles Basin. *A,* cross section of a normal 30-year-old pine; *B,* the same age tree further downslope, exposed to smog-polluted air.

3. The seasonal timing of rainfall may be as important as the annual total in determining the kind of vegetation present.

4. Slope aspect affects temperature in the microenvironment. It also affects the amount of net radiation that reaches the plants.

5. Light quantity and quality is changed as light passes through a leaf canopy or through water.

6. Soil formation proceeds through mechanical and chemical weathering and through plant activity. Colloidal soil particles are important as nutrient reservoirs, and soil pH affects their ability to retain nutrients. Through time, soils in different regions develop their own characteristic profiles. Soils can be classified by profile characteristics into 10 orders and thousands of soil types.

7. Fire is an important, natural environmental factor that affects the distribution and architecture of many vegetation types.

8. Forms of biological interaction include competition, amensalism, predation, and mutualism.

9. A taxonomic species includes genetically similar plants that are interfertile. Within any wide-ranging taxonomic species, however, are many ecotypes that have become adapted to different environments.

10. A plant community is composed of a group of associated species or populations that occupy a particular type of environment. One or more of the species dominate the community, but all of the species together combine to form a unique architecture of canopy layers. The structure and composition of communities can be objectively determined by the use of sampling methods such as line transects or quadrats.

11. Climax communities perpetuate themselves, but successional communities are transitory.

12. Vegetation types of the world include tundra, taiga, deciduous forest, tropical rain forest, grassland, and scrub.

13. An ecosystem consists of the plant and animal communities of a habitat, their abiotic environment, and all the interactions between these components. Organisms of an ecosystem are tied together by food chains. Damage done to one component of a food chain (as through the action of pollution) may cause indirect but serious damage to the entire ecosystem.

14. Important atmospheric pollutants include sulfur dioxide, ozone, and peroxyacetyl nitrate. The latter two are components of smog and cause millions of dollars of crop damage a year as well as ecological damage to natural vegetation.

Plant taxonomy deals with the identification, naming, and classification of plants. The term "identification" implies that many plant groups exist, and each group possesses unique characteristics so that it can be distinct from all other groups. These characteristics, as we shall see, involve morphology, anatomy, physiology, cytology, biochemistry, and geographic distribution. The term "classification" implies that plant groups can be ranked in some hierarchical relationship, based on similarities in their characteristics that reflect genetic relationships. Taxonomists are building a library of the world's plant resources; they are making a census of the unique characteristics of each of the half-million species that make up the world's **flora.*** This census is far from complete. Many areas of the world have not been botanically explored and their floras can only be estimated.

Taxonomists hope that, as the census becomes complete, patterns of similarity will become apparent within this maze of plant diversity, and that these patterns will enable them to reconstruct the evolutionary history of the present flora. They hope it will be possible to determine which characteristics—out of the hundreds available—are most reliable in showing pathways of evolution. Are they characteristics of flower morphology, wood anatomy, chemical composition, geographic distribution, breeding behavior, or chromosome number? Which traits in each category are **primitive** (imitating those of early, now-extinct plants)? Which traits are **advanced** (derived in time from primitive traits)? When the answers to such questions are known, it will be possible to construct a **phylogenetic** or **natural classification** based on genetic, evolutionary relationships. Such a classification, in addition to summarizing the past, might prove useful in predicting future pathways of evolution.

Historical Aspects

Pre-Linnaean Period

The history of taxonomy can be very roughly divided into a pre-Linnaean (300 B.C.–1753 A.D.) period and a post-Linnaean (1753–present) period.

Before the time of written history, human beings had learned to identify many plants. These plants were particularly useful as spices, medicines, foods, drugs, or religious symbols. The written history of botany begins with the Greek Theophrastus (ca. 300 B.C.), a student of Aristotle. He described and classified about 500 kinds of plants in the book *De Historia Plantarum.* He utilized morphological characters like habit (tree, shrub, herb), length of life (annual, biennial, perennial), corolla form (petals fused together or free), and ovary position (superior, inferior) to distinguish among his plants.

Many of the taxonomists who followed the lead of Theophrastus through the dark ages and Renaissance are typified by Otto Brunfels (ca. 1500 A.D.) of Germany. He too classified plants on the basis of gross morphology, but he contributed to the record by expanding the list of known plants and using drawings to accompany many of the descriptions. Hans Weiditz was Brunfels' illustrator, and he did a magnificent job of carving accurate and intricate woodcuts from which prints were made (Fig. 10.1). The descriptions and drawings of many species were compiled in books called **herbals.** Herbaceous plants used for medicinal purposes were often emphasized in the herbals. It was common to assign medicinal properties to a plant on the basis of its form. If the plant or parts of it imitated the shape of an organ, then it was used for correcting ailments of that organ. For example, the simple plant body of liverworts, which resembles the shape of a liver, was used to treat liver ailments.

Carolus Linnaeus (ca. 1750, Fig. 10.2) represents the culmination of pre-Darwinian thought. Born Carl Linné in southern Sweden, he attended the University of Uppsala and received an M.D. degree from the University of Harderwijk in the Netherlands. During his early days at Uppsala, however, he had become interested in the classification of plants and had developed his own system, called the sexual system. It was based almost entirely on the morphology and number of stamens and pistils, and proved very efficient in classifying the many new plants then being brought to Europe from Africa, America, and Asia and propagated in botanic gardens. After working on the collections of many large botanic gardens, Linnaeus published the book *Species Plantarum* in 1753. The work included about 6000 species in 1000 genera.

Each species was described in Latin by a sentence limited to twelve words that began with the genus name. Linnaeus considered this sentence (a **polynomial**) to be

---

* Another meaning of "flora" is a book that lists all the plant species found in a given region. Thus, floras have been published for countries, states, counties, or even more local areas.

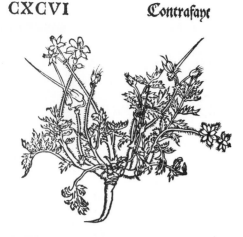

Storckensthnabel.
❡ Von dem Nammen.
Torckenschnabel/Gottes gnad/Kranch,

Figure 10.1   Part of one page from Brunfel's sixteenth century herbal, The plant illustrated is called stork's bill (probably *Erodium cicutarium*).

the official, scientific species name, but he also coined a shortened form consisting of the genus name and one additional word from the polynomial. This shortened form (a **binomial**) served as his common name for the plant. He described spiderwort by this polynomial: *Tradescantia ephemerum phalangoides tripetalum non repens*

Figure 10.2   Linnaeus (1707–1778), the Swedish botanist who initiated the binomial system of nomenclature.

Figure 10.3   *Tradescantia virginiana* (spider wort), ×2. Note the jointed staminal hairs, from which the common name derives.

virginianum gramineum, common name *Tradescantia virginiana.* Loosely translated, the polynomial means: The annual, herbaceous, upright Tradescantia from Virginia that has a grasslike habit, three petals, and [stamens with hairs that are] spider-like, common name Tradescantia of Virginia. The genus name was in honor of John Tradescant, a gardener to King Charles I.

Not surprisingly, the binomial became favored over the polynomial by later taxonomists, and was accepted as the official, scientific name. Common names were left to individual choice and not restricted to Latin. The usual English common name for *Tradescantia virginiana* is spiderwort (spider plant), which comes from the jointed hairs on stamen filaments that resemble the jointed legs of a spider (Fig. 10.3).

Post-Linnaean Period

After Darwin published *On the Origin of Species* in 1859, emphasis in taxonomy shifted from a description of species *per se* to: (a) a very selective description, based on characters that reflected genetic (evolutionary) relationships, and (b) the construction of a phylogenetic classification scheme. Construction of phylogenetic schemes demanded that decisions be made on which traits are primitive and which are advanced. As there was little or no fossil evidence to support any of several hypotheses, early schemes were both numerous and often in conflict. For example, the American C. E. Bessey

Table 10.1
Assumptions for the Besseyan Phylogenetic Scheme

Primitive Characters	Advanced Characters
Plants woody	Plants herbaceous
Flowers bisexual	Flowers unisexual
Floral axis elongated	Floral axis short
Floral parts spirally arranged	Floral parts whorled
Floral parts numerous	Floral parts few
Sepals or petals free	Sepals or petals fused
Floral symmetry radial	Floral symmetry bilateral
Fruit single	Fruit aggregated

started with the assumptions listed in Table 10.1. Although Bessey's assumptions are still widely accepted today, some 60 years after he proposed them, disagreements continue to arise over specific interpretations. Adolph Engler and Karl Prantl, living about the same time as Bessey, constructed another classification scheme from quite different assumptions.

Do species that now resemble each other represent progeny from a common ancestor, or is it possible that they started from quite different sources and simply evolved to look alike? These two possibilities are diagrammed in Fig. 10.4. It is seen that species A and B, which are very similar today, actually arose from quite different stocks. Similar environments, however, have shaped them along convergent paths. This pattern is termed **convergent evolution**. A striking example of convergent evolution is the present similarity between cactus plants of the southwestern United States and certain *Euphorbia* species of Africa. Both groups grow in hot, arid climates, and, as seen in Fig. 10.5, both have a similar morphology. They have no leaves, their stems are ribbed and swollen and contain water-storage tissue, and they are armed with spines. Yet, other traits, like flower morphology and geographic distribution, are so different that it is clear they must have arisen from different stocks.

On the other hand, B and C, which are today less similar than A and B, actually did evolve from a common ancestor. This pattern is called **divergent evolution**. As time passes, B and C will become more and more different. The same pattern will be repeated by C and D, but at this point in time they are still quite similar.

If a certain group of plants arose from a single ancestral type, the group is referred to as **monophyletic**. If a group arose from several different ancestral types, the group is referred to as **polyphyletic**. (Of course, the terms are relative, because ultimately all life can be traced back to some one, first cell that evolved in the primaeval seas.) Bessey considered each separate group of seed plants (e.g., the Anthophyta, Coniferophyta, Ginkgophyta) to be monophyletic. Some taxonomists considered all the seed plants together to be monophyletic. Still others did not think that even one group, the Anthophyta, was monophyletic. Unless fossil evidence is at hand, it is very difficult to separate the effects of convergent and divergent evolution. At present, most taxonomists agree that the

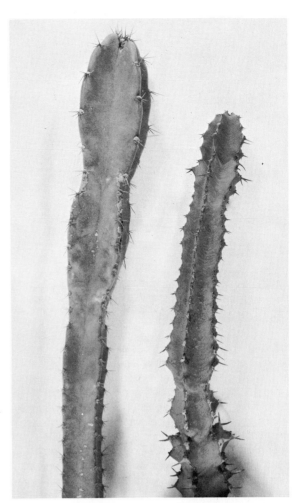

Figure 10.5 Convergent evolution of members of the Cactaceae (left) of the south-western United States and members of the Euphorbiaceae (right) of Africa, × 1/10.

largest group of plants that is definitely monophyletic is the family. (See the next section.)

At present, we still lack a classification scheme that all botanists accept. For convenience and consistency, this book follows a polyphyletic scheme by Scagel *et al.* (1967) and Bessey's views (Table 10.3).

How Are Plants Classified?

Over 550,000 species of plants clothe the earth and inhabit most of the waters to a depth of approximately 75 m. What is a species? A **species** consists of groups of morphologically and ecologically similar plants; these groups may be called **populations**. Since the populations within a species are reproductively isolated from the populations of other species, genetic information is not exchanged between them. Within a typical species, however, the plants are very similar in many aspects of their morphology and interbreed freely. Their genetic constitutions are similar, even identical; they are very closely related. Such populations of closely related, interbreeding individuals constitute a species.

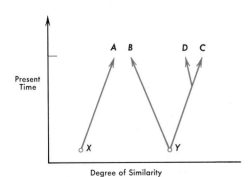

Figure 10.4 Hypothetical pathways of species evolution. X and Y represent ancestral forms from which A, B, C, and D have since evolved. The paths leading to A and B represent convergent evolution; the paths leading to B and C represent divergent evolution.

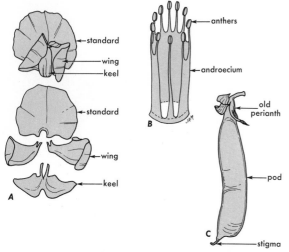

Figure 10.6 Diagram showing traits that are important in the classification of the sweet pea *(Lathyrus odoratus* L.). *A,* flower, exploded view; *B,* 10 stamens, 9 attached together and 1 free; *C,* pod with seeds.

The most basic requirement of a biological classification is that it show genetic relationships among the organisms classified. Several different species may have some characteristics in common, and an occasional mating may take place between members of two different but similar species. Such genetically related species are grouped together in **genera** (s. **genus**). A comparison of some genera shows that they have traits in common. Genera with similar traits are genetically related, but less closely than species within a genus; they are grouped in **families**. Plants in the same family have common characteristics that indicate they may all have evolved from the same ancestor. For instance, members of the pea family have flower parts in fives, pods like pea or bean pods, and dissected leaves like pea or locust leaves. Many members of the family have flowers resembling the sweet pea, with 10 stamens, nine of them grown together (Fig. 10.6).

Families are grouped together in **orders**, and orders are grouped in **classes**. At the top level it is customary to divide all plants into **divisions**. Differences in photosynthetic pigments, manner of leaf development, structure of the conducting (vascular) tissues, and methods of reproduction are the most important elements in separating plants into divisions.

Table 10.2
Sweet Pea

Classification Level	Name	Ending
Specific epithet	Odoratus	—
Genus	Lathyrus	—
Family	Fabaceae	aceae
Order	Rosales	ales
Subclass	Dicotyldeonae	ae
Class	Angiospermae	ae
Division (phylum)	Anthophyta	phyta
Kingdom	Plantae	not consistent

Thus, the word "similar," in a natural classification, means organisms with genetic constitutions indicating relationship. This system is exemplified in the complete classification of the sweet pea given in Table 10.2. Note the characteristic ending for each category.

Plant Groups

Broad groups of plants that include several divisions may be arranged. Although these categories are meaningful, they do not have taxonomic importance. For instance, all plants can be divided in two groups: those with nuclei (the **eukaryotes**, Fig. 10.7*B*) and those without nuclei (the **prokaryotes**, Fig. 10.7*A*). A large group of primitive plants, mainly aquatic and filamentous, may be set apart from other plants by a very distinctive characteristic: the sexual cells (eggs and sperms; that is, the gametes) are produced in structures not protected by a jacket of vegetative or sterile cells (Fig. 10.7*C*). These plants may be referred to collectively as **thallophytes**. The thallophytes constitute eight algal, one fungal, and a single bacterial division (Fig. 10.8). Fungi and most bacteria lack chlorophyll. The only unifying characteristics of all the remaining plants is that (1) the reproductive cells are protected by a jacket of sterile cells and (2) the embryo derives its first nourishment from the gametophyte. A small group of these plants typically lacks vascular conducting tissue. They are the **liverworts** and **mosses**; the **bryophytes** (Fig. 10.9). All other plants have developed a specialized conducting tissue and are known as **vascular plants**.

The less advanced vascular plants, called the **lower vascular plants**, do not produce seeds. Examples include the wisk fern *(Psilotum,* Fig. 10.10), horsetails *(Equisetum,* Fig. 10.11), club mosses *(Lycopodium,* Fig. 10.12), and true ferns (Fig. 10.13). These comprise four divisions.

The more advanced vascular plants bear seeds and are frequently referred to as the **seed plants**. These are separated into **gymnosperms** (Fig. 10.14), whose seeds are borne naked on a cone scale, and **angiosperms** (Fig. 10.15), whose seeds are protected within an ovary.

The most conspicuous division of gymnosperms are the **conifers** (Coniferophyta). Together with the angiosperms (division Anthophyta), they dominate the present flora. We may divide the angiosperms into two subclasses. The seeds of one subclass have two seed leaves, or **cotyledons**, as seen in the peanut or bean. These are the Dicotyledonae. The seeds of the other subclass (onions, lilies, grasses, orchids) have a single seed-leaf. These are the Monocotyledonae.

The relationship between the divisions and classes is shown by the key in Table 10.3.

Herbaria and Botanic Gardens

Several thousand new species of plants are discovered and described every year. The plant specimens may have been collected in parts of the world previously unexplored by botanists, although new species are being discovered every year in areas that botanists have traversed many

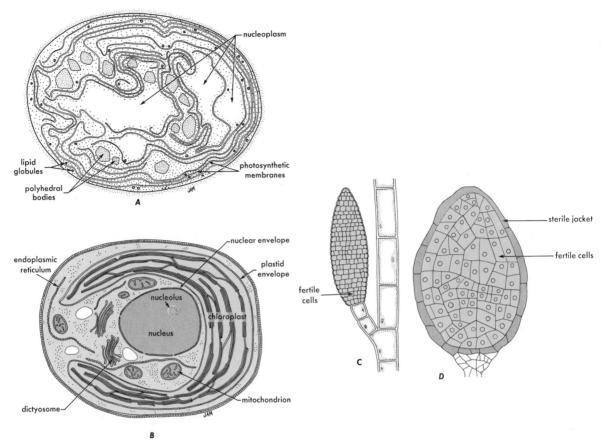

Figure 10.7 Diagrams to show differences between groups of plants. *A*, a prokaryotic cell (a blue-green alga); *B*, a eukaryotic cell (the green alga *Chlorella*); *C*, a group of reproductive cells lacking a protective jacket (the brown alga *Ectocarpus*); *D*, a group of reproductive cells provided with a protective jacket (the liverwort *Riccia*).

times. These specimens ultimately fall into the hands of a taxonomist who specializes in the group (family or genus) to which the plant in question belongs. This botanist sets out to identify the species by comparing it with allied forms, herbarium specimens, or published descriptions. He may find that the plant has characteristics similar to those of a species previously described, or he may find that the plant has characters so different from those of any known species that he concludes that it is a new species (*species novum,* sp. nov.) that has never been described. Accordingly, he describes this new species and gives it a name. The description is published in Latin in one of the many recognized botanical journals, and the specimen or specimens he used in making the description are properly labeled and placed in one or more of the herbaria of the world. This specimen is known as the **type specimen**.

Herbaria (s. **herbarium**) are storehouses of pressed, dead, preserved plants. In contrast, **botanic gardens**, or **arboretums**, "store" living plants. Some of the great botanic gardens of today, such as the Kew Gardens in England, have a long history of distributing valuable crop and ornamental species to areas far from their native habitats. Sometimes the species actually grow better in their new habitats because insect or fungal parasites are absent.

Methods of Taxonomic Research

Traditional (Morphological)

About 20 years ago, the geneticist C. H. Waddington said it is "an empirical fact that living organisms do not vary continuously over the whole range, but they fall into more or less well-defined groups, which are commonly called species." This philosophy is quite appropriate for traditional taxonomists. They realize that the love of classifying nature often puts boundary lines between things where boundaries simply do not exist; *but* when it comes to species, they are convinced from extensive field observations that *discrete species do exist* and that classification at this level is not arbitrary or artificial.

Traditional taxonomists primarily examine plant morphology, searching for a few traits that consistently enable them to separate plants into "well-defined groups." In pre-Darwin days, a variety of traits were utilized, but now the choice is limited to those which presumably are (a) genetically controlled (rather than environmentally

A

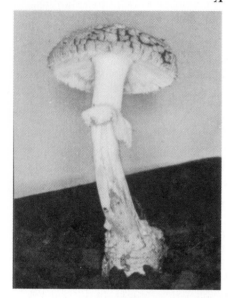

B

C

D

E

Figure 10.8 Pigments, hence color, are of importance in distinguishing divisions of thallophytes. *A, Amanita muscaria* is in the Mycota; it lacks chlorophyll but does have a bright red pigment in the cap. *B, Ulva* (sea lettuce) is in the Chlorophyta, has chlorophylls *a* and *b*, and is green. *C, Postelisa* (sea palm) is in the Phaeophyta, and has chlorophylls *a* and *c* but they are masked by a brown pigment. *D, Iridophycus,* and *E, Porphyra,* are in the Rhodophyta, and they have chrlorophyll *a* and a red pigment.

Figure 10.9   The hair cap moss, *Polytrichum,* is a representative of the Bryophyta.

Figure 10.11   Horsetail (*Equisetum telmateia* L.), the only genus in the division Sphenophyta.

Figure 10.10   *Psilotum nudum,* a living species closely resembling fossils of the earliest vascular plants, found in sedimentary rocks laid down about 400 million years ago.

Figure 10.12 *Lycopodium obscurum* L. (running pine), a representative of one of the five genera in the division Lycophyta.

Figure 10.13 The fern *Polypodium vulgare*, a representative of the division Pterophyta.

Figure 10.14 The Coniferophyta. *A, Pinus jeffreyi* (Jeffrey pine), closely related to *P. ponderosa* (ponderosa pine), the most abundant western states pine. *B*, an open cone of *P. monophylla* (single-leaf piñon pine), showing two seeds on a scale.

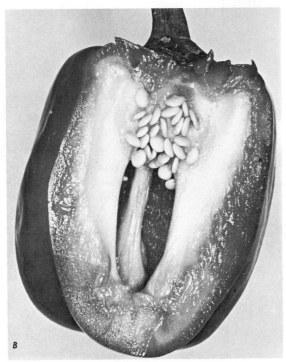

Figure 10.15 The angiosperms, or Anthophyta. *A,* the flower and fruit of *Citrus sinensis* (orange). *B,* the fruit of *Capsicum frutescens* var. *grossum* (bell pepper), showing seeds within the matured ovary.

controlled) and (b) **conservative** in the evolutionary sense. Flower and fruit morphology, for example, meet these criteria, but leaf size does not. Leaf size, and other traits that can be modified by the environment, are said to be **plastic** or **variable**.

It should not be supposed that traditional taxonomists are limited at the start of a study to the examination of only a few traits. They examine many characters, measure many plants, and keep records of all of them. They eventually, however, subjectively select only a few

characters to serve in determining the number of groups. The characters they select show discontinuities and are most helpful in separating related groups.

Anatomical

Anatomical studies of particular tissue and cell types within the vascular system reveal evolutionary patterns. Vessel elements, for example, show the changes summarized in Fig. 10.16. Evidence for this set of hypotheses comes from examination of fossils, of lower plants that have vessels (such as *Selaginella*), and of correlations between changes of vessel characters with changes in other characters such as flower morphology (i.e., plants with primitive flower morphology also show primitive vessel characters). Similar hypotheses of evolution have been applied to other cell types, such as fibers.

Biochemical

We have already seen (Table 10.3) that algal divisions are separated by biochemical characteristics, such as the type of chlorophyll or accessory pigments, the major component of the cell wall, or the principal form of food stored. Only within the last 20 years, however, has biochemistry been regularly applied as a taxonomic tool for higher groups of plants. Members of closely related taxa are chemically treated to extract particular compounds, and the amounts (or presence versus absence) of the compounds extracted are compared. The more chemically similar that taxa are, the closer their genetic relationship is presumed to be.

Almost every category of biochemical compound has been used at one time or another in these studies— sugars, amino acids, fats, oils, alkaloids, alcohols, terpenes, and phenols—and they have been isolated by paper chromatography, electrophoresis, fractionation or, more recently, gas chromatography. Many of these

Table 10.3
Synopsis of Plant Divisions Covered in This Book

	Divisions	Common names or examples
A. Plants without roots, stems, or leaves, and no protective jacket of vegetative cells around the developing reproductive cells		thallophytes
B. Plants lacking chlorophyll[a]		
C. Prokaryotes	Schizomycophyta	bacteria
C. Eukaryotes		
D. Vegetative stage of naked cytoplasm	Myxomycota	slime molds
D. Vegetative stage with a cell wall, mostly filamentous	Eumycota	fungi
B. Plants with chlorophyll		algae
C. Chlorophyll *a* alone		
D. Prokaryotes	Cyanophyta	blue-green algae
D. Eukaryotes		
E. With water-soluble blue and red pigments	Rhodophyta	red algae
E. With plastid pigment xanthophyll	Xanthophyta	
C. Chlorophylls *a* and *c*		
D. Large seaweeds with a cell wall of cellulose	Phaeophyta	brown algae
D. Small, mostly unicells, cell wall of cellulose, two anterior unequal flagella	Chrysophyta	golden algae
D. Small, mostly unicells, cell wall of silica	Bacillariophyta	diatoms
D. Small unicells lacking a cell wall or with cellulose plates, two lateral unequal flagella	Pyrrophyta	
C. Chlorophylls *a* and *b*		
D. Unicells without a cell wall	Euglenophyta	*Euglena*
D. Greatly diversified forms with cell walls	Chlorophyta	green algae
A. Protective jacket of vegetative cells present; roots, stems, and leaves may be present		
B. Plants lacking vascular tissue	Bryophyta	liverworts, mosses
B. Plants with vascular tissue		vascular plants
C. Plants without seeds		lower vascular plants
D. Roots absent	Psilophyta	*Psilotum*
D. Roots present		
E. Small leaves, no gap at union with stem, leaves do not unroll as they open		
F. Stems not jointed	Lycophyta	club mosses
F. Stems jointed	Sphenophyta	horsetails
E. Leaves well-developed, gap at union with stem, leaves unroll as they open	Pterophyta	ferns
C. Plants with seeds		seed plants
D. Seeds not covered		gymnosperms
E. Seeds in cones		
F. Trees palmlike	Cycadophyta	cycads
F. Trees conical	Coniferophyta	conifers
E. Seeds not in cones like those above		
F. Leaves fan-shaped and deciduous; trees	Ginkgophyta	*Ginkgo*
F. Leaves not fan-shaped; shrubs, vines, or prostrate perennials	Gnetophyta	*Ephedra, Gnetum, Welwitschia*
D. Seeds covered	Anthophyta	angiosperms
E. Two-seed leaves	Dicotyledonae[b]	dicotyledons
E. One-seed leaf	Monocotyledonae[b]	monocotyledons

[a] A few bacterial species do have chlorophyll.
[b] Classes.

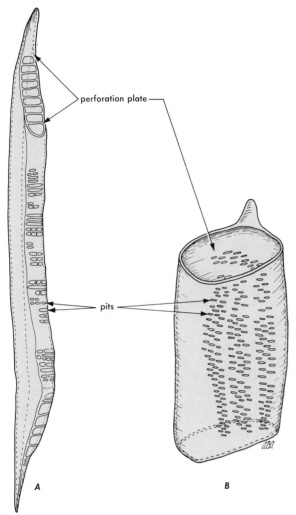

Figure 10.16 Vessel elements from members of the Anthophyta. A, primitive type from *Liriodendron tulipifera* (tulip tree); B, advanced type from *Quercus* (oak).

perforation plate

pits

A

B

turned off, and the gel is "developed." The gel is immersed in a solution that contains a substrate upon which the enzymes under study act. The solution also contains a stain that couples with the product of the enzyme–substrate reaction. The result is that bands of color on the gel mark the locations of very specific enzymes. The distance that the enzymes have moved from their origin depend on their charge and shape; the number of bands reveals the number of enzymes capable of reacting with the substrate (Fig. 10.17).

Biological

Biological taxonomists are interested in determining what the **natural biotic units** are. By natural biotic units they mean populations of plants that maintain their distinctiveness because of **biological barriers** that genetically isolate them from other populations. These isolating barriers may be due to breeding behavior (time of flowering, type of pollinator), habitat or geographic isolation, or to the inability to form fertile hybrids with closely related groups (sometimes because of differences in chromosome number or chromosome morphology).

Biological taxonomists often deal with populations at, and below, the rank of species because they are vitally concerned with detecting divergent evolution at an early stage. They are interested in discovering the ways in which species become distinct and in which they are able to maintain their distinctiveness even though living in close proximity or in similar habitats. Biological taxonomists are referred to as **biosystematists,** and their field as **biosystematics.**

Many species that are widespread and occupy a variety of habitats may be composed of several genetically distinct subspecies. The subspecies are all interfertile, but hybrids are not common because hybrids are less suited to the habitats than are parental types. Drs. Clausen, Keck, and Hiesey of the Carnegie Institution have shown the existence of subspecies in plants such as five-finger (*Potentilla glandulosa*), which grows in California from the coast near Stanford to high elevations near timberline in the Sierra Nevada mountains.

Clausen, Keck, and Hiesey dug up plants of five-finger that were growing all along this range and transplanted

compounds are high in molecular weight and are structurally complex. It is reasoned by biochemical taxonomists that these complicated molecules probably evolved at only one time and in only one original group of plants; therefore, species or genera or families that possess the same complicated molecules are undoubtedly related to this original group of plants.

One of the most widely used methods for comparing taxonomic similarity of organisms at the molecular level is **gel electrophoresis.** This method detects differences between enzymes (or other proteins) on the basis of differences in their electrical charge and shape. Such differences are the result of changes in the amino acid sequence of the molecule. Since amino acids are coded for by small segments of the DNA molecule, amino acid differences are thought to be equivalent to differences between single genes.

A small amount of liquid extract of crushed leaves (or of any other plant organ desired) is placed in a notch cut into a gelatinous material. This gel, often made from potato starch, either fills a tube or coats a glass plate. The extract is placed at one end of the gel, and an electric current is sent across it. After several hours, the current is

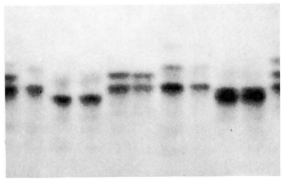

Figure 10.17 A "developed" gel of 11 *Clarkia* genotypes. The enzyme bands have migrated downward from the "top" of the gel.

Methods of Taxonomic Research

them together in a uniform garden at Stanford. In the uniform garden, any differences among the plants from different locations would be due to genetic control. They found morphological differences: timberline plants were much shorter than lowland plants. They also found physiological differences: the timberline plants flowered earlier in the summer than lowland plants. When five-finger plants from different locations were grown together in a garden near timberline, these differences proved critical to survival: lowland plants that did not become dormant were killed during the winter, and lowland plants that flowered late in summer were nipped by frost before seeds could be produced. Clausen, Keck, and Hiesey named the lowland group *P. glandulosa* ssp. (= subspecies) *typica* and the timberline group *P. glandulosa* ssp. *nevadensis* (Fig. 10.18).

Biosystematists are also interested in comparing **taxa** (**taxon**, singular; plants of any taxonomic rank) on the basis of their survival strategies.

One example of different strategies can be seen by comparing an oak with a dandelion. Oaks and other large perennial plants devote a lot of energy to the accumulation of great bulk that can withstand many environmental stresses. The price oak ''pays'' for its size is a long juvenile period of development when reproduction is nil, and even after that period the amount of energy put into flower and seed production is low, compared to the energy invested in the total mass of the plant. Seeds (fruits) are large and are disseminated by gravity, water, or animals. Seed production and seedling establishment can be very low for many years in succession because of poor environmental factors, or internal factors, but the population density remains constant because of the long life span of individuals. This is called a **K** strategy by ecologists.

Dandelions and herbaceous annuals, on the other hand, invest a high proportion of their energy in seed production and have a short life cycle. Each seed is small and often wind-disseminated, which means that the young seedling has very little food reserve to use while it establishes itself as a plant. The result is that most seedlings perish. But, because the reproductive capacity of the few survivors is so high, the population density is maintained. This is called an r strategy. The letters r and K are components in an equation for population growth rate. Obviously, both K and r strategies can lead to success in an evolutionary sense. The question is, under what set of environmental conditions will each strategy be more successful? The concept of r and K strategies was first proposed in 1962 by the very innovative ecologist Robert MacArthur.

Numerical

All the approaches to taxonomy just discussed share one feature in common: each subjectively selects one form of similarity on which to base conclusions, and each weights its own choice of evidence as more important toward understanding relationships than any other form of evidence. Traditionalists weight morphological evidence, biosystematists weight breeding behavior, and so on.

Another approach to taxonomy is to consider all forms of evidence with equal emphasis; all evidence has equal weight. This is a statistical approach and is called **numerical taxonomy**. A basic tenet of numerical taxonomists is that a great deal of evidence is required to objectively separate taxa from each other. They utilize 50 to 300 characteristics for each study.

The numerical taxonomist *does not admit that discrete species exist* in nature. He/she insists that when many traits instead of a few are considered, a continuum of

Figure 10.18 Growth of *Potentilla glandulosa* ssp. *typica* after one growing season at each transplant garden. This coastal subspecies, or ecotype, grew best in its normal environment at Stanford (left); it grew very poorly at timberline (right) and was killed by the following winter temperatures. Middle plant was grown at an intermediate elevation.

variation will emerge. Boundaries between species can be erected by taxonomists, and these boundaries may be objectively selected; but the boundaries are arbitrary and artificial in the final analysis, according to numerical taxonomists. The viewpoints of traditional and numerical taxonomists, then, are quite different. Interestingly, classification schemes arrived at by numerical means often show good agreement with those arrived at by traditional means. Can we conclude that experienced, traditional taxonomists are just as effective as the computers used in numerical studies?

Evolution

In a broad sense the term **evolution** refers to a process involving gradual changes. It is well known that neither animals, nor plants, nor cities, nor states remain the same through time. We may speak of the evolution of the means of transportation, the evolution of human clothing, the evolution of mountains and valleys, and the evolution of many other nonorganisms.

Organic evolution pertains to the gradual changes that have taken place in living organisms. Coal, for instance, is formed from the remains of plants that were different in appearance from those growing today. The student of organic evolution may be interested in, among other things, accounting for the disappearance from the earth's flora of the Coal Age plants and the appearance of the modern flowering plants.

The Geologic History of Plants

Our understanding of plant history comes from the examination of fossils. A fossil may be a shell or bone, little changed from its original form; it may be the microscopic cell wall of a spore or pollen grain, which is also very resistant. But most plant parts are not hard, and plant fossils seldom consist of unchanged structures. They often are simply the impressions of leaf or stem fragments that were trapped in mud—mud that later was compressed into sedimentary rock (Fig. 10.19). To yield a good fossil, plant parts must settle in quiet waters, then be buried under silt or volcanic ash to lie trapped in surroundings unfavorable to decay. They are seldom entombed where shed but, more typically, are first carried by water or wind for some distance. If the plant part is large enough, for example, a tree trunk, the interior of the cell may be replaced by silica quartz crystals. The tissue pattern and anatomical details are faithfully preserved (Fig. 10.20). In addition, degradation products from tissue decay may remain trapped as "chemical fossils." The porphyrins from chlorophyll degradation, for example, are very inert and long-lasting. Their presence in sediments indicates that photosynthetic plants were present at the time the sediments were laid down.

It is important to note that plant fossils consist almost entirely of plant remains that grew in or near water or were carried by streams into lakes, bays, or other sites of deposition. The remains of plants growing in arid or mountainous regions are rarely deposited where conditions are favorable for their preservation as fossils. Either it is too dry or the site is undergoing erosion and the deposits are of very short duration. This means that the geological record of plants must probably remain incomplete and that the records we do have mainly represent the floras of lowlands and moist habitats. Considerable evidence indicates that the great environmental extremes prevalent in arid or semiarid mountainous areas are especially favorably to rapid evolution. This factor, we shall see, may be very important

Figure 10.19 An imprint of a leaf of *Pecopteris*, a common plant of the Pennsylvania geologic period, about 300 million years ago.

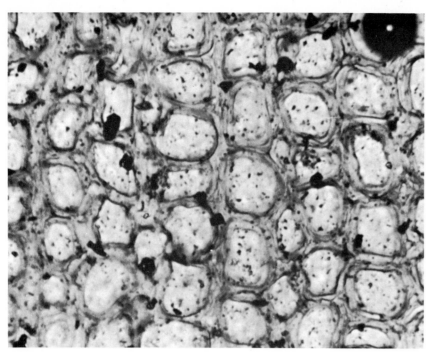

Figure 10.20 A thin section of petrified wood, as seen through a light microscope. Tracheid lumens are filled with silica stone.

in connection with the problem of determining the origin of the angiosperms and may serve to explain peculiarities that surround the first records of the angiosperms.

When Did the World Begin?

When did the world begin? Earth scientists have no sure answer, but the calculations of astronomers and physicists, coupled with known age of meteorite fragments and of rocks from the moon, lead us to estimate that the surface of the earth cooled to a crust about 4.5 to 5.0 billion years ago. This primordial crust probably contained uplands, ocean basins, plateaus, and mountainous volcanoes. Volcanic eruptions filled the atmosphere with carbon dioxide, methane, carbon monoxide, nitrogen, hydrogen, and strong-smelling ammonia and hydrogen sulfide. Free oxygen was absent. The first rains washed, dissolved, and eroded the uplands, carrying many nutrients to the young oceans.

Life, or some sort of half-way form of life, probably originated very early. It may have consisted of aquatic, small, self-replicating, heterotrophic cells that lived on the rich, anaerobic oceanic soup around them. However, our present methods of detecting remains of ancient life require that the fossils be preserved in rocks. So far, the oldest rocks found on earth are 3.4 billion years old. This means that we have no record of the first 1 to 1.5 billion years of earth's history (Table 10.4).

Aquatic life did exist 3 billion years ago, for we have fossil evidence of it. Some of the oldest rocks known lie exposed near the gold-mining town of Barberton, on the border between the Republic of South Africa and Swaziland. About 3.2 billion years ago, this area was a shallow, warm sea or embayment. Living things existed in thin sheets at the bottom of a silica-rich sea. Apparently,

conditions were perfect for preservation of the organisms. They were preserved in the sediment in a siliceous solution that later crystallized into rock called chert, much as a modern biological specimen is preserved by being embedded in plastic. The soft bodies of these early organisms were not distorted because the silica matrix of chert is incompressible. Today, these beds of chert, up to 130 m thick, are exposed. When thin sections are placed on a microscope slide, the perfectly preserved organisms (microfossils) can be seen. The chert itself has not been dated, but rock layers above and below it have. By inference, the age of the chert is estimated to be 3.2 billion years.

Many unusually shaped structures, which may or may not be organisms, have been seen in this chert (called the Fig Tree Formation), but only two are seen frequently enough and in great enough detail to leave no doubt that they once lived. One is a rod-shaped bacteriumlike cell, called *Eobacterium isolatum* (Fig. 10.21), the other is possibly a blue-green alga, *Archaeosphaeroides barbertonensis*. Organic residues were also found in the chert. Analysis of them showed the presence of certain hydrocarbons that can most reasonably be regarded as breakdown products of chlorophyll. Analysis also revealed that the ratio of carbon-13 to carbon-12, two natural isotopes of carbon, was lower than the ratio in today's atmosphere. When plants photosynthesize, they "prefer" to use carbon-12; this means that their tissue and fossil residues have a low carbon-13 to carbon-12 ratio, corresponding to the ratio in the chert. This combination of direct and indirect evidence strongly suggests that plant life existed 3.2 billion years ago.

A similar, but richer, collection of microfossils has been preserved in another chert deposit, the Gunflint Formation,

Table 10.4
Geologic Time and the Dominance of Different Plant Groups through Time[a]

Era	Period	Epoch or part	Began (millions of years ago)	Dominants
Cenozoic	Quaternary	Recent	Last 12,000 years	Flowering Plants
		Pleistocene	2.5	
	Teritiary	Pliocene	7	
		Miocene	26	
		Oligocene	38	
		Eocene	54	
		Paleocene	65	
Mesozoic	Cretaceous	Upper	90	
		Lower	136	Gymnosperms
	Jurassic	Upper	166	
		Lower	190	
	Triassic	Upper	200	
		Lower	225	
Paleozoic	Permian	Upper	260	
		Lower	280	Lower vascular plants
	Carboniferous	Pennsylvanian	325	
		Mississippian	345	
	Devonian	Upper	360	
		Middle	370	
		Lower	395	Algae
	Silurian		430	
	Ordovician		500	
	Cambrian		570	
Proterozoic	Precambrian		4,500–5,000	

[a] Shaded sections were times of great evolutionary changes in the plant kingdom. The time scale is not drawn to scale. All named epochs do not appear on this chart.

along the Ontario shore of Lake Superior. By dating layers above and below it, geologists estimate this deposit to be 2 billion years old. Filaments are most abundant, and some resemble modern blue-green algae (Fig. 10.22). Small spheres that might be colonial blue-green algae or spores are also common. Analysis of the organic residue and determination of the carbon isotope ratio show that photosynthetic organisms were present. The origin of photosynthesis was important; most scientists believe that the O_2 of the atmosphere resulted from it.

The first eukaryotic plants may have appeared 1 billion years ago. Fossils in the Bitter Springs Formation, about 65 km northeast of Alice Springs, in the heart of Australia, are rich and varied, and indicate that at least four groups of organisms existed at that time: (a) prokaryotic filamentous blue-green algae, akin to modern *Oscillatoria*

and *Nostoc;* (b) bacteria; (c) eukaryotic fungi; and (d) eukaryotic green algae. Cytological details are preserved so well that stages of cell division can even be detected (Fig. 10.23), but you should be aware that the interpretation of cell details is not uniform, and some biologists claim that the "nuclei" are artifacts and that the cells are those of prokaryotes.

If these are fossil eukaryotes, then perhaps they could reproduce sexually. Sexual reproduction brings with it variation in offspring and a more rapid rate of evolution. It enabled a greater diversity of forms to develop and proliferate. A spurt of evolution resulted. By the end of the **Proterozoic** era (Table 10.4), the seas had become crowded with life, including many forms of animals and plants. The rate of oxygen formation had increased; by the end of the Proterozoic era the level of oxygen in the air

Figure 10.21 Oldest known bacterium, *Eobacterium isolatum*, from the Fig Tree Formation, about 3.2 billion years old.

may have been 1% of its present level. Oxygen not only supports life; it also screens out ultraviolet radiation from the sun. The radiation can be damaging to cell activities. This amount of oxygen—1% of present level—produced a sufficient filter of radiation to permit life in all but the top 2 to 3 cm of water. Life on land, except in sheltered places, would still have been impossible.

A "sudden" evolutionary spurt 600 million years ago marked the beginning of the long **Paleozoic** era, a time when much of the land was low-lying, and inland seas moderated the climate so that seasonal and latitudinal

effects were minor. It was a time when a climate much like that in today's tropics dominated the entire world. It was a time of enormous evolutionary change in the plant kingdom. However, the rate of evolution did not proceed at a constant pace during the entire Paleozoic era. Most of the major changes in plant form appeared in only 25 to 50 million years, during the upper-Silurian to mid-Devonian periods. It was then that plants came to dominate the land and that the oxygen level of the air may have reached 10% of the present level.

Through the Cambrian, Ordovician, and most of the Silurian periods, algal forms continued to dominate the plant kingdom. Blue-green and green algae were most abundant. but lime-encrusted red algae and some brown algae resembling *Laminaria* and *Fucus* were also present. The lime-encrusted forms secreted calcium carbonate just as modern forms do.

Invasion of the Land: Silurian and Devonian Periods

What are some of the adaptations required for plant life on dry land? In water, plants need no complex structures for support nor for the uptake of nutrients. The surrounding water buoys them up and bathes them with soluble nutrients. In contrast, land plants must not only develop roots for the absorption of nutrients and the elaboration of stiff tissue for support, they must provide pathways for water and nutrient transport—xylem and phloem. Reproductive cells on land must be carried by agents

Figure 10.22 The filamentous blue-green alga *Animikiea septata* from the Gunflint Formation, 2 billion years old.

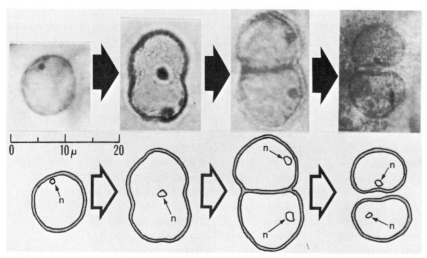

Figure 10.23 Sequence of fossil impression from the Bitter Springs Formation arranged to show cell division.

other than water, and they must be thick-walled and cutinized to avoid desiccation. All external surfaces of land plants, in fact, must be adapted to reduce water loss; these adaptations can take the form of stomata and cuticle.

Despite tempting fragments of evidence, such as cutinized spores and bits of xylem dating back to the Cambrian period, the first undisputed fossils of terrestrial, vascular plants do not appear until the upper-Silurian period. But what appeared at that time was not just one type of vascular plant. Many types suddenly occur in the fossil record: representatives of the Psilophyta, Lycophyta, Sphenophyta, and pre-ferns. Thus we can surely believe that the first vascular plant evolved many millions of years before these many types. Did they all evolve from a common ancestor or did they evolve from several separate lines? At this time there is not enough evidence to answer the question.

The early vascular plants consisted of slender forking stems, with or without leaflike appendages, and with sporangia either at the tips of the branches or along their sides. Perhaps the most primitive example is *Cooksonia,* discovered in 395 million-year-old fossil beds in Czechoslovakia and Wales, from the end of the Silurian period (Fig. 10.24). It was a plant probably less than 10 cm tall, made up of dichotomizing branches less than 4 mm in diameter that terminated with sporangia. The spores had a waxy cuticle, indicating that they were adapted for dissemination on land. The xylem in the stem was a solid, central core; there was no pith. We know nothing of the root system. Modern, living wisk fern (*Psilotum,* see Fig. 10.10) still retains many of these primitive features.

From this primitive, herbaceous start, evolution progressed rapidly to the first trees only 25 million years later. *Aneurophyton,* for example, was up to 12 m tall. *Archeopteris* was more than 30 m tall and up to 1.5 m in diameter at the base (Fig. 10.25). Both had evolved pith, cambium, and the capacity to produce considerable secondary xylem. Since the petrified wood samples lack annual rings it is not possible to determine their age.

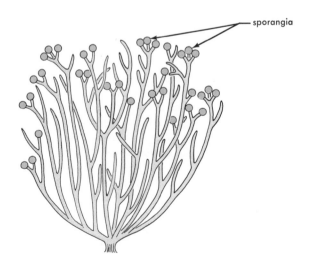

Figure 10.24   A reconstruction of *Cooksonia,* one of the first vascular plants. The above-ground portion, shown here, was less than 10 cm tall.

Leaves had also evolved, changing from very small epidermal outgrowths of the primitive vascular plants, to larger true leaves having a branching vascular system, as found in *Aneurophyton* and *Archeopteris.* The "true" leaves are called **megaphylls**; the epidermal outgrowths are called **microphylls**.

Coal Age Forests and Seed Plants

Plant development during the Devonian period not only modified the vegetative size of plants, it affected their reproduction as well. The first seed plants developed in the upper Devonian period, but it was not until the Coal Age (Mississippian and Pennsylvanian epochs, also known as the Carboniferous period) that seed plants became abundant.

Extremely lush, swamp forests dominated the **Coal Age** landscape (Fig. 10.26A). Because the land was low, minor changes in sea level successively inundated, buried, then supported one forest after another. The buried organic remains have become compressed and changed through time. Today, they form the coal, gas, and some of the oil reserves of the world. These sources of energy are the chemically changed, fossil remains of Coal Age forests and, for that reason, they are referred to as **fossil fuels**. Burial of plants and transformation into fossil fuel have continuously taken place throughout time, but not at the rate it did during the Coal Age (Fig. 10.26B).

These forests were dominated by Lycophyta. In the Devonian period, this division evolved rapidly into both herbaceous and woody types; in the Coal Age, the woody types reached tree stature. One example is *Lepidodendron* (Fig. 10.27), which is up to 35 m tall and with a trunk 1 m across. Straplike leaves and sporangia occurred near the ends of the branches. A cross section of the trunk reveals pith, primary and secondary xylem, cambium, and an enormous amount of cortex and cork. Very little of the stem area served for conduction or strengthening. The roots were dichotomously branched.

Second in abundance were giant horsetails, such as *Calamites* (Fig. 10.28). *Calamites* was smaller than *Lepidodendron,* but was still 10 m tall and had a trunk about 30 cm in diameter. Whorls of branches developed, from which formed smaller branches; and from these branches leaves developed at the nodes. The upright stems arose from a horizontal rhizome system.

Third in abundance—not tall, but dominating the forest floor—were ferns and seed ferns. Seed ferns were fernlike, but in addition bore seeds on their large leaves instead of spores. Exactly from what plant group the ferns evolved is not clear, for they appear rather suddenly in the fossil record in the Coal Age.

Gymnosperms (other than seed ferns) were fourth in abundance, but these were forms quite unlike modern gymnosperms and were not abundant. Gymnosperms did not become abundant until 100 million years later.

Herbaceous forms of Lycophyta and Sphenophyta were common, but members of the Psilophyta—direct descendants of the first, weak, vascular plants to climb onto land only 70 million years earlier—were rare, although we have only a few fossils from that time. The lower vascular plants dominated world vegetation during

Figure 10.25 Reconstructions of large Devonian trees. *A, Aneurophyton*, 7 to 13 m tall; *B, Archeopteris*, about 30 m tall.

the Coal Age just as thoroughly as the algae dominated the world in previous ages. The dominance of the lower vascular plants was short, however, and great extinctions lay ahead.

Permian Extinctions and on to the Mesozoic Era

The 375 million-year-long Paleozoic era was marked by two concentrated periods of evolution: a 25 million year period of innovation in the lower-Devonian period during which weak vascular herbs led to forest trees; and a 50 million year period of major extinctions in the Permian period. These changes were associated with increasing oxygen in the air; the extinction may have been caused by a cooling and drying climate and an uplifting of the land. In many ways, the sudden rise and fall of the Coal Age forests are just as striking and mysterious as the rise and fall of the dinosaurs many years later. The Permian extinctions marked the start of the **Mesozoic** era.

The plants that replaced the lower vascular plants in dominance were gymnosperms: the true conifers (Coniferophyta), Cycadophyta, and Ginkgophyta. Most of

the Mesozoic era was the age of the gymnosperms. However, the group that dominates the world today, the flowering plants, must have been evolving all through the Mesozoic era. Pieces of wood, leaf impressions, and pollen scattered through the geologic record as far back as the Triassic period seem to be angiosperm in nature. But the first uncontestable appearance of fossil angiosperms is in the Cretaceous period.

The Cretaceous period was a third interval of rapid evolution, undoubtedly because of major climatic and geologic changes that took place. Two major biological changes at the close of the Cretaceous period were (a) a spread of the flowering plants, and (b) extinction of the dinosaurs. A number of other changes in the plant world accompanied all this (e.g., the extinction or near-extinction of several gymnosperm groups), but the rise of the angiosperms was the greatest change. Where had they come from, and why did they evolve so quickly? We may never know the answers. At present, one leading theory is that flowering plants evolved over a long period of time. This evolution took place in tropical uplands where fossil

preservation in sediments is rare. These early angiosperms may have occupied warm, seasonally wet, rocky slopes with great variations in microhabitats resulting from differences in exposure, elevation, drainage, and soil type. This environmental variability could have been a stimulus to evolution. As the land lifted and Cretaceous seas withdrew, new lowlands could have been rapidly invaded by diverse forms of flowering plants that had evolved on mountain slopes. By the time the **Cenozoic** era began, 65 million years ago, the conifers dominated only the cold temperate and polar regions, and to the angiosperms belonged everything else.

Climatic Change in the Cenozoic Era

The gradient of temperature with latitude that we have today apparently did not exist at the beginning of the Cenozoic era. Looking at fossil deposits of marine plankton, we can estimate the ocean surface temperatures from the equator to the north pole at past times. In the mid-Cretaceous period, the difference from equator to pole was only 11°C, but by the end of the Cretaceous period, it

"I HAVE A FEELING IT'S TOO SOON FOR FOSSIL FUELS AROUND HERE."

Figure 10.26 Coal Age forests, and the fossil fuel results. *A,* reconstruction of such a forest, courtesy of the Field Museum of Natural History. Note the seeds attached to fernlike leaf in left center. *B,* cartoon relating the age of dinosaurs wrongly with the coal age. Dinosaurs flourished 100–200 million years after the coal age.

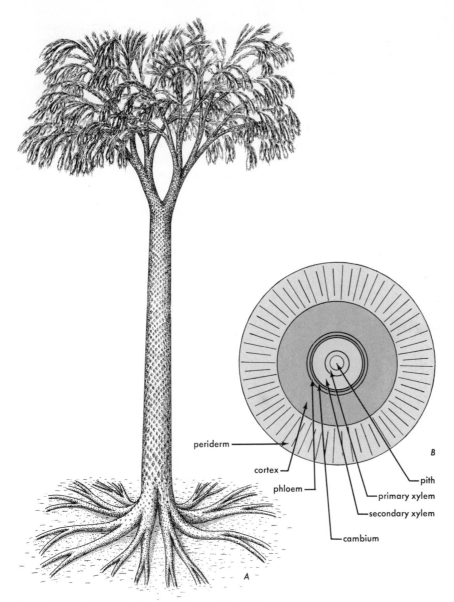

periderm

cortex

phloem

pith

primary xylem

secondary xylem

cambium

B

A

Figure 10.27 *Lepidodendron,* a dominant of the Coal Age forest. *A,* reconstruction of the entire plant, about 30 m tall; *B,* cross section of the trunk.

had grown to 15°C. The difference continued to grow during the Cenozoic era. We can guess that similar temperature changes occurred on land. In the past, the climate over the globe was broadly zoned and without temperature extremes, but today many climatic zones exist between the pole and the equator. Figure 10.29 shows how these zones may have formed and shifted during the last 60 million years. During Pleistocene glaciation, the temperature gradient was at a peak.

The Pleistocene **Ice Age,** which ended only a moment ago in geologic time—about 12,000 years—produced another period of extinctions. Glaciated areas were scraped clean of plants, and they are still today being slowly revegetated.

In the Recent period, conifers have become more and more restricted in the land area they dominate. The flowering plants, however, continue to evolve, especially in harshly cold or dry environments. Today, only two groups

of vascular plants appear to be expanding in terms of diversity and abundance: ferns and angiosperms. Both groups have survived great physical changes of the earth, yet their evolution seems to have been stimulated by these stresses, while many other forms have become extinct or survive as tenuous remnants.

By the end of the Ice Age the human species began to play a role in the evolution and distribution of plants. Although the earliest archeological evidence for seed agriculture (cultivation of annuals such as grain, beans, and squash) takes us back to 9000 B.C., people may have been cultivating **root crops** (perennials propagated by cuttings and harvested for starch in the ''roots'', such as sweet potato and taro) in southern Asia as long ago as 13,000 B.C. Some of these root crops have been asexually propagated for so many years that they have lost or nearly lost the capacity for sexual reproduction, and it is doubtful that they could survive in nature.

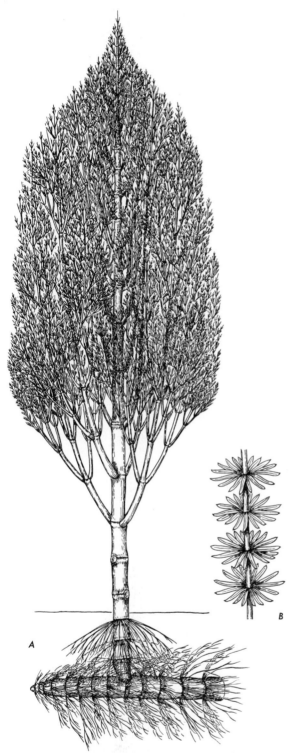

Figure 10.28 *Calamites,* another Coal Age forest tree. *A,* reconstruction of an entire plant, including rhizome and roots, of *C. carinatus. B,* detail of whorled leaves of the form species *Annularia radiata,* which are very similar to the leaves of *C. carinatus.* The leaves are about 1 to 2 cm long.

Annual grains have been selected for productivity, rather than natural survival, and the result is they have become so changed from their wild relatives that it is difficult to determine where, and from what stock, they were first domesticated.

People have also accidentally domesticated and favored the evolution of certain other plants, the **weeds**. These plants grow well in disturbed or trampled soil, in waste areas rich in nitrogen, or in cropland. Some of them have evolved seeds that imitate the size of crop seeds, so they may not be separated during the threshing or sieving of crop seeds. When the next season's crop is sown, the weeds are sown inadvertently along with it.

Charles Darwin and Evolution

Many biologists since the time of the Greek philosophers have attempted to explain the mechanism of evolution. Great progress has been made during the nineteenth and twentieth centuries. Our better understanding was made possible chiefly because the works of Charles Darwin and Gregor Mendel provided a basis that enabled biologists to approach the problem of evolution in an entirely different way.

Darwin, as a result of many years of careful study and observation of a large number of plants and animals, emphasized the following points:

1. The numbers of plants or animals may increase in a geometric ratio. For example, a given plant may produce 1000 seeds. If each seed grows into a new plant and each new plant produces 1000 seeds, 1 million new plants could result. A third generation from the one original plant would result in 1 billion plants. This example shows that plants (and animals also) have the potential to increase their numbers at a tremendous rate.
2. Actually, the number of individuals of a given plant or animal remains fairly constant. There is no such tremendous increase in the number of individuals that seed production seemingly makes possible.
3. No two individual plants or animals are identical; there is variation.

Reasoning from these observations, Darwin arrived at these conclusions:

1. Any given population is usually able to reproduce many more young individuals than can adequately be raised in the region it occupies. Therefore, a struggle for existence occurs among the individuals.
2. In the struggle for existence, only those individuals survive that, because of their particular variation, are best adapted to their immediate environment. Thus, a natural selection takes place; the unfit do not survive or do not reproduce.
3. These selected variations may be inherited, that is, passed from one generation to another and, thus, may gradually give rise to new species.

The Sources of Variation

Since variation plays such an important part in plant evolution, let us examine the sources of variation a little more closely.

One source of variation is **mutation,** the spontaneous transformation of a gene (Fig. 10.30). Mutation does not create new genetic material (i.e., new DNA); it simply

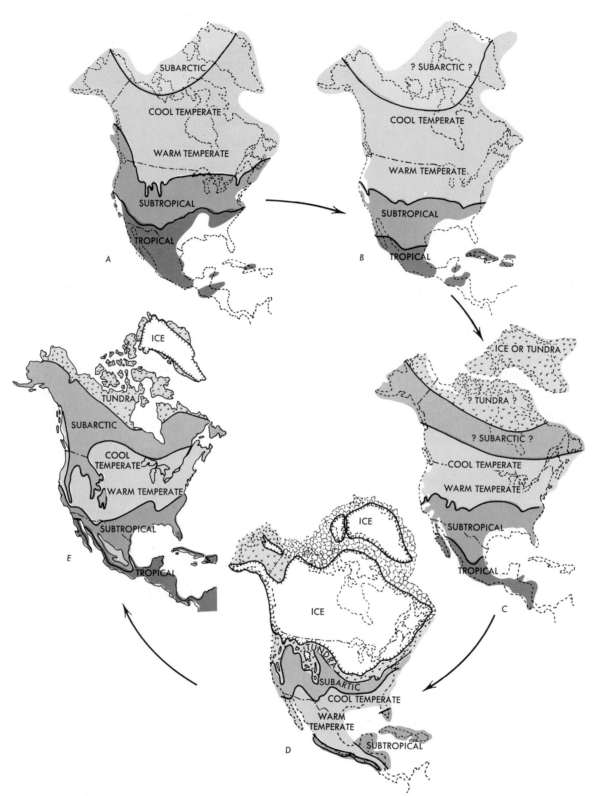

Figure 10.29 Climatic and vegetation zones in North America during the past 40 million years. *A*, Oligocene; *B*, Miocene; *C*, Pliocene; *D*, at the peak of the Pleistocene Ice Age; *E*, present.

Figure 10.30 Examples of mutations. *A*, common wild sunflower; *B*, mutant sunflower, known as sun gold; *C*, variation in the seed coats of beans.

changes the arrangement of the existing material so that the enzymes and other proteins encoded by the DNA are no longer quite the same. A mutated gene can back-mutate to the original condition also. Although several environmental factors, such as radiation, can cause a mutation, most mutations seem to result from cellular "mistakes" in copying the DNA molecule during cell division.

Mutations are a universal fact of life. They are known to occur in every plant and animal that has been studied. This does not mean that a given gene is relatively unstable and is likely to mutate very often. Rates of mutation for an individual gene, or locus, probably average one for every 100,000 cells. The mutation rate does vary from gene to gene, however, as shown for corn (Table 10.5), where one gene may have a mutation rate 500 times that of another.

Other sources of variation include chromosome aberrations such as deletions, duplications, inversions, translocations. A final powerful source of variation is **recombination**, the reshuffling of chromosomes during sexual reproduction (see chapter 3).

Table 10.5
Mutation Rates of Different Genes in Corn (*Zea*)

Gene	Number of Mutations per 1,000,000 Gametes
Seed color, not purple	492
Seed color inhibitor	106
Purple seed color	11
Sugary seed	2.4
Shrunken seed	1.2
Waxy seed	Less than 0.1

The Role of Natural Selection

Mutation and recombination produce new patterns of heredity, which result in variation among individuals. However, for an entire population of individuals in a species to progressively change into something new, more than this raw material is needed. Some force or pressure must be exerted on the population so that the abundance of certain mutations becomes higher and higher, until all or most members of the species possess it. How does this happen? The presently accepted answer had its own evolution, over the course of 150 years of debate and experimentation. The answer is based on the idea of **natural selection**.

At the beginning of the nineteenth century, the French naturalist Jean Baptiste Lamarck (1744–1829) was a liberal—for his day—in his thoughts on evolution. Lamarck did not believe that all species were of the same age, that all were created together at one time. Neither did he believe that the same species always existed. He believed that new ones were forming all the time as a result of changing environments. This stand was quite heretical, considering that Linnaeus 50 years earlier had practically founded ''modern'' taxonomy on the principles that there was only one time of creation, and that all the species in the world from that time on were fixed and constant. Lamarck thought that new traits and new species could evolve from old ones if a species were placed under stress. For example, if a tall plant at sea level were transplanted to a severe, timberline habitat, the climate would stunt it. This stunted plant would shed seeds and the new seedlings would also be stunted, even if grown back at sea level. Characters acquired during the lifetime of one plant would be passed on to succeeding generations.

Lamarck himself did no experimentation to bolster his theory. Other botanists did, however. Bonnier in France made reciprocal transplants of alpine and sea-level species. He grew a number of plants of each species to a convenient size at Paris, then cloned them (split them into pieces) and planted them in plots in the Alps, the Pyrenees, and Paris. They were not planted in gardens, but were put in among natural vegetation. The plots were sometimes fenced, sometimes not; watering and fertilization were not practiced. The plots were periodically visited for a number of years (1884–1920) and any changes in the plants noted. He concluded that a number of lowland species were transformed into related alpine or subalpine species during the years of the study. Also around 1920, the American ecologist Frederic Clements established a similar series of ''gardens'' from Pike's Peak to the California shore. He too concluded that some lowland species were transformed to related high-elevation species, and vice versa.

However, the most carefully documented and controlled transplant experiments were last conducted, between 1920 and 1940, by Clausen, Keck, and Hiesey in California, as mentioned earlier in this chapter. Their detailed study of some 60 taxonomically diverse species showed not one case of a lowland species becoming an alpine species, or the reverse. A lowland species at timberline might become dwarf or prostrate or stunted, but if its seeds were collected and sown back at sea level, normal tall plants would result. There had been no genetic change—only a temporary, plastic response to a harsh environment. Such temporary changes are called **phenotypic** changes. In contrast, **genotypic** changes (changes in the genes) are passed on to offspring.

Darwin's concept of evolution, expressed over a hundred years ago, is the theory accepted today, not Lamarck's. Its differences with Lamarck's are illustrated in Fig. 10.31. Basically, Darwin's concept is this: Variation exists in the initial population; an environmental stress places certain individuals at an advantage; because those individuals survive or reproduce more successfully, they leave more offspring that carry the same genetic traits; the abundance of the advantageous traits in this way increases in every generation, but variation still persists. This process of directed change is called **natural selection**.

How Species Remain Distinct

Evolution proceeds most rapidly in a population of organisms if it is ''isolated'' from other populations. But we mean a very special kind of isolation: **reproductive isolation**. If plants in one isolated population cannot breed with similar plants in other populations, then natural selection will produce a distinctive species in the shortest amount of time. This species will remain distinct from all others, once it has formed (Fig. 10.32). Without isolation, crossbreeding would dilute the abundance of new genes that are of most value in one particular environment; all members of a species would continue to be more or less alike and not as well adapted to extremes within their range, as they could be if isolated. Mediocrity—a population of generalists—would result. One large population of generalists is not always as successful a strategy for survival as several small populations of specialists.

How can reproductive isolation be established? Table 10.6 lists the most important isolating mechanisms. The prezygotic mechanisms have to do with preventing pollen of one population from reaching the stigmas of another. This can be accomplished by separating the populations in space (isolated valleys or separating continents), time (different seasons of flowering), or biology (different pollinating insects, which do not visit both flowers). On a world scale, **continental drift** (the gradual spreading apart of continents over the past 200 million years) may have been the most important isolating mechanism. Postzygotic mechanisms prevent normal offspring from developing, even though pollination occurs. The hybrids may be weak or sterile, for example, or the developing embryos can abort.

Hybrids are often weak in nature, or at least they are less well-adapted to a given habitat than either parent. So the most common type of hybrid in nature is not a ''pure'' hybrid, halfway between each parent. Instead, the successful hybrids are often the result of backcrossing between the original pure hybrid and either or both parental types (Fig. 10.33). The few pure hybrids in this way produce an entire series of partial hybrids between

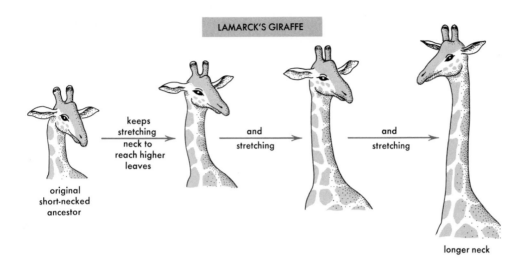

LAMARCK'S GIRAFFE

original
short-necked
ancestor

keeps
stretching
neck to
reach higher
leaves

and
stretching

and
stretching

longer neck

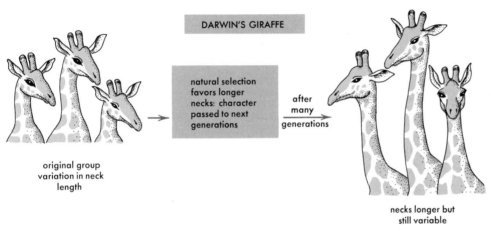

DARWIN'S GIRAFFE

original group
variation in neck
length

natural selection
favors longer
necks: character
passed to next
generations

after
many
generations

necks longer but
still variable

Figure 10.31   Comparison of Lamarck's and Darwin's concepts of evolution.

themselves and original parental types. This is called **introgressive hybridization,** and it creates great diversity (hybrid swarms) that may spell success for the group as a whole.

Summary

1. As a floral survey of the world becomes more complete, the goal of taxonomy becomes the construction of a phylogenetic classification scheme that summarizes the evolutionary history of our flora.

2. The history of taxonomy can be divided into a pre-Darwinian period and a post-Darwinian period. Linneaus culminated the pre-Darwinian period (ca. 1753) and organized plant nomenclature. During the post-Darwinian period there were several attempts to classify the plant kingdom on a natural (phylogenetic) basis; schemes by Bessey and Engler and Prantl are examples. At present, we still lack a classification scheme accepted by all botanists because the evolutionary history of plants is not completely known.

3. A species is a convenient unit of information. Species that show a close genetic relationship are grouped into a genus. Related genera are grouped into a family, families into orders, orders into classes, and classes into divisions. For the sake of worldwide uniformity, each species is described in Latin and its name is written in Greek or Latin as a binomial: the genus (capitalized) followed by the specific epithet (not usually capitalized).

4. The 21 plant divisions covered in this book can be grouped into several categories, which include prokaryotes, eukaryotes, thallophytes, liverworts and mosses, lower vascular plants, higher vascular plants, seed plants, gymnosperms, and angiosperms. Two classes of angiosperms are the Monocotyledonae and the Dicotyledonae.

5. Dried, pressed, and labeled specimens of all named species of plants are kept in herbaria. Living examples are grown in botanic gardens (arboretums).

6. Taxonomic research may take any of several directions, utilizing patterns of morphology, anatomy, biochemistry, and breeding behavior, any of which

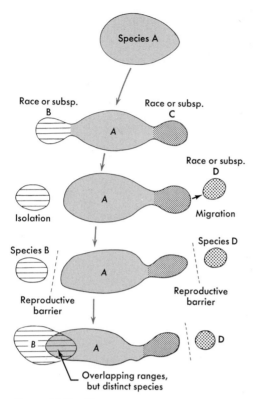

Figure 10.32 Diagram showing the sequence of events which leads to the formation of new races, subspecies, and species. One species (green, at top) extends into new habitats, and populations at the extremes become modified into races or subspecies. If the races or subspecies become isolated, they may evolve further and become incompatible with the original species; at this point they are recognized as distinct species. The reproductive barrier will keep the species distinct, even if later migration brings them back into contact (bottom).

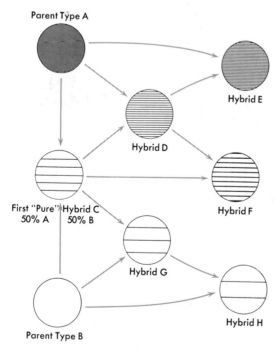

Figure 10.33 Introgressive hybridization. The degree of shading (horizontal lines in the circles) shows the degree of similarity of hybrids to parental types A and B. Introgressive hybrids D, E, F, G, and H would be more numerous in nature than "pure" hybrid C.

Table 10.6
Summary of Some Important Reproductive Isolating Mechanisms

A. Prezygotic mechanisms: prevention of pollination or fertilization

1. Separation in space: two populations live far from each other so that pollen cannot be transferred, or they live close but are separated by barriers (mountains, bodies of water), or they occupy different habitats (lowlands versus uplands), or drifting continents move them apart.

2. Separation in time: two populations flower at different times of the year, or if they flower at the same time, the pollen is shed before the stigmas are receptive.

3. Biological separation: because the pollinating insect or animal is different for each population, it is unlikely that cross-pollination will occur, even if the plants are close to each other; or the flowers may have the same pollinator but open at different times of the day (morning versus evening).

4. Physiological differences: the pollen of one is unable to grow through the style of the other, or it grows more slowly than pollen of the other.

B. Postzygotic mechanisms: prevention of normal offspring development

1. Incompatibility of zygotic or embryonic tissue with that of the mother plant produces seed abortion.

2. Hybrids (F_1) are completely inviable, or considerably weakened.

3. Hybrids are vigorous but sterile.

4. F_1 is vigorous, but successive generations (F_2, etc.) are weak or sterile.

may be used to show genetic relationships between taxa. The numerical taxonomist utilizes information from all fields and does not give extra weight to any particular type of information. This is in contrast to the work of other taxonomists who choose subjectively to weight particular types of information. Biosystematists, for example, emphasize the importance of breeding behavior; they often deal with populations at the species and subspecies levels.

7. Plants can also be characterized or described by their life cycle pattern. For example, herbaceous annuals may exhibit an "r" type life cycle "strategy," while long-lived woody perennials may exhibit a "K" type "strategy."

8. Organic evolution is the gradual change that has taken place in living organisms.

9. The fossil record reveals that prokaryotic life began more than three billion years ago, in the Proterozoic era. Eukaryotic life began more than one billion years ago. The fossil record became richer, more varied, and more complex in the early part of the Paleozoic era, as algal forms and the first terrestrial plants evolved. The Cambrian period saw the evolution of

tree forms and seed plants. Later in the Paleozoic era, Coal Age forests were dominated by lower vascular plants. Climatic and geological changes next brought dominance of the world's vegetation to the gymnosperms and then to the angiosperms in the Mesozoic era. The angiosperms have dominated most terrestrial habitats, in all kinds of climatic extremes, during the Cenozoic era (the last 65 million years).

10. A modern theory of the mechanism of evolution can be outlined as follows:

   a. Genes, contained in the chromosomes, are largely responsible for the development, structure, and metabolism of plants and animals.

   b. The complement of genes does not remain absolutely constant. Mutations (changes in chromosomes and genes) occur, which modify the structure and metabolism of the individuals containing them.

   c. Mutations causing considerable phenotypic change are likely to kill or to greatly weaken the plant because they upset the delicate equilibrium existing between the plant and its environment.

   d. Hybrids differ from their parents because of the resulting new combinations of genes. Hybrid swarms, resulting from introgressive hybridization, are common in some taxa.

   e. If the variants produced by mutations or hybridization are better adapted to the environment than the parent plants, the parent plants may be replaced by the new forms. Many complicated factors are involved in this replacement.

   f. New species are formed by processes that divide a population, ultimately leading to reproductive isolation.

11 CHAPTER

algae

Algae are photosynthetic, nonvascular plants. They include some of the most primitive plants on earth. They may exist as single cells, as loosely organized clumps of cells (**colonies**), as flat, leaflike sheets, or as intricately branched and intertwined **filaments** (Fig. 11.1). Only rarely, in the kelps, do algal bodies become large and complex enough to differentiate their tissues into organs that resemble roots, stems, and leaves (Fig. 11.2). Approximately 25,000 species of algae have been identified, but many more, especially those in the oceans, remain to be described for the first time. It is estimated that there may be 10 times this number of species living today.

Grouping all these species under the general classification of algae is highly artificial because there does exist a far greater diversity of form and metabolism among them than among the bryophytes and the vascular plants. It is difficult to find definitive traits that are common to all algae yet absent from all other divisions. Possibly the only structure possessed by all higher plants that rarely occurs in the algae is a protective jacket of sterile cells surrounding the developing gametes. In addition, most algal bodies are relatively undifferentiated. Their aquatic and semi-aquatic environments lack survival pressure for the evolution of vascular and supporting tissue. All the cells of most algae can carry on photosynthesis and

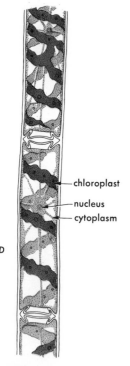

Spirogyra

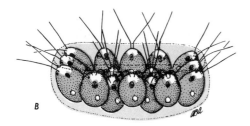

chloroplast
nucleus
cytoplasm

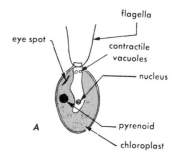

flagella
eye spot
contractile vacuoles
nucleus
pyrenoid
chloroplast

Chlamydomonas

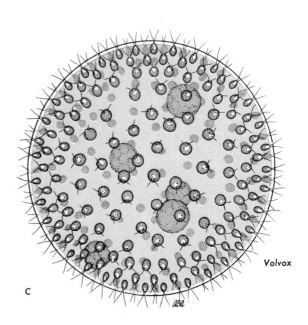

Volvox

Volvolcine Line

222

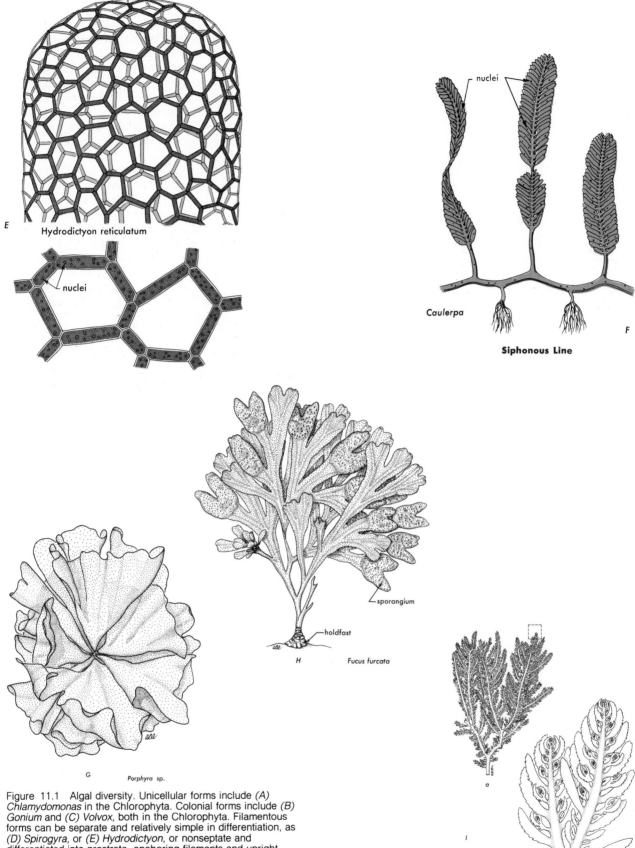

E

Hydrodictyon reticulatum

nuclei

nuclei

Caulerpa

F

Siphonous Line

sporangium

holdfast

H   Fucus furcata

G   Porphyra sp.

a

I

b   Ptilota filiciana

Figure 11.1   Algal diversity. Unicellular forms include *(A)*
Chlamydomonas in the Chlorophyta. Colonial forms include *(B)*
Gonium and *(C) Volvox,* both in the Chlorophyta. Filamentous
forms can be separate and relatively simple in differentiation, as
(D) Spirogyra, or *(E) Hydrodictyon,* or nonseptate and
differentiated into prostrate, anchoring filaments and upright,
photosynthetic filaments as in *(F) Caulerpa,* all in the
Chlorophyta. Sheetlike forms include *(G) Porphyra* in the
Rhodophyta. More differentiated, larger forms with leaflike fronds
include seaweeds and kelps such as *(H) Fucus* and *(I) Ptilota
filiciana.*

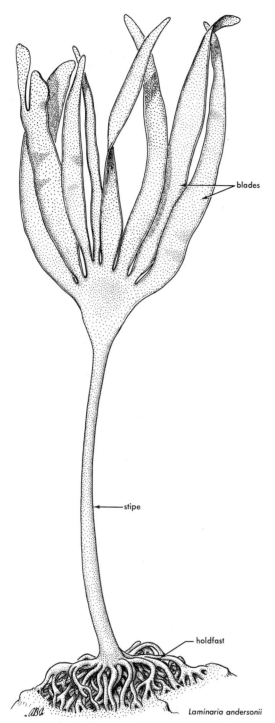

blades

stipe

holdfast

Laminaria andersonii

Figure 11.2 Example of a kelp, *Laminaria andersonii*, in the Phaeophyta.

obtain water and nutrients directly from their surroundings. Algae lack true roots, stems, and leaves (with vascular tissue); such a body is called a **thallus**. Plants with such a body are called **thallophytes**, as described in Chapter 10.

Algae can be found nearly everywhere. They float in air or water. They are attached to tree trunks or branches, to the bottom of streams, to soil particles, or to rocky intertidal cliffs battered by surf. They are also found

growing symbiotically with other plants or animals. Algae occur in the most severe habitats on earth. Some species grow on snow in perpetually freezing temperatures; others thrive in hot springs at temperatures of 70°C or more; a few live in extremely saline bodies of water, such as the Great Salt Lake in Utah; yet others survive the pressure and low light intensity conditions of 100$^+$m below the surface of lakes or seas. They have even been detected 1 km from ground zero, 6 months after a 20 kiloton atomic bomb tower test in Nevada, and they were considered to be survivors of the explosion.

Probably the most commonly noticed natural urban habitats for algae are the sides of glass fish tanks. They may also generally be found around leaking faucets and in garden or park pools that are not kept "pure" with chemicals. The "bloom" occurring during the summer on many lakes or the scum found on ponds is actually algae. Microscopic forms occur in most natural waters, including the top 75 m of the ocean. Here they constitute a primary food source for marine animals of all types and account for about 50% of the oxygen released into the atmosphere through photosynthesis.

Economic Importance of Algae

Algae are important in two basic, but quite different ways. They are important to the entire biosphere because of the ecologically vital functions they perform. These functions include the production of carbohydrates, which places the algae at the base of food chains, the fixation of nitrogen, and reef building. They are also economically important to people because they serve as food, fodder, and fertilizer, and also have many industrial and pharmaceutical uses.

Ecological Functions

Algae as Producers

In aquatic ecosystems, algae are the major producers. Shallow bodies of water may depend on productivity by higher plants such as rushes, water lilies, duckweed, or the like, or they may be fueled by a constant supply of decomposing plant material (**detritus**) washed down from the surrounding land. About 70% of the globe is nonterrestrial and most of that is covered by deep water. Here algae are the only producers. They have been called the grasses of the oceans, converting solar energy into chemical energy that is passed up through the rich marine food chain, and releasing oxygen as an important by-product to both water and atmosphere.

Algae (and other life) occupy several distinct habitats in aquatic ecosystems. One habitat, along rocky oceanic coasts, is the **intertidal** zone (Fig. 11.3). This zone is especially severe for the growth of algae, because the plants are alternately exposed by low tide and inundated by high tide. When exposed, the plants are beset by desiccation, high temperatures, high light intensity, and increased salinity. When the tide comes in, all environmental factors change abruptly. In winter, the plants may be locked in ice. The **neritic** zone is below the intertidal, but is still relatively shallow and near shore. Species of attached algae in the intertidal and neritic zones have their own particular distribution: some occur at the upper end, others occur in the middle, and others in the lower region (Fig. 11.4). Their distribution probably reflects different degrees of adaptation to the stresses of exposure, for those species at the upper part of the intertidal zone are exposed more frequently and for longer periods of time than species farther down. Experiments tend to confirm this hypothesis: algae that grow in the upper part of the intertidal zone are actually more resistant to drying, and to fluctuations in temperature, light intensity, and salinity.

Gradually, with increasing depth, the quality and quantity of penetrating light becomes less and less favorable for photosynthesis. The **compensation depth** or **compensation intensity** is the point where positive growth is no longer possible (Fig. 11.5). In deep bodies of water, the habitat between the surfaces and the compensation depth is dominated by floating algae, the **phytoplankton**.

Phytoplankters are usually microscopic, but they do include the large, floating *Sargassum* seaweed, namesake of the Sargasso Sea off Bermuda. The microscopic forms are typically unicellular and maintain buoyancy by storing oil droplets or by developing fine projections that extend out from the cell wall (Fig. 11.6). The average life span of a given cell is probably measured in hours or days; if it does not reproduce the protoplast dies and the cell wall remains will sink to the bottom. Most phytoplankters are members of the Bacillariophyta (diatoms) and Pyrrhophyta (dinoflagellates) divisions, but the Chlorophyta (greens), Cyanophyta (blue-greens), and Euglenophyta are also represented.

The quality of light changes with depth. Since red light is completely absorbed in the upper layers, a blue-green twilight prevails further down. Experiments show that aquatic algae have adjusted their metabolism to the light at this depth. Figure 11.7 shows the action spectrum of photosynthesis for the green alga sea lettuce (*Ulva taeniata),* which grows high up in the intertidal zone; superimposed on the same graph is the action spectrum for the red alga *Myriogramme spectabilis,* which grows at much greater depths. You can see that the peak has shifted and condensed to center around the blue-green region, 440 to 680 mμ. Special accessory pigments (phycobilins) present in the red algae, trap the light of blue-green wavelengths, eventually transferring this energy to chlorophyll.

Productivity and Pollution

The density of phytoplankton is not very high in the open ocean, perhaps a few thousand cells per liter, but the oceanic expanse is enormous and this results in a high annual rate of gross productivity: 32.6×10^{16} kcal (3 quintillion, 260 quadrillion). This is about three times the annual productivity of all the world's grassland and pasture, and four times that of all cropland.

As great as this productivity is, however, it is limited by many environmental factors such as light, temperature, and nutrients. If these limiting factors are ameliorated, productivity will increase manyfold. A natural process of increasing productivity and a changing environment takes place very slowly in most bodies of water because silt accumulates, making the depth shallower and the water uniformly warmer, and vegetation encroaches along the edges and contributes increasing amounts of detritus. This process is called **eutrophication**, and it ordinarily takes centuries, millenia, or eons. Human activities, however, mainly through the disposal of industrial, agricultural, and residential wastes, can telescope eutrophication into a matter of decades. This is called **cultural eutrophication**.

As nutrients increase, algal and bacterial activity increase, resulting in more turbidity and lower light. Oxygen becomes limiting near the bottom; this limitation plus siltation prove fatal to fish eggs of some species. Algal species replace one another in an aquatic

Figure 11.3   The intertidal zone, exposed at low tide, along the central coast of California.

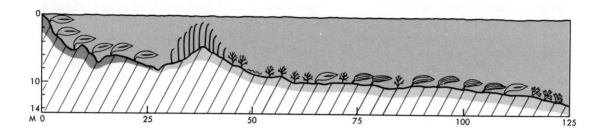

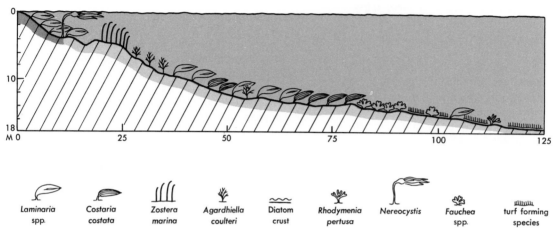

Figure 11.4 Representative transects of benthic (bottom attached) algae in the near-shore region of Puget Sound, Washington. Dark substrate is rocky.

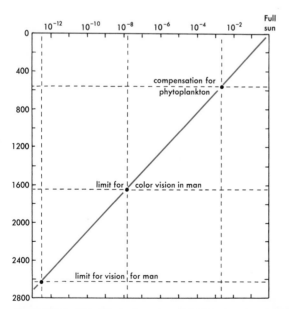

Figure 11.5 The penetration of sunlight into the clearest ocean water. The compensation intensity for most phytoplankton is 1/100th to 1/1000th the intensity of full sunlight at the ocean surface; compensation depth is about 550 ft (170 m).

succession: generally greens are replaced by blue-greens that may impart disagreeable odors or tastes to the water. As the base of the food chain changes, so does the entire chain. After 100 years of accelerating industrial and residential development along its shores, Lake Erie had changed from a relatively clean lake with important fishing and recreational values to an aesthetic and economic loss. Plankton and other algae increased six- or sevenfold in some places, and the composition shifted toward the noxious blue-greens. Oxygen content of bottom waters declined 67%. Commercial catches of trout, whitefish, herring, sauger, walleye, and blue pike declined by 99.9%, while such fish as carp, shad, alewives, and smelt increased. Recently, the rate of pollution has declined and water quality has improved.

Red Tides and Blooms

One of the results of cultural eutrophication is the episodic appearance of **red tides** or **blooms**: the rapid growth of phytoplankton populations until they become dense enough to color the water. This phenomenon can occur in bodies of water that range from small ponds to lakes to large coastal regions. Marine occurrences are often reported along the southwest coast of India, southwest

Africa, southern California, Texas, Florida, Peru, and Japan. Factors associated with red tides or blooms are long hours of sunlight, shallow, warm, offshore water, and high levels of nitrogen and phosphorus.

Red tides are most often caused by dinoflagellate species of *Gymnodinium* or *Gonyaulax*. Both species produce a water-soluble toxin of high potency that affects animal nervous systems. It is related to curare poison, extracted from certain tropical flowering plants, and it is 10 times as effective as cyanide. Massive fish kills and the poisoning of shellfish result from red tides. A 1947 episode off Florida killed an estimated 500 million fish.

Other blooms are caused by the chrysophyte *Prymnesium parvum* and blue-green species of *Microcystis, Anabaena, Nostoc, Aphanizomenon, Gloeotrichia,* and *Oscillatoria*. The blooms are not necessarily blue-green in color, however; a species of *Oscillatoria* causes red blooms, which give the Red Sea its name. Some of the plants above also produce toxins. A mere 72 g of *Anabaena* toxin would cause the death, within two minutes, of an adult human.

Nitrogen Fixation

Some bacteria and about a fourth of all blue-green algal species are able to assimilate or "**fix**" elementary nitrogen, N_2. The details of this metabolic pathway are still not clear, but this overall reaction can be written:

$$N_2 + 3H_2 \longrightarrow 2NH_3$$

The NH_3 can then be taken up by organisms, whereas the N_2 cannot. All forms of biological nitrogen fixation, bacterial as well as algal, contribute on the average 11 kilograms of nitrogen per hectare (ha) per year to terrestrial surfaces around the world. In exceptionally favorable areas, such as tropical grasslands, the amount added may be as high as 227 kg/ha. Blue-green algae are cultivated in rice paddies and in some other kinds of agriculture as a form of free fertilizer. Living cells secrete free amino acids and peptides into the soil or water medium, and dried algal crusts can be plowed into the land. Specialized, thick-walled cells called **heterocysts** conduct most of the fixation; the resulting organic nitrogen then diffuses to other cells in the filament (Fig. 11.8) or out to the environment.

Economic Uses

Algae as Food or Medicine

Seaweed—marine algae of moderate size, which usually grow in the intertidal or neritic zones—forms an important part of the human diet and medicine chest in several parts of the world. In the orient, seaweed harvesting has been known for 5000 years. Shen Nong, the legendary Chinese "father of medicine," prescribed seaweed for certain ailments in 3600 B.C. Some 3000 years later, Confucius praised its curative value. For centuries, the Japanese have used algae as a healthful, tasty supplement to their rice diet. The demand for *nori* (the red alga *Porphyra*) has

grown to such an extent in Japan that it is cultivated in shallow, intertidal bays (Fig. 11.9). The Polynesians in Hawaii were known to have utilized and named at least 75 species of *limu* (algae) as food sources. Some rare species were cultivated for the nobility in marine fish ponds. Dulce (the red seaweed *Rhodymenia palmata*) has been known as a food for 12 centuries in the British Isles. It was the Irish who discovered that small quantities of Irish moss (the red alga *Chondrus crispus*), when boiled with milk, would produce a jelly dessert that the French later called *blanc mange*. Brown algae off the coast of California (mainly *Macrocystis pyrifera*, Fig. 11.12) have been harvested for their content of iodine, which is added in trace amounts to the diet to prevent goiter, an enlargement of the thyroid gland in the neck.

It is not as food or medicine, however, that algae are most important today. With some exception, they do not have much nutritive value—in fact, their major constituents are largely indigestible. Algae are used more as condiments, garnishes, or desserts than as staple foods, much as we use lettuce, watercress, celery, or whipped cream. Iodine now is regularly obtained from other sources and added to table salt. Claims as to the health-giving value of seaweed do not have much foundation in fact.

Fodder and Fertilizer

Seaweeds not only contain such important trace elements as iodine, but they contain large amounts of potash (potassium), nitrogen, phosphorus, and other characteristics of good fertilizer or cattle feed supplements. In historic times, the North American Indians and the Scotch-Irish used Irish moss as a fertilizer to build up poor soils for such diverse crops as corn and potatoes. Seaweed has more recently been shown to compare favorably in nutrition to barnyard manure: it enhances germination, increases the uptake of nutrients in plants, and seems to impart a degree of frost, pathogen, or insect resistance. *Macrocystis* (Fig. 10.12) continues to be harvested off the California coast for use as a cattle feed supplement.

Cell Walls and Cell Wall Extracts

Peculiar characteristics of the cell walls of diatoms, browns, and reds have led to many recent industrial, pharmaceutical, and dietary advances. It is in these areas that the algae have their greatest economic value. The characteristics result in products such as diatomite, agar, carrageenan, and algin, each of which we shall consider individually.

Diatomite. Diatomite, also called diatomaceous earth, is a chalky, sedimentary rock composed of the cell wall remains of unicellular algae called diatoms. Diatoms, you recall, are important members of the phytoplankton (Fig. 11.6). One of the richest deposits of diatomite is located near Lompoc, California (Fig. 11.10). About 15 million years ago, that area was submerged beneath a warm, shallow sea. As diatoms flourished and died, their remains accumulated in bottom sediments. These particular remains persisted intact because the wall is not composed

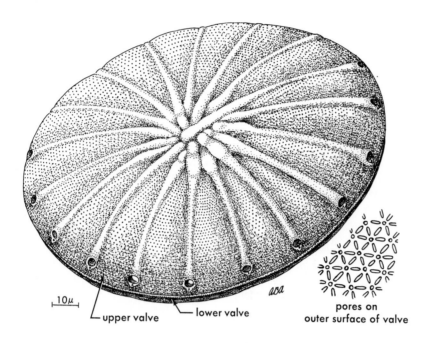

10μ

upper valve ⎯⎯ lower valve

pores on
outer surface of valve

Asteromphalus elegans

A

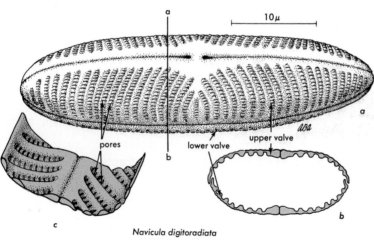

10μ

pores

lower valve

upper valve

Navicula digitoradiata

B

c

b

Figure 11.6 Examples of phytoplankters. *A* to *D*, diatoms *Asteromphalus elegans*, *Navicula digitoradiata*, *Asterionella formosa*, and *Biddulphia biddulphia*. *E* and *F*, dinoflagellates *Ceratocorys aultii* and *Ceratium* sp.

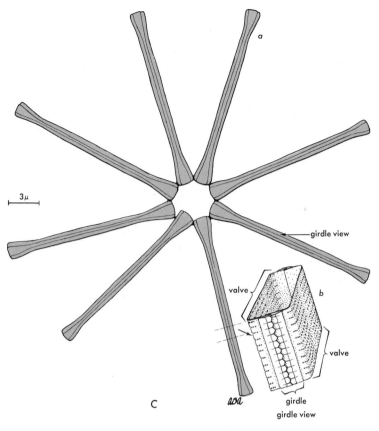

C

Asterionella formosa

girdle view

valve

valve

b

a

girdle

girdle view

3μ

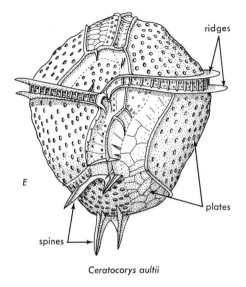

E

ridges

plates

spines

Ceratocorys aultii

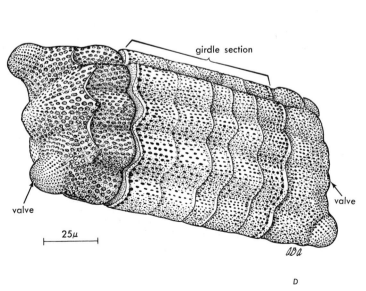

girdle section

valve

valve

25μ

D

Biddulphia biddulphia

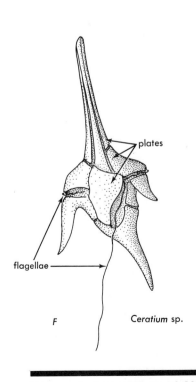

plates

flagellae

F

Ceratium sp.

Economic Importance of Algae

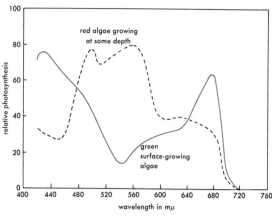

Figure 11.7 Action spectrum for photosynthesis in a green, surface-growing alga (solid line) and in a red alga growing at some depth (dashed line).

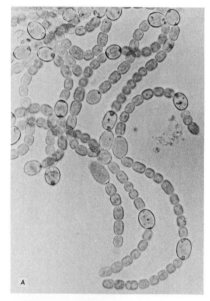

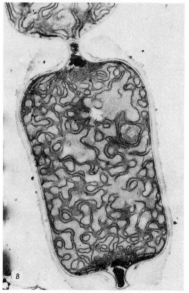

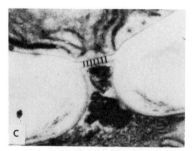

Figure 11.8 Ultrastructure of the heterocyst of the blue-green alga *Anabaena*. *A*, filaments of *A. azollae*, ×880, showing enlarged heterocysts in different stages of development. *B*, electron micrograph of a heterocyst of *A. cylindrica*, ×22,000, showing constrictions at heterocyst poles, three layers outside the cell wall, and contorted thylakoids (photosynthetic membranes) in the cytoplasm. *C*, an enlargement of *B*, showing microplasmodesmata connecting the cytoplasm of the heterocyst with that of the adjoining vegetative cell.

Figure 11.9 Nori (*Porphyra tenera*, Rhodophyta) culture, Sendai Prefecture, Honshu Island, Japan. *A*, distance view, showing the extent of the hibi nets in January, about the time of harvest. *B*, closer view, showing the hibi nets about 30 cm above mean low water in September, at the beginning of nori cultivation.

of carbohydrates such as cellulose; rather it is 95 percent silica.

The walls of a diatom fit together like two halves of a Petri dish and the glasslike case is perforated with hundreds of microscopic pores (Fig. 11.11). Each half is called a **frustule** and the region of overlap is the **girdle**. Under the electron microscope, it is possible to see that even the pores have pores. The perforations form exquisite, symmetrical designs, each design peculiar to a given species. Diatoms extract the opaline silica from the surrounding water by a process that is still not understood. In more recent geologic time, the Lompoc area was uplifted above sea level and the diatom deposit was revealed by erosion. Today, such companies as General Refractories and Johns-Manville mine the diatomite for industrial and pharmaceutical use.

Diatomite makes a superior filter or clarifying material because the microscopic pores create a large surface area (0.5 lbs—230 g—contain the area of a football field), and the rigid walls are incompressible. Diatomite is inert and can be added to many materials to provide bulk,

improve flow, and increase stability. In those ways it is used in cement, stucco, plaster, grouting, dental impressions, paper, asphalt, paint, and pesticides. Diatomite is also used as an abrasive.

Agar. Less than 100 years ago, a physician's wife, Frau Fanny Eilshemius Hesse, discovered the use of agar as a bacterial culture medium. She passed the information on to her husband, and he to Robert Koch, and Koch to the world via his important scientific writings in the late nineteenth century.

Agar (or agar–agar) is a polysaccharide, analogous to starch or cellulose but chemically quite different. It is found in the walls of certain red algae. Mainly species of *Gelidium* and *Gracilaria* seaweeds are commercially harvested, divers descending 3 to 12 m in the warm near-shore waters off Japan, China, California, Mexico, South America, Australia, and the southeastern United States.

In addition to its bacteriological use, agar is important in the bakery trade. When added to icing, it retards drying in the open air or prevents running in cellophane packages.

Economic Importance of Algae

231

Figure 11.10   Aerial view of diatomite deposits near Lompoc, California.

Because it is virtually indigestible, agar is also used medically as a bulk-type laxative.

Carrageenan.   This is another polysaccharide from the walls of red algae. It is mainly harvested from Irish moss and the substance takes its name from the town of Carragheen, County Cork, along the south shore of Ireland, where its properties were first discovered.

Carrageenan reacts with the proteins in milk to make a stable, creamy, thick solution or gel. Consequently, it is used in ice cream, whipped cream, fruit syrups, chocolate milk, custard, evaporated milk, bread, and even macaroni. It is added to dietetic, low-calorie foods to bring back the appropriate mouth "feel" of nondietetic foods. Carrageenan is also used in toothpaste, pharmaceutical jellies, lotions of many sorts and as a whiskey chaser for a cold "cure." Irish moss is commercially harvested in this country in Maine, principally by two companies: Kraft Foods and Marine Colloids.

Algin.   Since this long-chain polymer is made up of repeating organic acid units, it is also called alginic acid. It is the principal wall component of brown algae, constituting up to 40% of the middle lamella and primary

wall by weight. Water is strongly adsorbed to algin, resulting in a thick solution. For example, one tablespoon of algin added to a quart (one liter) of pure water will increase the viscosity to approximately that of honey. In nature, algin may be valuable to intertidal browns during exposure by enabling water to be retained in and on the plant. Commercially, however, it has many uses (Table 11.1).

Hundreds of species possess algin, but only a few are commercially harvested: *Macrocystis pyrifera* along the California coast, and species of *Laminaria, Fucus,* and *Ascophyllum* off England. *Macrocystis* is the largest **kelp** (a general term for large, marine, brown algae) known, and it has the fastest growth rate of any multicellular plant in the world. From a single-celled zygote, it grows into a young plant rooted on a rocky bottom 6 to 30 m below the surface, and then into a 60 m long mature giant—much of its shoot floating on the surface—in the course of a single growing season (Fig. 11.12). At this stage it is differentiated into a rootlike region (the **holdfast** or **hapteron**) that anchors the plant to bottom substrate, a stemlike region (the **stipe**), and numerous leaflike **fronds** that arise all along the stipe and possess gas-filled **bladders** at their base that increase buoyancy. Each stipe,

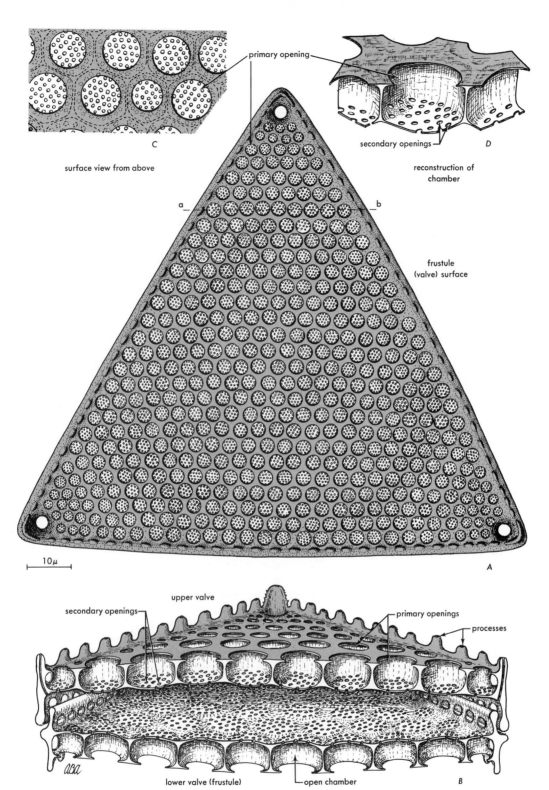

surface view from above

primary opening

secondary openings

reconstruction of chamber

frustule (valve) surface

a b

10μ

A

C

D

upper valve

secondary openings

primary openings

processes

lower valve (frustule)

open chamber

B

Figure 11.11   Cell wall details of the triangular diatom *Triceratium favus*.

Table 11.1
Some of the Products in Which Algin Is Used

Food Products	*Industrial*
Bakery icings and meringues	Water base paints
Salad dressings	Wall joint cements
French dressings	Welding rod coatings
Dietetic dressings	Textile print paste
Pickle relish	Textile sizing
Meat sauces and pepper sauces	Latex creaming and thickening
Orange concentrates	Adhesives
Fruit drinks	Paper coatings
Dietetic beverages and drinks	Corrugated paper
Beer	Paper sizing
Ice cream and related frozen desserts	Paper cartons for foods, soaps, and detergents
Egg nog	Food wrappers
Creamed cottage cheese	Waxed cardboard cartons
Cream cheese	Boiler compounds
Pasteurized cheese spreads	Can sealing compounds
Canned buttered vegetables	Battery plate separators
Canned chow mein	Mold release coatings
Canned meat stews	Wax cleaner polishes
Cake mixes	Ceramic glaze
Puddings, pie and cake fillings	Sugar beet clarification
Fountain syrups	Finger paints
Buttered pancake syrups	
Berry syrups	*Pharmaceutical*
Chip dips and mixes	Antibiotic tablets and suspensions
Instant dessert mixes	Dental impression compounds
Chocolate drink	Toothpaste
Delicatessen salads	Surgical jellies
Candy	Mineral oil emulsions
Dessert and salad gels	Medicated rubbing ointments
Breading batters	Tranquilizing tablets
	Hand lotion
	Facial beauty masque

growing as much as 30 cm a day, may live only a year or less, even though it has the potential for continuous tip growth. Other stipes continue to arise from the holdfast, however, and an average plant might live 4 to 10 years.

Macrocystis has been called the sequoia of the sea; indeed, it grows in such towering, dense stands or ''beds,'' that a scuba diver may feel he is swimming through a forest. It is the fast growth of *Macrocystis,* combined with its occurrence in dense beds, that make it commercially harvestable. Barges visit the beds every six weeks, cutting the tops one meter below the surface, then taking them to shore for processing.

Algal Classification

As shown in Table 11.2, there are many algal divisions that differ in biochemistry, cytology, and habitat.

Differences in cell wall construction, number and placement of flagellae, and type of chlorophyll are quite fundamental and significant. We shall briefly survey seven of the divisions.

Cyanophyta: the Blue-Green Algae

Unique among all the algal divisions, the blue-greens are prokaryotic and are thus most closely related to bacteria. In fact, some taxonomists place them in the bacteria, calling them the **Cyanobacteria.** They lack an organized nucleus, membrane-bound organelles, and endoplasmic reticulum. They do, however, contain gas vacuoles, photosynthetic membranes, and storage granules of material that has been likened to lipid, glycogen, and protein (Fig. 11.13). The wall resembles that of certain bacteria, being composed of muramic acid, glucosamine, glutamic acid, glucose, xylose, mannose, and several other mureins and sugars. Cellulose is not present. Virtually all free-living blue-green algae are surrounded by a mucilaginous sheath (Fig. 11.14).

Morphological diversity is limited in the Cyanophyta to unicellular, colonial, and simple filamentous forms (Fig. 11.14). There are no flagellated cells, and sexual reproduction has never been documented. Instead, asexual reproduction by fragmentation of filaments, the formation of resting cells, or simple fission predominates. Movement is passive except in some filamentous forms, which rotate and glide by an unexplained mechanism. Some forms, such as *Nostoc* or *Anabaena,* possess scattered specialized cells called heterocysts, which are the site of nitrogen fixation, as mentioned earlier (Fig. 11.7).

Rhodophyta: the Red Algae

Cyanophyta and Rhodophyta have a few characteristics in common. Both possess only chlorophyll *a* and identical accessory pigments called phycobilins, neither possesses flagellated cells, and both form a nonstarch storage product related to amylopectin. The differences, however, are much more striking. The Rhodophyta are eukaryotic, possess cellulose walls, and exhibit some of the most complex life cycles, incorporating both sexual and asexual reproduction, of any group of plants. Their range of morphological diversity is also greater than that of the blue-greens, including sheetlike forms and complex, finely branched filamentous forms with holdfast and stipe (Fig. 11.1). In some species, pits connect the cytoplasm of adjacent cells.

The reds are almost exclusively marine, and are most abundant in warm water, often extending to some depth. Again like the Cyanophyta, some forms are lime-encrusted and participate in tropical reef building.

Euglenophyta

This division is a small one, consisting mainly of unicellular, motile species (Fig. 11.1). Members of this division have several protozoan features, such as the lack

young lateral blades

terminal blade

D

mature blade

C

holdfast

haptera

5 mm

B

A

Macrocystis pyrifera

Figure 11.12 The kelp *Macrocystis pyrifera*, including details of the blades (C and D) and holdfast (B).

Table 11.2
Major Characteristics of 10 Algal Divisions[a]

Division	Chlorophylls and Accessory Pigments	Cell Wall	Storage Product	Flagella	Habitats	Number of Species	Notes
Cyanophyta (blue-greens)	a + phycobilins (phycocyanin and phycoerythrin)	Mureins and sugars + sheath	Cyanophyte starch (= amylopectin)	none	All, but ofen polluted water; 75% marine	1500	Asexual reproduction only; prokaryotic; prominent in blooms; some fix N
Rhodophyta (reds)	a (possibly + d) + phycobilins	Cellulose + sometimes agar or carrageenan	Floridean starch (= amylopectin)	none	98% marine; often warm, deep water	4000	Includes some seaweeds; complex life cycles
Chrysophyta (golden algae, class Chrysophyceae only)	a + c + fucoxanthin	Cellulose	Fats, oils, chrysolaminarin	2, unequal, anterior	Mainly fresh water	300	
Xanthophyta (yellow-greens)	a (+ c in some)	Cellulose or none; sometimes with Si and diatomlike	Oil or chrysolaminarin	2, various	Mainly fresh water	400	Sometimes placed in Chrysophyta
Euglenophyta	a + b	None	Paramylon (paramylum)	1–3, equal, anterior	Mainly polluted fresh water	450	Many animal-like properties
Chlorophyta (greens)	a + b	Cellulose	Starch	2, equal, anterior	Widest distribution of any division	7000	Possible precursor of higher plants
Charophyta (stoneworts)	a + b	Cellulose	Starch	2, equal, anterior	Often bottom dwellers in fresh water	250	Differentiated into nodes and internodes; gametangia with sterile jacket, sometimes put in Chlorophyta
Bacillariophyta (diatoms)	a + c + fucoxanthin	Silica + pectin	Fats, oils, chrysolaminarin	generally none	Mainly aquatic	8000+	Prominent in phytoplankton; sometimes put in Chrysophyta
Pyrrhophyta (dinoflagellates, class Dinophyceae only)	a + c + peridinin	None or plates of cellulose	Starch(?)	2, unequal, lateral	93% marine	1000	Prominent in phytoplankton
Phaeophyta (browns)	a + c + fucoxanthin	Cellulose + algin	Laminarin, mannitol	2, unequal, lateral	99.7% marine; often, shallow water	1500	No unicellular forms; includes the kelps (another name for brown seaweeds)

[a] A Few Small Groups Have Not Been Included.

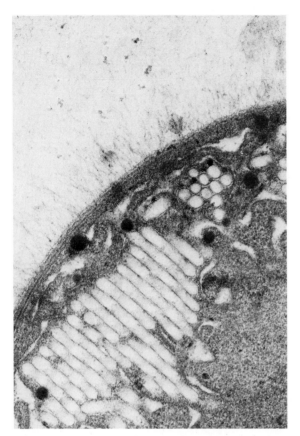

Figure 11.13 Ultrastructure of the blue-green alga *Anabaena flos-aquae*, ×73,500, showing microfibrils in the outersheath, a four-layered cell wall, gas vesicles in longitudinal section (rods) and cross section (circles), photosynthetic membranes, dark-staining lipid bodies, and many granular ribosomes.

of a cell wall (although the outer cytoplasm may be modified for rigidity), rapid movement, and the ability to produce a nonstarch food reserve called paramylon. Some Euglenophyta, in addition, lack chloroplasts and either engulf or absorb food. Exposure to ultraviolet radiation can halt chloroplast division but not cell division; the result is that within a few generations some progeny cells will lack chloroplasts. These cells will continue to live as heterotrophs and produce progeny just like themselves, revealing what a minor genetic difference there sometimes is between "plants" and "animals."

Chlorophyta: the Green Algae

The green algae are predominantly fresh-water forms, but they also exist in salt water, on snow, in hot springs, on soil, on branches, and on the leaves of terrestrial plants. They form the second-largest division within the algae. (Diatoms are the largest.)

Cytologically, this group shares several important traits with higher plants: chlorophyll *a* and *b,* similar accessory pigments, starch as a storage product, and cellulose cell walls. For this reason, some scientists hypothesize that ancient Chlorophyta served as the ancestors of all higher plants. The most complex greens have well-developed prostrate portions and upright portions, indicating a high degree of differentiation. This type of habit is called **heterotrichy.** Other forms as shown in Fig. 11.1, are unicellular, filamentous (with or without cross-walls), colonial, and sheetlike.

Bacillariophyta: the Diatoms

We have already discussed these organisms earlier in the chapter as one of the major components of phytoplankton. They exist as single cells (Figs. 11.6, 11.11)—sometimes stalked and sedentary, sometimes floating free—or as filaments. As seen in a face view, their silica walls are either circular or elongate, giving rise to two main orders, Centrales and Pennales. Some diatoms are capable of a gliding motion even though cilia and flagella are absent.

Diatoms contain chlorophyll *a* and *c,* store carbohydrate as oil, and their unusual silicate wall is embedded in a pectin matrix. They are mainly aquatic, but are also found in soil.

Pyrrhophyta: the Dinoflagellates

Together with the diatoms, this group dominates the phytoplankton. Most of its species are unicellular, flagellated, and marine (Fig. 11.6), but a few are colonial or filamentous. These organisms in great density cause "red tides." The cells are either naked, that is, with only a thickened membrane around them as in Euglenophyta, or they are enclosed by a unique cellulose wall that resembles patches or armor plating stuck together. Usually two flagella are present, both emerging from the same pore, but otherwise different. One flagellum is flat and ribbonlike and encircles the cell in a transverse groove; it is responsible for rotation and some forward movement. A second flagellum trails behind and provides forward movement while at the same time acting as a rudder.

Dinoflagellates contain chlorophyll *a* and *c* and a brown pigment, peridinin, which gives the group its green-brown or orange-brown color. True starch appears to be stored. Some forms are luminescent and contribute to the glow of water when it is disturbed at night, as in the wake of a ship.

Phaeophyta: the Brown Algae

The browns include filamentous forms, sheetlike forms, and large kelps with more complex differentiation in anatomy and morphology than any other algae (Figs. 11.1, 11.2, 11.12, 11.15). Unicellular forms are unknown. Chlorophyll *a* and *c* are present, plus the accessory pigment fucoxanthin, which gives these plants their characteristic brownish color (though color can range from olive-brown to golden-brown to practically black). Carbohydrate is stored as mannitol or laminarin, not as starch. The cell walls are composed of cellulose and often a great deal of algin as well.

Kelps such as *Macrocystis* (Fig. 11.12) are complex anatomically as well as morphologically. If the stipe is

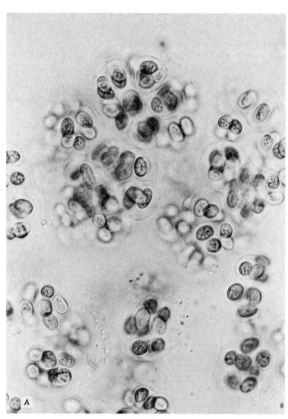

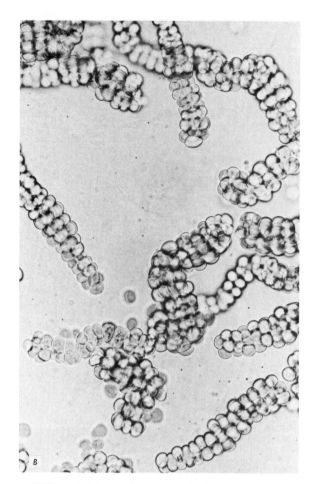

Figure 11.14 Diversity of form in the Cyanophyta. *A* and *B*, loosely organized colonies of *Gloeocystis* sp. and *Fisherella musicela*, both ×800; note the pronounced gelatinous sheath around *Gloeocystis*. *C*, a colony of the filamentous *Nostoc* sp., ×880, with enlarged heterocysts visible, all embedded in a gelatinous matrix. *D*, filaments of *Oscillatoria* sp. (large cells) and *Anabaena* sp. (small cells, with terminal and intercalary heterocysts), ×880. *E*, the spiral filamentous *Arthrospira* sp.; two filaments are intertwined along one portion of the figure, ×880.

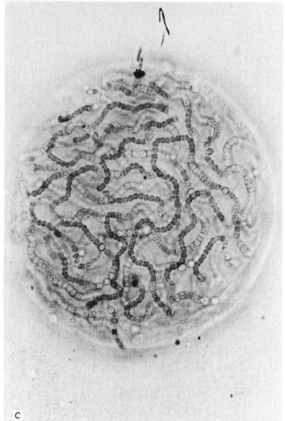

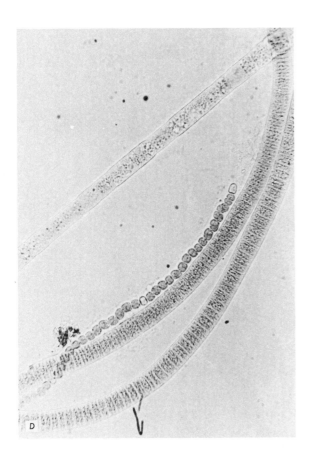

sectioned and examined under the microscope, several regions are apparent (Fig. 11.16). Cells in the outermost layer not only are protective, but they remain meristematic and contain chloroplasts as well. To distinguish this diverse tissue from epidermis, it is given the name **meristoderm**.

A broad cortical region is composed of parenchyma-like cells. Mucilage-secreting cells form definite canals through the cortex. Loosely packed filaments of cells fill the central region, the **medulla**. Some inner cortex cells next to the medulla appear to function as sieve-tube members: they have sieve plates, form callose, adjoin one another to form continuous tubes, and are known to translocate mannitol by a mechanism resembling that found in vascular plants (Fig. 11.17). However, these cells contain no tissue that resembles xylem. The value of a photosynthate-conducting system in these large plants is easy to perceive. The mass of floating fronds considerably reduces the penetration of light into the water below; consequently the lower part of the stipe and the holdfast may be below the compensation depth. They must be nourished by food translocated from above.

Unlike many of the other algae, the dominant generation of the kelps is diploid. The dominant generation is also larger and more complex than any other alga.

Reproduction

Knowledge of reproductive processes in the algae is incomplete and largely confined to the Rhodophyta, Phaeophyta, and Chlorophyta. In some other divisions, detailed knowledge is known only for a very few species, genera, or families. There is no typical life cycle for the algae as a group.

Asexual Reproduction: Vegetative

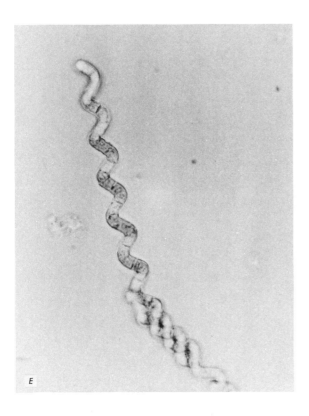

Vegetative reproduction is widespread in many forms and is probably used in all divisions. Indeed, in those algal groups where sexual reproduction is unknown or rare, asexual reproduction may be the only form of reproduction.

The longitudinal splitting of a biflagellate motile *Euglena* into two, still motile, but uniflagellate daughter cells is simple vegetative reproduction. The diatoms may divide vegetatively for up to five years. Recall that the cells have two sections that fit into each other like the top and bottom of a Petri dish. After division, a new wall forms within the old wall (Fig. 11.18). This means that the new cell receiving the lower and smaller wall will be smaller than the parent cell. Eventually, a minimum size is reached and division stops. Full size is regained at the time of sexual reproduction (Fig. 11.18).

Vegetative reproduction in filamentous forms occurs by a simple fragmentation of the filament. In some species, specialized cells seem to be associated with the breaking of the thallus. In species of the Cyanophyta, notably *Oscillatoria* (Fig. 11.15), the point at which fragmentation occurs is marked by a dead cell. Separation at these dead

Reproduction

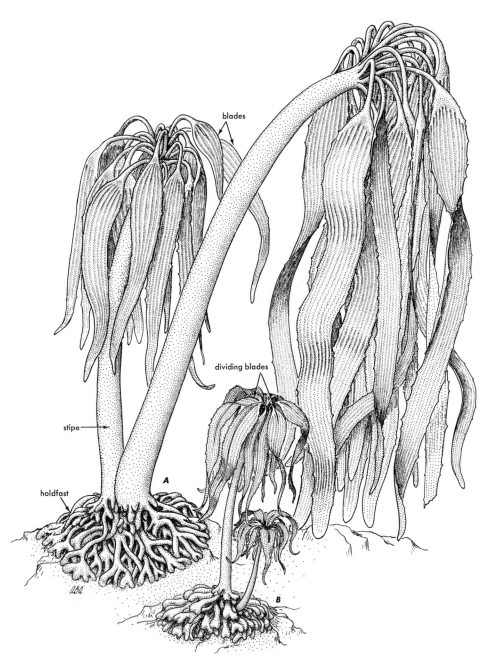

blades

dividing blades

stipe

holdfast

A

B

Figure 11.15 *A*, an example of a kelp, or brown seaweed:
Postelsia palmaeformis (sea palm), showing organs that resemble
roots, stems, and leaves. *B*, young plant.

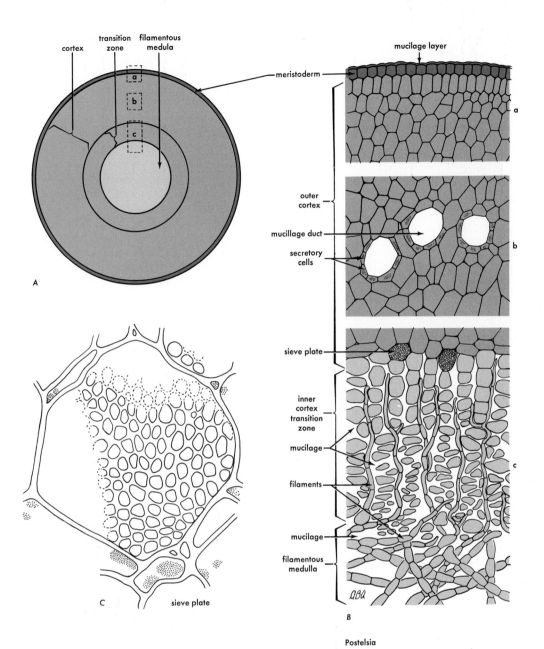

cortex · transition zone · filamentous medula

A

meristoderm

mucilage layer

a

outer cortex

mucillage duct

secretory cells

b

sieve plate

inner cortex transition zone

mucilage

filaments

mucilage

filamentous medulla

c

B

Postelsia

C · sieve plate

Figure 11.16   Anatomical details of the stipe of *Postelsia*. *A*, diagrammatic cross section, showing the major regions and the location of detail drawings *B* and *C*. *B*, details of the meristoderm region, cortex with mucilage canals, and inner cortex with sieve-tubelike cells in a transition region near the central medulla. *C*, detail of the end wall of a sieve-tubelike cell, showing pores through which mannitol and other substances can flow.

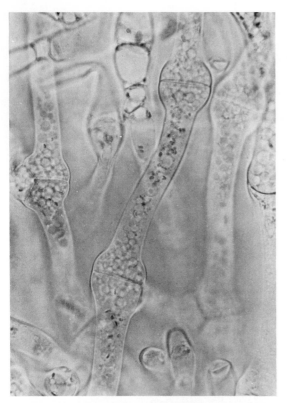

Figure 11.17 Photograph of sieve-tubelike cells, with swollen junctures, from the kelp *Laminaria groenlandica*, about ×900.

Most zoospores are pear-shaped or spherical. Depending on the species, they have two, four, or many flagella. After liberation from the sporangium, zoospores are actively motile for as short a time as a few minutes or for as long as three days. Their movement and periods of activity are frequently affected by light. After their period of activity they settle to the bottom of a pond or culture tank, lose their flagella, and secrete a cell wall, and by cell division begin to develop a new thallus.

Sexual Reproduction

Sexual reproduction is responsible for variation within a population. Since sexual reproduction is frequently associated with resistant cells that carry the plants over seasons unfavorable to growth, all of the new plants in a given population formed by sexual reproduction will have different genotypes when growth is resumed; those plants best adapted to the new growing conditions will become established.

The cells that fuse in sexual reproduction are known as **gametes**. Gametes have the **haploid** or *n* number of chromosomes. The single cell resulting from the fusion is a **zygote**, and has a **diploid** or *2n* number of chromosomes. By cell division, the originally one-celled zygote can produce a diploid plant (the **sporophyte** generation). The number of chromosomes is returned to the haploid, or *n*, number by meiosis. Reproductive cells resulting from meiosis, with the *n* number of chromosomes, may be called **meiospores**. Meiosis occurs at different places in the sexual cycle, and the cells in which it occurs have been called **meiocytes**. Meiospores can be motile or nonmotile. If they lodge in an appropriate environment, they germinate and by cell division produce a haploid plant (the **gametophyte** generation). The gametophyte generation produces the sexual cells called gametes.

Gametes are produced in cells called **gametangia**. A gametangium, in the most unspecialized case, may be indistinguishable from an ordinary vegetative cell, except that several additional mitoses occur within it, giving rise to 16 to 32 cells. In the unicellular, motile green alga *Chlamydomonas*, these cells strongly resemble the vegetative cells, but they may be smaller (Figs. 11.19, 11.21). In the simple filamentous species of the green alga *Ulothrix* (Figs. 11.20, 11.21), an ordinary vegetative cell gives rise to 8 to 64 flagellated cells. When only eight motile cells are formed, they germinate directly, acting simply as zoospores (mitospores), and the cell in which they are formed is called a sporangium. However, when a greater number of cells is formed, the cells are smaller and they behave as gametes that fuse in pairs. When gametes resemble each other, they are called **isogametes**, and the species that produces them is **isogamous** (Fig. 11.21). Thus, *Chlamydomonas* and *Ulothrix* are isogamous.

In some species, the gametes differ in size, one being slightly smaller than the other. Gametes with only a slight difference in size are said to be **anisogametes** (heterogametes), and species expressing this state are **anisogamous species** or exhibit **anisogamy** (heterogamy) (Fig. 11.21).

cells results in short filaments that have a gliding motion. This enables them to change their position in the mucilaginous mass in which the filaments grow, and gradually to enlarge the mass. Other species may modify somatic cells by adding a thick wall, and these cells may function as **resting spores**, living through a hostile period of time in dormancy when all other somatic cells die.

Asexual Reproduction: Mitospores

Asexual spores are always preceded by mitosis and may be called **mitospores** to distinguish them from spores preceded by meiosis, called **meiospores**. The latter constitute a stage in sexual reproduction and will be considered later. Since only mitotic divisions are involved, the genotypes of all asexual spores arising from one and the same parent are identical with each other and with the parent plant. They will produce a population of genetically identical individuals. When a population has the genotype best fitted to a given growing condition, or locality, it will multiply by asexual reproduction and populate that locality.

Mitospores may be either motile or nonmotile. Motile spores are called **zoospores** and nonmotile spores are frequently called **aplanospores** (or, in *Polysiphonia*, carpospores). Mitospores are produced in specialized cells called **sporangia**. In many algae, the sporangia show little, if any, difference in appearance from ordinary vegetative cells. The number of spores produced depends on the species, but it is typically 16 to 64.

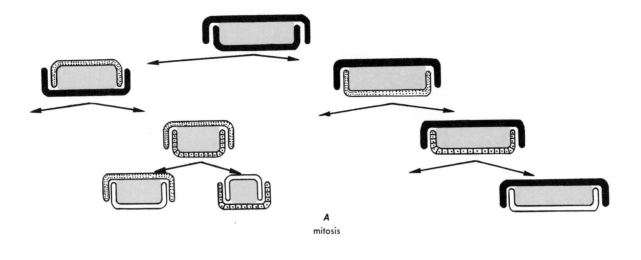

A
mitosis

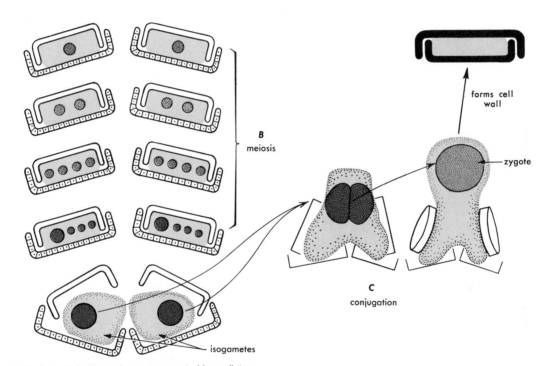

B
meiosis

forms cell wall

zygote

C
conjugation

isogametes

Figure 11.18   A gametic life cycle as represented by a diatom. *A*, the progeny cells continually become smaller if, after cell division, the new cell forms within the silica shell of the old cell. *B*, meiosis occurs; three of the nuclei with the *n* number of chromosomes degenerate and the single remaining haploid cell becomes a gamete. *C*, fusion of the two gametes to form a zygote.

In other species, the gametes not only differ in size but in degree of motility. One type of gametangium produces many small, motile gametes called **sperm**. A sperm-producing gametangium is known as an **antheridium**. Another vegetative cell may become quite enlarged, sometimes becoming flask-shaped. The protoplast retracts slightly from the cell wall. It has become an **egg** cell, and the cell in which it is produced is known as an **oogonium**.

The sperms are liberated from the antheridium and swim toward the oogonium. One sperm enters the oogonium, either through the funnel end or by a breaking down of the oogonial cell wall. Union of egg and sperm now ensues; this process is known as **fertilization**. When gametes differ in size and activity as they do in *Laminaria, Fucus,* and *Polysiphonia,* the plants exhibit **oogamy** (Fig. 11.21).

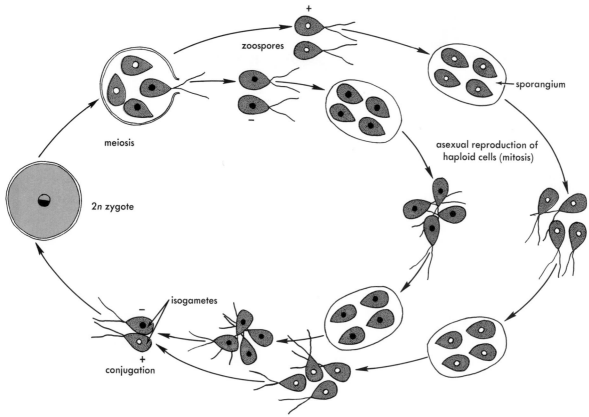

Figure 11.19 A zygotic life cycle is represented by the green alga *Chlamydomonas*. Meiosis occurs during the germination of the zygote; thus the zygote is the only diploid (sporophytic) cell.

How does a sperm cell find an egg cell? It could happen by chance, much as wind-blown pollen lands on stigmas of flowers of its own species. However, there is considerable evidence that the female gametangia or gametes in plants produce chemical agents that attract the sperms to the close proximity of the egg cells. For instance, a solution prepared from the mature female filaments of the alga *Oedogonium* may be drawn up into a capillary tube. When the tube is placed in a suspension of sperms, the sperms will collect at the tip of the capillary tube and some of them will eventually enter the tube. This attraction seems to be species-specific for *Oedogonium*. Once the gametes fuse, a diploid zygote cell is formed. Further cell division will produce the sporophyte generation.

Alternation of Generations

Fertilization constitutes one of the two critical steps of a complete sexual life cycle; meiosis is the second critical step. We may distinguish three types of life cycles, depending on the relation of meiosis to fertilization.

In most animals, meiosis results directly in gamete formation. In a sense, the entire gametophyte is composed of gametes. In this case, the mature individuals are diploid and the haploid stage is limited to the gametes. This is a **gametic life cycle**. It occurs in the diatoms and in the order Codiales of the Chlorophyta.

In many of the more primitive algae, meiosis accompanies the germination of the zygote. In this case, the mature plants are haploid and the diploid stage is limited to the zygote. This is a **zygotic life cycle**. It occurs in many Chlorophyta and in a few primitive Rhodophyta.

In most plants, meiosis and fertilization take place in distinct generations. One generation is haploid and produces gametes, the other is diploid and produces meiospores. This is a **sporic life cycle**. The plants of the sporophyte and gametophyte generations may be identical in appearance; if so, there is an **alternation of isomorphic generations**, as occurs in some Chlorophyta, primitive Phaeophyta, and all higher Rhodophyta. In **alternation of heteromorphic generations**, the haploid plants of the gametophyte generation have different forms from the diploid plants of the sporophyte generation. This occurs in a few Chlorophyta and in all of the more advanced Phaeophyta, and in the rest of the plant kingdom beyond the algae.

Gametic Life Cycle. The diatoms are unicellular forms that may divide by cell division for periods as long as five years. With each division, one of the cells is smaller than the parent cell (Fig. 11.18). At the end of a period of mitotic divisions, meiosis occurs and four gametes are produced, not all of which may function (Fig. 11.18). Depending on the species, these most generally are

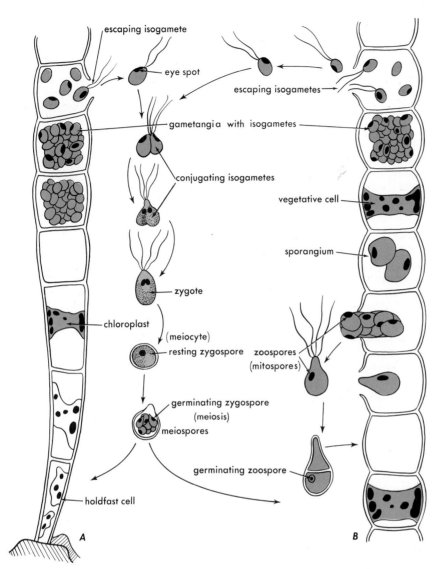

Figure 11.20 A zygotic life cycle as represented by the filamentous green alga *Ulothrix*. As with *Chlamydomonas*, only the zygote is diploid.

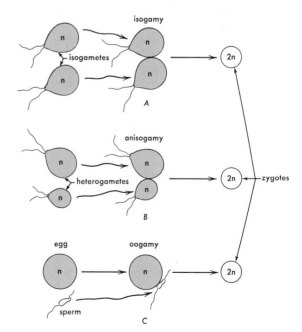

Figure 11.21 Differences in gametes. *A*, in isogamy the gametes are identical; *B*, in anisogamy the gametes are slightly different in size or activity; *C*, in oogamy, the gametes are very different in size and activity and are called sperms and eggs.

isogametes, but they may be anisogametes or even eggs and sperm. After fusion, the zygote increases greatly in size and, again depending on the species, may begin a rest period. The germination of the zygote is by mitotic cell division. Note that in diatoms, the products of meiosis are gametes, not meiospores. The gametes themselves are the only haploid cells in the life cycle.

Zygotic Life Cycle. *Chlamydomonas* and *Ulothrix* are examples of a zygotic life cycle (Figs. 11.19, 11.20). In each of them, two haploid gametes fuse to produce a diploid zygote. Meiosis occurs in the germination of the zygote, which is the only diploid cell in the life cycle.

Sporic Life Cycle: Alternation of Isomorphic Generations. *Ectocarpus* (Phaeophyta) and *Polysiphonia* (Rhodophyta) will serve to illustrate alternation of isomorphic generations. In *Ectocarpus* (Fig. 11.22), specialized reproductive cells are formed. The gametangia are grouped together, forming elongated, slightly curved structures. Each cell in this structure is a gametangium that has resulted from mitotic divisions and will produce one motile gamete with the haploid chromosome number. They conjugate to form the usual diploid zygote. The zygote, upon germination, will grow into a diploid sporophyte (Fig. 11.22C) whose thallus is identical in appearance to that of the gametophyte thallus. Two different kinds of sporangia are formed on this sporophyte plant. One of them resembles exactly the group of gametangia found on the gametophyte plant. Like the gametangia, all sporangium cells have been formed through mitotic divisions, and each cell will produce a motile diploid zoospore that can germinate directly into a new diploid plant (Fig. 11.22B). The other type of sporangium found on the sporophyte plant is spherical; after meiosis, it will produce numerous motile zoospores or meiospores that have the haploid chromosome number. These meiospores grow into new gametophyte plants (Fig. 11.22A). Haploid gametes, haploid meiospores, and diploid zoospores are all identical in appearance.

Polysiphonia. In the Rhodophyta, an alternation of isomorphic generations is complicated by a second, distinctive sporophyte phase. The delicate feathery thalli of male and female gametophytes and one sporophyte of *Polysiphonia* are identical in appearance and growth (Fig. 11.23). No motile cells are produced. The nonmotile sperm (spermatia) are produced in great profusion near the tips of male gametophyte filaments (Fig. 11.23C). The oogonia (carpogonia) form near the tips of the female gametophyte. The oogonia are flask-shaped cells with long slender necks (trichogynes) that protrude from a protecting envelope formed from the cells at the bases of the oogonia (Fig. 11.23D). Spermatia are carried by water currents to make contact with the necks of the oogonia. Nuclei enter the necks and pass downward to unite with the female nuclei.

Fertilization stimulates the development of an array of short diploid filaments (the **carposporophyte**), which will eventually produce diploid mitospores (**carpospores**). The original protective envelope of haploid filaments is also

stimulated to grow. It enlarges and gives rise to a surrounding protective case formed of haploid filaments (**cystocarp**, Fig. 11.23E). Thus, the first sporophyte generation (carposporophyte) in *Polysiphonia* and many other Rhodophyta is composed of short diploid filaments that produce diploid spores. This sporophyte is protected by an envelope of gametophyte cells. When discharged, the diploid carpospores give rise to a second sporophyte (**tetrasporophyte**), identical in appearance to the two gametophyte plants. Meiosis occurs in sporangia in the sporophyte plant in cells formed between rows of axial and pericentral cells (Fig. 11.23G). Each sporangium produces four meiospores (**tetraspores**) that on germination give rise to male or female gametophyte plants similar in appearance to the sporophyte plants.

Sporic Life Cycle: Alternation of Heteromorphic Generations

Kelp. Alternation of heteromorphic generations occurs in both the Chlorophyta and Phaeophyta but is most highly developed in the Phaeophyta, particularly in the kelps of the order Laminariales. The sporophyte generation of the kelps is large and well known; its vegetative characteristics have been discussed in some detail earlier in the chapter. The sporangia arise from meristematic cells, or the meristoderm, that form the outermost layer of cells on the fronds (Fig. 11.24B). The sporangia usually arise in groups (**sori**) and are accompanied by an elongation of adjacent cells that serve to protect the developing sporangia (Fig. 11.24). Depending on the species, each sporangium produces 8 to 64 motile meiospores.

The meiospores germinate into male or female gametophyte plants, both of which consist of a small branched filament (Figs. 11.24C,D). The antheridia are produced in large numbers, either singly or in groups, at the ends of short branches. There appear to be no specialized oogonia, as eggs may be produced in any of the cells of the female gametophyte (Fig. 11.24E). The eggs are usually discharged from the oogonium before fertilization, but remain attached to it for some time. The egg is thus fertilized outside of the oogonium in open sea water.

Fucus. Alternation of heteromorphic generations has reached an extreme in *Fucus* (Fig. 11.25). The diploid generation is the conspicuous one (Fig. 11.1), and the haploid phase has been reduced to a few haploid cells. These do not develop into free-living vegetative gametophytes but become transformed into sperm and eggs and fuse to produce a new diploid generation (Fig. 11.25).

The ends of the diploid strap-shaped thallus of *Fucus* are swollen and notched. They bear small raised areas (Fig. 11.25A). A section through these areas shows them to be small cavities (**conceptacles**) that open by small conical pores to the sea water (Fig. 11.25B). Unilocular sporangia arise in these cavities; and they are of two types. Microsporangia are formed profusely at the ends of short branching filaments, and megasporangia are large spherical cells separated from the wall of the cavity by a

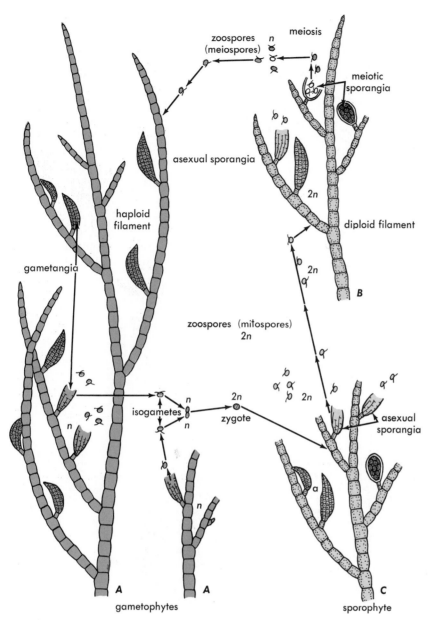

Figure 11.22 Sporic life cycle, alternation of isomorphic generations differing only in having diploid and haploid chromosome numbers, as represented by the filamentous brown alga *Ectocarpus*.

single cell (Fig. 11.25C). Sterile hairs are numerous, surround the sporangia and may protrude outward through the pore. In the Pacific Coast species of *Fucus*, both types of sporangia occur in the same cavity, but in the Atlantic Coast species they occur on separate plants.

The microsporangium contains a single large cell—a microsporocyte (Fig. 11.25F). Meiosis takes place, and four haploid cells are produced. Each can be simultaneously considered a meiospore and a microgametophyte. Each microgametophyte will rapidly divide by four mitoses, giving rise to 16 cells (a total of 64 cells within the old sporangium walls). Each of the 16 (64)

cells will differentiate into a sperm and be discharged into the cavity.

The megasporangium also contains a single large cell—a megasporocyte. Meiosis produces four meiospores, and each meiospore undergoes a mitosis without cell division; the result is four binucleate cells, each one of which is a megagametophyte. Ultimately, cell walls do form and an additional mitosis occurs, resulting in eight cells that further differentiate into eight eggs (Fig. 11.25E,O). The unfertilized eggs are discharged from the old megasporangium walls and they pass, with the sperms, outside the cavity into the open sea where fertilization takes place.

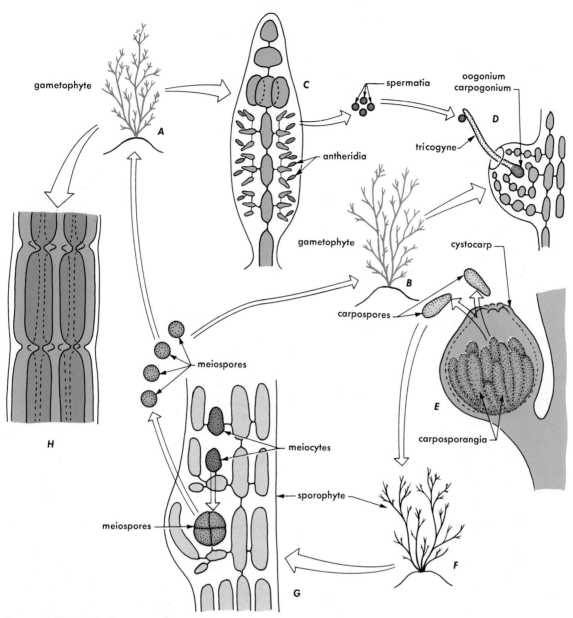

Figure 11.23 Sporic life cycle, alternation of isomorphic generations, as represented by the red alga *Polysiphonia*. *A,* male gametophyte plant. *B,* female gametophyte plant. *C,* production of spermatia at the tip of male gametophyte. *D,* the carpogonium (oogonium) is protected by filaments and spermatia become attached to the protruding trichogyne. *E,* the zygote produces a body of 2n carpospores, and the gametophytic cystocarp enlarges and surrounds the carposporophyte. *F,* a sporophyte, identical in appearance to the gametophyte, arises from the carpospores. *G,* some pericentral cells give rise to meiocytes which, through meiosis, produce four meiospores (tetraspores). *H,* the vegetative structure of both gametophytes and the sporophyte are identical.

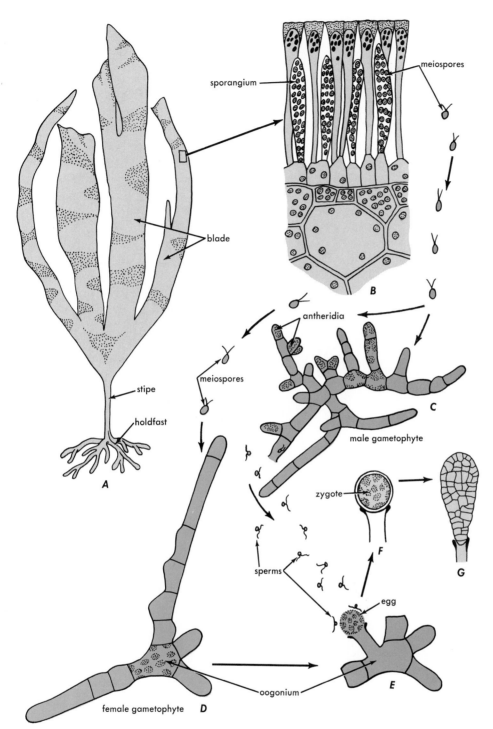

Figure 11.24 A sporic life cycle, with alternation of heteromorphic generations, as shown by the brown alga *Laminaria*.

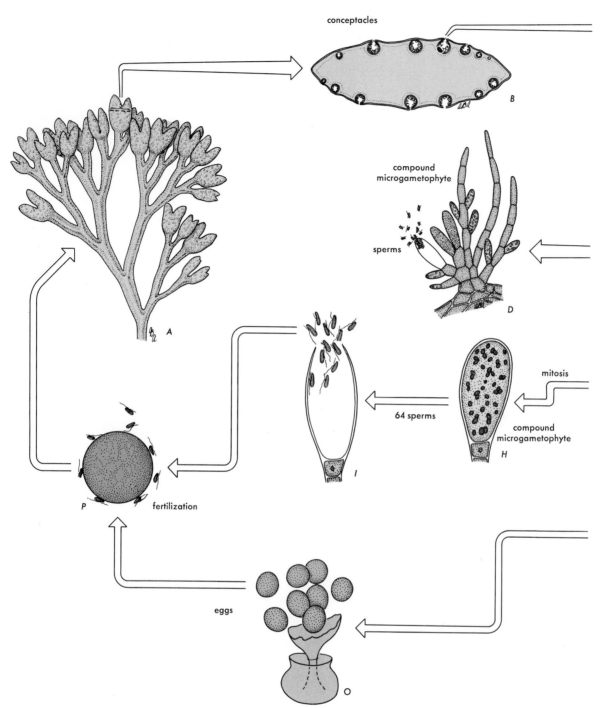

conceptacles

compound
microgametophyte

sperms

mitosis

64 sperms

compound
microgametophyte

H

I

P

fertilization

D

B

eggs

O

Figure 11.25   A sporic life cycle, as shown by *Fucus*. *A*,
dichotomously branching blade with swollen tips. *B*, cross section
of tip showing cavities (conceptacles). *C*, the conceptacles
contain sterile hairs (paraphyses), shorter filaments bearing many
small antheridial meiocytes, and large oogonial meiocytes on
short stalks. *D*, detail of filaments bearing sperm. *E*, detail of
oogonial meiocyte after meiosis. *F* and *G*, meiosis in male
meiocyte. *H*, mitotic division resulting in a compound
microgametophyte. *I*, each cell functions as a sperm, shown
leaving the microsporangium, *J*, *K*, and *L*, meiosis in the oogonial
meiocyte. *M*, a single mitosis following meiosis. *N*, an eight-celled
megagametophyte forms. *O*, the eight cells leave the
megasporangium and function as eggs. *P*, fertilization takes place
in the open water.

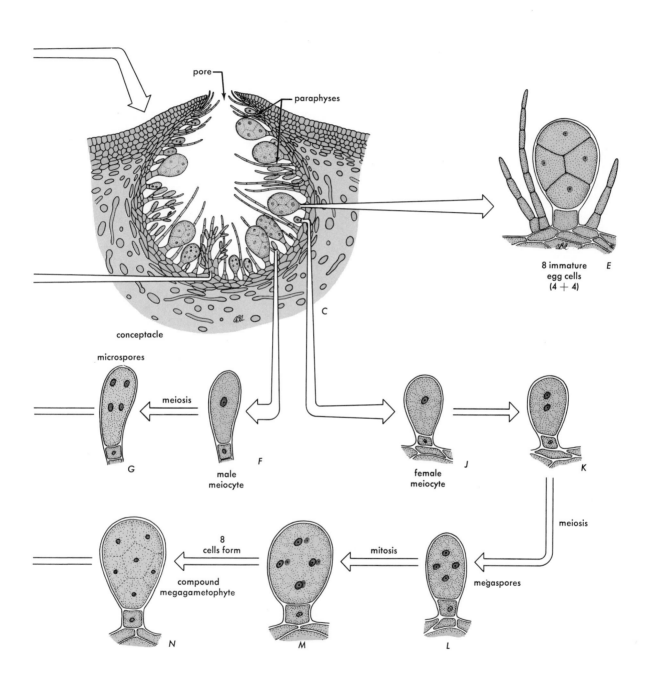

pore

paraphyses

8 immature
egg cells
(4 + 4)

E

conceptacle

C

microspores

meiosis

G

male
meiocyte

F

female
meiocyte

J

K

meiosis

8
cells form

mitosis

megaspores

compound
megagametophyte

N

M

L

Summary

1. Algae are photosynthetic, nonvascular, relatively undifferentiated organisms. Their gametangia are unicellular—that is, they lack a jacket of sterile protecting cells around the developing gametes.

2. The algae consist of more than 25,000 species belonging to 10 divisions. Although most species are aquatic, the algae as a group are found in a great diversity of habitats including some extreme environments.

3. Ecological importance of the algae centers around their role as phytoplankton at the base of aquatic food chains. Water pollution can lead to cultural eutrophication, red tides, and blooms. Some blue-green algae are also important because they fix atmospheric nitrogen into forms that can be utilized by other organisms.

4. The algae are economically important mainly because of their cell wall materials (agar, algin, carrageenan), but they have some medical, food, fodder, and fertilizer uses.

5. Algal classification is based on biochemical and cytological traits, as well as gross morphology. The Cyanophyta are the only prokaryotic algae. The Rhodophyta share some biochemical traits with the Cyanophyta but, in contrast, exhibit complex life cycles and a greater range of morphological diversity. The Euglenophyta are mainly unicellular and have some protozoan traits, such as the absence of a cell wall. The Chlorophyta are thought to be precursors of higher plants because they share the same chlorophylls and storage products as higher plants. Their species exhibit a great range of morphological diversity: unicellular motile, unicellular nonmotile, colonial, filamentous without cross-walls, filamentous with cross-walls, heterotrichous forms, and sheetlike forms. The Pyrrhophyta include the organisms which, in great density, cause certain kinds of red tides. The Phaeophyta include the largest, most anatomically complex algae, called kelps. Photosynthate is translocated in some kelps through cells that closely resemble sieve cells. Tissue resembling xylem, however, is absent.

6. Reproduction in the algae includes asexual and sexual types. There is no typical life cycle for the algae as a group. Asexual reproduction may involve cell division in unicellular forms, fragmentation of filaments, or the production of mitospores. The Cyanophyta only reproduce asexually. Sexual reproduction may involve isogamy, anisogamy, or oogamy, and the alternation of generations that results may be of identical-looking generations (isomorphic generations) or of different-looking generations (heteromorphic generations). In any case, the gametophyte generation is haploid and the sporophyte generation is diploid.

7. In a gametic life cycle, mature plants are diploid and the haploid stage is limited to the gametes. In a zygotic life cycle, mature plants are haploid and the diploid stage is limited to the zygote. In many algae, however, there are two distinct, independent generations that alternate in a sporic life cycle.

12 |

the mycota (fungi)

We commonly meet the fungi as mushrooms, mildews, molds, and the organism that causes athlete's foot. A few fungi provide food for humans, but many more compete with us for food sources and several of them use us as their own living food supply. The baking and brewing industries have an ancient partnership with the fungus known as yeast (*Saccharomyces*), and the medical profession has the makings of another long partnership with fungi such as *Penicillium notatum* for their production of penicillin and other antibiotics. It is doubtful whether any other group of organisms is associated with human beings in as widely varied an array of relationships.

Classification of the Fungi

The fungi are eukaryotic organisms that reproduce by means of spores and that cannot perform photosynthesis.

There are more than 200,000 known species of fungi. They are classified as a single division, the **Mycota**, which is subdivided in different ways by different taxonomists. The scheme shown below will be used in this text.

Division Mycota (the fungi)

I. Subdivision Myxomycotina (slime molds)
II. Subdivision Eumycotina (true fungi)
  A. Class Oomycetes (egg fungi)
  B. Class Zygomycetes (zygote fungi)
  C. Class Ascomycetes (sac fungi)
  D. Class Basidiomycetes (club fungi)
  E. Class Fungi Imperfecti (imperfect fungi)

All the fungi are **heterotrophs**: they must obtain food from the environment. Most are **saprophytes**. That is, their food is taken from nonliving sources such as plant and animal wastes and dead bodies. The rest are **parasites**, organisms that take organic materials directly from other living organisms. There are even a few fungi that are parasites on other fungi!

The Myxomycotina are separated from the rest of the fungi because their active cells lack cell walls. The Eumycotina have cell walls, giving the organisms a more or less rigid body shape and plantlike habits.

Mycologists (those who specialize in the study of fungi) subdivide the Eumycotina on the basis of the structures they produce for sexual production. Since there is no simple way to summarize the differences between the various groups without discussing their life cycles, we shall survey each group individually.

Subdivision Myxomycotina

The slime molds, of which there are some 300 known species, have a vegetative body that consists of a slimy mass of naked protoplasm without an external wall. The body has no definite shape; it creeps slowly by amoeboid movement on rotting tree trunks or dead leaves, engulfing bacteria and the spores of other fungi, or absorbing nutrients from dead organic matter. The absence of cell walls in most stages, the amoeboid movement, and the ability to take in solid food are characteristics that are usually associated with animals. A good case may be made for the idea that the slime molds are really animals that belong to the same group as the amoeba. It is in regard to reproduction that they are most like plants: they produce spores with cellulose walls.

There are two major groups of myxomycete fungi. The **acellular slime molds** have a vegetative body called a **plasmodium** (Fig. 12.1). It is a large, easily visible mass of protoplasm with no internal division into cells (*ie*, it is **coenocytic**). The plasmodium originates as a single cell, an amoeba or a flagellated cell. Feeding, growth, and nuclear division proceed without cytokinesis to produce the large plasmodium. Sometimes individual plasmodia fuse and sometimes two amoebae fuse to form a diploid zygote that develops into a plasmodium.

Eventually the plasmodium moves to a drier location and puts forth stalked masses of protoplasm (**sporangia**) in which the contents become divided by cellulose walls to form spores. The spores are discharged from the sporangia and spread by the wind. In the presence of water, they germinate; the wall is ruptured, and the contents escape in the form of a flagellated or amoeboid cell to repeat the cycle.

The **cellular slime molds** differ from the acellular types in that they are not coenocytic. A **pseudoplasmodium** (pseudo = false) forms from many separate amoebae. These amoebae originally live separately, feeding and dividing, until eventually one or more of them begin to emit a chemical signal (cyclic AMP) that diffuses into the surroundings and induces neighboring amoebae to migrate toward them. When the amoebae meet, they stick

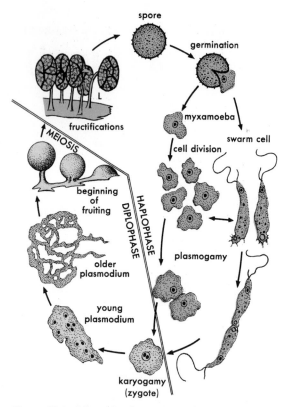

Figure 12.1 Life cycle of a myxomycete.

together; however, they retain their plasmalemmas and do not fuse. This mass of separate cells, the pseudoplasmodium, now moves about as a unit. As in the acellular forms, the cellular slime molds reproduce by means of spores in stalked sporangia.

Subdivision Eumycotina

All but the simplest of the true fungi (Eumycotina) have a vegetative body composed of branching tubes (Fig. 12.2) called **hyphae** (**hypha**, singular). True cell walls, produced by the protoplast, surround the hyphae. The fungi are eukaryotes whose cell contents are similar to those of plants except that plastids and usually dictyosomes are missing. Most of the space within the hypha is occupied by large vacuoles.

Hyphae grow only at their tips; there the walls are soft and extensible under the turgor pressure from the protoplast. As a region of wall is left behind the growing tip, it becomes thicker and thus highly resistant to further stretching. Branches originate as bulges in the sides of the growing hyphal tips, where the walls are still soft. The branches behave exactly as their parent hyphae. Hence the system of hyphae progressively extends and ramifies as it absorbs nutrients and water from the substratum. The hyphae possess a "chemical sense" and can orient their growth and branching toward sources of raw materials such as sugars, amino acids, water, and minerals. The orientation of growth by chemical signals is called **chemotropism**. Its effect here is to cause the fungus to

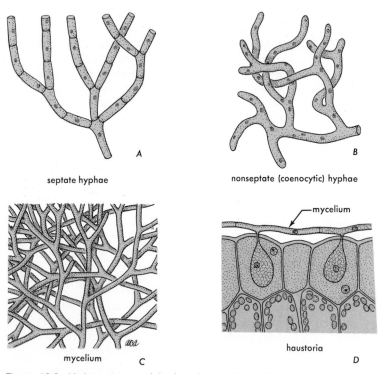

septate hyphae

nonseptate (coenocytic) hyphae

mycelium   C

haustoria   D

Figure 12.2 Various aspects of the fungal vegetative body.

progressively invade food sources such as moist slices of bread and decaying stored fruits.

The system of connected hyphae is the fungus organism. Mycologists call it a **mycelium** (Fig. 12.2). It is comparable to the plasmodium of the slime mold. There is some cooperation between neighboring hyphae, but the degree of cooperation rapidly declines with distance. Hence as the mycelium grows and spreads, its advancing hyphae tend to form local groups that function independently of one another, as if they belonged to entirely separate mycelia. Older parts of the mycelium may be abandoned, their usable contents withdrawn, and their empty walls sealed off by cross-walls that resemble abandoned tunnels. Such an event, or the accidental shearing of mycelia into parts by outside forces, readily converts a single mycelium into many mycelia. Thus, **fragmentation** is a real and useful mode of vegetative reproduction that is available to nearly all fungi.

Mycelia often show sophisticated responses in which development is keyed to environmental limitations and opportunities. These responses use light, gravity and chemical signals to direct growth and to select the best times for reproduction.

True fungi are solution feeders: they can take in only small molecules. In many cases their food requirements are very simple, consisting of no more than an organic carbon compound such as a sugar, plus water, minerals, and perhaps a few vitamins (enzyme cofactors, required only in small amounts, which the fungus cannot build for itself.) The vitamin most often needed by fungi is the same Vitamin B_1 that is essential in human nutrition. Despite the simplicity of their needs, mycelia can exploit a wide range of organic materials as food sources if they are available. Food sources are often hard materials such as wood, but also include insoluble storage polymers such as seed proteins and starches. These materials must be rendered soluble before they can diffuse through the walls of the hyphae and reach the protoplast. Hyphae build and secrete enzymes that hydrolyze proteins to amino acids and polysaccharides to free sugars. Working outside the hyphae, these enzymes require water as a reactant and as a medium for support and transport. Hence mycelia tend to be active only in moist places.

Class Oomycetes

The Eumycotina include some fungi that have swimming cells and some that do not. This distinction seems fundamental, and mycologists group all fungi with motile cells into a class that in this book we call *Oomycetes* (other names are available; e.g., Mastigomycotina).

Many Oomycetes cause severe plant diseases; others infect fish, insects, or even humans. Many members are saprophytes or weak parasites. Some are minute forms of one to several cells, and a few of these smaller sorts are reported to attack small aquatic animals. Most representatives of the Oomycetes develop a more or less extensive mycelium of indefinite form (Fig. 12.3).

The hyphae of actively growing Oomycetes are generally coenocytic (Fig. 12.2). The whole mycelium consists of a single, very highly branched multinucleate

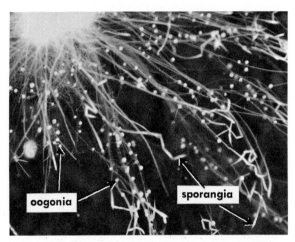

Figure 12.3 Water mold growing on a hemp seed, ×3.

cell. **Septa**, or crosswalls (**septum**, singular), may form occasionally in old hyphae of some species, and normally the reproductive cells are cut off from the vegetative hyphae by crosswalls.

The Oomycetes are divided into several orders. Specialists do not agree completely and the number of recognized orders depends on the authority one is consulting. We shall consider the orders Saprolegniales and Peronosporales. Note that the names of orders end in -ales.

Saprolegniales

Fungi of this order usually live in fresh or salt water; hence the common name **water molds**. Most are saprophytic. Even the species that attack the gills of fish grow on tissues that have been weakened or suffered injury. They may be easily cultivated by placing small pieces of meat, egg albumin, radish seeds, or dead flies in a dish of pond water (Fig. 12.3). After the mycelium has ramified through the substratum, hyphae grow outward into the water. Reproductive cells are produced by these hyphae.

Asexual Reproduction

With ample food supply, there is little tendency to produce reproductive cells. If a well-developed mycelium is transferred to distilled water in which a food supply is lacking, asexual sporangia will usually appear. Sporangia are formed by a cross-wall cutting off the tip of a hypha from the rest of the mycelium (Fig. 12.4A). The sporangium is multinucleate. The protoplasm of the sporangium divides into a large number of spores, each with one nucleus (Figs. 12.4B,C). Upon maturity, the spores are discharged from the sporangium. In most species these spores have flagella that permit swimming movements, and are therefore known as **zoospores**. In *Saprolegnia*, each zoospore has two flagella attached to its anterior end that enable it to swim actively (Fig. 12.4D). After a time, the zoospores settle down, lose their flagella, develop a cellulose wall, and pass through a resting period. When they have resumed activity, they escape from the wall; the two newly developed flagella are now

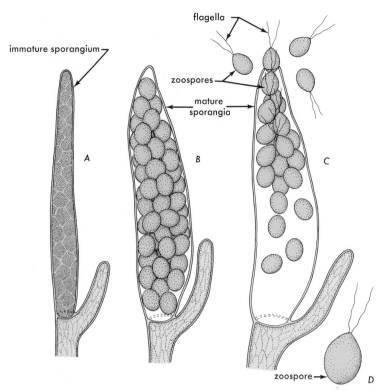

Figure 12.4 Zoosporangia of *Saprolegnia. A,* immature sporangium; *B,* mature sporangium; *C,* discharge of zoospores; *D,* a zoospore.

attached laterally, and zoospores swim about for a period. If they come to rest on a suitable substance, each sends out a hypha that penetrates this substance.

In a few species of water molds, one or even both zoospore stages are suppressed. When zoospores emerge from the sporangia, they may germinate directly into a new mycelium. In certain other species, the spores never leave the sporangia but germinate while still enclosed within it, and the germ tubes pierce the old sporangial wall. In still other species, spores are not even formed; the sporangia germinate directly into a coenocytic mycelium.

Sexual Reproduction

When stimulated to reproduce sexually, the mycelium produces specialized hyphae called **oogonia** and **antheridia** in which gametes are formed. The nuclei in the vegetative mycelium are diploid; meiosis occurs in the specialized sexual hyphae to produce the haploid nuclei needed for gametes.

Egg cells are formed inside the oogonia, which are spherical cells at the tips of short side hyphae (Fig. 12.5). When mature, each oogonium is three to four times the diameter of an ordinary hypha and may contain from one to twenty eggs. Each spherical egg contains one nucleus.

The antheridia are also formed at the tips of hyphae (Fig. 12.5). Each is separated from the rest of the hypha by a cross-wall. The antheridia are usually similar in diameter to ordinary hyphae.

The antheridium grows toward an oogonium; it is directed in its growth by a chemical signal (a *sex hormone*) released by the oogonium. When the

antheridium touches an oogonium, it produces a short hypha called a *fertilization tube* that punctures the oogonial wall, and comes in contact with one or more eggs. If the oogonium contains several eggs, the fertilization tube usually branches, sending a branch to each egg. Nuclei (male gametes) from the antheridium migrate into each egg, and fertilization ensues. The zygote develops a heavy wall, becoming an **oospore,** and usually will not germinate for several months, even under favorable conditions. Upon germination, the oospore sends out new hyphae, which rapidly grow into a typical mycelium.

Some oomycete species are bisexual, both kinds of gametangia being borne on the same mycelium. But others have separate sexes. In these species, the antheridia and oogonia do not form unless a suitable partner is available. The system that signals the presence of a partner has attracted considerable scientific interest because it is a clear-cut case where environmental signals cause major developmental changes. By placing male and female mycelia in the same tank of water but separated by a membrane, it has been found that the female constantly emits very small quantities of a compound that functions as a sex hormone. The male sensitively detects this hormone and responds by producing antheridia. The antheridia in turn give off a second hormone that diffuses to the female mycelium and induces it to produce oogonia. Additional hormones induce the antheridium to grow toward the oogonium, cling to it, and produce the fertilization tubes. It has not yet been discovered how the hormones produce these responses, but special sets of genes are probably called into play.

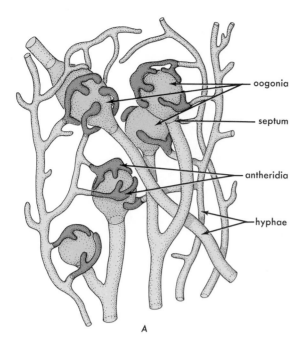

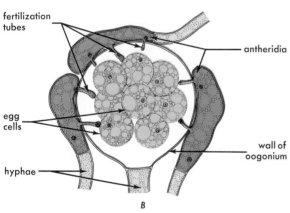

Figure 12.5 Gametangia in *Saprolegnia. A,* oogonia and antheridia with associated hyphae; *B,* antheridia, oogonium with many eggs, fertilization tubes in place.

Peronosporales

Nearly all the Peronosporales are obligate parasites. In general, heavy-walled oospores carry them over unfavorable periods, and various sorts of asexual spores bring about rapid multiplication under suitable conditions.

The most famous member of the Peronosporales is *Phytophthora infestans,* (Fig. 12.6), a fungus that indirectly caused the deaths by starvation of at least 250,000 people in Ireland between 1843 and 1847, and induced another million Irish people to move to the United States. This fungus causes a disease of the potato called **late blight.** In those years Ireland had virtually a one-crop economy, featuring the potato. *Phytophthora* was present in the potato fields for many years prior to the starvation years, gradually spreading, but conditions were never right for maximum growth. Then, starting with 1843, there followed a period of unusually warm, humid summers during which the fungus nearly annihilated the potato crops. The disease kills foliage and rots the potato tubers. Farmers of

the day had no idea of the cause and were skeptical when botanists discovered that a fungus was the culprit. Since then it has been found that *Phytophthora infestans* can be controlled (it still exists in potato fields) by spraying infected fields with fungicide, as well as by carefully disposing of infected tubers.

The mycelium of *Phytophthora* grows between the cells within a leaf or stem (Fig. 12.6). Parasitic hyphae **(haustoria)** penetrate the cells. Aerial sporangia arise on long aerial hyphae called **sporangiophores** that extend out through the stomata of the infected plant. The sporangia break loose and are disseminated by air currents.

Some of the sporangia eventually come to rest on the leaves of susceptible plants. When moisture on the leaf surface is sufficient, zoospores escape from them. They soon send out small hyphae called **germ tubes** that penetrate the host tissues and bring about new infections.

Downy mildews are a group of fungi in the Peronosporales that cause infections on many cultivated and wild plants (Fig. 12.7). They are easily identified by the sporangiophores that may, in severe cases, nearly cover the leaf. In the laboratory, the sporangia can usually be induced to germinate by floating them on cool water.

A small group of Peronosporales, the "white rusts," develop sporangia beneath the epidermis of such crop plants as mustards and spinach. We mention them because their sporangiophores do not grow out of the stomata as in the downy mildews. Instead, they collect in pustules under the epidermis of the stem or leaf. Sporangia are cut off in chains from the tips of the sporangiophores. They accumulate in large numbers and finally rupture the epidermis, forming creamy-white pustules (Fig. 12.8).

Sexual reproduction in the Peronosporales is similar to that observed in the Saprolegniales.

Class Zygomycetes

The Zygomycetes are the simplest group of fungi that lack motile cells. They owe their name to the trait of converting the zygote into a resting spore called a **zygospore** (Fig. 12.10). These are chiefly terrestrial fungi that form an extensive coenocytic mycelium. In contrast to the Oomycetes and the green plants, the walls of the zygomycete mycelium have **chitin** as their fibrous framework. Chitin is a polysaccharide in which the sugar subunits have been modified by the addition of a nitrogenous group. The same substance is responsible for the strength of insect exoskeletons.

The hyphae of the vegetative mycelium invade the food substrate and also grow on the surface, forming a cottony mass (Fig. 12.9). When the fungus is well established, it commences sexual reproduction with the formation of specialized sexual hyphae (Fig. 12.10). These hyphae grow toward one another, touch, and fuse at the tips. Special enzymes dissolve the wall at the contact point so that the protoplasm of the two hyphae flows together in a single mass. At the same time, a crosswall forms behind the contact point on each side. This walls off a single large cell with many nuclei from each parent. The nuclei fuse in pairs to form diploid zygote nuclei. The

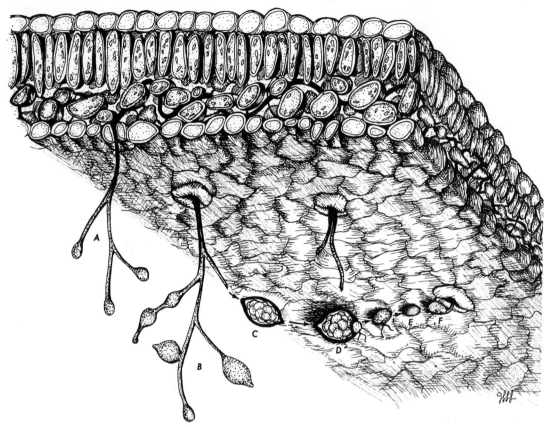

Figure 12.6 Asexual reproduction of *Phytophthora infestans*. The vegetative mycelium invades the spongy mesophyll of a potato leaf. Sporangiophores *(A, B)* grow through stomata. Multicellular sporangia *(B, C)* form. Each sporangium breaks off as a unit *(C)* and is carried by wind to another leaf surface *(D)*. Zoospores emerge from the sporangium *(D)* and swim in the film of moisture on the leaf surface before encysting and becoming dormant *(E)*. The dormant spore later germinates *(F)* and produces a germ tube that may penetrate the epidermis through cell walls or stoma, to produce a new mycelium within the leaf. By repeating these events the spores can quickly infect a whole potato field.

surrounding cell wall thickens and the cell becomes a dormant spore (a zygospore). The diploid nuclei undergo meiosis to produce a new generation of haploid nuclei before the zygospore germinates. In all the life history of the zygomycete fungus, only the zygospore is diploid.

In some species of Zygomycetes, a single mycelium can produce zygospores without a partner. But many species require a mating between sexual hyphae of two different mycelia of opposite "sexes." The partners look alike and contribute equally to mating, so mycologists designate them as "+" and "−" **mating types** rather than "male" and "female." The mycelia advertise their mating type to one another by emitting sex hormones, as in the oomycetes.

The Zygomycetes reproduce asexually by means of sporangia, which form at the tips of aerial sporangiophores (Fig. 12.11). The sporangiophore swells at the tip and a crosswall forms below the swelling. The multinucleate protoplasm in the swollen sporangium then divides into many spores. Members of the class differ greatly in the details of asexual reproduction: for instance,

Figure 12.7 Downy mildew *(Bremia)* on lettuce leaf.

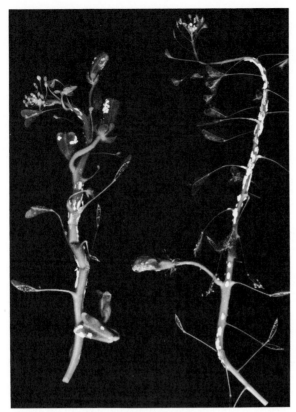

Figure 12.8 *Albugo* on shepherd's purse.

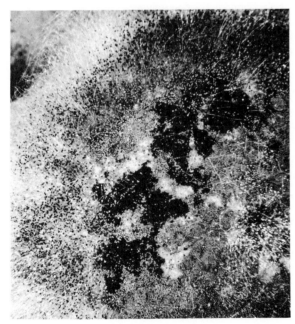

Figure 12.9 *Rhizopus* on peach, ×4.

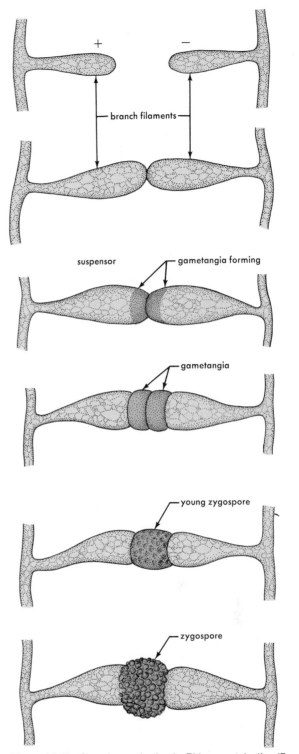

Figure 12.10 Sexual reproduction in *Rhizopus stolonifer*. (From *Cryptogamic Botany*, vol. I, *Algae and Fungi*, by G. M. Smith. Copyright 1955, McGraw-Hill Book Company. Used with permission of McGraw-Hill Book Company.)

the common bread mold, *Rhizopus nigricans*, forms special, thick hyphae **(stolons)** that arch through the air and are anchored to the substrate at intervals by finer branch hyphae or **rhizoids** (Fig. 12.11). Clumps of sporangiophores arise at each anchor point. Other members of the Zygomycetes produce solitary

sporangiophores, or sporangiophores that branch and bear several sporangia.

Many of the Zygomycetes have sophisticated mechanisms that use light and gravity to direct sporangiophore growth so that the spores will be brought to locations suitable for dispersal. Other mechanisms may

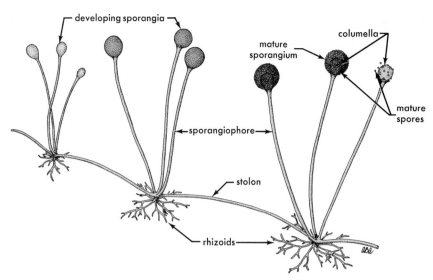

Figure 12.11 Asexual reproduction in *Rhizopus stolonifer.*

time the formation of sporangiophores on the basis of temperature, illumination, and nutritional conditions.

Spores are dispersed by various means. In some species the sporangial wall breaks down at maturity and the light, dry spores are spread by wind. In others the spores form a sticky mass that is carried by animals. Members of the genus *Pilobolus* build up pressure in the sporangiophore until the sporangium as a whole is discharged explosively and travels several meters through the air (*Pilobolus* literally means "hat thrower.")

Most of the Zygomycetes are saprophytes, though a few are parasitic on other members of the same class. If left uncontrolled some, such as *Rhizopus*, cause major losses of stored foods. Few of them cause human diseases.

Class Ascomycetes

The *Ascomycetes* are named for their trait of producing sexual spores inside a sac (**ascus**) like marbles in a transparent bag (Fig. 12.14). Ascus is derived from the Greek word for sac (plural asci).

The life of an ascomycete typically begins with the germination of a haploid spore to produce a vegetative mycelium. As in the Zygomycetes, the walls contain chitin instead of cellulose. But in the Ascomycetes the mycelium is divided into cells by cross-walls. The cross-walls have a hole in the center (Fig. 12.12) that allows the passage of organelles from one cell to the next; however until sexual reproduction occurs, each cell usually maintains just one haploid nucleus.

Sexual reproduction begins with the fusion of two cells (Fig. 12.13). Many ascomycete fungi are self-incompatible; fusion will occur only if the cells come from mycelia of opposite mating types. In some species, fusion can occur between any two hyphae, while in others, fusion involves specialized multinucleate hyphae called **ascogonia** and **antheridia,** and still others produce small free cells called **spermatia** instead of antheridia.

When contact is made, nuclei pass from the antheridium

to the ascogonium. The (+) and (−) nuclei pair but do not fuse. The fertilized ascogonium produces hyphae into which the pairs of nuclei migrate. Hyphae with the paired nuclear arrangement are not truly diploid ($2n$), nor are they haploid ($1n$). Rather, their nuclear condition is symbolized as ($n+n$). An ($n+n$) hypha is known as a *dikaryon* (di = two; karyon = nucleus). Since these $n+n$ hyphae produce asci, they are called **ascogenous hyphae.**

The ascogenous hyphae may grow and branch before they develop into asci. The formation of the ascus is outlined in Fig. 12.14.

Most ascomycetes concentrate their asci within a complex reproductive structure called an **ascocarp** (Fig. 12.15).

The ascocarp, composed of both vegetative and ascus-bearing hyphae, is characteristic of the species. It may be microscopic or as much as 15 cm in diameter. There are three general types of ascocarps:

1. **Cleistothecium:** hollow, completely closed sphere (Figs. 12.15*A,B*).
2. **Perithecium:** hollow, flask-shaped body with narrow opening (Figs. 12.15*C,D*).
3. **Apothecium:** open, cup-shaped body (Figs. 12.15*E,F*).

The asci line the inner surface of the ascocarp. This surface layer is the **hymenium** or fertile layer. Sterile haploid cells, called **paraphyses,** also arise in the hymenium (Fig. 12.16).

The hymenial layer is surrounded and protected by haploid hyphae that form a layer called the **peridium.**

The fertilization of the ascogonium is usually the signal that starts the formation of an ascocarp. Therefore in the ascomycetes, $n+n$ hyphae are usually confined to the reproductive body and do not contribute to the vegetative mycelium. This contrasts with the Basidiomycetes, to be discussed later.

The ascocarp is composed of both haploid and $n+n$ hyphae. The result is a multicellular body that is considerably more complex in structure than anything

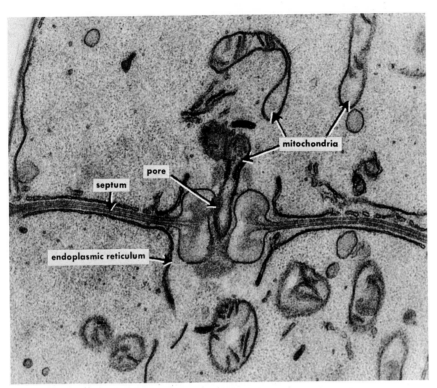

Figure 12.12   A small portion of a hypha of *Rhizoctonia solani* showing a septum, with pore, through which a mitochondrion appears to be passing, ×24,000.

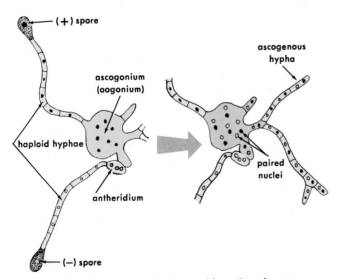

Figure 12.13   Diagram of conjugation and formation of *n+n* ascogenous hyphae.

found in the Oomycetes and Zygomycetes. For this reason the Ascomycetes are often called **higher fungi** and the former groups **lower fungi**.

Although sexual reproduction is important for producing new genetic combinations in the Ascomycetes, asexual reproduction is more important as a means of multiplication and dispersal. Most species reproduce asexually by pinching off the end cells of aerial hyphae (Figs. 12.17). These cells develop resistant walls and are

light enough for long-distance travel on the wind. Spores produced by such a pinching process are called **conidia**. The hyphae that produce them are **conidiophores** (-phore = bearer). Various species differ greatly in the branching of the conidiophores. Conidia are usually produced in immense numbers. Cold generally kills them. But within a single summer, the growth of mycelia is so fast that dozens of generations of conidia and mycelia can be formed and epidemics of fungal growth can occur.

Subdivision Eumycotina

261

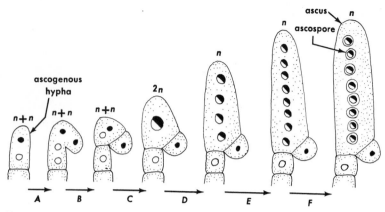

Figure 12.14   Ascus formation. *A*, paired nuclei divide by mitosis; hypha develops J shape. *B*, crosswalls form; penultimate cell is *n+n*. *C*, nuclear fusion (karyogamy) in penultimate cell. *D*, meiosis divides 2*n* zygote into four haploid nuclei. *E*, each nucleus undergoes mitosis. *F*, walls form around haploid nuclei, yielding eight ascospores inside ascus (old hyphal wall.)

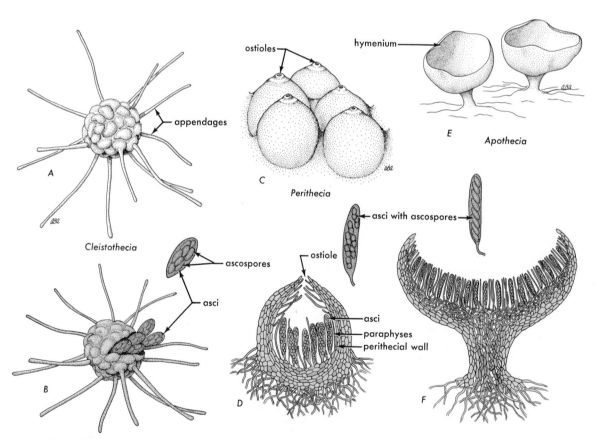

Figure 12.15   Diagram showing three types of ascocarps. *A* and *B*, cleistothecia, ×500; *C* and *D*, perithecia, ×150; *E* and *F*, apothecia, *E* ×¼, *F* ×10. (From *Comparative Morphology of Fungi*, by E. A. Gäuman; copyright 1928, McGraw-Hill Book Company. Used with permission of McGraw-Hill Book Company. *D*, redrawn from E. A. Gäuman, *The Fungi. A Description of Their Morphological and Evolutionary Development*. Hafner Publishing Company, New York, 1952.)

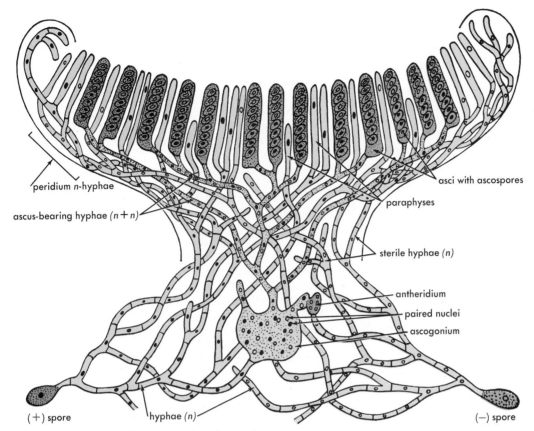

Figure 12.16 Diagram of a cross-section of an apothecium.
(Redrawn from *Fundamentals of Cytology,* by L. W. Sharp.
Copyright 1943 by McGraw-Hill Book Company, New York. Used
with permission of McGraw-Hill Book Company.)

Labels in Figure 12.16:
peridium *n*-hyphae
ascus-bearing hyphae (*n* + *n*)
asci with ascospores
paraphyses
sterile hyphae (*n*)
antheridium
paired nuclei
ascogonium
(+) spore
hyphae (*n*)
(−) spore

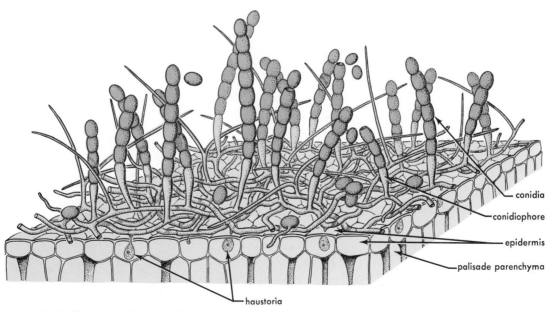

Figure 12.17 Powdery mildew on leaf surface.

Labels in Figure 12.17:
conidia
conidiophore
epidermis
palisade parenchyma
haustoria

Examples of Ascomycetes

There are about 15,000 species of Ascomycetes, and so many variations that even an introductory survey of them would require many pages. Thus in the present book we will consider only a few illustrative examples.

Yeasts. The simplest ascomycetes are the yeasts. *Saccharomyces cerevisiae* is the yeast of baking and brewing. This fungus is important because it converts sugar (glucose) into alcohol and carbon dioxide. This trait is useful in gaining energy from glucose when molecular oxygen is not available (Chapter 2). In the presence of oxygen, yeast (like other organisms) can completely oxidize glucose to carbon dioxide and water; although it gains much extra energy this way, it does not produce the economically valuable alcohol.

Yeasts are normally single-celled. After cell division the new cells separate; thus ordinary mitotic division is a form of asexual reproduction (Fig. 12.18). In older cultures the cells may hang together and form short chains. Some yeasts double in size and pinch in half. More often, though, division is a process of **budding**: new cells grow out from the mother cell much as a small bubble would form if a piece of thin rubber were made to expand through a small opening in some heavier material. The small "bubbles" are called **buds**. They enlarge and finally separate from the parent cell.

Ascus formation in yeasts is very simple. Two compatible yeast cells fuse, the nuclei fuse, and meiosis follows. The original wall forms the ascus. There is no ascocarp.

Most Ascomycetes do nothing more with the diploid zygote nucleus than to divide it by meiosis, producing haploid nuclei. Some yeasts follow the same pattern. But in others, the diploid cell may reproduce asexually by mitosis and budding (Fig. 12.18). Later, meiosis finally intervenes to restore the haploid condition. Yeasts differ greatly in the relative time spent in the diploid and haploid states. Some strains are diploid most of the time, a highly unusual trait among the sac fungi.

Powdery mildews. These are a group of ascomycete genera that owe their common name to the development of a powdery appearance on the leaves of infected plants. All the powdery mildews are **obligate parasites**; they can only grow on living plants. The genus *Erysiphe* is representative (Fig. 12.17).

The mycelium is generally confined to the surface of the leaves, flowers, or fruits. Haustoria penetrate epidermal and parenchyma cells, from which they secure nourishment. At first the mycelium appears like a delicate cobweb. Eventually, it assumes a white powdery appearance because of the development of numerous conidiospores that serve to spread the fungus from one plant to another during the growing season.

Sulfur dusted on the host plants at this stage prevents infections from these spores.

For sexual reproduction, small black cleistothecia are formed, each enclosing one or several asci (Figs. 12.15A,B). The ascospores are usually discharged by force from the cleistothecium.

Although the powdery mildews may not kill their host, they weaken it and greatly reduce the crop yield. Powdery mildews cause diseases of apples, grasses, grains, grapes, cherries, and many other plants.

Monilinia. Brown rot of stone fruits, a very severe disease of peaches, cherries, plums, apricots, and nectarines, is caused by a disk fungus (*Monilinia*, Fig. 12.19, Color Plate 3), which infects mainly blossoms and fruits. The ascocarps of the order to which *Monilinia*

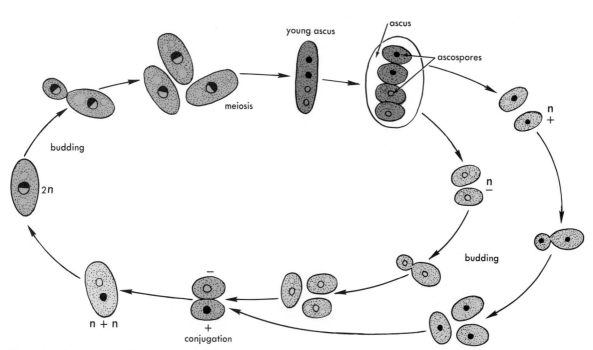

Figure 12.18 Life cycle of *Saccharomyces cerevisiae*.

Figure 12.19 *A*, apothecia of
the brown rot fungus
(Monilinia fructicola), × ½;
B, brown rot on cherry
compared with a healthy fruit,
× ½.

A

B

A

Figure 12.33 Various types
of lichens. *A*, the conspicuous green foliose lichen is
Peltigera aphthosa and it is
surrounded by the white
branches of the fruticose
lichen *Cladonia* sp.; *B*, crustose lichen *Acarospora flava*
colors a rock on the skyline
of a Sierra Nevada ridge; *C*,
Ramalina reticulata, a fruticose lichen, clothes the
branches of an oak in the
foothills of the Coast Ranges
along the Pacific Coast.

B

C

Figure 13.11 Habit of *Marchantia. A,* × ⅛; *B,* × 1. Note gemmae cups. *C,* antheridial heads, × 2. *D,* archegonial heads, × 1.5. *E,* under surface of an archegonial head; the sporophytes are enveloped by a protective perianth. Immature sporophytes are green; ripening sporophytes show yellow, × 3.

belongs are cup- or disk-shaped (apothecia) (Fig. 12.15*E,F;* 12.19*A*—Color Plate 3). They may be as much as 15 cm in diameter, depending on the species, and are sometimes brilliantly colored. Members of a related genus, *Morchella* (the morels), are edible.

Claviceps. The genus *Claviceps* is parasitic on grasses, including grains. A dormant mycelium that replaces the mature grain is known as **ergot**. It possesses several alkaloids that have medicinal properties. Ergot constricts the blood vessels, particularly those that pass into the hands and feet, thus depriving the extremities of a normal blood supply. During humid summers in Central Europe, *Claviceps* may infect rye heavily. In centuries past, before the nature of the fungus was understood, ergot would be milled along with the grains of rye. The contaminated flour, which might contain as much as 10% of powdered mycelium, would be baked in bread. A continued diet of bread from this flour resulted in much misery. Because of the contraction of the blood vessels and the limited supply of blood reaching feet or hands, gangrene sets in. Hands, arms, and legs die and finally drop off. The disease was known as "Holy Fire" because of a burning sensation in the extremities. Today, ergot is a valued drug used to control hemorrhage, particularly during childbirth, and to treat migraine.

With the knowledge of the poisonous nature of the ergot, diseased grain is no longer milled and *Claviceps* is not of concern in human diet. However, cattle may occasionally feed on grasses infected by ergots and thus be poisoned.

Class Basidiomycetes

Based on the complexity and size of their reproductive structures, the most advanced of the higher fungi fall in the class *Basidiomycetes*. These include all the mushrooms, puffballs, and the bracket fungi that grow on the trunks of trees. These elaborate structures are called **basidiocarps.**

But not all the Basidiomycetes have such elaborate reproductive bodies. The real feature that determines whether a fungus belongs to this class is the precise way in which sexual meiospores are formed. Rather than being enclosed in a common sac, here the spores formed after meiosis are carried on separate stalks called **sterigmata** (**sterigma** singular) (Fig. 12.22). The cell that undergoes meiosis and produces the spores is called a basidium, because in some cases it is club-shaped (*basidium* is Greek for "club"). The meiospores are called **basidiospores.**

The Basidiomycetes consist of two main groups or subclasses, which will be considered in turn.

Subclass Homobasidiomycetidae

The Homobasidiomycetidae comprise all the species that form elaborate fruiting bodies or basidiocarps. Some of these are parasitic but most of them are saprophytic. Some form mycorrhizae, symbiotic associations with the roots of higher plants, previously discussed in Chapter 5.

Others obtain their carbohydrates from wood and may even cause extensive decay of timbers in mines and buildings. Specialized tissues and organs may be formed, some of the *n + n* hyphae becoming specialized to transport food and water, and others becoming tough and woody.

The vegetative body of such a fungus is a mycelium that starts from a haploid spore and invades its substratum as in previous groups. The walls of the mycelium contain chitin rather than cellulose, and the hyphae are almost completely septate. There is a pore in the center of each cross-wall but an elaborate caplike structure covers the pore so that organelles probably cannot pass between cells.

These fungi are not noted for their asexual reproduction. Instead, sexual reproduction is highly developed.

Sexual reproduction begins with the conjugation or fusion of hyphae to produce a dikaryon with the paired, *n + n* nuclear condition. Some species are **homothallic**, conjugation taking place between any two cells of any two hyphae or even between two cells of the same hypha. Other species are **heterothallic** and conjugation will only occur between hyphae from mycelia of different mating types.

Once formed, the dikaryon may grow and branch to form a long-lived *n + n* vegetative mycelium. The dikaryon may persist for many years, perhaps even for centuries. If conditions are not right, the production of sexual spores and basidiocarps may be postponed indefinitely.

In the *n + n* mycelium, division occurs in such a way that all the cells maintain the *n + n* condition. This requires a special set of events (Fig. 12.20). Growth of the tip cell of the hypha is accompanied by division of both nuclei. The tip cell develops a J-shape and one of the nuclei migrates into the end of the J. Cross-walls are laid down as shown in Fig. 12.20. The lateral cell bends around, fuses with the trailing cell, and the *n + n* condition is restored in that cell. This happens each time a new cell is formed at the tip of the hypha. The curved and fused lateral cell is termed a **clamp connection.** Of all the fungi, only the Basidiomycetes form a dikaryotic vegetative mycelium with clamp connections.

Although both haploid and *n + n* mycelia develop extensively, the *n + n* phase is of special interest because it gives rise to the spore-bearing body, or **basidiocarp,** in which the basidia are developed. The hyphae form a tangled mass in the substrate or host and emerge to form

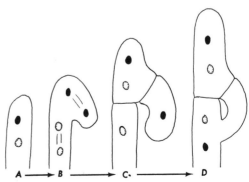

Figure 12.20   Formation of a clamp connection, which maintains the dikaryotic condition in the basidiomycete hypha.

Figure 12.21 *Agaricus brunnescens,* the common edible mushroom, ×1.

a basidiocarp—the mushroom, puffball, or bracket fungus. The basidiocarp of a typical mushroom such as *Agaricus brunnescens* (Fig. 12.21) consists of a short upright stalk or **stipe**, attached at its base to a mass of mycelium and expanding on top into a broad cap or **pileus**. The underside of the pileus of *Agaricus* is formed by thin **gills** radiating outward from the stipe. These gills are lined with a hymenium or spore-bearing layer.

Basidia develop from the tipmost cells of hyphae in the hymenium (Fig. 12.22). Here the paired nuclei finally fuse to give a diploid ($2n$) nucleus, which immediately undergoes meiosis. Sterigmata form at the tip of the basidium. The haploid nuclei that result from meiosis are squeezed through the narrow necks of the sterigmata

along with cytoplasm. The formation of a resistant wall completes each spore.

If the gill is very young, developing basidia on which basidiospores have not yet been formed may be observed; Figure 12.23 shows a section through three gills. Note two stages of development of basidia and the loose tangle of hyphae that form the center of the gill. When a basidiospore is discharged, it is shot horizontally from its position on the basidium straight into the space between two gills. A basidiospore resembles a toy rubber balloon, having a large volume for its weight. It is shot from the basidium with a force sufficient to carry it about midway between the two gills. It then falls straight downward. It has been estimated that some basidiocarps may discharge

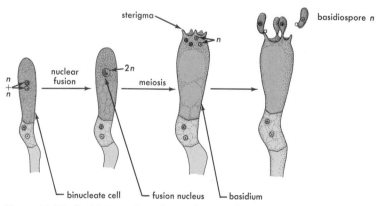

Figure 12.22 Basidiospore formation as it occurs in a mushroom (Homobasidiomycetidae). (From *Cryptogamic Botany:* Vol. I Algae and Fungi, by G. M. Smith; copyright 1955, McGraw-Hill Book Company; after H. Kniep, *Sexualität der Niederen Pflanzen,* copyright G. Fischer Verlag, Jena 1928; and after *Researches on Fungi,* Vol. VII *The Sexual Process in the Uredinales,* by A. H. R. Buller; University of Toronto Press 1922. Used with permission from McGraw-Hill Book Company, Gustav Fischer Verlag, the Royal Society of Canada, and the University of Toronto Press.)

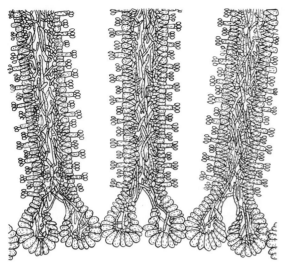

Figure 12.23   Section showing three gills of a mushroom.

as many as a million spores a minute for several days.

The *Agaricales* include not only edible fungi but also some of the most deadly of poisonous mushrooms. Those of the genus *Amanita* (Color Plate 1) are especially dangerous; it is said that the Eskimos used to pay high prices for dried specimens that, when taken in extremely small quantities, induce hallucinations. Experimentation is ill-advised, however, because small dosages can be toxic. Although poisonous mushrooms are much less common than benign ones, mycologists usually advise enthusiasts to consult an expert before sampling mushrooms they have gathered from nature.

Not all Agaricales have gills. In some, the lower surface of the cap or bracket has many small pores that extend upward to form small tubules (Fig. 12.24). The hymenium lines the tubules. Many bracket fungi found on the trunks of forest trees are such **pore fungi.**

Some bracket fungi cause serious damage to forest trees and lumber in mines and wooden buildings. During World War II, much green lumber had to be used; because fungi were able to feed on the improperly cured lumber, much loss resulted. One of the best-known fungi in this group is *Fomes applanatus* (Fig. 12.25). It causes a disease of forest trees known as *white-mottled rot,* to which many hardwoods and conifers are susceptible.

The basidiocarp of *Fomes applanatus* is a very elegant shelf-shaped bracket or **conk.** The conks are perennial and grow to a large size. They are gray on top, and the lower surface has millions of pores lined with a creamy-white hymenium.

Most of the bracket fungi of the forest normally live in the dead heartwood, to which they gain entrance by wounds. They are, however, able to invade the living cells of sapwood, which they may destroy.

Other types of basidiocarps also occur, such as those in puffballs, and bird's-nest fungi.

Subclass Heterobasidiomycetidae

The fungi of this subclass are much less familiar to the average person than those of the group just described, because they do not produce elaborate basidiocarps. But economically they may be much more important. This subclass includes some of the most virulent crop destroyers: the **rusts** and **smuts,** which attack cereal grains such as wheat and corn and cost millions of dollars

Figure 12.24   The underside of the pore fungus *Polyporus brumalis,* × ⅓.

Figure 12.25 Basidiocarp of *Fomes applanatus*, × ⅓.

Puccinia graminis has two hosts: a grass, and a shrub known as barberry *(Berberis)*. Various strains of the fungus attack different grasses. Variety *avenae* attacks oats and variety *tritici* attacks wheat. These host choices are hereditary with the fungus.

The two hosts have different roles in the life of the fungus (Fig. 12.26). Roughly speaking, the grass serves to multiply and spread the existing genetic types of the fungus population. This is achieved by asexual reproduction using food reserves that are drained from the leaves and stems of the grass plant. The grass is usually not killed, but its vitality is impaired and its seed production reduced.

The barberry host is the sexual ground for the rust fungus. Here, matings between different mycelia produce new genetic combinations. These in turn move out to invade the grain fields. The barberry plant is not killed by the infection.

To survey the life cycle of wheat rust (Fig. 12.27), let us begin with the basidiospores. These spores are carried by air currents in the spring, and they can only attack barberry shrubs. They germinate on barberry leaf surfaces, and form a germ tube that penetrates the epidermis.

A mycelium develops in the tissues of the host. Pustules, called **spermagonia**, appear on the upper surface of leaves. A spermagonium is pear-shaped and has a small opening through the upper epidermis to the exterior of the leaf (Fig. 12.27A). Hyphae in the spermagonium cut off large numbers of spermlike cells, the **spermatia**. These are forced out of the spermagonia through the pores.

Puccinia graminis is heterothallic; both plus and minus mating types appear on barberry leaves. These develop from plus or minus basidiospores. Spermatia are transferred by wind or insects to adjacent spermagonia, where they come in contact with **receptive hyphae**.

in lost harvests and preventive costs. Most members of this subclass are parasites on plants.

Some of the Heterobasidiomycetidae may have two hosts; that is, separate phases of their life history are passed on different plants. Such forms are **heteroecious**. If only one host is required to complete their life history, they are said to be **autoecious**.

The rust fungi owe their name to the rusty color of some of their spores, which form in abundance on the leaves of infected grasses. An example of this group is *Puccinia graminis*. It causes common wheat rust, and is a persistent and costly pest.

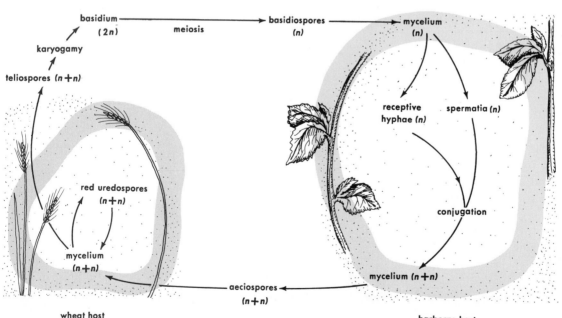

Figure 12.26 Summary of *Puccinia graminis* life cycle.

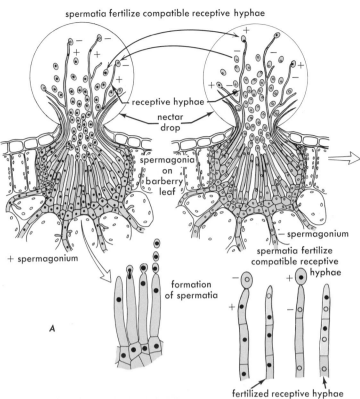

spermatia fertilize compatible receptive hyphae

receptive hyphae

nectar drop

spermagonia on barberry leaf

— spermagonium

+ spermagonium

formation of spermatia

spermatia fertilize compatible receptive hyphae

fertilized receptive hyphae

A

aecia

B

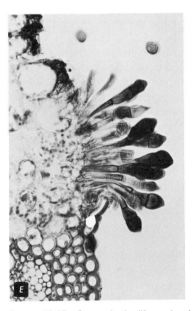

D

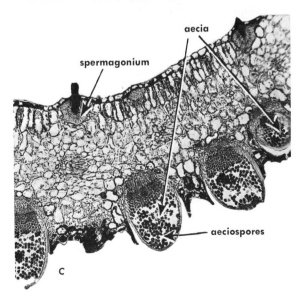

aecia

spermagonium

aeciospores

C

E

Figure 12.27 Stages in the life cycle of wheat rust, *Puccinia graminis. A,* formation of receptive hyphae and their conjugation on barberry; *B,* pustules (aecia) on barberry leaf; *C,* section through aecia and spermagonia on barberry; *D,* pustule (sorus) on wheat stem, producing red uredospores; *E,* the same mycelium on wheat later produces black teliospores as seen in this section view.

Subdivision Eumycotina

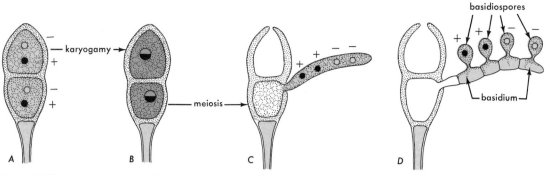

Figure 12.28 Germination of a teliospore, *Puccinia graminis*. A, teliospore with binucleate cells; B, 2n cells; C, meiosis and germination of a cell produces a hypha with four haploid nuclei; D, crosswalls form and basidiospores are extruded.

A plus (+) spermatium fuses with a cell of a minus (−) receptive hypha, and vice versa. An n+n mycelium results. In its formation, the process of genetic recombination has been completed.

How do mycelia get from barberry to wheat? The newly formed n+n mycelium on the barberry leaf invades the spongy mesophyll. Close to the lower epidermis the mycelium forms pustules called **aecia** (Figs. 12.27B,C). Here n+n spores (**aeciospores**) are formed asexually. They are released when the pustule breaks through the leaf epidermis. Aeciospores, borne by the wind, attack only wheat plants.

Hyphae from the germinating aeciospores enter the wheat plant through its stomata. A delicate mycelium forms with haustoria that penetrate the cells of the leaf and extract nutrients. About 10 days after infection, red n+n **uredospores** form. The pustules or **uredia** containing these spores open up (Fig. 12.27D). Carried on wind currents, the uredospores attack other wheat plants at a rate that can attain epidemic proportions. Wheat rust gets its name because of these red lesions (uredia) in the stems.

As the wheat begins to mature, the production of uredospores gives way to the production of black teliospores by the same mycelium (Fig. 12.27E). Whereas uredospores and aeciospores are killed by winter cold, teliospores are not. They are overwintering spores. Each cell of the teliospore begins with the dikaryotic (n+n) nuclear condition, but during winter, fusion occurs to form a zygote. Meiosis occurs in the warm, moist weather of spring. Then each cell of the teliospore puts out a single short hypha, the basidium (Fig. 12.28), and four basidiospores are extruded. No host is involved at this stage; germination occurs at any moist surface where the spore happens to land. With this event the fungus has completed its cycle.

The smuts are another group of basidiomycetes that attack grain crops. Their hallmark is the replacement of grains by black masses of spores that have a sooty look; hence the name.

Smuts vary in their method of attacking the host. *Ustilago maydis* spores cause local infection of corn plants (Fig. 12.29). The mycelium stays near the point of infection and produces tumors where spores are formed. The spores are n+n teliospores that survive the winter to germinate and produce basidiospores as in *Puccinia*. Haploid mycelia from these spores develop on the corn plants and conjugate to restore the n+n condition that leads to teliospores. Resistant strains of corn effectively limit this pest.

Another group of smut fungi can invade only the pistils of flowers in grain plants. These are the blossom-infecting smuts. *Ustilago tritici* and *U. avenae* attack wheat and oats, respectively. The germinating spore builds a small mycelium that gently invades the grain seed. When the infected seed is planted, the contained mycelium grows in pace with the plant, even accelerating its growth. But when flowers are formed masses of smut spores form in place of the seeds and are spread by wind to the slower-growing normal plants in the neighborhood.

Still other smuts infect only young seedlings. This is true of *Tilletia tritici*, which invades wheat plants. These also replace the mature grain with masses of spores.

Class Fungi Imperfecti

About 24,000 species of fungi, in some 1200 genera, are known only by their asexual stages. The class name, Fungi Imperfecti, arises from the custom of calling the sexual stages of fungi **perfect stages** and the asexual stages **imperfect stages**. Since only the imperfect stages of this large group of fungi are known, they are called the Fungi Imperfecti.

The structure of the hyphae, which are septate, and of spores, suggest that many imperfect fungi may be Ascomycetes; others may be Basidiomycetes. The Fungi Imperfecti may be considered to be Ascomycetes or Basidiomycetes whose sexual stages have not been observed or no longer exist. Occasionally, sexual stages are discovered in species that have previously been classified as imperfects. When this happens, the fungus is renamed, but usually the old name for the imperfect stages is kept too. For instance, the imperfect fungus *Penicillium vermiculatum* is now known to form asci similar to those of the ascomycete genus *Talaromyces*; hence this fungus is also called *Talaromyces vermiculatum*.

Since the classification of imperfect fungi is very doubtful, the groupings are designated as **form genera** or **form families** to show that the members do not

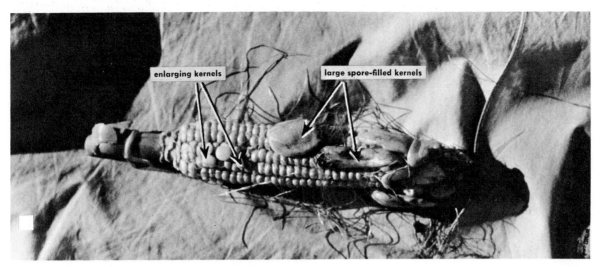

Figure 12.29 Smut disease of corn caused by *Ustilago maydis*.

necessarily have a natural or family relationship. For instance, some species have been named simply on the basis of the host upon which they were found.

Only a single form group will be mentioned, the form order Moniliales; it is the largest in the class and has over 10,000 form species.

Penicillium and *Aspergillus* are probably the most widespread fungi. They are the common blue, green, and black molds that occur on citrus fruits, jellies, and preserves. Their conidia are in the air and soil everywhere.

In *Penicillium,* the spores are formed on profusely branched conidiophores (Fig. 12.30*A*). In *Aspergillus,* the tip of the conidiophore swells and conidia form in long chains radiating from this swollen tip (Figs. 12.30*B,C*).

Enzymes secreted by these fungi are particularly active in digesting starch and other carbohydrates. When purified, these enzymes are important industrial preparations. *Aspergillus oryzae* is used in the preparation of rice wine and soybean sauces. Several species of *Aspergillus* are important in cheese manufacture. A disease resembling tuberculosis is caused by *Aspergillus fumigatus,* and several other *Aspergillus* species cause diseases in plants. Strains of *Aspergillus flavus* form compounds called aflatoxins that are toxic to animals.

Penicillium notatum has won deserved fame because **penicillin**, a drug derived from it, will inhibit the growth of bacteria without injuring human tissue. Substances such as penicillin, formed by one organism and inhibiting the growth of other organisms, are called **antibiotics**. Our knowledge of them dates back to 1870, but their significance was not fully appreciated until the discovery of penicillin.

In 1928, Alexander Fleming investigated extracts of *Penicillium;* after demonstrating their antibiotic properties,

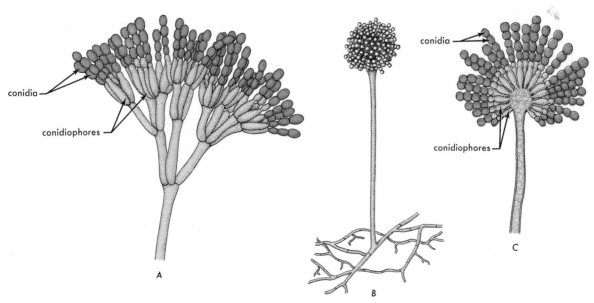

Figure 12.30 Diagrams of different forms of conidiophores. *A, Penicillium; B, C, Aspergillus; B,* three-dimensional appearances; *C,* optical section through *B.*

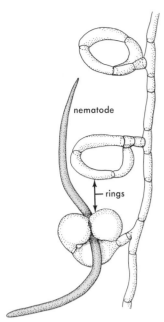

Figure 12.31 Nematode trapping loops of *Dactylaria*. (Redrawn from E. A. Gäuman, *Fungi, A Description of Their Morphological Features and Morphological Development,* Hafner Publishing Company, New York, 1950.)

he gave the antibiotic the name penicillin. It was not until after 1940 that a concentrated effort was made to produce penicillin on a commercial scale, for use in World War II.

Many thousands of pounds of mycelium of *Penicillium notatum* have been grown, and the species has been subjected to intensive study, yet no sexual stages have ever been observed. The absence of sexual reproduction presents the geneticists who are studying *Penicillum notatum* with a difficult breeding problem. Nevertheless, they have succeeded in growing strains that produce many times more penicillin per pound than do the original strains.

Several forms in the Moniliales are adapted to capture and destroy microscopic animals. A few trap nematode worms by forming small rings that quickly constrict when stimulated by the contact of a nematode crawling through them. A nematode-trapping form is shown in Fig. 12.31.

Since the sexual process has not been observed to occur in the Fungi Imperfecti and genetic variation is not thought to be a part of asexual reproduction, the question may be asked: Is genetic variation possible in the Fungi Imperfecti? The answer is, apparently, yes, at least in the laboratory. Careful genetic analysis has produced evidence that mycelia of imperfect fungi may sometimes fuse and bring together nuclei of different origin. By unknown mechanisms, nuclei may be formed of genes from each of the original kinds of nuclei. Meiosis appears not to be involved; hence this way of forming recombinant nuclei is called the parasexual cycle. It is well-established in laboratory cultures where it may account for the great variability of some Fungi Imperfecti. It has not been established that the process is significant in wild populations.

The Lichens

The lichens are composite plants composed of algae and fungi. The algal components are generally single-celled forms belonging to either the Chlorophyta or the Cyanophyta. When free from the fungus, the algae may exist normally.

The fungal component is generally an Ascomycete, but algal associations with Basidiomycetes are also known. The fungal component of the lichen cannot live separated from the alga unless supplied with a special nutritive medium.

The association of a fungus and an alga in the lichen is generally believed to be **symbiotic**; that is, both the fungus and the alga derive benefit from the association. The alga furnishes food for the fungus, which supplies moisture, shelter, and minerals for the alga.

A section through a lichen thallus reveals four distinct layers. The top and bottom layers consist of a compact mass of intertwining fungal filaments. The algal cells form a green layer beneath the top mass of fungal filaments, and a loose layer of hyphae lies directly below the algal cells (Fig. 12.32).

Lichens are widespread. They form luxuriant growths on the frozen, northern tundras, where they supply feed ("reindeer moss") for animals. In the United States they frequently cover rocks, trees and boards exposed to sun and wind (Figs. 12.33,*A,B,C,* Color Plate 3). They are slow but efficient soil formers; rocks are disintegrated slowly by their action. Lichens are pioneer plants, appearing before any other plants on recent lava flows. Lichens and mosses accumulate soil and organic matter sufficient to allow herbs and later shrubs and trees to become established.

For convenience in classifying, lichens may be divided into three groups simply on the basis of their vegetative body. This is a highly artificial grouping. Probably the most conspicuous lichens are members of the **fruticose** type. They are formed by small tubules or branches. Fruticose lichens may make extensive and attractive growths on oaks or the dead branches of certain pines (Fig. 12.33*C*). Some lichens have a leaflike plant body.

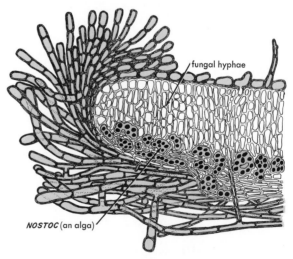

Figure 12.32 Cross section through a lichen thallus.

They constitute the **foliose** type and are common in moister climates, where their growths on tree trunks or on fallen logs are extensive (Fig. 12.33A). However, the **crustose** lichens are the most common type, as they grow extensively on rocks, bark, poles, boards, and other hard surfaces. Indeed, any bright coloring on the dark rocks (Fig. 12.33B) and gray granites of our northern mountains, on unpainted wood, or on bark of trees is quite likely to be caused by a growth of crustose lichens.

Asexual Reproduction

Lichens may multiply simply by the distribution of small pieces of the vegetative thallus. Many lichens produce special powdery bodies, composed of both a fungus and an alga, that serve as a means of vegetative reproduction. These bodies are called **soredia** (singular, **soredium**).

Both the algal and the fungal components may produce spores. The algae may grow independently, but if the young hyphae do not find the required algal cells when fungal spores germinate, they soon die.

Sexual Reproduction

Sexual reproduction is characteristic of the type of fungus present in the lichen. Since the fungal component is an Ascomycete in the great majority of lichens, ascocarps with asci and ascospores are formed. Upon the germination of the ascospore, the germ tube must find its proper algal associate in order to survive.

Summary

1. The fungi (Division Mycota) are heterotrophic eukaryotes that lack photosynthesis; typically reproduce by spores; and have cell walls during at least part of the life cycle. About 200,000 species are known.
2. Slime molds (Subdivision Myxomycotina) lack cell walls except in spores. The vegetative body has amoeboid motion and engulfs solid food particles.
3. True fungi (Subdivision Eumycotina) have cell walls throughout the life cycle and feed only by absorbing dissolved solutes from solution. Some are unicellular but most form a vegetative body (the mycelium) consisting of branched filaments (hyphae). Five Classes of true fungi are discussed.
4. The egg fungi (Class Oomycetes) have flagellated male reproductive cells and may have flagellated spores. The mycelium may be diploid. Many species are aquatic. Some are important plant pathogens.
5. The remaining classes of true fungi lack motile cells and are classified by their mode of sexual reproduction.
6. The zygote fungi (Class Zygomycetes) mate by fusing special hyphae. The zygote becomes a spore. Sporangia are produced for asexual reproduction. The mycelium is haploid, coenocytic, and is usually terrestrial. Most are saprophytes.
7. The sac fungi (Class Ascomycetes) form meiospores inside a sac (the ascus). Vegetative body may be unicellular (yeasts) or a multicellular mycelium that is divided into cells by perforated cross-walls. The vegetative mycelium is haploid. Fusion of sexual hyphae produces dikaryotic ($n+n$) hyphae that produce asci. Most species form their asci within an ascocarp composed of many haploid and dikaryotic hyphae. Asexual reproduction is usually by conidia. Most species are terrestrial saprophytes but some are dangerous plant pathogens.
8. The club fungi (Class Basidiomycetes) produce meiospores on the surface of special cells, often club-shaped, called basidia. Mushrooms and brackets are the fruiting bodies (basidiocarps) of fungi in the subclass Homobasidiomycetidae. The basidiocarps consist entirely of dikaryotic hyphae. Dikaryons form by fusion of haploid hyphae. The dikaryotic condition is maintained by clamp connections. The vegetative mycelium may be haploid or dikaryotic. Most fungi in this subclass are terrestrial saprophytes. The rusts and smuts (Subclass Heterobasidiomycetidae) lack basidiocarps. They are plant parasites and may be serious pathogens. Some of them use two plant species as alternate hosts at different stages of the life cycle.
9. The imperfect fungi (Class Fungi Imperfecti) have no known sexual reproduction. Reproduction is asexual, usually by conidia. Most form a septate terrestrial mycelium. This class includes the most common household molds and the fungi that produce penicillin and fine cheeses.
10. The lichens are symbiotic associations in which a fungus and an alga produce a joint body. Widely varied forms occur. Reproductive structures include separate fungal and algal spores as well as bodies (soredia) that include both alga and fungus. Some lichens are important pioneers on bare rock.

13 CHAPTER

bryophytes

The division Bryophyta includes hornworts, mosses, and liverworts. Perhaps the two greatest differences between algae and bryophytes are the general structure of the plant bodies and the occurrence of multicellular gametangia. It will be recalled that the plant body of thallophytes is generally composed either of single cells (or filaments of cells), or of intertwining filaments, resulting in a more or less simple body. Layers of

Table 13.1
Tabulation of Differences between Thallophytes and Bryophytes

Thallophytes	Bryophytes
1. Mostly aquatic	**1.** Mostly terrestrial, but moist habitat
2. A few of the kelps have elements that resemble sieve cells	**2.** Some mosses have cells suggesting sieve cells; otherwise food conducted in relatively undifferentiated cells
3. None have water-conducting elements	**3.** Simple water-conducting cells (not vessels or tracheids)
4. In general, no specialized water- and nutrient-absorbing tissue; rhizoids and haustoria may occur in some fungi	**4.** Rizoids anchor and absorb water and mineral nutrients
5. Mostly filamentous or a lacework of intertwining filaments; parenchyma in a few	**5.** Only one stage of mosses is filamentous; all others are formed of parenchyma cells
6. A definite alternation of generations in many forms	**6.** All have an alternation of heteromorphic generations
7. Both sporophytes and gametophytes nutritionally independent	**7.** Gametophyte independent; sporophyte nutritionally dependent
8. Gametangia are either single cells or groups of single cells not accompanied by a jacket of sterile vegetable cells	**8.** Gametangia are always composed of gamete-producing cells covered by a jacket of sterile vegetative cells
9. Spores often water-dispersed	**9.** Spores wind-dispersed

parenchyma tissue occur in some forms. With the exception of one stage in the life history of mosses, the bryophyte plant body is never filamentous. It is composed of blocks or sheets of cells that form a parenchymatous tissue. Another important difference is that the sporophyte generation in the bryophytes is attached to the gametophyte and it remains there throughout its life time.

General Characteristics

The comparative characteristics of the thallophytes and bryophytes are shown in Table 13.1 The multicellular female gametangia characteristic of all members of the Bryophyta produce but a single gamete, the egg, which is surrounded by a layer of protecting cells. Male gametangia of the bryophytes produce numerous sperms, but they are always surrounded by a protective layer of parenchyma cells, called a sterile jacket.

Water is required by bryophytes, as in the algae, to effect fertilization. A definite alternation of generations occurs in all members of the Bryophyta, with the sporophyte (diploid) generation always more or less dependent on the gametophyte (haploid) generation. In all Bryophyta, an **embryo sporophyte**, consisting of a spherical or elongated mass of tissue, is formed directly after the germination of the zygote.

Asexual spores are not formed in Bryophyta. Some members of this division reproduce asexually either by fragmentation or by multicellular units called **gemmae**. A haploid plant, producing gametes, alternates with a diploid plant; meiosis occurs in the diploid plant, and haploid spores containing $1n$ sets of chromosomes in each nucleus are formed. Thus, a haploid gamete-producing plant—the **gametophyte**—regularly alternates with a diploid spore-producing plant—the **sporophyte**, as shown below.

Economic Importance

Bryophytes occur on soil, trunks of trees, and rocks; many species are truly aquatic. They are not extremely important economically, but they do have an interesting if short, ethnohistory. Native American Indians in Utah, for instance, ground up mosses, such as *Mnium* and *Bryum,* into a paste and applied it as a poultice to treat burns and bruises. In Europe, the growth of mosses such as *Dicranoweisia* was encouraged on shingle roofs to make them watertight.

Sphagnum is by far the most important moss economically. The surface layers of stems and "leaves" of *Sphagnum* are composed of alternating files of living green cells and large empty hyaline cells (Fig. 13.2). The hyaline cells are capable of absorbing water or other fluids in great quantities as high as 20 times their own weight. In the 1880s in Germany it was discovered, quite by accident, that *Sphagnum* moss pads make excellent absorbent bandages. A workman in Kiel lacerated his arm while working in a peat moor. Since he had no other bandage material, he packed his wound with dry peat *(Sphagnum)*. Ten days later, when he was able to obtain treatment, the wound had healed. After that, disinfected bags filled with dry *Sphagnum* became a common wound dressing in Germany. The use of *Sphagnum* as wound dressings was not popular outside Germany until World War I. At that time they were used on a large scale by the Allied Armies as well. The British, for example, used 1,000,000 *Sphagnum* dressings each month. *Sphagnum* dressings were used very little in the United States, but the American Red Cross did publish inquiries and instructions on finding useful *Sphagnum*. More recently, *Sphagnum* is still used as packing for vegetables, in potting mix to increase the water holding capacity of the soil; and old compressed *Sphagnum* is used as a fuel in the form of a low grade coal known as peat.

Classification

The Bryophyta are divided into three classes: **Hepaticae** (liverworts), **Anthocerotae** (hornworts), and **Musci** (mosses). The Hepaticae are the simplest and perhaps the most primitive bryophytes. They consist of a flat, ribbonlike, green thallus that produces gametes and a meiospore-producing capsule or sporangium usually embedded within the gametophyte thallus. The name liverwort is very old, having been used in the ninth century. It was probably applied to these plants because of their fancied resemblance to the liver and the belief (the "Doctrine of Signatures") that plants resembling human organs would cure diseases of the organs they resembled. At any rate, a prescription for a liver complaint in the 1500s called for "liverworts soaked in wine."

Mosses are the best-known class of Bryophyta. They have simple "stems" and "leaves," and sporophytes are borne on green gametophytes. They have either an upright or prostrate leafy gametophyte, and may form extensive mats on moist, shaded soil. The sporophyte is raised above the leafy gametophyte.

Class Musci

General Characteristics

Although the mosses are small plants, they are nevertheless conspicuous. They frequently cover rather large areas of stream banks. They grow on rocks and trees, and are sometimes submerged in streams. Although some mosses are able to resist considerable drought, (Figs. 13.1, 13.2*A*, 13.5) all require moisture for active growth and reproduction.

Mosses as a class show great structural uniformity. *Funaria* and *Mnium* and *Sphagnum* are examples (Figs. 13.1, 13.2*A*, 13.5). Gametophytes of nearly all species have two growth stages: (a) a creeping, filamentous stage (the **protonema**) (Fig. 13.1), from which is developed (b) the moss plant with an upright or horizontal stem bearing small, spirally arranged green leaves (Fig. 13.5).

Gametophyte

Germinating spores of mosses do not develop directly into a leafy gametophyte but first become a filamentous structure. This early stage of gametophyte is called the protonema (Fig. 13.1). The protonema is not a permanent structure, although it may branch considerably under favorable conditions and cover a rather large area of soil, sometimes forming a green coating resembling algal growth. Cells composing the protonema contain numerous chloroplasts (Fig. 13.1). At various locations, cell divisions along the protonema produce swollen regions called **buds**. From them develop the upright gametophyte stage.

The mature stem of the leafy gametophyte shows a wide range of differentiation, depending on the species, age, and environment. Elongate cells called **rhizoids** anchor the gametophyte to the substrate (Figs. 13.1, 13.5, 13.7). These structures are not roots, and apparently not involved in absorption. The simpler mosses, like *Tetraphis* (Fig. 13.2*E*) have an outer layer of thick-walled epidermal cells (stereids), surrounding an undifferentiated cortex made up of parenchymalike cells. Epidermal cells of the peat moss *Sphagnum* (Figs. 13.2*B* to 13.2*D*) are large and empty, with pores that open to the outside.

Some mosses, such as *Mnium* (Figs. 13.3*A*, 13.3*B*, 13.4) have a central conducting strand in their stems. It is made up of elongated, thin-walled **hydroids,** which are dead, empty cells that conduct water (Figs. 13.3*B*, 13.4). Their end walls are oblique and sometimes very thin, perforated with pores, or partly dissolved. Experiments with dyes show that translocation of water can occur in these cells. Some moss "leaf" midribs also contain hydroids, but only in one genus are these known to connect with the hydroids of the stem. Hydroids show some resemblances to tracheids, but lack specialized pitting and lignified walls. No lignin has been detected in bryophytes.

A few of the most specialized mosses also contain cells resembling the sieve cells of vascular plants. Between the

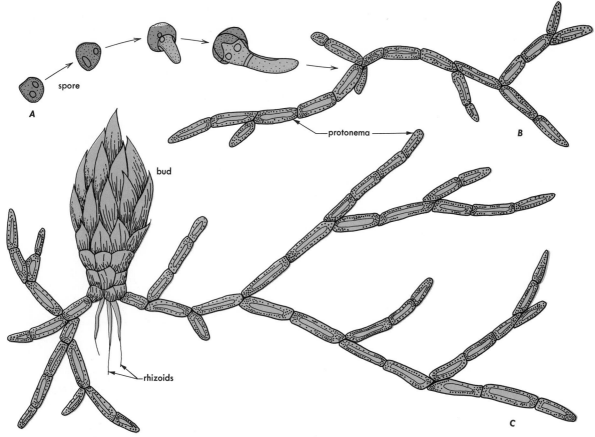

Figure 13.1 *Funaria. A,* germination of spores; *B,* protonema; *C,* protonema with bud.

epidermis and central strand, elongated cells with oblique end walls **(leptoids)** may occur (Fig. 13.4). These cells are alive, but their nuclei are degenerate and inactive. They have many plasmodesmata in their end walls, and callose may be present. Studies show that sugars may be translocated through these cells.

It is important to emphasize that: (1) most mosses do not contain leptoids; (2) there are important differences between hydroids and tracheids, and between leptoids and sieve cells; (3) there is as yet no firm evidence to show these cells are phylogenetically related to conducting cells of vascular plants.

Sexual Reproduction

The gametophytes of many mosses are **monoecious** (homothallic); that is, both antheridia (sperm-producing) and archegonia (egg-producing) are produced by the same gametophyte (Fig. 13.5). Some mosses are **dioecious** (heterothallic); in other words, antheridia and archegonia are produced on separate gametophytes. In *Mnium,* shoots bearing antheridia are easily recognized, by the surrounding leaves that spread around them somewhat like petals of a flower. The group of antheridia appears as an orange spot in the center of the terminal cluster of leaves (Figs. 13.5, 13.6).

Antheridia of most true mosses (Fig. 13.6*B*) are enclosed by cells forming a **sterile jacket.** These cells

contain chloroplasts that become orange-red when the antheridium ripens. The sperms consists mainly of an elongated nucleus and each has two flagella. The antheridia are surrounded by club-shaped, multicellular, sterile hairs **(paraphyses)** with conspicuous chloroplasts.

The archegonia have a long **neck,** a thickened middle portion called the **venter,** which contains the egg, and a long **stalk** (Fig. 13.6*C*). When sufficient moisture is present, sperms are extruded from the antheridium. The neck cells of the archegonium separate, opening a passage to the egg. Sperms swim down, possibly responding to a chemical signal, and one fuses with the egg. Mosses are dependent on free water for fertilization.

Sporophyte

Soon after fertilization, the zygote begins to develop into a spindle-shaped embryo that differentiates into a sporophyte consisting of a **foot, seta,** and **sporangium** or **capsule.** The foot penetrates the base of the venter and grows into the apex of the leafy shoot. It absorbs water and nourishment from the gametophyte for its growth and development. The seta elongates rapidly, raising the sporangium 1 cm or more above the top of the leafy gametophyte (Fig. 13.7*A*). The old archegonium increases in size as the sporophyte enlarges. When the seta elongates, the top of the expanded archegonium is torn

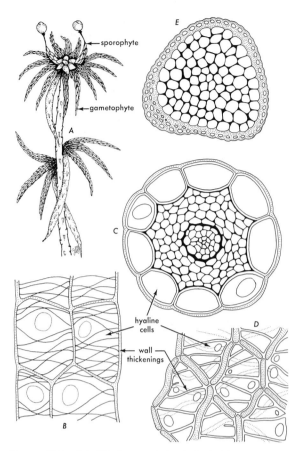

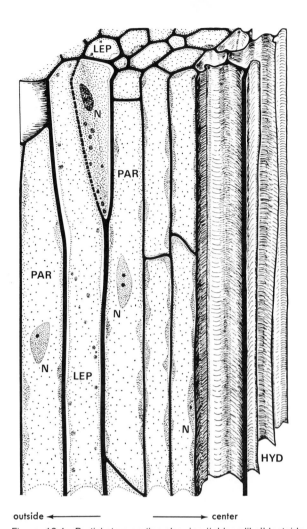

Figure 13.2 *A*, habit of sporophyte plant of *Sphagnum papillosum*, ×2. *B*, external view of stem showing large hyaline cells, ×450. *C*, cross section of branch stem showing external hyaline cells and thick-walled cells of central axis, ×590. *D*, leaf cells; note the narrow chlorophyll containing cells and the larger hyaline cells with peculiar bandlike thickenings, ×330. *E*, cross section of *Tetraphis pellucida* stem to show its simple structure, ×70.

outside ←——————→ center

Figure 13.4 Partial stem portion showing "phloemlike" leptoids and "xylemlike" hydroids.

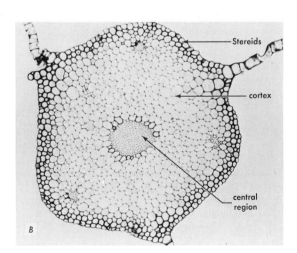

Figure 13.3 Longitudinal sections through gametophyte stems of mosses. *A*, stem of *Mnium* showing central strand of sclerenchymalike cells, ×100; *B*, cross section of *Mnium* seta (1) thick-walled stereids; (2) thin-walled cortex; (3) central region of partially collapsed hydroids, ×72.

Class Musci

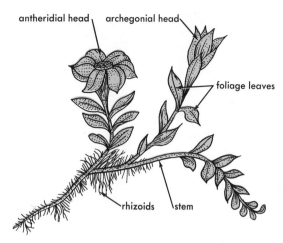

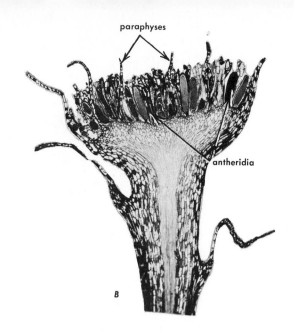

Figure 13.5 *Mnium,* showing location of gametangia, ×1.

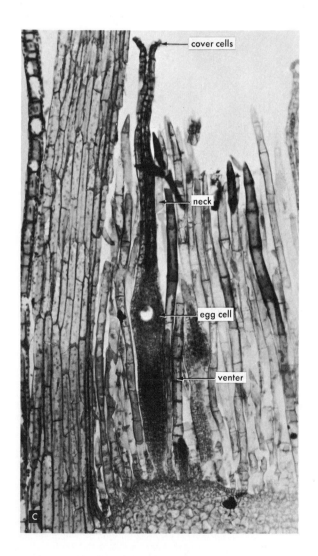

Figure 13.6 Microscopic views of moss gametangia. *A,* antheridial heads of *Polytrichum,* ×1; *B,* antheridial head of *Mnium,* ×35; *C,* archegonium surrounded by paraphyses (*Mnium*), ×50.

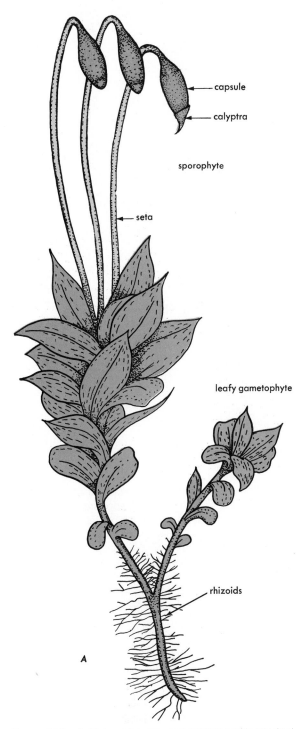

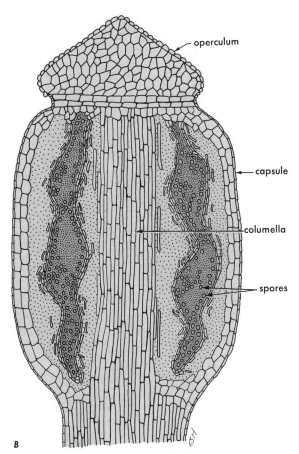

B

Figure 13.7   A, Mnium, showing gametophytes with attached sporophytes. B, median section through a mature but unopened capsule of Mnium, ×25.

from its point of attachment to the gametophyte and elevated with the sporangium. The upper end of the old archegonium, now known as the **calyptra**, remains for a time as a covering for the sporangium (Fig. 13.7A).

The sporangium of mosses (Figs. 13.7, 13.8) may measure 1 to 3 mm in diameter and 2 to 6 mm long. It is surrounded by an epidermal layer composed of cells similar to those in the epidermis of higher plants. Stomata

occur in the epidermis covering the lower half of the sporangium. The sterile tissue forming the inner portion of the sporangium may conveniently be divided into three regions, each of which may be recognized by the type of cells comprising it. The fourth region, composed of sporogenous cells, forms a layer around the columella (Fig. 13.7B). The cells of the sporogenous tissue (sporocytes) may increase in number by mitosis; each cell contains the diploid number of chromosomes. Meiosis takes place next, and haploid spores containing 1n chromosomes result. They are the first cells of the gametophyte generation.

The columella projects upward, forming a small dome above the main mass of the sporangium. The four or five outer layers of cells of this dome differentiate into a dry, brittle cap called the **operculum** (Figs. 13.7B, 13.8). The cells immediately beneath the operculum form a double row of triangular **peristome teeth** (Figs. 13.8B,C). The broad bases of the teeth are attached to the thick-walled deciduous cells that form the **annulus** around the upper end of the sporangium. When the sporangium matures and becomes dry, thin-walled cells of the annulus break down, allowing the operculum to fall away and expose the peristome teeth. By this time, most of the thin-walled cells within the columella have collapsed and the cavity thus formed is filled with a loose mass of spores. Peristome teeth are rough and are very sensitive to the amount of moisture in the air (Figs. 13.8B,C). When they are wet or

Class Musci

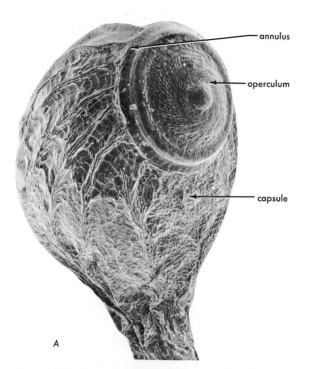

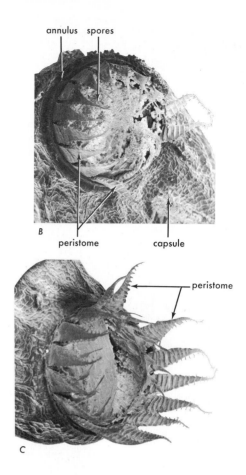

Figure 13.8 Scanning micrograph of a capsule of *Funaria*. *A*, capsule with operculum in place; *B*, operculum has been shed, peristome teeth partially open, spores attached to them; *C*, peristome teeth fully open, capsule empty, ×300.

the atmospheric humidity is very high, they bend into the cavity of the sporangium; when dry, they straighten and lift out some of the spores, which are then disseminated by air movements. If a spore comes to rest on moist soil and if illumination and temperature are favorable, it germinates and grows into a protonema.

Summary of Moss Life History and Characteristics

1. The gametophyte consists of (a) a filamentous, branched, algal-like structure called a protonema, and (b) leafy shoots that develop from buds on the protonema. The shoots consist of a stalk bearing rhizoids at its lower end and leaves throughout its length. The gametophyte is green and able to synthesize food. All gametophyte nuclei are haploid.
2. Antheridia and archegonia develop at the apex of the leafy gametophyte.
3. Each antheridium produces hundreds of sperms.
4. A single egg is formed in the venter of each archegonium.
5. Gametes (eggs and sperms) each contain $1n$ chromosomes.
6. When sufficient moisture is present, mature sperms are extruded from the antheridium. The neck of the mature archegonium is open and sperms swim down it to the egg in the venter of the archegonium.
7. The egg is fertilized by one sperm.
8. The zygote, as a result of the union of gametic nuclei, is diploid.
9. An embryo sporophyte, consisting of a spindle-shaped mass of undifferentiated cells, partially dependent on the gametophyte, develops from the zygote.

10. The embryo develops into a sporophyte consisting of a foot, seta, and sporangium. The sporangium contains sporogenous tissue and several types of sterile tissues, among which are the operculum, annulus, and peristome.
11. Sporocytes, each containing $2n$ chromosomes, form from sporogenous cells.
12. Sporocytes undergo meiosis. Four spores result from each sporocyte
13. When spores germinate, they form the protonema.
14. In the central strand of some mosses, *eg, Sphagnum* and *Mnium,* there are specialized cells called hydroids that may be involved in water conduction.
15. Cells specialized for sugar transport, called leptoids, are also found in some mosses.

Class Hepaticae

General Characteristics and Distribution

Hepaticae are commonly known as liverworts. There are some 8500 species of liverworts. The great majority of them grow in moist, shady localities. The gametophyte is the prominent plant, and the sporophyte is partially or wholly dependent on the gametophyte. The gametophyte is green and may grow either as a flat ribbon or as a leafy shoot. In either event, the plant body is frequently called a thallus, even though its internal structure does not

Figure 13.9   Habit of *Riccia*, ×1.

correspond to that of the thallophytes. Of the four orders in the Hepaticae, we shall consider briefly the characteristics of two genera, *Riccia* and *Marchantia*, both belonging to the order Marchantiales.

Gametophytes of the Marchantiales are small, green, ribbon-shaped plants (Figs. 13.9, 13.11). They branch regularly by a simple forking at the growing tip, resulting in a number of Y-shaped branches. In some species, a rosette may be formed. The upper surface of the thallus is composed of cells adapted for photosynthesis and arranged to form air chambers that open to the surface by a pore (Figs. 13.10A, 13.12C,D). Several types of storage cells generally make up the lower surface. **Rhizoids,** specialized elongated cells, extend downward from the lowermost layer of cells and anchor the gametophyte to the substrate. The thallus is thus differentiated into distinct upper and lower portions. Scales, frequently brown or red, are also formed on the lower surface (Figs. 13.10A).

Riccia

Riccia is a widely distributed genus, and although it requires water for active growth, most species can tolerate considerable drought. Several species are aquatic, growing either on mud or on the surface of small ponds.

Gametophyte

The gametophyte is a small green thallus, frequently forming a rosette (Fig. 13.9). Tissue on the lower side is composed of colorless cells that may contain starch. Tissue on the upper side consists of vertical rows of chlorophyll-bearing cells, between which are air chambers (Fig. 13.10A). The gametangia are embedded in deep, lengthwise depressions or furrows, on the upper surface of the thallus. Antheridia and archegonia are usually found on the same gametophyte (homothallic).

Antheridia.   Antheridia of different representatives of the Bryophyta are similar, though they may vary somewhat in shape. In *Riccia,* they are pear-shaped and composed of two types of cells, (a) fertile and (b) sterile (Fig. 13.10B). Fertile cells give rise to sperm, which are relatively numerous, small, and dense with protoplasm. Sterile cells form a protective jacket, one cell in thickness, around the fertile cells. The antheridia are connected to the gametophyte by a short stalk (Fig. 13.10A).

Mature sperm consist mainly of an elongated nucleus with two long flagella. They may be shot with considerable force from the mature antheridium or they may be extruded slowly in a single mucilagenous mass. In any event, they do not leave the antheridium until enough moisture is present to allow them to swim about.

Archegonia.   The archegonium of *Riccia* is a flask-shaped structure consisting of two parts, (a) an expanded basal portion, the **venter,** and (b) an elongated **neck** (Figs. 13.10C,D). Four **cover cells** are located at the top of the neck. Each archegonium contains a single egg cell, which is located in the venter. A short stalk attaches the archegonium to the gametophyte. Archegonia of a similar structure are found in other Bryophyta.

Shortly before the egg cell is mature, the cover cells separate. At the same time, the cells in the center of the neck dissolve, so that an open canal connects the venter with moisture outside the archegonium. Free-swimming sperms move toward certain chemical substances formed by the archegonium. Several sperms may enter the archegonium, but only one sperm fertilizes the egg in the venter.

Sporophyte

As a result of fusion of sperm and egg nuclei, the zygote contains the diploid, or $2n$, set of chromosomes. Mitosis now proceeds in a more or less orderly manner until a spherical mass of some 30 or more similar cells is formed (Fig. 13.10D). These cells are partially dependent on the gametophyte. This mass of undifferentiated cells comprising the young sporophyte is an **embryo.** It is located within the venter of the archegonium.

Further development of the embryo involves differentiation of an outer layer of cells to form a protective jacket of sterile tissue surrounding a mass of cells that are capable of forming spores. This spore-forming tissue is called **sporogenous tissue.** The sporogenous cells continue to divide by mitosis until many have formed. These cells, now known as sporocytes, will divide by meiosis to produce four spores with the haploid chromosome number. Remember that meiosis involves two cell divisions and results in a reduction in the chromosomes by one half (see Chapter 3).

The mature sporophyte of *Riccia* (Fig. 13.10E) is called a **capsule.** It is composed of a jacket of sterile cells enclosing a mass of spores. The sporophyte wall breaks down when the spores are mature. These spores remain embedded in the gametophyte thallus and are not released until after the death and decay of the gametophyte. As each spore germinates, it grows into a typical gametophyte plant.

Class Hepaticae

281

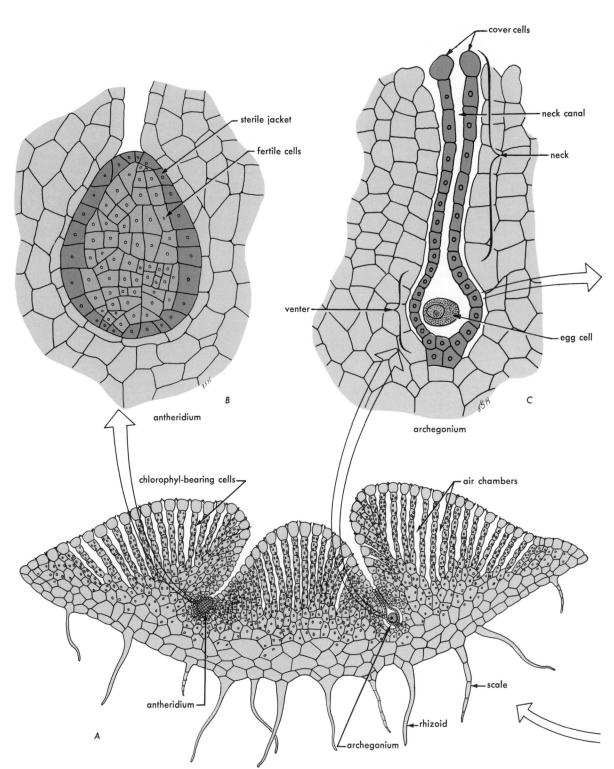

Figure 13.10 *Riccia*. A, cross section through a thallus: gametangia are on the upper surface of the thallus between the photosynthetic filaments or in a notch in the thallus, ×8. B, an antheridium; note jacket of sterile cells surrounding the developing sperm cells. C, archegonium on thallus, neck canal open, egg cell in venter. D, young sporophyte developing in venter of old archegonium. E, sporophyte with spore mother cells still enclosed in venter of old archegonium, F, meiosis occurs within the sporophyte. G, the venter now contains fragments of the old spore case (sterile cells of the sporophyte) and spores. H, a spore. B through G, about ×50; H, ×100.

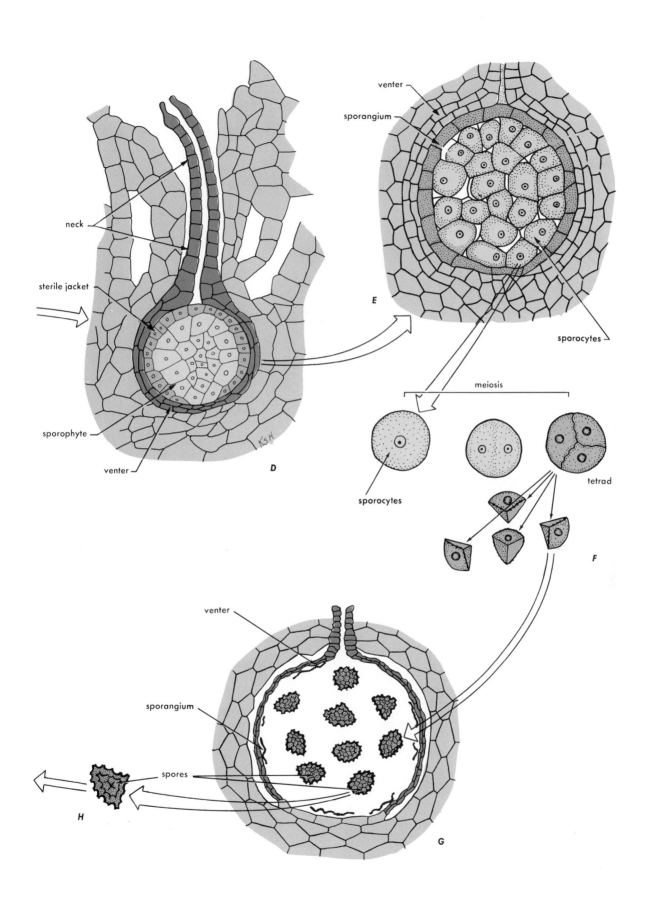

neck

sterile jacket

sporophyte

venter

D

venter

sporangium

E

sporocytes

meiosis

sporocytes

tetrad

F

venter

sporangium

spores

G

spores

H

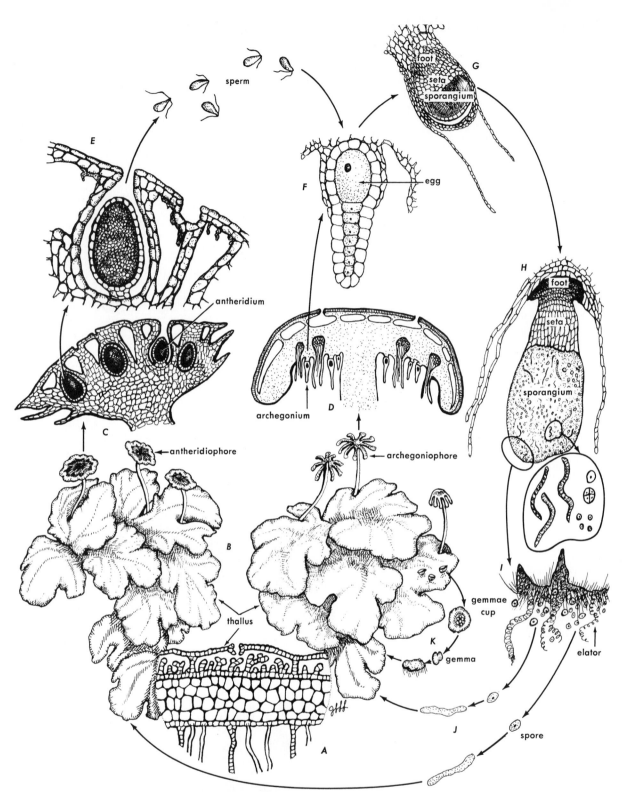

Figure 13.12 Life cycle diagram and habit of *Marchantia polymorpha*. *A*, cross section view of thallus. Note the pores and air chambers containing columns of chlorophyllous cells. *B*, habit of male (left) and female (right) thalli with antheridiophores and archegoniophores respectively. *C*, section through head of antheridiophore to show antheridia. *D*, section through head of archegoniophore to show archegonia. *E*, enlarged view of antheridium. *F*, enlarged view of an archegonium. *G*, immature sporophyte plant. *H*, mature sporophyte composed of foot, seta, and sporangium. *I*, sporangium releasing spores and elaters. *J*, spore germination to form new gametophyte thallus. *K*, asexual reproduction by gemmae.

Significant Steps in the Life History of *Riccia*

1. The gametophyte, a small, green, flat plant, absorbs water and mineral salts from the soil and carbon dioxide from the air. It contains chlorophyll and can synthesize food. All gametophytic nuclei are haploid (contain $1n$ set of chromosomes).
2. Gametangia (antheridia and archegonia) develop in a furrow on the upper surface of the gametophyte.
3. Each antheridium produces thousands of sperms.
4. A single egg is formed in the venter of each archegonium.
5. Gametes (eggs and sperms) contain $1n$ set of chromosomes.
6. When sufficient moisture is present, mature sperms are extruded from antheridia. The neck of the mature archegonium is opened and sperms swim down to the egg in the venter.
7. Each egg is fertilized by a single sperm.
8. The zygote, as a result of the union of two haploid gametic nuclei, is diploid.
9. An embryo sporophyte, consisting of a spherical mass of undifferentiated cells, partially dependent on the gametophyte, develops from the zygote. Each nucleus of the embryo sporophyte is diploid ($2n$).
10. The cells of the embryo differentiate into a sporangium that consists of a jacket of sterile cells enclosing a mass of fertile cells.
11. Sporocytes, each with $2n$ nuclei, form in sporangia.
12. Sporocytes undergo meiosis. A quartet of spores forms from each sporocyte. All sporocytes are $2n$ and spores are $1n$.
13. Upon death and decay of old gametophytes, spores are liberated from the capsules and new gametophytes from them.

Marchantia

Marchantia may grow in large mats on moist rocks and soil in shady locations. It is widely distributed and fairly common on banks of cool streams. It is somewhat better adapted to growing on land than is *Riccia*, but considerable moisture is still required for active growth and for fertilization. For all references below to Figure 13.11, see Color Plate 3.

Gametophyte

The gametophyte of *Marchantia* differs from that of *Riccia* in that gametangia on special disks are raised some distance above the vegetative thallus (Figs. 13.11*C,D,* 13.12*A,E*). The thallus of *Marchantia* is similar to that of *Riccia,* but much coarser. It is strap-shaped, with a prominent midrib and shows dichotomous branching (Figs. 13.11*A,B*). The tips of the branches are notched. An individual thallus may be 1 to 1.5 cm broad. Their length (1 to 1.5 cm) and the degree of branching depend on growth conditions. It is usual for new growth to overlap the older decaying ends of adjacent thalli. On its upper surface are polygonal areas, with a small but conspicuous pore in the center (Figs. 13.12*L*). These areas, with their air pores, mark the outlines of air chambers, each of which is filled with short filaments of cells containing chloroplasts. As in *Riccia,* the lower surface of the thallus is composed of colorless cells, some of which are modified for storage. Rhizoids, which anchor the thallus, grow from the cells covering the lower surface. Several rows of scales are also attached to this lower surface (Fig. 13.12*L*).

Asexual Reproduction. *Marchantia* reproduces asexually in two ways: (a) Older parts of the thallus die and younger portions, no longer attached, develop into new individual plants; and (b) small cups, known as **gemmae cups,** form on the upper surface, and small disks of green tissue, called gemmae, grow from the bottom of these cups (Figs. 13.11*B* 13.12*M*). Gemmae, when mature, break off from the thallus and are distributed into nearby areas. Raindrops are often the agents that break off gemmae and scatter them away from the thallus. New gametophyte plants grow from gemmae.

Sexual Reproduction. *Marchantia* is heterothallic; that is, gametophytes have either antheridia or archegonia. The gametangia, however, are very similar in structure to those of *Riccia.* Antheridia are pear-shaped bodies composed of a jacket of sterile tissue surrounding sperms (Fig. 13.12*B,C,D*). They are borne on disks raised above the thallus on slender stalks, called **antheridiophores** (Figs. 13.11*C,* 13.12*A*). Antheridiophores are modified portions of the thallus having furrows, rhizoids, and air chambers. Antheridia develop in cavities on the upper surface of the disk, the youngest antheridia being close to the outer margin of the disk (Fig. 13.12*B*). Mature sperms are extruded in a mucilagenous mass.

Archegonia are flask-shaped and have the same structures as those of *Riccia*—venter, neck, and cover cells (Figs. 13.11*D,* 13.12*G,H*). They are borne on specialized branches called **archegoniophores.** The disk at the top of the archegoniophore typically has eight or more lobes. In development, the lobes bend downward and then grow inward toward the stalk. Thus, the youngest portion of an older disk bearing archegonia is on the underside and close to the stalk (Figs. 13.11*E,* 13.12*E,F*). Archegonia develop in the lower surface of the disk but in tissue similar to that found on the upper surface of the thallus. Fingerlike processes grow out from between the lobes. The archegonia mature, and fertilization takes place when the disk of the archegoniophore is but slightly elevated above the thallus.

Sporophyte

Zygotes develop, as in *Riccia,* into diploid embryos, which are spherical masses of undifferentiated, dependent tissue (Fig. 13.12*H,I*). Subsequent growth, however, is more complicated than in *Riccia.* Some cells of the embryo divide to form a mushroomlike growth that becomes embedded in gametophyte tissue. This growth is called the **foot.** Cells in the central portion of the embryo form a seta. The remaining lower cells develop into a **sporangium** (Fig. 13.12*H,I*). Lengthening of the seta suspends the sporangium below the disk. While this development is taking place, the stalk of the archegoniophore elongates, lifting the disk with its archegonia well above the thallus (Figs. 13.11*D,E,* 13.12*I*).

Fertilization stimulates enlargement of the old archegonium, which keeps pace with the enlargement of the developing sporophyte. As a result, the sporophyte is continually enclosed within the archegonium, which, because of its increase in size and change in function and shape, is now known as the **calyptra**. In addition, the surrounding gametophyte tissue produces two other envelopes that protect the sporophyte.

The mature sporophyte (Fig. 13.12*I*) is composed of three parts, **foot, seta,** and **sporangium.** The foot is an absorbing organ. The seta serves to lower the sporangium away from the archegonial disk and thus to facilitate distribution of spores. Before meiosis, the sporangium or capsule is composed of a jacket of sterile cells surrounding a mass of sporocytes, among which are a number of sterile elongated cells (Figs. 13.12*I,J*). Four spores, each containing 1*n* chromosomes, develop from each sporocyte. The elongated sterile cells are transformed into spiral elements, called **elaters**, that change shape under the influence of varying moisture conditions. Their twisting motion as they imbibe and lose water aids in dispersal of spores (Figs.13.12*I,J,K*).

Dissemination of spores is aided by two structural features not found in *Riccia:* (a) the sporangium of the sporophyte hangs from the lower side of the raised archegonial disk (Figs. 3.11*E*), and (b) elaters help to empty the spore case. Spores develop immediately into new gametophytes if they land in an appropriate habitat.

Significant Steps in the Life History of *Marchantia*

1. The green gametophyte thallus absorbs water and mineral salts from soil and carbon dioxide from air.
2. Gametangia (antheridia and archegonia) are borne on upright branches called antheridiophores and archegoniophores.
3. The zygote is diploid (2*n*).
4. The embryo sporophyte, consisting of a spherical mass of undifferentiated cells, is formed from the zygote.
5. Cells of the embryo sporophyte differentiate into foot, seta, and sporangium.
6. The foot is an absorbing organ. It is embedded in the gametophyte, from which it receives nourishment.
7. The sporangium consists of a jacket of sterile cells surrounding a mass of sporogenous tissue interspersed with elongated sterile cells.
8. The seta is a stalk that, in lengthening, lowers the sporangium below the surface of the archegonial disk.
9. Repeated mitotic cell divisions in sporogenous tissue result in sporocytes each one of which contains 2*n* chromosomes.
10. Sporocytes divide by meiosis into tetrads of spores; each spore contains 1*n* chromosomes.
11. The elongated sterile cells are transformed into spiral elaters.
12. The presence of elaters and the hanging position of the sporophyte aid in disseminating spores.
13. Spores germinate immediately into new gametophyte plants.

Class Anthocerotae

The Anthocerotae have the simplest gametophytes of the Bryophyta. They are small, green thallus plants with little internal differentiation of vegetative tissues. They are slightly lobed, with numerous rhizoids growing from the lower surface (Fig. 13.13). The antheridia are similar in structure to those encountered among the Hepaticae (Fig. 13.14*A*). They are located in roofed chambers in the upper portion of the thallus. The archegonia are embedded within the thallus and are in direct contact with the vegetative cells surrounding them (compare Figs. 13.10*C*, 13.14*B*).

The sporophyte (Figs. 13.13, 13.15) of the Anthocerotae is in striking contrast to those of the Hepaticae. The subepidermal cells contain chloroplasts, and typical stomata are found in the epidermis. A foot embedded in the thallus serves as an absorbing organ. The sporangium is an upright elongated structure. Sporogenous tissue forms a cylinder parallel with the elongated axis of the sporangium. Spores mature in progression from the top down. A meristematic region (i.e., region of continuous cell division) lies just above the foot and continually adds new cells to the base of the sporangium. Its presence means that spores may be produced over long periods.

Under exceptionally favorable growing conditions, the sporophyte may lengthen greatly. Some sporogenous tissue at the base of the sporangium may be replaced by a conspicuous conducting strand. The foot enlarges and, through decay of the gametophyte, comes into more or less direct contact with the soil. Such sporophytes are capable of surviving independently for some time.

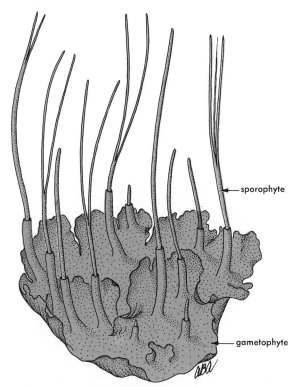

Figure 13.13 *Anthoceros* gametophyte with upright, dependent sporophyte, ×3.

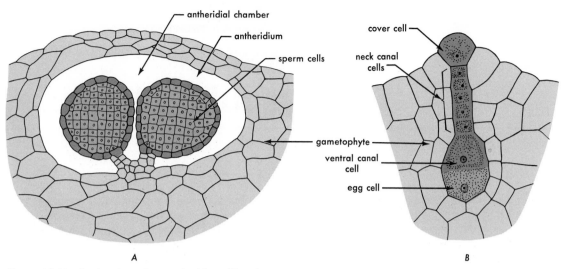

Figure 13.14 Section through an antheridium *(A)* and an archegonium *(B)* of *Anthoceros*, ×50.

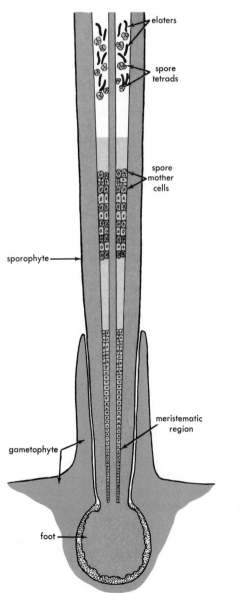

Figure 13.15 Longitudinal medium section through sporophyte of *Anthoceros*.

lower vascular plants

In contrast to algae, fungi, liverworts, and mosses, the sporophytes of vascular plants possess a well-developed vascular system that serves to conduct water, mineral salts, and food. There are 10 divisions of vascular plants; six are seed-bearing and four do not bear seeds. The latter four divisions, which together make up the lower vascular plants, are the **Psilophyta, Lycophyta, Sphenophyta,** * and **Pterophyta.**

The number of families and genera in these divisions is small, and their members do not form a very conspicuous part of the current land flora. The Pterophyta is currently the largest of the four divisions; it comprises the ferns, which are prominent in certain cool, moist habitats. However, plant fossils indicate that there was a period in the earth's history when the members of the Lycophyta, Sphenophyta, and Pterophyta formed a large and dominant flora (Chapter 10).

Although they are land plants, they still require free water for fertilization. The sporophyte is the dominant generation and—except for the youngest stages in the formation of an embryo—it is independent of the gametophyte. Gametophytes are small, and some may lack chlorophyll. In its simplest form, the plant body of the sporophyte is an axis. Meristematic tissue terminates the shoots and roots. Most lower vascular plants are herbaceous perennials, but some ferns reach tree stature and have some woody tissue.

Division Psilophyta

This division is represented in the existing flora by a single family, the Psilotaceae, with two genera, *Psilotum* and *Tmesipteris.* They are rare plants found mainly in the tropics, one form of *Psilotum* growing as far north as Florida. Figure 14.1 shows the simple plant body of *Psilotum nudum.* Roots are not present and leaves (microphylls) are small and scalelike. These are not "true" leaves (megaphylls, see Chapter 10); they are instead more like epidermal outgrowths. The branching, upright stem is slightly flattened and contains chlorophyll. A fungus is always associated with the branched rhizome, which is clothed with many rhizoids. There are no true

* We choose to use the old designation, Sphenophyta, for the division to which the horsetails belong, because the new proposed designation, Arthrophyta, might be confused with Anthophyta or Arthropoda.

roots. The vascular tissue is simple (Fig. 14.2), consisting of poorly developed phloem and xylem. An endodermis, with a conspicuous Casparian strip, separates the central vascular tissue from the outer cortex. The xylem contains spiral and scalariform tracheids.

Sporangia are borne in axils of some of the scalelike leaves (Fig. 14.1). Meiosis occurs in the sporangia, forming tetrads of meiospores. When the meiospores are ultimately released from the sporangia, they may land in a suitable habitat, germinate, and divide by mitosis repeatedly to form a gametophyte.

Psilophyta gametophytes are little more than a nonphotosynthetic axis of parenchymatous tissue several millimeters long (Fig. 14.3). Xylem and phloem are rarely present. Gametophytes may grow on the surface of soil, rock, or bark, or may grow within the soil. Nutrients are provided via a symbiotic or parasitic relationship with a fungus, which ramifies throughout the tissue.

Gametangia are scattered over the surface of the gametophyte: flask-shaped archegonia with eggs, and spherical antheridia with spirally coiled, multiflagellate sperm (Fig. 14.4).

After a sperm swims through free water and reaches an egg, fertilization occurs and a diploid zygote cell begins dividing. The young sporophyte (embryo) consists of a shoot-root axis attached to the gametophyte by a **foot.** While the shoot-root portion is assuming form, the foot enlarges by repeated cell divisions, sending haustorial outgrowths into the gametophytic tissue. The foot, by virtue of its position and organization, is well suited for the functions of anchorage and absorption until the shoot-root axis becomes physiologically independent and a mature sporophyte results.

Division Lycophyta

The Lycophyta are well-represented in the northern hemisphere by three genera: *Lycopodium, Selaginella,* and *Isoetes.* We shall consider only the first two genera.

There are many fossil Lycophyta belonging to several orders, Lepidodendrales being one of the best known. Now-extinct representatives of this order were forest trees, some of which bore seeds and possessed megaphylls (Chapter 10).

The leaves of the living members of this division are generally small and usually arranged in a spiral. Leaf gaps are never formed at the junction of stem and leaf in the

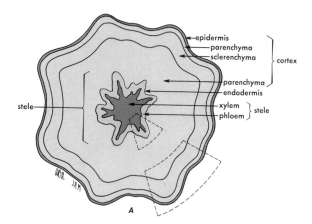

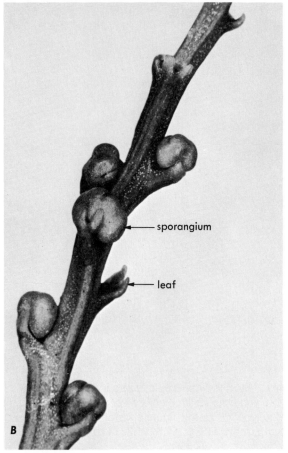

Figure 14.1 *Psilotum nudum. A,* plant; note sporangia on branches at left. *B,* end of a branch, showing scale leaves and sporangia at nodelike swellings.

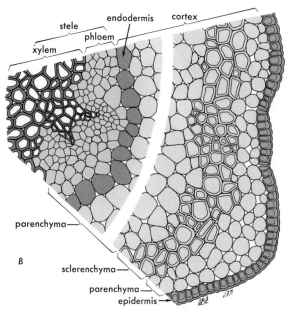

Figure 14.2 Cross section of *Psilotum* stem. *A,* diagram to show the arrangement of tissues. *B,* enlarged sector showing cellular detail.

Lycophyta, so that a living stem of a member of this division has a core of xylem. The leaves are microphylls; the central vein does not branch.

Lycopodium

Approximately 400 species are in this genus. Most are trailing plants, many forming short, upright branches that loosely resemble pine seedlings (Figs. 14.5, 14.6). They are frequently called "ground pine" or "club moss." Although widely distributed, they are most abundant in subtropical and tropical forests. They cannot grow in arid habitats. Several species grow in the eastern and northwestern United States, but none occur in the more arid states of the southwest. Some eastern species are in danger of extinction because they are popular as Christmas decorations.

Mature Sporophyte

The main stem branches freely and is prostrate. Upright stems, approximately 20 cm in height, grow from the horizontal stem. Both types of stems are sheathed with small green leaves. Small but well-developed adventitious roots arise irregularly from the underside of the horizontal stem. As in many higher plants, the primary root, which grows from the embryo, is not long-lived.

The xylem in a *Lycopodium* stem occurs as a complex system of anastomosing strands, so that its distribution in the stem varies greatly in different cross sections of the stem. The phloem is present between the strands of xylem and thus, too, forms a complex system of anastomosing

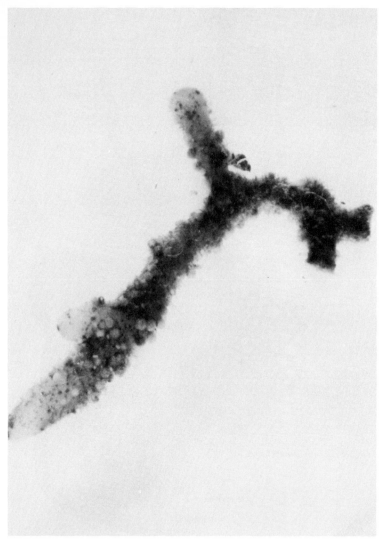

Figure 14.3 Gametophyte of *Psilotum.*

strands. The xylem is composed of tracheids, whereas the phloem contains sieve cells and some parenchyma cells. An endodermis encircles the vascular cylinder. *Lycopodium* stems also lack pith, as do roots of most higher vascular plants.

Leaves of *Lycopodium* show a well developed epidermis with stomata, a mesophyll with many air spaces, and a midvein (Fig. 14.7).

Reproduction

Lycopodium may reproduce asexually or through an alternation of generations involving distinctive gametophytic and sporophytic generations. Both generations are independent and the sporophyte generation is the dominant generation. Meiospores are borne in sporangia in or near the leaf axils (Figs. 14.5, 14.6), sheathing the stem. Spores (Fig. 14.8) and sporangia of a given species are all alike. Leaves bearing sporangia are called sporebearing leaves, or **sporophylls**. In some species, they closely resemble ordinary nonspore-bearing leaves in their structure and appearance. In other

species, they differ from sterile leaves in size, shape, position, and color. Such modified sporophylls are grouped together closely at the ends of stems, forming a **cone** or **strobilus** (Figs. 14.5, 14.6, 14.8).

Gametophyte. The meiospores germinate and develop into gametophytes. Gametophytes in some species grow above ground and are green; in other species, gametophytes are subterranean and lack chlorophyll. The gametophytes are always associated with a fungus. Both male and female gametangia are found on the same gametophyte. Gametangia are borne on the upper portion of the gametophyte (Fig. 14.8). Fertilization occurs when sufficient free water is present to allow sperms to swim to mature archegonia.

Sporophyte Embryo. The embryo sporophyte possesses (a) a well-developed foot, (b) rudiments of a short primary root, (c) leaf primordia, and (d) a short shoot apex (Fig. 14.8). The embryo grows directly into the mature sporophyte plant.

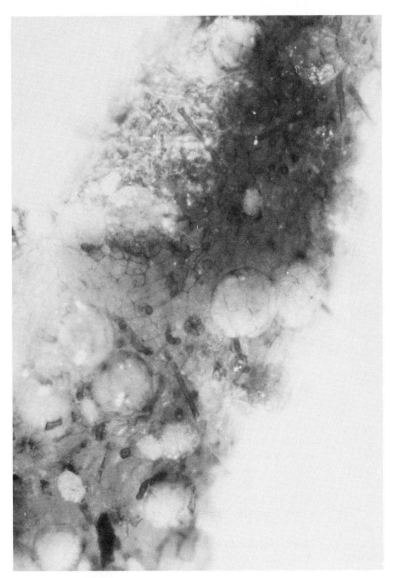

Figure 14.4 Closer view of *Psilotum* gametophyte, showing globular antheridia (male gametangia) and cross-shaped neck cells of archegonia (female gametangia).

Selaginella

Selaginella species resemble those of *Lycopodium* in their general appearance but are smaller. They number more than 700. Although they are widely distributed, most of them are tropical; a few grow in temperate zones. Some species are adapted to withstand periods of drought and hence may grow in relatively dry localities.

Mature Sporophyte

As in *Lycopodium*, the sporophyte of *Selaginella* generally consists of a branched, prostrate stem with short, upright branches, usually only 10 cm high. In some species, the stem is upright and slender and taller. Both horizontal and upright stems are sheathed with small leaves in four longitudinal rows or ranks.

Two species of *Selaginella* are shown in Fig. 14.5. One of them, *Selaginella watsonii*, grows in exposed rock

crevices in the higher Sierra Nevada and is able to withstand periods of drought. The more delicate *Selaginella emmeliana* grows best in a humid environment; it is frequently grown as an ornamental plant.

The vascular system in stems of *Selaginella* consists of one to several branching strands, each having a central core of xylem surrounded by phloem. The cross sections in Fig. 14.9 show a single strand in a stem of *Selaginella*. It occurs in a large air space and is supported by strands of radially arranged endodermal cells. Vessels are present in the xylem of several species of *Selaginella*. Parenchyma and sclerenchyma tissue form a cortex, which is bounded externally by an epidermis.

Reproduction

As in *Lycopodium*, spores are borne in sporangia, which grow in or near axils of sporophylls. Although sporophylls do not differ greatly in appearance from sterile leaves,

Figure 14.5 Two species of *Selaginella*. A, *S. watsonii*, a montane species growing in crevices in granite; B, *S. emmeliana*, native to tropical America and requiring a moist, warm climate.

they are always grouped to form cones, or strobili, at the ends of upright branches.

Two types of sporangia are formed: **megasporangia** (Greek, *mega,* large) and **microsporangia** (Greek, *micro,* small). As the names indicate, megasporangia produce larger spores. A single strobilus usually contains both types of sporangia.

Within a developing megasporangium, all but one of the **megasporocytes** (potential spore-producing cells) degenerate. This remaining megasporocyte, nourished in part by fluid resulting from the degenerating spores and in part by cells surrounding the spore cavity, increases greatly in size. During meiosis, four large spores, called **megaspores**, are formed from the megasporocyte. Each megaspore may germinate and give rise to a **female gametophyte** (megagametophyte; Fig. 14.10).

Only a few microsporocytes within the developing microsporangium degenerate. The approximately 250 microsporocytes that remain undergo meiosis, each forming four small spores, or **microspores** (Fig. 14.10).

The production by a given species of two distinct types of meiospores—megaspores and microspores—is called **heterospory**. Thus, *Selaginella* is heterosporous. *Lycopodium,* which produces but one spore type, is said to be **homosporous**.

Female Gametophyte. Repeated cell divisions within the megaspore result in the female gametophyte, which is contained within the megaspore wall until it nears maturity. Its increase in size eventually ruptures the megaspore wall, and a small cushion of colorless gametophytic tissue protrudes from the megaspore along the lines of rupture (Fig. 14.10). Archegonia develop on the protruding cushion.

The megaspore (containing the female gametophyte) may be shed from the cone or strobilus at almost any stage of the development of the gametophyte. In some species, it may be retained in the strobilus until well after fertilization. In any event, fertilization occurs only when sufficient water, either from rain or dew, is present, allowing sperms to swim to archegonia.

Male Gametophyte. Upon germination, microspores divide into two cells. One of them, the **prothallial cell,** is small and does not divide further; it represents the vegetative portion of the male gametophyte. The other cell, by repeated divisions, develops into an antheridium composed of a jacket of sterile cells enclosing 128 or 256 bilflagellated sperms. This development takes place within the microspore wall. Microspores are shed from the microsporangium midway in the development of the male gametophyte and grow to maturity without direct connection with parent sporophyte or soil. Usually, microspores sift down to the bases of the megasporophylls below the microsporophylls (Fig. 14.10). In this position, they are close to the developing female gametophytes. The sperms escape when the microspore wall ruptures.

Sperms swim to the archegonia, which grow on that portion of the female gametophyte protruding from the megaspore (Fig. 14.10). Fertilization ensues, and the resulting zygote initiates the diploid or sporophyte generation.

Sporophyte Embryo. Of the two cells formed by the first division of the zygote, only one develops into an embryo. The other cell grows into an elongated structure, the **suspensor,** which pushes the developing embryo into

sporophylls

strobilus

Figure 14.6 Sporophylls in *Lycopodium*. *A* and *B*, sporophylls similar to vegetative leaves in *L. selago*. *C*, sporophylls different from vegetative leaves and grouped in terminal strobili in *L. obscurum*.

leaves

sporangia

sporophylls

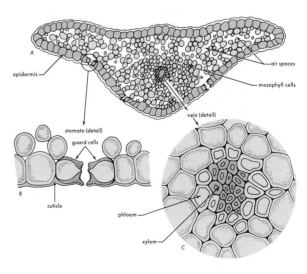

A

epidermis

air spaces

mesophyll cells

stomate (detail)

vein (detail)

guard cells

B

cuticle

phloem

xylem

C

Figure 14.7 *A*, cross-section of *Lycopodium* leaf. *B*, stomate detail. *C*, vein detail.

Division Lycophyta

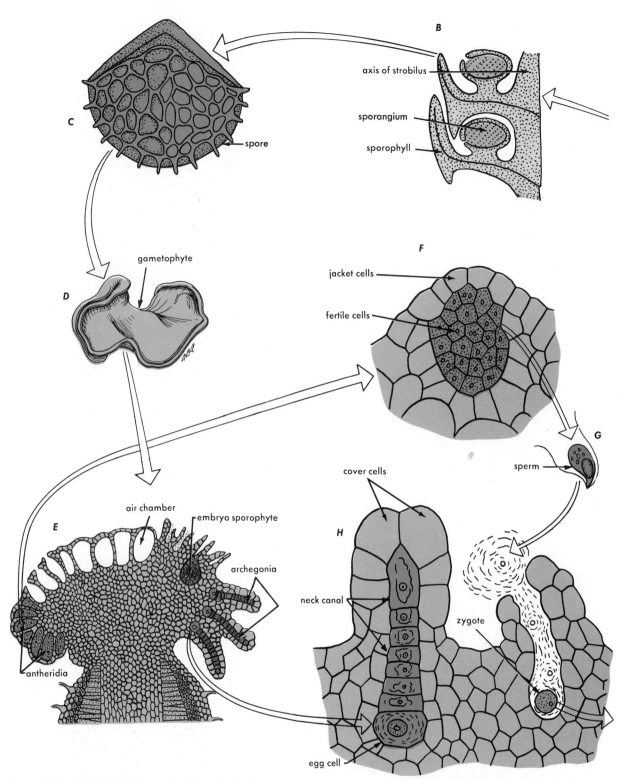

B

axis of strobilus

sporangium

sporophyll

C

spore

gametophyte

D

F

jacket cells

fertile cells

G

sperm

E

air chamber

embryo sporophyte

archegonia

antheridia

cover cells

H

neck canal

zygote

egg cell

Figure 14.8 Stages in the life cycle of *Lycopodium. A, L. annotinum; B,* section of sporophyll showing location of sporangia; *C,* spore; *D,* gametophyte; *E,* section of gametophyte showing location of gametangia; *F,* antheridium; *G,* sperm; *H,* archegonium; *I,* section of embryo; *J,* underground gametophyte with young sporophyte.

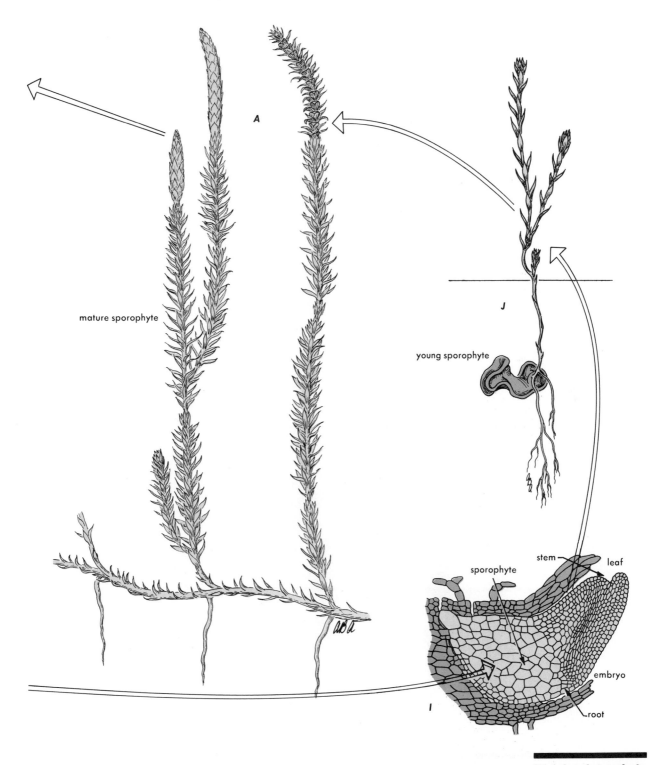

mature sporophyte

young sporophyte

A

J

stem

leaf

sporophyte

embryo

root

I

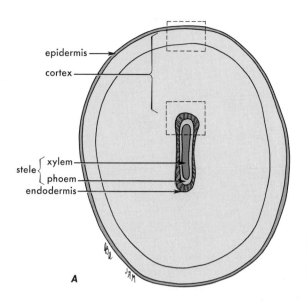

A

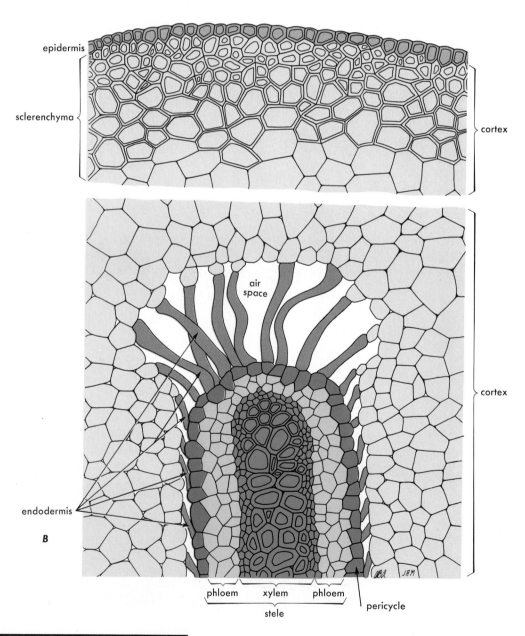

Figure 14.9 Cross section of a stem of *Selaginella*. *A*, diagram showing distribution of stem tissues; a single, flattened cylinder of vascular tissue is supported in an air space by filaments of endodermal cells and the cortex is composed of parenchyma and sclerenchyma. *B*, detail showing cellular structure.

gametophytic tissue, where there is a food supply (Fig. 14.10).

The embryo is a structure with (a) a foot, (b) a root, (c) two embryonic leaves, and (d) a shoot (Fig. 14.10). In certain species of *Selaginella,* the embryo is held by the megaspore and retained within the strobilus. It does not pass into a dormant state, as do embryos of seeds, but continues to grow. Young sporophytes may be found extending from the strobilus of parent sporophytes. If the developing embryo were to pass into a period of dormancy while being held by the parent sporophyte, a condition would arise that approaches the seed habit.

Division Sphenophyta

Members of this division once grew very abundantly, as their good representation in the fossil record shows. Today, the division is represented by a single genus (*Equisetum*) of about 25 species (Fig. 14.11). Many species inhabit cool, moist places, but *Equisetum arvense* also grows in dry habitats. Most species are characterized by the presence of silica in the epidermis of stems. Because of this trait, they were used in colonial days to scour pots and pans and hence were called scouring rushes. The genus is also commonly known as the horsetails.

Mature Sporophyte

The sporophyte of *Equisetum* is the dominant phase of its life history. In one tropical species, the sporophyte is vinelike and may reach 7 m in length. Usually, 1 m represents the maximum upright growth. All species are perennials and have a branched rhizome from which upright stems arise. Stems, depending on the species, may branch either profusely or sparingly. In either case they are straight and marked by vertical ridges and distinct nodes (Fig. 14.11). Bases of nodes are sheathed by whorls of small simple leaves. Many branches may also form at each node. The microphyllous leaves are much reduced in size, nongreen and, in many species, short-lived. The stems are green and are, therefore, the organs of food manufacture. Roots occur at the nodes of the rhizomes, but they may also develop from stem nodes if the stem becomes procumbent on a moist surface.

Except at nodes, stems of *Equisetum* are hollow and the ribs make prominent markings on their outer circumferences. The ridges are formed of sclerenchyma tissue and not only project outward, but extend inward almost to the small vascular bundles (Fig. 14.12). There are air spaces, or canals, between vascular bundles. Photosynthetic tissue lies between these air canals and the thin, outer layer of sclerenchyma tissue and epidermis. Vascular strands are also marked by a canal (Fig. 14.12). Arms of xylem extend outward from this canal, and phloem tissue lying outward from the canal is present between radial arms of xylem. Some species contain vessels in the xylem. An endodermis is present, but its location varies from species to species; it may surround each vascular bundle, or encircle a ring of vascular bundles. In some species, there is also an inner endodermis.

Reproduction

In all species of *Equisetum,* the sporangium-bearing organs, **sporangiophores,** are specialized structures very different from ordinary leaves. They are grouped together in strobili, or cones (Fig. 14.11) at the summit of main upright branches and occasionally on lateral branches. In most species, strobili are borne on ordinary vegetative shoots; in a few species, they are formed only on special fertile shoots.

The sporangiophores are stalked, shield-shaped structures borne at right angles to the main axis of the cone. The cone may be compared to a pole to which open umbrellas have been fastened, the handles of the umbrellas being at right angles to the pole. Sporangia are attached to the underside of the shield (umbrella), close to its edge. They extend horizontally inward, toward the axis of the cone. Four meiospores with the haploid number of chromosomes are formed from each sporocyte. All meiospores are morphologically alike. *Equisetum,* therefore, is homosporous. Upon maturity the meiospores are discharged from the sporangia. The spores are fragile and normally live but a few days.

Gametophytes

Gametophytes are small green bodies about the size of a pinhead and consist of a cushionlike base with many erect, delicate lobes (Fig. 14.13). They are easily cultured in the laboratory.

Gametangia are similar to those of *Lycopodium.* Fertilization takes place only when there is free water. Sperms are spiral shaped and multiciliate.

The zygote develops directly into an embryo. No suspensor is formed. The embryo is similar to those already described, except that the foot is small or lacking.

Division Pterophyta

The division Pterophyta comprises the ferns, some 10,000 species, most of which are shade-loving plants of small size, their upright leaves, or **fronds,** generally being their most prominent feature. Most species native to temperate zones have an underground rhizome with leaves and roots arising at nodes (Fig. 14.14); a few are small, floating aquatics without rhizomes. Some ferns of tropical regions do grow into fairly large trees. All Pterophyta have a definite alternation of generations, with both sporophyte and gametophyte being autotrophic plants.

The Pterophyta comprise four orders, the largest of which, the Filicales, includes the **true ferns.** The Filicales are divided into 11 or more families, of which the largest and best-known in the United States is the Polypodiaceae. Most of the discussion below deals with representatives of this family.

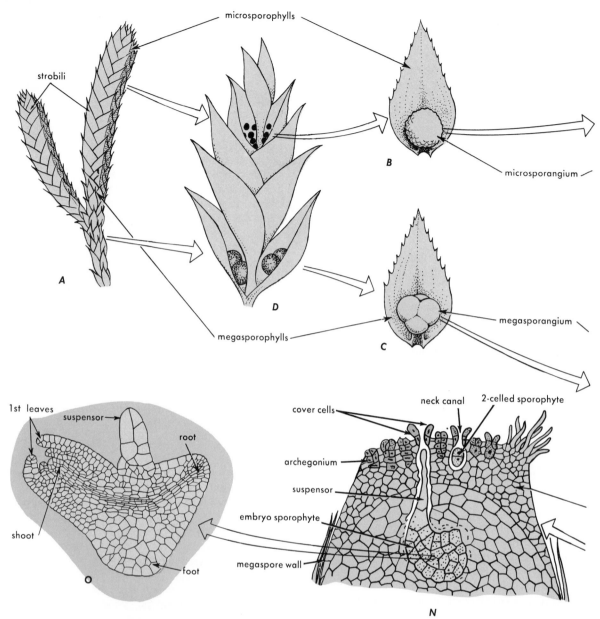

Figure 14.10 Stages in the life cycle of *Selaginella*. *A*, shoot of
S. watsonii showing two strobili; *B*, microsporophyll with
microsporangium; *C*, megasporophyll with megasporangium; *D*,
strobili showing sporophylls with sporangia; *E*, microsporangium
discharging microspores; *F*, megasporangium discharging
megaspores; *G*, microspores sift downward and lodge in axils of
megasporophylls close to megaspores; *H*, a microspore; *I*, section
of a microspore; *J*, male gametophyte with microspore; *K*,
antheridium with developing sperm cells; *L*, mature sperm cells;
M, female gametophyte within megaspore; *N*, section of female
gametophyte showing location of archegonia and a young embryo
pushed by its suspensor into the vegetative tissue of the female
gametophyte; *O*, the young embryo sporophyte generally remains
embedded in vegetative gametophyte tissue.

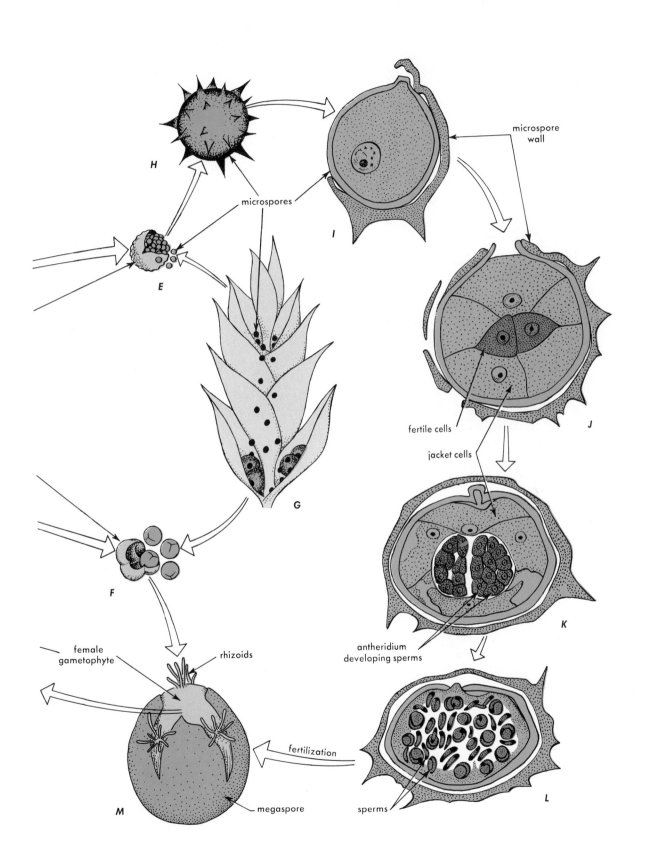

microspores

microspore wall

H

I

E

G

fertile cells

jacket cells

J

F

antheridium
developing sperms

K

female
gametophyte

rhizoids

fertilization

M

megaspore

sperms

L

Figure 14.11 Stems of *Equisetum telmateia*. *A,* fertile stems with strobili; note much-branched, adjacent sterile stems. *B,* strobilus with sporangiophores. *C,* enlarged view of the stem, showing three nodes, vertical ridges, and reduced leaves.

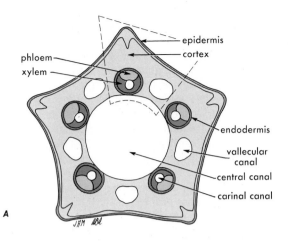

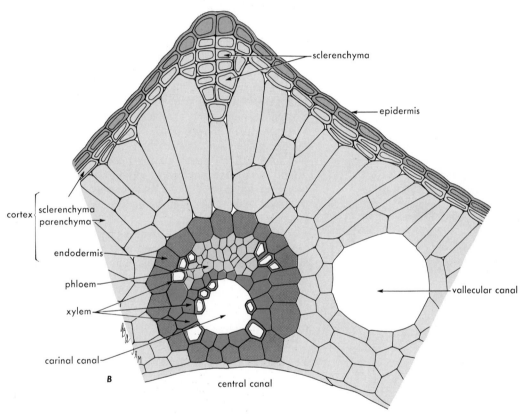

Figure 14.12 Cross section of *Equisetum* stem. *A*, diagram
showing arrangement of tissues; *B*, enlarged view showing details
of cellular structure.

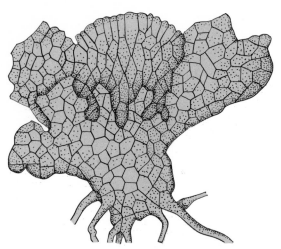

Figure 14.13   Gametophyte of *Equisetum*.

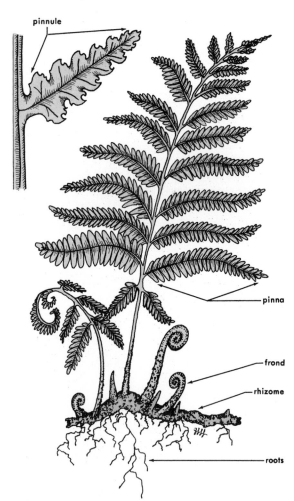

Figure 14.14   A typical, mature fern sporophyte, showing rhizome below ground and fronds above ground. Lady fern, *Athyrium felix-femina*.

Mature Sporophyte

The sporophyte, which is the dominant generation of all ferns, possesses an underground stem or rhizome from which leaves and adventitious roots arise (Fig. 14.14). Leaves are the most prominent part of the fern plant and they vary greatly in size and form. Young fern leaves are rolled into tight spirals and consist chiefly of meristematic tissue (Fig. 14.15). As the leaf matures, it unwinds from the base upwards. The upright, expanded portion is mature, but the leaf continues to grow by cell division at its coiled tip. In certain species, the apical meristem may remain active for years, resulting in leaves that are nearly 3 m long. The uncoiled, fully expanded fern leaf lacks any residual meristematic tissue.

Most fern leaves are compound, although simple types exist in all groups. The most common type of fern leaf has a stout or rigid petiole, which is prolonged to form a rachis from which leaflets (pinnae and pinnules) arise (Fig. 14.15).

The vascular tissue of ferns is organized into one or more vascular strands, each having a core of xylem, surrounded by phloem and separated from the ground parenchyma by an endodermis. These vascular bundles are interconnecting, so that, as in *Lycopodium*, the precise arrangement will vary from section to section. However, there are arrangements characteristic of various genera. For instance, in *Polypodium* there is a ring of small bundles placed well out from the center of the rhizome (Fig. 14.16). Each strand is formed of a central core of xylem, surrounded by phloem, and the whole has a bounding endodermis. The xylem consists largely of tracheids, vessels being known in only two genera. Sieve cells occur in the phloem.

A typical fern leaf has a well-developed epidermis with chloroplasts in the epidermal cells and stomata on the lower surface. The mesophyll may be relatively undifferentiated or divided into palisade and spongy parenchyma layers (Fig. 14.16).

Reproduction

Vegetative Reproduction

Vegetative reproduction may occur in one of two ways: (a) by death and decay of the older portions of the rhizome and the subsequent separation of the younger growing ends; and (b) by the formation of deciduous leaf-borne "buds," which become detached and grow into new plants. Such buds occur in only a few genera. Some clumps of ferns are thought to be hundreds of years old.

Sexual Reproduction

In the sexual life cycle, independent sporophyte and gametophyte generations alternate with each other. The vegetative structure of the sporophyte has been described above. Spores are borne in sporangia, which ordinarily develop on the lower surface or margins of fronds (Fig. 14.17). Not all leaves are fertile (i.e., spore-producing), and the fertile leaves in some species are dissimilar in structure to the sterile (non-spore producing) leaves. The distribution of sporangia on the leaf surface varies

Figure 14.15 Diversity of form in fern fronds. *A*, tightly rolled tip of a young frond of the tree fern *Cyathea; B*, adder's tongue *(Ophioglossum vulgatum); C*, moonwort *(Botrychium lunaria); D*, grape fern *(Botrychium multifidum); E*, golden back fern *(Pityrogramma triangularis); F*, spleenwort *(Asplenium pinnatifidum); G*, leather leaf *(Polypodium scouleri); H*, holly fern *(Polystichum lonchitis; I*, five-finger fern *(Adiantum pedatum); J*, lip fern *(Cheilanthes covillei);* and *K*, cliff brake *(Pellaea compacta).*

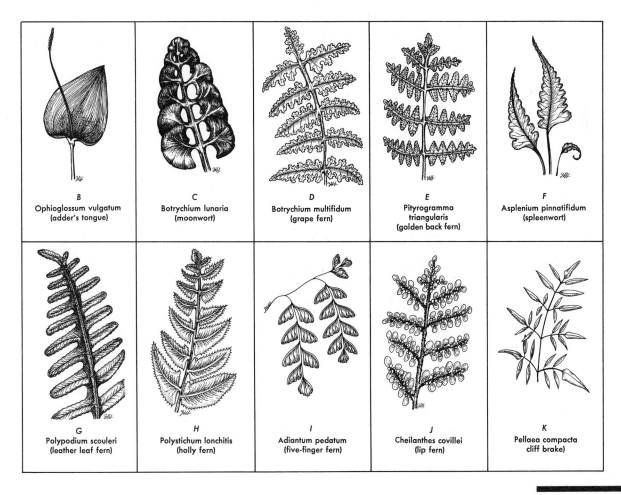

B Ophioglossum vulgatum (adder's tongue)	*C* Botrychium lunaria (moonwort)	*D* Botrychium multifidum (grape fern)	*E* Pityrogramma triangularis (golden back fern)	*F* Asplenium pinnatifidum (spleenwort)
G Polypodium scouleri (leather leaf fern)	*H* Polystichum lonchitis (holly fern)	*I* Adiantum pedatum (five-finger fern)	*J* Cheilanthes covillei (lip fern)	*K* Pellaea compacta cliff brake

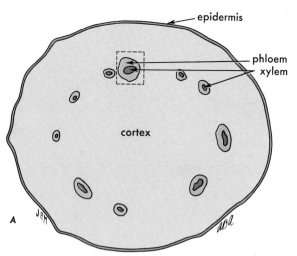

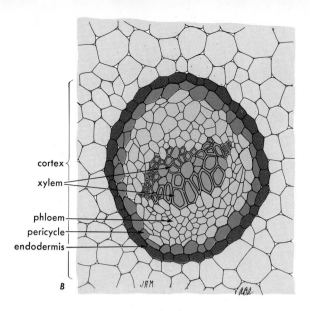

Figure 14.16 Fern anatomy. A, Polypodium stem cross section, showing relationship of tissues; B, a higher magnification of A, showing cellular detail; C, leaf cross section of Polypodium.

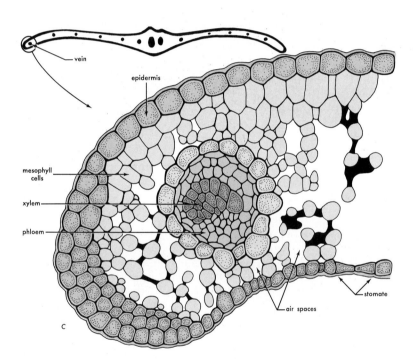

considerably in different genera and species (Fig. 14.17). Sporangia may (a) cover much of the lower surface, (b) be grouped in **sori** (singular, **sorus**) and grow in a definite relationship with veins, or (c) grow only along margins of the leaf.

When sporangia are grouped together in sori, a structure called the **indusium** (Fig. 14.18), which is sometimes umbrella-like, may be present, thereby protecting young and developing sporangia. Frequently, marginal sporangia are protected by the curled edge of the leaf itself, which forms a **false indusium**.

The sporangium is a delicate, watch-shaped case, consisting of a single layer of epidermal cells, only one row of which possesses heavy walls. This row, which nearly encircles the sporangium in the Polypodiaceae, is the **annulus**; it functions in opening dried mature

sporangia and aids in dispersal of ripe spores (Fig. 14.18).

The young sporangium is filled with sporocytes. Meiosis results, as always, in spores with a reduced number of chromosomes. Ferns, with the exception of two families, the Marsileaceae and the Salviniaceae, are homosporous. These two families are widely distributed, small, aquatic, and unfernlike in appearance.

Gametophyte. Gametophytes, or **prothallia**, of the Polypodiaceae are small, flat, green heart-shaped structures (Fig 14.19) with rhizoids on their lower surface (Fig. 14.18). In most species, they apparently mature rapidly and are not long-lived. Antheridia and archegonia are borne on the same prothallus. Antheridia are formed when the prothallus is very young and are scattered over its lower surface. Normally, 32 sperms develop within

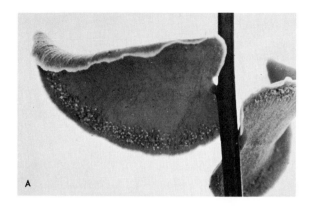

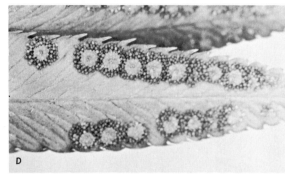

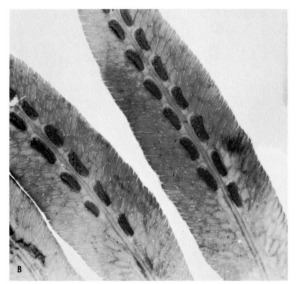

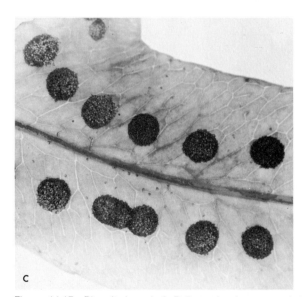

Figure 14.17   Diversity in sori. *A, Pellaea,* showing sporangia in a band close to leaflet margin; *B, Woodwardia,* showing sori parallel to midrib and covered with indusia; *C, Polypodium,* showing round sori; *D, Polystichum,* showing round sori with centrally attached indusia; *E, Asplenium,* showing sporangia in long rows on veins.

each antheridium.

Archegonia form later than antheridia and are usually clustered close to the notch, also on the undersurface of the gametophyte (Fig. 14.18).

Fertilization occurs when free water is present and sperms are thus able to swim to the neck of the archegonium. The resulting zygote is diploid and rapidly develops into an embryo sporophyte comprising a foot, root, stem, and leaf (Fig. 14.18). The embryo develops directly into a young sporophyte, without a dormancy period.

Summary

1. Lower vascular plants is a general name for vascular plants that do not produce seeds. This group consists of four divisions, the Psilophyta, Lycophyta, Sphenophyta, and Pterophyta. The latter division, the ferns, contains the most species and is the most widespread.

2. The Psilophyta is represented by only two genera, *Psilotum* and *Tmesiptris.* Sporophytes of *Psilotum* consist of a branched, underground rhizome with rhizoids (no true roots) and an upright, green, branched stem with scalelike microphylls. It is homosporous and

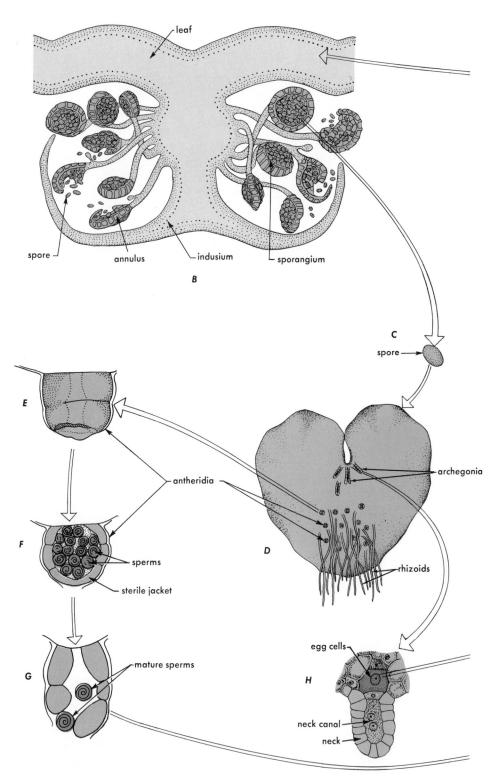

leaf

spore
annulus
indusium
sporangium

B

C
spore

E

antheridia

archegonia

F
sperms
sterile jacket

D

rhizoids

G
mature sperms

egg cells

H

neck canal
neck

Figure 14.18   Stages in the life cycle of a fern. *A,* fern frond; *B,*
section through a sorus, showing indusium, sporangia, and
spores; *C,* spore; *D,* gametophyte; *E,* antheridium; *F,* section of
antheridium showing sperm; *G,* open antheridium discharging
sperm; *H,* archegonium; *I,* archegonium receptive to sperm; *J,*
section of gametophyte showing a zygote with the venter of the
archegonium; *K,* section of gametophyte and embryo; *L,*
gametophyte with a young sporophyte still attached.

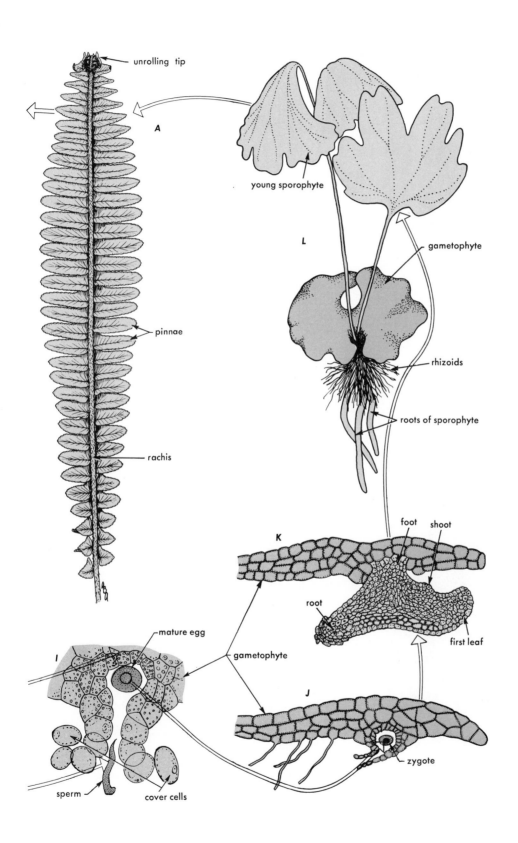

unrolling tip

A

pinnae

rachis

young sporophyte

L

gametophyte

rhizoids

roots of sporophyte

foot   shoot

K

root

first leaf

mature egg

gametophyte

I

J

sperm

cover cells

zygote

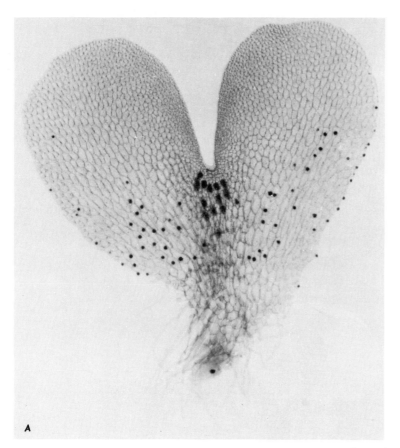

Figure 14.19 Photographs of fern gametophytes. *A*, a view from beneath, showing rhizoids, antheridia (dots), and archegonia (near notch); *B*, a gametophyte (G) with attached young sporophyte (S) emerging and overtopping the gametophyte.

the meiospores germinate to produce nonphotosynthetic, small gametophytes. Free water is required for fertilization. The embryo sporophyte is at first attached to the gametophyte by an absorbing organ (the foot), and it depends for nourishment on the gametophyte.

3. The Lycophyta is represented by only two genera in this book: *Lycopodium* and *Selaginella. Isoetes* (the quillworts) was not discussed, nor were two other genera, *Stylites* and *Phylloglossum.*

4. Sporophytes of *Lycopodium* (club mosses) are less than 20 cm tall and exhibit true roots, stems, and leaves. The stems are sheathed with small leaves that contain an epidermis with stomata, a mesophyll with intercellular air spaces, and a central vein. Sporangia are borne in the axils of leaves; they may be clustered near the tips of branches in a strobilus (cone). The genus is homosporous, and the meiospores produce small gametophytes that may or may not have chlorophyll. Free water is required for fertilization, and the embryo sporophyte is at first attached to the gametophyte by a foot, as in *Psilotum.*

5. Sporophytes of *Selaginella* are similar in appearance to those of *Lycopodium.* Vessels are present in the xylem of some species. Sporangia are borne in the axils of leaves and they are clustered into strobili. Sporangia and spores are of two kinds. Megasporangia produce megaspores that germinate and grow (within the ruptured spore wall) into female gametophytes. Microsporangia produce microspores that germinate and grow (within the spore wall) into male gametophytes. Microspores may sift down the strobilus

to lie near megaspores, then they rupture to release sperm; if free water is present fertilization can be accomplished in the strobilus. The embryo is at first pushed deep within the female gametophyte by a suspensor, and nutrition is gained from the gametophyte through a foot. *Selaginella* is heterosporous.

6. Sporophytes of Sphenophyta (horsetails) are about 1m tall and consist of an underground rhizome, with true roots, and an upright, jointed, silica-rich stem. Microphylls and branches arise in whorls at the nodes. Vessels are present in some species. Sporangia are grouped into complex strobili. The division is homosporous. Meiospores germinate and grow into small, photosynthetic gametophytes. Free water is required for fertilization. The embryo stage is similar to that of *Psilotum*.

7. Sporophytes of Pterophyta (ferns) are typically below 1 m in height, but some reach tree stature. In most cases, the stem is completely underground as a rhizome, and only the large leaves and petioles are above ground. A few ferns are small, floating aquatics. Vessels are present in some species. Leaves may exhibit palisade and spongy parenchyma. Sporangia are produced on specialized leaves in some species, or on the undersides of all leaves in other species. The sporangia may be clustered into compact groups called sori and covered with an indusium. Most ferns are homosporous and, if so, the meiospores germinate and grow into small, photosynthetic, heart-shaped gametophytes (prothallia). Free water is necessary for fertilization, and the embryo sporophyte is much like that of *Psilotum*.

Table 14.1

Comparison of Psilophyta, Lycophyta, Spenophyta, and Pterophyta

Generation	*Psilotum*	*Lycopodium*	*Selaginella*	*Equisetum*	Ferns
Sporophyte	A branching stem, scale leaves, no roots	Prostrate branching stem with upright branches, roots, and leaves	Prostrate branching stem with upright branches, roots, and leaves, also scales	Rhizome, upright, jointed stem, small leaves	Rhizome, roots, and leaves
	Herbs or epiphytes	Herbs	Herbs	Herbs	Herbs or trees
	Simple vascular system	Simple vascular system	Simple vascular system, but vessels are present	Simple vascular system but vessels may be present	Stem: several strands of vascular tissue each with xylem surrounded by phloem; Vessels: only in two genera; Sieve cells: no companion cells in phloem
	One type of sporangium	One type of sporangium	Megasporangia and microsporangia	One type of sporangium	One type of sporangium (mostly)
	Homospory	Homospory	Heterospory	Homospory	Homospory (mostly)
	No sporophylls	Sporophylls present	Sporophylls present	Sporangiophores	Leaves bear spores in sporangia
	No cone	A cone in some species	Cone	Cone composed of sporangiophores	No cone
	Embryo develops attached to gametophyte	Embryo develops attached to gametophyte	Embryo develops within female gametophyte on sporophyll	Embryo develops attached to gametophyte	Embryo develops attached to gametophyte
Gametophyte	Irregular subterranean structure, associated with fungus	Irregular to tapered subterranean structure, associated with fungus	Male gametophyte: one prothallial cell, and an antheridium within microspore	A single green thallus resembling *Anthoceros*	Heart-shaped, small, green, completely independent thallus
	Gametangia embedded in thallus	Gametangia embedded in thallus	Female gametophyte: small amount of tissue within megaspore, cushion protrudes in which gametangia are embedded	Gametangia embedded in thallus, neck of archegonium protruding	Antheridium of approximately 6 to 10 cells, archegonia partially embedded in gametophyte
	Motile sperms	Motile sperms	Motile sperms	Motile sperms	Motile sperms

Summary

309

gymnosperms

The higher vascular plants possess both vascular tissue and seeds. **Gymnosperms** (conifers and their relatives) and angiosperms (flowering plants) together make up the higher vascular plant group. These plants dominated the world's vegetation during the Cenozoic era (Chapter 10).

Gymnosperms are a major source of lumber and paper pulp in North America. This group is represented by such common trees as pines *(Pinus)*, Spruces *(Picea)*, firs *(Abies),* and true cedars *(Cedrus),* all of which possess well-developed seed-bearing cones (Fig. 15.1). Other gymnosperms, such as Ginkgo, yews *(Taxus),* and Mormon tea or joint-fir *(Ephedra),* do not bear cones.

A trait common to all gymnospermous plants is the absence of a protecting case, that is, an ovary wall, around the seeds. In pines and other cone-bearers, seeds are borne on the surface of scales that comprise the cone and, though well-protected by the scales, they are not surrounded by floral parts. In yews, seeds are partially surrounded by a red berrylike structure (Fig. 15.2). In *Ephedra,* ovules are borne in axils of short bracts (Fig. 15.2) and in *Ginkgo,* naked ovules are attached to ends of short branches (Fig. 15.2). By contrast, the seed of an angiosperm is surrounded by a matured ovary wall.

Seeds lacking protection of an ovary wall are said to be "naked"—thus the name gymnosperm, derived from two Greek words *gymnos* (naked) and *sperm* (seed). Fossil Lepidodendrales, belonging to the Lycophyta, bore seeds more than 200 million years ago, thus indicating that the seed habit is neither of recent geological origin, nor indicative of a single evolutionary line.

There are four divisions of living gymnosperms, encompassing more than 600 species. Of these divisions, we shall only briefly consider: (a) the Cycadophyta, or cycads; (b) the Ginkgophyta, composed of only one species, the maiden-hair tree; and (c) the Gnetophyta, an artificial grouping of three unusual genera, *Gnetum, Ephedra,* and *Welwitschia.* We will discuss a fourth division, the Coniferophyta or conifers, in much more detail, and we begin with that group.

Division Coniferophyta

Classification

Without exception, the well-known and economically important gymnosperms of temperate zones belong to the Coniferophyta. There are nine families and 550 species. They comprise pines, hemlocks, firs, spruces, junipers, yews, redwoods, and many others. Not all of them bear cones, yet the cone is such a conspicuous feature of many of them that the division has been named the Coniferophyta or cone bearers. Most conifers form true cones. Juniper berries are in reality cones with fleshy adhering scales. In yews, seeds are surrounded at the base by a more or less pulpy, berrylike body and are not borne in cones. In either case, the seeds are naked, not being surrounded by an ovary wall.

The following brief descriptions of several common genera enable us to introduce the more important conifers in some of the seven families.

Family Pinaceae

Pinus (Pines).   Pines are usually large trees. Some members of bristlecone pine *(Pinus longaeva)* are the oldest living things, over 5000 years old. The leaves are needlelike, two or more growing together (except in *Pinus monophylla) in a* **fascicle** or group, which is sheathed at the base (Fig. 15.1). The cones vary greatly in size and shape and are very characteristic of the species to which they belong.

Abies (Firs).   Firs are stately trees of a symmetrical, cylindrical or pyramidal shape. The leaves are flat and linear; in cross section they are relatively broad, without marked angles. The cones are erect on the branches and shatter at maturity (Fig. 15.3).

Picea (Spruces).   These trees closely resemble the firs, but they can be distinguished by the position of the leaves on the branchlets, by the angular appearance of the leaves in cross section, and by the cones that are hanging and do not shatter at maturity (Fig. 15.4).

Tsuga (Hemlocks).   The trees of this genus are pyramidal with slender horizontal branches. The leaves are usually two-ranked, linear, flat, and with a short petiole. They resemble the leaves of firs but are much shorter. The cones are small.

Pseudotsuga (Douglas Fir).   Only two species occur in the United States. One of them, the Douglas fir *(Pseudotsuga menziesii),* is the most important timber tree of the United States. It may grow to a height of 60 m, with a trunk diameter of 3 m. Its leaves are flat, like those of the true firs, but have white lines on either margin and a

Figure 15.1   Pines. *A, Pinus jeffreyi* (Jeffrey pine), showing the trunk of one tree and the crown of a second tree. *B,* mature ovulate cones of *P. ponderosa* (ponderosa pine), a closely related species. *C,* fascicle of ponderosa pine needles.

Figure 15.2   Gymnosperms without typical cones. *A*, branches of *Taxus* with seeds; note berrylike aril surrounding seed. *B*, branch and seeds of *Ginkgo biloba. C*, a cluster of female strobili of *Ephedra;* note "pollination drops" at the trip of two cones.

groove along the upper surface. The cones are 5 to 11 cm long, pendulous, and easily recognized by bracts extending outward below each scale.

Larix (Larches).   Trees belonging to this genus grow in the cooler portions of the Northern Hemisphere. They differ from most other members of the Pinaceae in that they are deciduous. The American larch, or tamarack, is a tall tree frequently found in bogs. The needles are short, linear, and grouped in crowded clusters on short spurs. On leading shoots, however, the needles are arranged spirally. The cones are small and persistent (Fig. 15.5).

Family Cupressaceae

This family contains the junipers, cypresses, and false cedars (true cedars, *Cedrus* species, are in the Pinaceae). The leaves are borne singly, but are usually

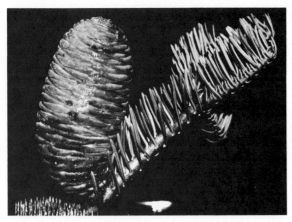

Figure 15.3 *Abies*. Ovulate cone and branch. Note arrangement of needles.

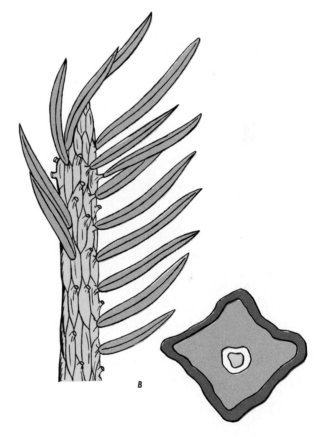

Figure 15.4 Picea. *A*, branch of *P. alba* with ovulate cones; *B*, leaves of *Picea*.

small and scalelike and are crowded closely around the stems (Fig. 15.6). The cones may be woody, as in *Cupressus,* or fleshy, as in *Juniperus*. The cone of juniper is often called a ''berry,'' but it is a modified cone, composed of fleshy scales that completely enclose the seeds (Fig. 15.6).

Family Taxodiaceae

This family contains the bald cypress, redwood, and Sierra bigtree. Redwood *(Sequoia sempervirens)* may be the tallest tree in the world, individuals over 60 m being

common and the record height being 112 m. Redwood has outstanding lumber qualities, not the least of which is its resistance to decay. Consequently, logging has removed 90% of all virgin redwood forest in the past 150 years. Only half of the remaining acreage is protected from future logging. Sierra bigtree *(Sequoiadendron giganteum)* is the most massive of any plant or animal species in the world; its trunk at breast height may reach 9.8 m in diameter and its height above 82 m. The trunk alone of the General Sherman tree weighs 625,000 kg (over 680 tons).

Figure 15.5 *Larix lyallii A*, branch with needles; *B*, mature ovulate cones.

Figure 15.6 *Juniperus occidentalis* growing at 2200 m in the Sierra Nevada Mountains of California. *A*, tree; *B*, branch with staminate cones; *C*, branch with ovulate cones.

Family Taxaceae

With two exceptions, ovulate cones are not borne by members of the yew family. The seed of this family so strongly resembles a drupe or a nut that it is usually called a fruit. The embryo is protected by an outer flesh called an **aril**, and an inner hard pit. Both these tissues may be derived from the integuments of the ovule; hence, the structure is morphologically a naked seed. As shown in Fig. 15.2, the outer red aril of ground pine (*Taxus canadensis*) almost encloses the seed. It drops away when the seed is mature.

Native yews are common only in a few places in the United States. Possibly the best-known yew is the English yew, which, because of the excellent bows that were made from its wood, is closely linked with English history and folklore. Yews are now widely cultivated.

Life History of a Conifer

The following outline of a gymnosperm life cycle is based largely on the genus *Pinus*. It differs from life cycles of other genera mainly in requiring three summers for completion. Significant stages of pollination, fertilization, and maturation of the seed occur within a decimeter of each other on a given branch, thus making it an ideal subject for class study.

Sporophyte

All species of pines are trees. They range in size and age from scrubby, 70-year-old fire-adapted species such as knobcone pine (*Pinus attenuata*) to large, straight trunked ponderosa pine (*Pinus ponderosa,* Fig. 15.1) 500 years old, to twisted, slow-growing bristlecone pine (Fig. 15.7) 5000 years old.

Conifer wood has no vessels. A three-dimensional reconstruction from a small piece of *Sequoia* is diagrammed in Fig. 15.8. It is extremely regular. Tracheids are elongated cells that appear in cross section as square, hollow cells. Those formed in the spring are largest in diameter; as the season progresses, newly formed tracheids are smaller in diameter with thicker walls. The heavy-walled cell is called a **fiber-tracheid**. Note xylem rays are composed of brick-shaped parenchyma cells, sometimes with dark-staining contents. Other parenchyma cells may run in vertical columns; these are called axial parenchyma.

Many gymnosperms produce resin. Typically, resin occurs in long **resin ducts** that are surrounded by parenchyma cells (Fig. 15.9). Resin may play a role by inhibiting the activity of wood-boring insects. *Sequoia* does not produce resin.

In the phloem, companion cells are missing, but otherwise, **sieve cells**, fibers, and parenchyma cells are present.

Figure 15.7 *Pinus longaeva* (bristlecone pine) in the White Mountains of California, growing near timberline.

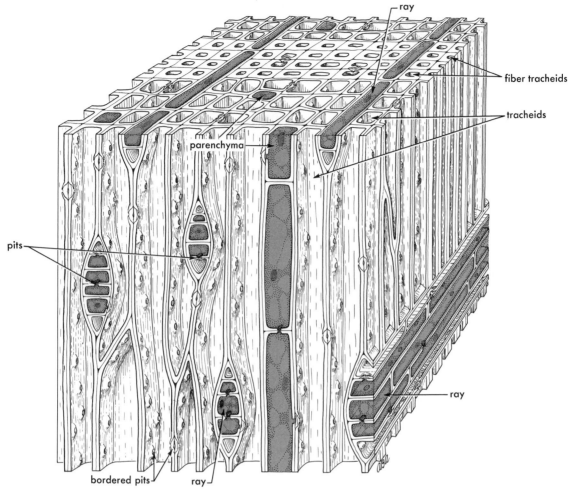

Figure 15.8 Block diagram of secondary xylem of redwood (*Sequoia sempervirens*).

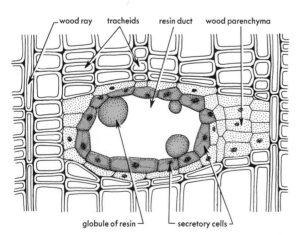

Figure 15.9 Resin duct in pine wood, as seen in cross section.

The needle leaves of *Pinus* show considerable adaptation to drought (Fig. 15.10): sunken stomata, thick cuticle, fibrous epidermis, close-packed mesophyll, and veins only in the center of a thick leaf. Such modifications are important in winter, when soil moisture is frozen and the evergreen leaves are subjected to drying winds.

All conifers produce two kinds of spores—**microspores** and **megaspores**—borne in cones that are morphologically distinct. The two types of cones are known, respectively, as **staminate** (or pollen) and **ovulate** (or seed) **cones**.

Staminate (Pollen) Cone

Staminate cones average 10 mm or less in length, by 5 mm in diameter. They are borne in groups, usually on lower branches of trees (Fig. 15.11). Each cone is composed of a large number of small scales (microsporophylls) arranged spirally on the axis of the cone. Two microsporangia develop on the under surface of each scale (Fig. 15.11).

Stages of microspore development are quite similar to those of *Selaginella* and to spores of mosses and ferns. Microsporocytes, which are surrounded by a nutritive cell

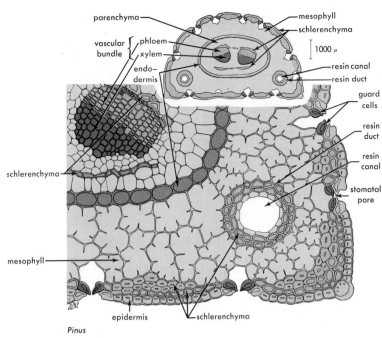

parenchyma
mesophyll
schlerenchyma
vascular bundle { phloem
xylem
endo-dermis
resin canal
resin duct
guard cells
resin duct
resin canal
stomatal pore
schlerenchyma
mesophyll
epidermis
schlerenchyma

1000 μ

Pinus

Figure 15.10 Cross section of a two-needled pine leaf. The inset (above) shows the entire cross section, with major tissues outlined; the large drawing shows cellular detail of a wedge from the epidermis in to one of the vascular bundles at the center.

layer, the **tapetum**, undergo meiosis, and four microspores result. As usual, each microspore contains the haploid number of chromosomes. The nucleus within the newly formed microspore divides several times by mitosis, forming a **pollen grain** that contains two viable, haploid nuclei and vestiges of several vegetative cells (Figs. 15.11, 15.12). Enormous numbers of pollen grains are finally shed from the microsporangium. They are light in weight and bear two wings that may facilitate dispersal by wind.

Ovulate (Seed) Cone

The ovulate cone, when mature, is the well-known cone of pines and other conifers. Each cone is composed of an axis upon which are borne several woody scales in a spiral fashion. Ovulate cones develop at tips of young branches in early spring (Fig. 15.13). Two ovules, each enclosing a single megasporangium, develop on the upper surface of **ovuliferous scales** (Figs. 15.12, 15.13).

The ovules first appear as small protuberances on the upper surface of this scale, close to the axis of the cone. A protective layer of cells, the **integument**, develops early on the outer surface of the ovule. In the end of each ovule near the axis of the cone, there is a small opening, the **micropyle**, through which pollen grains may enter.

In pine, one megasporocyte lies in the center of each ovule (Fig. 15.12); several are contained in ovules of redwoods and cypresses. Like microsporocytes, the megasporocyte is surrounded by a nutritive tissue called the **nucellus** (Fig. 15.12). The nucellus is in reality a megasporangium because it surrounds the area where megaspores arise.

Female Gametophyte. The megasporocyte soon divides by meiosis. Four megaspores, usually arranged in a single row of four cells, result from each megasporocyte. The nucleus within each megaspore has the haploid number of chromosomes. Generally, only one of the four megaspores develops into a female gametophyte; the other three degenerate. Germination of the remaining megaspore and growth of the female gametophyte progresses very slowly. Several months are required in most conifers, and 13 months are required in pine.

The development of the female gametophyte takes place entirely within the ovule. There are approximately 11 mitotic divisions of the megaspore before cell walls begin to appear between the newly formed nuclei. Walls do gradually form, however, resulting in a small mass of gametophytic tissue, completely enclosed by the diploid cells of the ovule. The adjacent sporophytic tissue of the ovule, the nucellus, is in part digested by the developing female gametophyte.

While cell walls are being laid down in the developing female gametophyte, two or more archegonia differentiate at its micropylar end. Figs. 15.12 and 15.14 show that the ovule at this stage consists of integuments, nucellus, and female gametophyte containing several archegonia, each with its enclosed egg. Directly beneath the micropyle is a space, the **micropylar chamber**. Nucellar tissue lies between the micropylar chamber and the archegonia.

Pollination

It will be recalled that **pollination** in flowers is the transfer of pollen from anther to stigma; in conifers, it is the

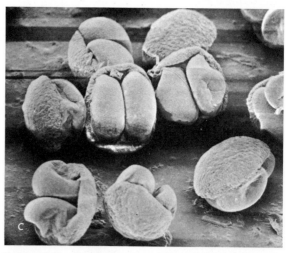

Figure 15.11 Staminate cones of *Pinus*. *A*, end of a branch with a cluster of staminate cones holding ripe pollen. *B*, longitudinal section of one staminate cone, showing microsporangia attached to the underside of each scale with pollen grains inside. *C*, scanning electron micrograph of pollen grains (probably shrunken from treatment in the SEM).

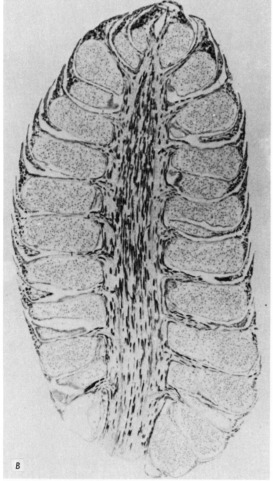

transfer of pollen from staminate cone to ovulate cone. Conifer pollen is windblown. In many species, ovulate cones are borne on higher branches of the tree and staminate cones are concentrated on lower branches; since pollen does not blow directly upward, cross pollination is usual.

Pollination occurs in most conifers in early spring soon after the ovulate cone emerges from the dormant bud (Table 15.1)—about the time of meiosis. At this age, scales of young cones turn slightly away from the axis so that pollen grains can sift down to the axis of the cone. Here they come in contact with a sticky substance secreted by the ovule. As this material dries, it draws some pollen grains through the micropyle into the micropylar chamber where they lodge close to the developing female gametophyte (Fig. 15.13). The pollen grain germinates and develops slowly into a tubular male gametophyte as it grows through the nucellus. Thus, male and female gametophytes of conifers develop to maturity in close proximity within the ovule. They are both dependent on nucellar tissue for nourishment and protection.

Several nuclear divisions occur in the tube, but no cell walls are formed. Two of the last-formed nuclei are **sperm nuclei**. This branched **pollen tube**, containing two sperm nuclei and several vegetative nuclei, is the male gametophyte (Fig. 15.12).

Fertilization and Embryo Formation

Development of male and female gametophytes is so coordinated that the egg is formed and ready for fertilization when the pollen tube, containing two sperm nuclei, has reached the archegonium. In pine, this development requires about one year. At the time of fertilization, pine cones are generally green and the scales tightly closed, showing a spiral pattern of arrangement (Fig. 15.13). The sperm nuclei, together with other protoplasmic contents of the pollen tube, are discharged directly into the egg cell. Sperm nuclei do not possess cilia and hence are not actively motile. One sperm nucleus comes in contact with the egg nucleus and unites with it. The nonfunctioning sperm nucleus and the other protoplasmic material discharged into the egg cells soon undergo disorganization.

The formation of the embryo is preceded by the development of a relatively elaborate **proembryo**. This structure consists of four tiers of four cells each (Fig. 15.12). The four cells farthest from the micropylar end of the proembryo may each develop into an embryo. The intermediate cells are the suspensor cells; they elongate greatly and push the embryo cells deep into the female gametophyte (Fig. 15.12). While this development is taking place, the female gametophyte continues to grow, digesting most of the rest of the nucellus. It enlarges and

becomes packed with food to be used not only for the growth of the embryo but also as a reserve in the seed.

It will be recalled that the female gametophyte usually contains two or more archegonia. Since the egg in each archegonium may be fertilized and since each of the four embryo-forming cells may give rise to an embryo, eight or more embryos may develop in every seed. Normally, however, only one embryo survives, but seeds with two well-formed embryos are not rare.

The mature embryo consists of several **cotyledons** or seed leaves, **epicotyl**, **hypocotyl**, and **radicle** or rudimentary root (Fig. 15.12). The embryo is embedded, as previously mentioned, in the enlarged female gametophyte. Both embryo and female gametophyte are surrounded by a papery shell and a hard protective **seed coat**, both formed from the integument of the ovule. The

Table 15.1
A Generalized Life History of a Pine

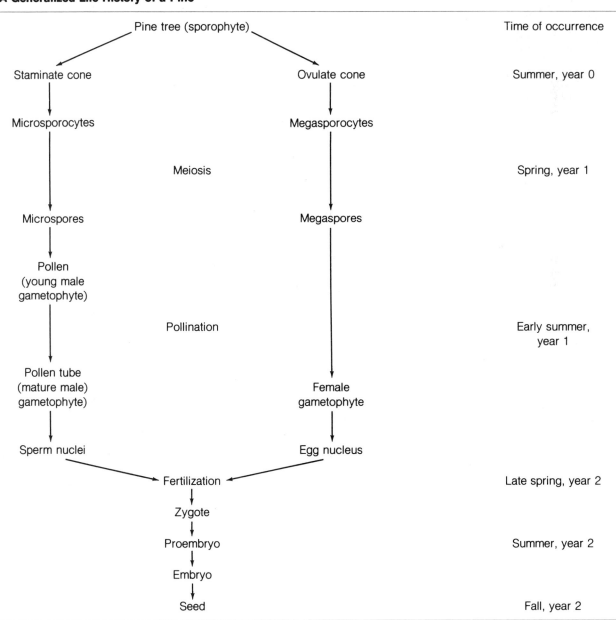

	Time of occurrence
Pine tree (sporophyte)	
Staminate cone → ← Ovulate cone	Summer, year 0
Microsporocytes — Megasporocytes	
Meiosis	Spring, year 1
Microspores — Megaspores	
Pollen (young male gametophyte)	
Pollination	Early summer, year 1
Pollen tube (mature male) gametophyte) — Female gametophyte	
Sperm nuclei — Egg nucleus	
Fertilization	Late spring, year 2
Zygote	
Proembryo	Summer, year 2
Embryo	
Seed	Fall, year 2

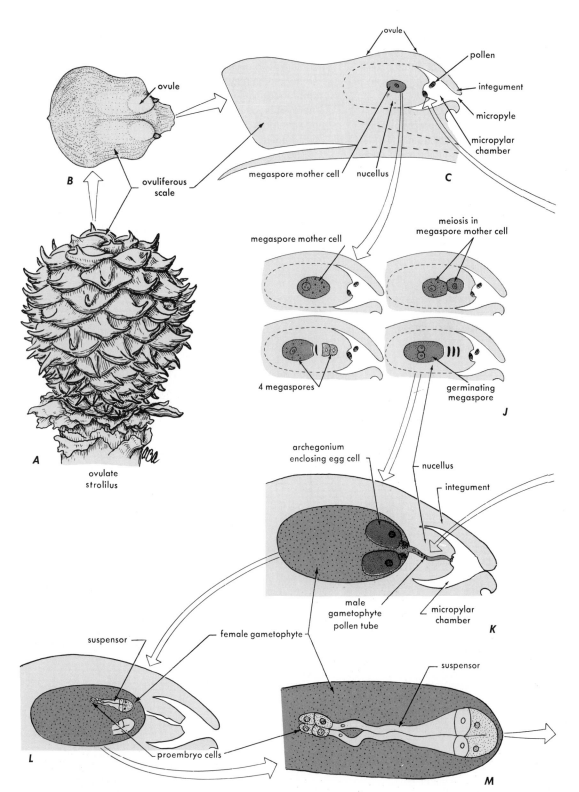

Figure 15.12 States in the life cycle of a pine, *A*, young ovulate cone; *B*, scale from ovulate cone showing two ovules on upper surface; *C*, section of ovulate cone showing megasporocyte cell and pollen in pollen chamber; *D*, staminate cone; *E*, scale from staminate cone; *F*, cross section of staminate scale; *G*, winged pollen grain; *H*, development of male gametophyte in pollen grain; *I*, pollen tube penetrating nucellus; *J*, formation of linear tetrad of megaspores; *K*, female gametophyte with egg cell and pollen tube; *L*, suspensor and proembryo sporophyte within female gametophyte; *M*, proembryo; *N*, section of seed showing seed coats, female gametophyte, and embryo sporophyte; *O*, winged seed; *P*, seedlings.

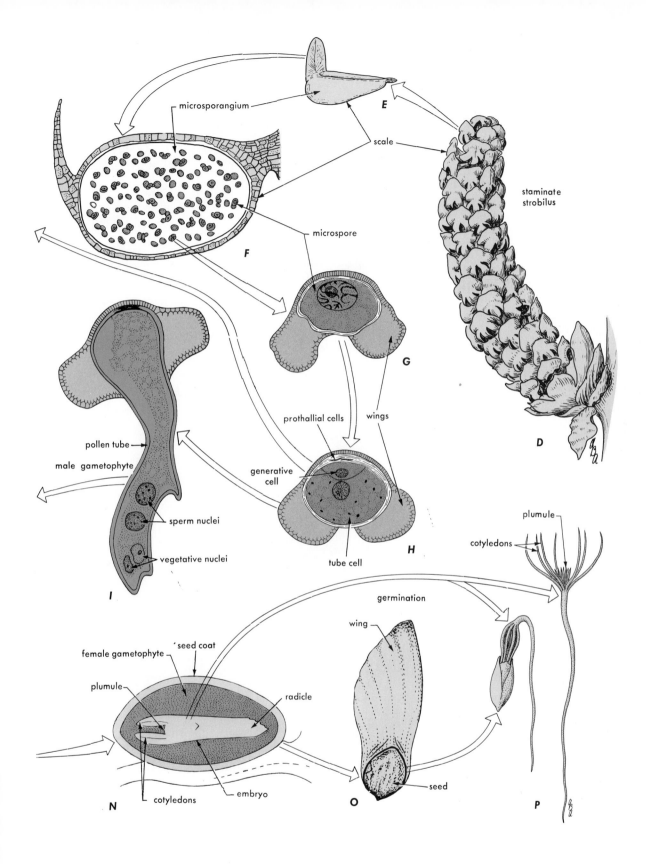

microsporangium

scale

E

staminate
strobilus

microspore

F

D

prothallial cells

wings

G

pollen tube

male gametophyte

generative
cell

sperm nuclei

vegetative nuclei

tube cell

H

I

plumule

cotyledons

germination

wing

female gametophyte

seed coat

plumule

radicle

embryo

cotyledons

seed

N

O

P

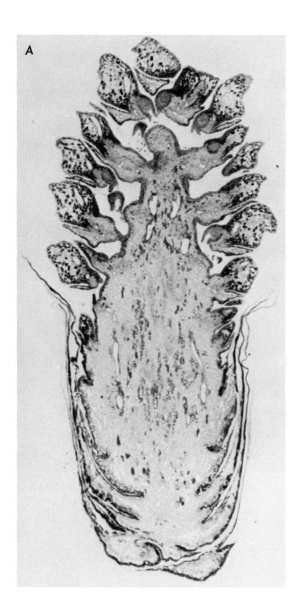

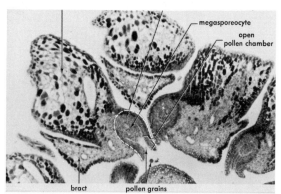

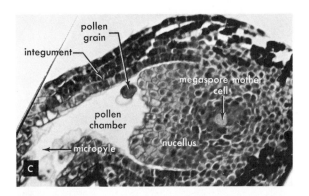

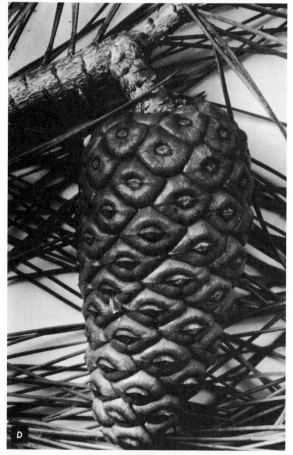

Figure 15.13 Ovulate cones of *Pinus*. *A,* longitudinal section of entire cone at the time of pollination, less than 1 cm long; note the pollen chambers with open micropyles. *B,* close-up view of the one ovule from *A. C,* an ovule with a pollen grain in the pollen chamber. *D,* cone at the time of embryo development, scales closed.

whole structure is the **seed** (Fig. 15.12). In many pines, the seeds are winged.

Normally, in pines, the seed matures some 12 months after fertilization; two years intervene between initiation of ovules and formation of seeds. Young pine seedlings (Fig. 15.12) have several cotyledons. A generalized life history of a pine is shown in Table 15.1

Gymnosperm seeds may remain dormant for many years, and some may remain embedded in the old mature cones for six years or more. Heat causes the resin coating the cones of some species to melt, allowing the cones to open and release the seeds. This behavior has considerable survival value, since large numbers of seeds are released after fires, and injured trees are thus replaced by the young ones. In many species, however, seeds are shed soon after they are mature.

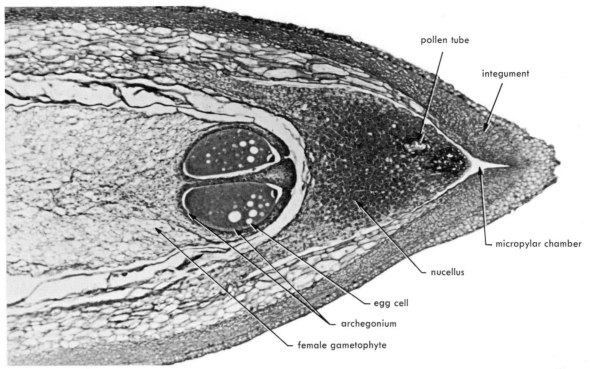

pollen tube

integument

micropylar chamber

nucellus

egg cell

archegonium

female gametophyte

Figure 15.14 Section through tip of ovule after pollination but before fertilization.

The Conifer Life Cycle in Perspective

The pine life cycle, representative of the Coniferophyta, is part of a general life cycle trend from Bryophyta to Anthophyta. Quite simply, that trend is: an increasing importance and complexity of the sporophyte generation and at the same time a decreasing size and degree of independence of the gametophyte generation.

As a generalization, thallophytes are part of this trend. In the algae and fungi, the gametophyte generation tends to be dominant; that is, a given species tends to spend most of its active life span in the haploid state. Furthermore when sperm and egg (or + and − gametes) join to produce a zygote, that cell is not retained or protected within parental tissue. Recall the life cycle of *Fucus,* where the eggs and sperm are released from gametangia and meet in the open water; the zygote then begins dividing into a young plant while still in the open water.

In the Bryophyta, the gametophyte generation is still dominant and it tends to be more anatomically and morphologically complex than the sporophyte. The sporophyte is still not an independent entity, because it partly relies on the gametophyte for nutrition during its entire life. In contrast, the gametophyte is photosynthetic and completely independent and can reproduce asexually to continue growth indefinitely (indeed, in some mosses, the sporophyte generation is rarely seen). Cells that resemble tracheids and sieve cells are present in some species. There is more protection given to the young sporophyte (embryo) than in thallophytes: the zygote is retained in the tissue of the gametophyte parent and its early growth takes place within this tissue. There is also additional protection given to developing gametes: the gametangia have a sterile jacket of cells around the gametes.

In the lower vascular plants, the sporophyte becomes the dominant generation—dominant in terms of life span, size, and anatomical complexity. It also becomes an independent phase. During the embryo stage, the sporophyte is still attached to and nourished by the gametophyte, but ultimately it becomes established in the soil and is from then on independent of the gametophyte. Multicellular gametangia are retained, as is the requirement for free water for fertilization.

In the conifers, the size, complexity, and longevity of the sporophyte generation become even more extreme, and the gametophyte generation becomes dependent on the sporophyte for nutrition and protection. Both the male gametophyte (pollen tube) and the female gametophyte are parasitic on the nucellus ($2n$ tissue) and incapable of manufacturing their own food or of existing in the open. They exhibit very little differentiation, and antheridia are not recognizable. The gametes are protected to a greater degree than in lower vascular plants in that the egg and the pollen tube that carry sperm to egg always lie within surrounding tissue.

Pollination is traditionally regarded as an adaptation to a dry, terrestrial habitat. Furthermore, free water is not required for fertilization. Another innovation is additional protection for the embryo by its enclosure in a dispersal package (the seed) complete with stored food and a hard seed coat.

Division Coniferophyta

323

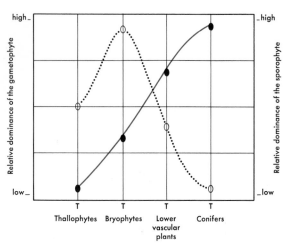

Figure 15.15 Life cycle trend from thallophytes to conifers, highly diagrammatic. Sporophyte, solid line; gametophyte, dotted line.

Figure 15.15 summarizes the shift in life cycle patterns that we have seen from thallophytes to conifers.

Division Cycadophyta

Some 200 million years ago members of the Cycadophyta formed an extensive portion of the earth's flora, probably constituting the food supply for at least some of the herbivorous dinosaurs. Today, there remain 9 to 10 genera of cycads with about 100 species growing in widely separated areas of the earth's surface but largely confined to the tropics. Only one genus, *Zamia,* occurs naturally in the continental United States; it is found in southern Florida. Various genera are, however, cultivated outdoors in warmer regions, and in greenhouses elsewhere.

Two cycads, *Cycas revoluta* and *Dioön spinulosum,* are shown in Fig. 15.16. Both were grown in the conservatory at Golden Gate Park, in San Francisco. These trees are palmlike in appearance; in fact, *Cycas revoluta* is known as Sago palm. In *Zamia,* the genus native to Florida, the trunk is largely subterranean (Fig. 15.16), but in *Cycas* it forms a single straight bole. The trees grow slowly, a 2 m high specimen being perhaps as much as 1000 years old.

That these trees are really gymnosperms and not palms is readily evident from their large cones (Fig. 15.16). A primitive feature of the division is the presence of flagellated sperm.

Division Ginkgophyta

Only one living representative, the maiden-hair tree *(Ginkgo biloba),* remains of this very ancient division of plants. It has been reported as growing wild today in forests of remote western China. *Ginkgo* has, however, been grown for centuries on Chinese and Japanese temple grounds and is now cultivated in many countries. It is a large tree with characteristic small, fan-shaped leaves

(Fig. 15.2) that are divided into two lobes. The sperms are flagellated. In the United States the *Ginkgo* is commonly planted as a city sidewalk tree.

Division Gnetophyta

This division consists of only three genera, and in some respects their traits place them as intermediate between the gymnosperms and the angiosperms. For example, they contain vessels in the xylem, the ovules appear to be surrounded by two integuments, and the pollen-producing structures superficially resemble stamens. All these traits are shared by angiosperms and absent in other gymnosperms, yet true flowers and fruit are absent— seeds are still borne naked.

Only one genus, *Ephedra* (joint-fir, Mormon tea), is found in North America. Its 40 species are distributed in warm temperate parts of the Mediterranean region, India, China, the southwestern deserts of the United States, and mountainous parts of South America. It is a shrubby xerophyte, with whorls of small deciduous leaves at prominent joints; most photosynthesis is accomplished in the green stems (Fig. 15.17).

The other genera look quite different from *Ephedra* and exhibit important life cycle differences. Their inclusion into one division is a matter of convenience for classficiation and the resulting division is quite artificial. *Welwitschia* is another xerophyte, distributed along the southwest coast of Africa. It has two long, straplike leaves that trail along the soil surface (Fig. 15.18) *Gnetum* is a tropical genus of 30 species, mainly climbing lianas. The leaves look very much like typical dicot leaves.

Summary

1. The gymnosperms are higher vascular plants and possess both vascular tissue and seeds. Seeds aid plant dispersal, protection of the embryo, and establishment of the young sporophyte. A second advancement is pollination, which permits sperms to reach eggs while surrounded by protecting tissue and in the absence of free water.
2. Of the four divisions of gymnosperms, Cycadophyta, Ginkgophyta, Gnetophyta, and Coniferophyta, the latter contains the most species, is the most widespread, and contributes the most to modern vegetation types. The Coniferophyta contains seven families, several of which were discussed. The Pinaceae contains pines, firs, spruces, hemlocks, Douglas fir, larch and true cedars (*Cedrus* spp.). The Cupressaceae contains junipers, cypresses, and false cedars. The Taxodiaceae contains bald cypress, redwood, and Sierra bigtree. The Taxaceae contains the yews. Other families are better represented in the southern hemisphere.
3. *Pinus* (pine) life cycle was examined as representative of the Coniferophyta. The sporophyte is large and anatomically complex. Vessels are absent, but considerable secondary xylem is produced. Leaves are modified for arid conditions. Pine is heterosporous, the microspores being produced in small (staminate) cones

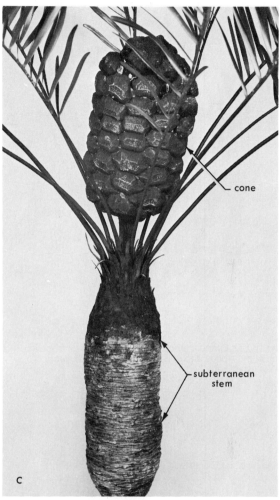

Figure 15.16 Cycads. *A, Cycas revoluta* showing a microsporangiate cone. *B, Dioön* spinulosum showing ovulate cone. *C, Zamia* showing ovulate cone and subterranean stem.

on lower branches. Megaspores are produced in other (ovulate) cones on upper branches; these cones are at first small but over a two year period mature into large, woody cones.

4. The microspores begin dividing while still within the spore wall. After two divisions, and the elaboration of air-filled wings on the outside of the thickened spore wall, growth ceases and the immature male gametophytes are liberated to the wind. These are pollen grains. Some, by chance, sift down into young ovulate cones and are sealed in when the cone scales close.

5. Two ovules are present on each scale in the ovulate cone. An ovule consists of an outer integument, nucellar tissue, and the female gametophyte. One nucellar cell functions as a megasporocyte, and, of the four resulting megaspores, only one survives. It begins to divide to form the female gametophyte at the same time that pollination (the transfer of pollen from male to female cone) occurs. At maturity, the female gametophyte is a relatively undifferentiated, multicellular, nonphotosynthetic thallus embedded in the nucellus. At one end are several archegonia.

6. If the pollen grain is drawn through the micropyle (an opening in the integument), it comes to lodge against the nucellus. It germinates and a pollen tube begins to parasitically grow toward the female gametophyte through the nucellus. Finally, a third cell division produces two sperm and the formation of the male gametophyte is complete. No antheridium is recognizable.

Summary

Figure 15.17 *Ephedra viridis*, a member of the Gnetophyta. *A*, a portion of the woody shoot; *B*, a close-up of *A*, showing two scalelike leaves attached to a young, green stem.

7. Fertilization is accomplished when the pollen tube ruptures near an archegonium and one sperm unites with an egg. The resulting zygote divides and differentiates into an embryo. A mature embryo lies surrounded by female gametophyte tissue, some remnants of the nucellus, and a seed coat (from the integument). The embryo is dormant until the seed is shed and comes to lie in an appropriate microenvironment.

8. In general, life cycles from thallophytes to conifers have shown a definite trend or pattern: the dominance and complexity of the sporophyte have increased and the dominance and complexity of the gametophyte have declined.

9. The Cycadophyta contain about 100 species of tropical, slow-growing, palmlike plants. The Ginkgophyta consists of a single species that may now be extinct in the wild but is widely cultivated. The Gnetophyta exhibit some traits that link them more closely to the flowering plants than to other gymnosperm groups. Only one of its three genera, *Ephedra*, is found in North America. This latter division is quite artificial in terms of classification.

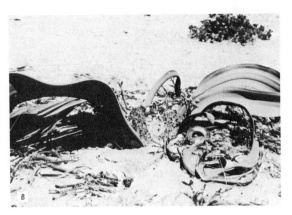

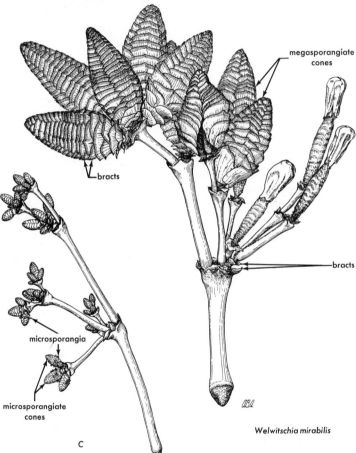

Figure 15.18 *Welwitschia*, growing in the Namib Desert of southwestern Africa. *A*, one of the larger plants, about 1.5 m tall and carbon dated at 1500 years old; *B*, closer view of smaller plant, showing the two leaves emerging from the nearly buried stem; *C*, male (left) and female (right) cones borne near the growing base of the long, leathery leaves. The plants are unisexual.

angiosperms

The angiosperms, or flowering plants, are the dominant plants of the world today. They include nearly all the crop plants of orchard, garden, and field. Hardwood forests, shrublands, grasslands, and deserts are composed chiefly of flowering plants. They show great variation in form, from simple stemless, free-floating duckweed (*Lemna*) through a whole series of herbaceous types to shrubs and trees such as oaks and beeches (*Fagus*). We have already studied the structure and physiology of the angiosperm plant body in considerable detail. Angiosperms constitute the Anthophyta, which is divided into two classes, Monocotyledonae and Dicotyledonae.

The outstanding and unique structure of the angiosperms is the *flower* (Fig. 16.1). The flower is a shoot bearing floral leaves. In a complete flower (Fig. 16.1), the floral leaves are sepals, petals, stamens, and carpels. Some flowers have only the essential reproductive structures, stamens and carpels. Other flowers are unisexual (Fig. 16.7): they have either stamens or carpels.

All flowering plants produce seeds except some that human beings have modified so that they are now seedless (e.g., seedless grapes and bananas). In angiosperms, seeds are borne within a closed structure, the ovary, which eventually becomes the fruit.

Review of Life Cycle

The life cycle of the angiosperms was thoroughly discussed in Chapter 7. At this point let us simply review its most important aspects.

Male Gametophyte

The anther is the part of the stamen responsible for production of pollen. Microsporocytes form within the pollen sac (Fig. 16.1). They divide by meiosis to form haploid (1*n*) microspores, which in turn divide mitotically to form a generative and a tube nucleus (Fig. 16.1*C*). A heavy, sculptured wall forms about the microspore and a mature pollen grain results. The pollen is now shed and conveyed in one fashion or another (Chapter 7) to a stigma where it germinates. A pollen tube penetrates the stigma (Figs. 16.1*A,D*) and grows through the style, down to the ovule within the ovary. The generative nucleus usually divides within the pollen tube into two sperm nuclei

(Fig. 16.1*D*). Each sperm nucleus, with its associated cytoplasm, is a sperm cell.

Free water is not needed for fertilization in angiosperms; motile, ciliated sperms are not produced by any members of this division. The pollen tube with sperm cells and tube nucleus, if present, constitute the mature male gametophyte.

Female Gametophyte

An enlarged cell, the megasporocyte, within nucellar tissue of a young ovule undergoes meiosis and forms four megaspores arranged in a row. While this process is occurring, integuments form about the nucellar tissue, resulting in the formation of an ovule. The three megaspores closest to the micropyle generally disintegrate. The remaining megaspore undergoes three successive mitotic divisions, resulting in seven cells of the embryo sac or female gametophyte: one egg cell, two synergid cells, one endosperm mother cell with two polar nuclei, and three antipodal cells (Fig. 16.1*A*).

Fertilization and Seed Development

With penetration of the pollen tube into the mature embryo sac, the stage is set for fertilization. In angiosperms, fertilization involves not only the union of the egg cell with the sperm cell but, in addition, the union of a second sperm cell with the endosperm mother cell to form the primary endosperm cell. Since there are two nuclei in the endosperm mother cell, the nucleus of the primary endosperm cell will have three sets of chromosomes. The fusions of two sperm cells, one with the egg, the second with the endosperm mother cell, is **double fertilization**.

The resulting zygote generally develops directly into a small proembryo, from one end of which a typical embryo develops. The fate of the primary endosperm cell depends on the species. This cell gives rise to an endosperm that plays some role in nourishing the developing embryo. In a few genera, notably the grasses, the endosperm enlarges and persists in the seed to form the source of nourishment for the young seedling. A seed always includes an embryo and a food supply to enable the young seedling to establish itself, except in rare cases like orchids where the embryo remains quite rudimentary. Food is generally stored either in cotyledons of the embryo or in endosperm.

With the germination of a seed and the development of

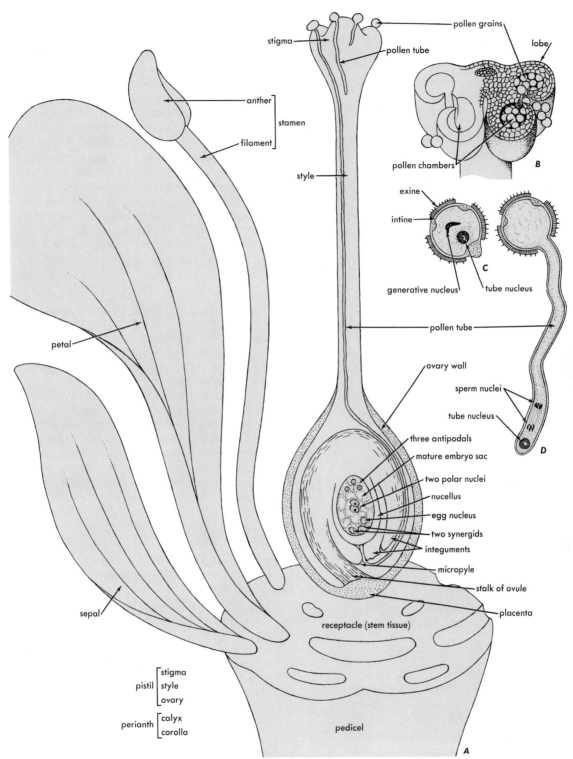

Figure 16.1 *A*, diagram of a flower; *B*, cross section of an anther; *C*, mature pollen grain; *D*, germinating pollen grain.

Table 16.1
Generalized Life History of an Angiosperm

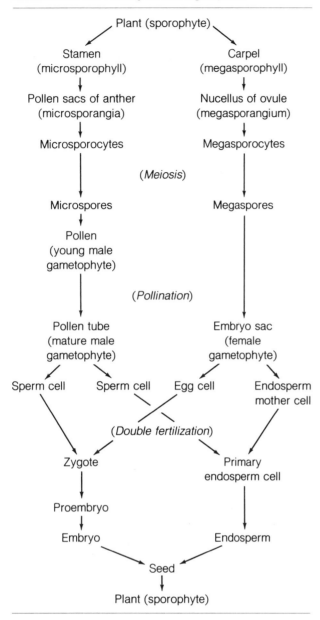

the seedling into a flowering plant, the life cycle of an angiosperm is completed. The essential steps are shown in Table 16.1.

Comparison of Angiosperm Life Cycle with more Primitive Plants

Certain comparative details between structures and life cycles of members of the gymnosperms and angiosperms are given in Table 16.2. Details concerning the Psilophyta, Lycophyta, Sphenophyta, and Pterophyta are given in Table 14.1. A general comparison of life cycles in

thallophytes, bryophytes, lower vascular plants, and seed plants appears in Fig. 15.15.

In all Pterophyta the sporophyte is the dominant generation, and both gametophyte and sporophyte generations are independent. In gymnosperms and angiosperms, there occurs a further reduction in size and complexity of the gametophyte to a condition of complete dependency on the parent sporophyte. This is accompanied by an overall increase in the general complexity of the sporophyte. It can be stated that, as a guiding principle, the more advanced forms in these divisions are more highly adapted to grow and flourish in a dry land habitat than are the more primitive forms. The most obvious developments designed to accomplish this are: (a) adaptations that remove any dependence of the plants on free moisture for fertilization, and (b) adaptations, both vegetative and reproductive, that enable the plant to grow and reproduce on dry land. Examples are given below.

Protection of the Female Gametophyte

The gynoecium of the angiosperm flower is essentially a modified leaf, or leaves, enclosing an ovule, or ovules, in which the female gametophyte is to be found. The ovary, as we know, eventually develops into a fruit. This enclosure of ovules and, subsequently, seeds by an ovary wall is a new development that sets angiosperms apart from all other groups of plants. The female gametophyte (embryo sac) has gained, in the ovary wall, an added protective barrier. However, this has necessitated further adaptations to enable sperms to pass through this added protection: (a) development of a receptive stigma and (b) growth of a pollen tube through the style.

Size of Gametophytes

Reduction in size of gametophytes has proceeded to a point in which the male gametophyte, the pollen tube, consists of one vegetative nucleus and two sperm cells (Fig. 16.1*D*). The female gametophyte is a seven-celled structure, one of whose cells is an egg cell. A second cell, the endosperm mother cell, is an innovation in that it, too, is receptive to a sperm cell. The remaining five cells are vegetative cells (Fig. 16.1*A*). It should be pointed out that embryo sacs of some angiosperms have more cells, while a few of them have less, but all have an egg cell and an endosperm mother cell.

Nourishment for the Developing Embryo

Double fertilization is still another difference between the life cycle of the angiosperms and more primitive forms. In the latter, nourishment for the young developing sporophyte is generally provided by the female gametophyte (Figs. 14.10*M,N*) or, in the case of many ferns, by the simple prothallus, or gametophyte (Fig. 14.18*J*). In angiosperms, food for the developing seedling is stored either in the endosperm or in the cotyledons of the embryo itself.

Table 16.2
Comparison of the Conifers and Flowering Plants

Generation	Gymnosperms	Angiosperms
	Roots, stems, and leaves present	Roots, stems, and leaves present
	Trees and shrubs, no herbs	Trees, shrubs, and herbs
	Stem: one core of xylem surrounded by phloem, vessels in one small division only	Stem: one core of xylem surrounded by phloem, or several vascular strands with phloem exterior
	Sieve cells without companion cells	Companion cells and sieve-tube members
	Pith present in stems, not in roots	Pith in stems, not in roots, of dicots; in some monocot roots
Sporophyte	Cambium develops in all species	Cambium in perennial forms; absent generally in annuals
	Staminate and ovulate cones, generally, with staminate and ovulate scales	Flowers, stamens, and carpels
	Heterospory	Heterospory
	Microspores in microsporangia	Microspores in anther sacs (megasporangia)
	Megaspores or nucellus	Megaspores in nucellus
	Ovules present, exposed on scales	Ovules present, enclosed within carpel
	Embryo develops within a seed	Embryo develops within a seed
Fertilization	Double fertilization absent; no endosperm	Double fertilization, resulting in zygote and primary endosperm cell
	Pollen tube: male gametophyte	Pollen tube: male gametophyte
	Female gametophyte within ovule	Female gametophyte (embryo sac) within ovule
	Antheridium absent	Antheridium absent
	Sperm cells formed in pollen tube	Sperm cells formed in pollen tube
Gametophyte	Motile sperms in several primitive genera, but free water not needed for fertilization	
	Nonmotile sperms in all other forms	Nonmotile sperms
	A very greatly reduced archegonium completely embedded in the female gametophyte	Archegonium only suggested by synergid cells of the embryo sac
	Embryo within female gametophyte and seed coat	Embryo within seed coats, remains of nucellus, and endosperm

Fruit

Fertilization stimulates tissues around the zygote to begin further development. In angiosperms, the ovary wall is stimulated to produce a fruit, and the fruit is a development originating with, and characteristic of, angiosperms. The protection of the seed from its surroundings by the ovary wall is accompanied by various devices for bringing it under conditions favorable for germination.

Evolution in the Angiosperms

If angiosperms represent the culmination of an extensive evolutionary development, it is likely that within the division anthophyta itself, evolution has been and still is active. In other words, families of angiosperms must be related and it should be possible to arrange the families in a sequence that would give some indication of relationships. Some families will be found to be more primitive than others; and from the more primitive types

the more advanced and specialized forms have presumably arisen (Chapter 10).

The Besseyan System

Many systems of classifying plants have been proposed. A system that has found much favor with American botanists was developed by Charles E. Bessey (1900s). His system regards the order Ranales as being the most primitive. In this order, the Magnoliaceae is one of the most primitive families, with the Ranunculaceae somewhat more advanced. The Christmas rose (Helleborus), magnolias (Fig. 16.3), and buttercups (Ranunculus) are representatives of these families.

If we consider the arrangement of flower parts in magnolias and buttercups to be most primitive, then it is possible to compare them to other families to determine their apparent relationships. The principal tendencies in evolution of the flower, according to the Besseyan system, are as follows:

1. From an elongated to a shortened floral axis.

2. From a spiral to a whorled condition of floral parts.
3. From numerous and separate stamens and carpels to few and connate stamens and carpels.
4. From numerous and separate sepals and petals to few and connate sepals and petals.
5. From complete and perfect flowers to incomplete and imperfect flowers.
6. From regular to irregular flowers.
7. From hypogyny to epigyny.

According to the Besseyan view, there were at least three main lines of advance from the primitive Ranalian type of flower. These lines culminated in (a) mints (Lamiaceae), (b) asters (Asteraceae), and (c) orchids (Orchidaceae).

Selected Families of Angiosperms

Let us now examine selected families of angiosperms with two objectives in mind: (a) to learn the characteristics of some important angiosperm families and (b) to discover how these families fit into the Besseyan system of angiosperm classification. In our discussion of these matters so far, floral characteristics have been emphasized. The emphasis is placed on floral parts because their form and structure are little influenced by the external environment. They also are structurally more constant from generation to generation than are vegetative parts. Because of this constancy in structure, floral parts have a much greater value in both plant identification and in tracing evolutionary relationships than do vegetative characters. However, the latter cannot be neglected.

Since the magnolia and buttercups are thought to be primitive forms, we shall begin our discussion with these families, then proceed to follow through a line of ascent ending in the mint family. We shall consider the line culminating in the Asteraceae (Compositae)* and

* According to rules of plant nomenclature family names should end in -ceae. Traditionally, the names of certain families such as Cruciferae and the Compositae have not ended this way. In this chapter the currently more acceptable synonym names ending in -ceae will be used.

complete our discussion with the line of monocotyledonous families. It should be further noted that lines of development are not straight but branched and treelike. The arrangement of the families discussed here as they occur in the Besseyan system is shown in Fig. 16.2.

Magnoliaceae to Lamiaceae

In this line of ascent (Fig. 16.2) all flowers, in all families are hypogynous. Syncarpy (the fusion of carpels) occurred very early, as did a reduction in number of floral parts. Other types of connation and the change to irregular flowers appeared somewhat later but still early in the line of ascent. Based on their floral characteristics, primarily, it is possible to show an evolutionary connection between the following families: Magnoliaceae, Ranunculaceae, Malvaceae, Brassicaceae, Solanaceae, Salicaceae, and Lamiaceae (Fig. 16.2).

The Malvaceae is a clearly marked family with relatively primitive characters as shown by its regular, perfect, and hypogynous flowers. There are numerous stamens with connate anthers, and the five or more carpels are connate. In Solanaceae, the flowers, while perfect and regular, show connation of sepals, petals, and carpels, with stamens adnate to the corolla tube. The Salicaceae is a family in which reduction in both number and size of floral parts has resulted in its advanced condition. The flowers are imperfect and grouped in catkins, and both calyx and corolla are lacking. The pistillate flowers are hypogynous and, in staminate flowers, the number of stamens has been reduced to one or two. The Lamiaceae, or mints, represent the most advanced family in this line of ascent. The flowers are hypogynous, with an irregular symmetry; there is a reduction in number of parts to four or two; and sepals, petals, and carpels are connate. The stamens are not only adnate to the corolla tube but are also highly specialized for insect pollination. Let us consider some of the general characteristics of these families.

Magnoliaceae

This group shares with the Ranunculaceae the distinction of being very ancient. The tulip tree, *Liriodendron,* as

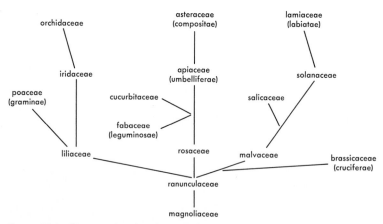

Figure 16.2 Diagram showing the evolutionary tendencies in the Angiospermae according to the Besseyan system.

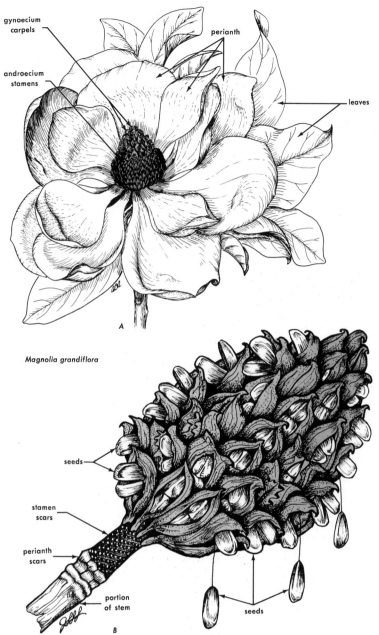

gynoecium
carpels

perianth

androecium
stamens

leaves

Magnolia grandiflora

A

seeds

stamen
scars

perianth
scars

portion
of stem

seeds

B

Figure 16.3 *Magnolia grandiflora* flower and fruit with hanging seeds.

shown by fossil remains, once had a very wide distribution. Most species of the Magnoliaceae are trees or fairly large shrubs. Magnolias themselves are magnificent trees, *Magnolia grandiflora* grows to 30 m or more with stiff, large, evergreen leaves and large white blossoms 18 to 20 cm across. The magnolias are largely warm-climate species, but tulip trees will tolerate northern winters (Fig. 16.3). There are some 12 genera and 230 species* in the family.

---

* The estimates of genus–species numbers were taken from Willis, J.C., *A Dictionary of the Flowering Plants and Ferns*, 8th edition, revised by H.K. Airy Shaw, 1973, Cambridge University Press. Note that species numbers from other authors may differ depending on the system used to calculate their estimates.

Ranunculaceae

Most of these are herbs but there are a few small shrubs and woody climbers. Leaves are frequently compound, a feature that has led to the common name of crowfoot family. There are about 50 genera and approximately 1900 species growing largely in the North temperate and arctic regions. Many of our common garden flowers belong to this family. Among these are *Delphinium, Aquilegia, Paeonia, Ranunculus, Anemone,* and *Clematis.* The marsh marigold *(Caltha)* and *Hepatica,* common spring flowers, are members of this family. One species, *Aconitum napellus,* yields the drug aconite, a cardiac and respiratory sedative, and some species of *Delphinium* are poisonous to livestock.

Selected Families of Angiosperms

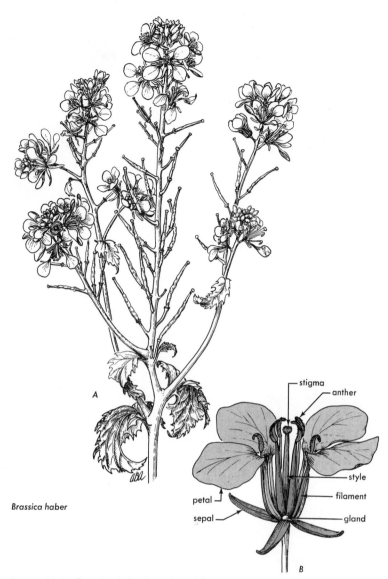

Figure 16.4 *Brassica haber* branch and flower.

Brassicaceae (Cruciferae)

This is a large, distinct family of many cultivated forms, as well as some that are noxious weeds. Cabbage, cauliflower, broccoli, brussels sprouts, kohlrabi, and kale are all horticultural varieties of a single species, *Brassica oleracea* (Fig. 16.4). Wild mustard, another *Brassica* species, is a sometimes noxious weed in grain fields, though sometimes used as cover crop in orchards. Radishes, turnips, and stocks are other members of this family, as are many garden herbs. The family has a worldwide distribution in temperate and subarctic zones; all are herbs. There are about 375 genera and 3200 species.

Malvaceae

This is a large family of 95 genera and 1000 species. Cotton, taken from seed coats of various members of the genus *Gossypium,* makes this family important from the viewpoint of politics, agriculture, and industry (Fig. 16.5). The cotton of commerce occurs as long hair or fuzz on seeds, which are borne in large capsules. Herbs, shrubs and trees, and such ornamentals as *Hibiscus,* okra, hollyhocks, and numerous weeds are members of this family.

Solanaceae

A large family with many tropical forms, it is also well-represented in temperate regions. There are about 90 genera and over 2000 species, some 1700 of the latter being in a single genus, *Solanum.* To this family belong tomatoes, potatoes, tobacco, eggplant, peppers, *Petunias,* and *Salpiglossis.* Although the family is of worldwide distribution, most of the cultivated forms were brought under domestication in the Western Hemisphere. There are many poisonous and drug plants in the family, such as belladonna and atropine. Even such a common plant as the tomato was long supposed to be poisonous, as its

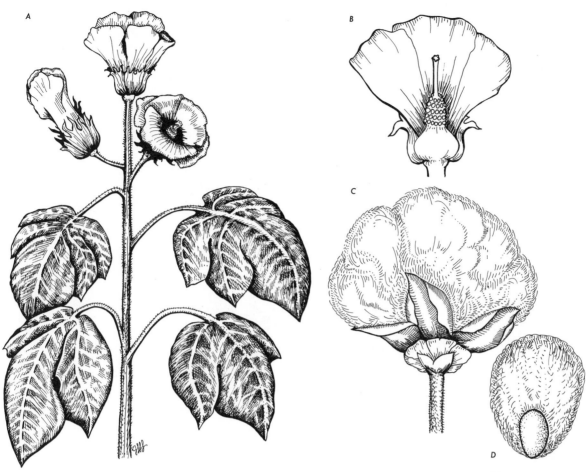

Figure 16.5 *A, Gossypium* sp. (cotton) shoot with flowers. *B*, split flower showing carpels. *C*, dehisced capsule (bole) showing cotton "fiber," which is actually made of seed coat hairs. *D*, cotton seed with attached "fibers."

scientific name *Lycopersicon,* which means "wolf peach," seems to indicate (Fig. 16.6). There are many erect and climbing herbs in the family, also some shrubs and small trees.

Salicaceae

The willow family is comprised of three genera (*Salix,* willows; *Populus,* poplars; *Chosenia,* an asiatic shrub), with about 530 species. They are mostly trees, but some shrubby forms are known. They are very abundant in the Northern Hemisphere, mostly in temperate zones. Baskets are woven from branches of the basket willow, and paper pulp is obtained from trunks of one species. Willows are common along water courses (Fig. 16.7), frequently overhanging or even choking mountain streams, to the ill comfort of fishermen. Pussy willow is a familiar example of this family.

Lamiaceae (Labiatae)

Mints (Fig. 16.8) represent the supposed highest advance of one of the three lines of angiosperm evolution. It is a large family of considerable economic importance, largely because of volatile oils produced by certain of its

members. Peppermint, spearmint, thyme, sage, and lavender are examples. There are about 180 genera with 3500 species well-distributed over the surface of the earth. There are herbs and shrubs in the family, generally with characteristic square stems and opposite leaves.

Magnoliaceae to Asteraceae

The second line of ascent involves the Fabaceae, Rosaceae, Apiaceae, and Asteraceae, with an offshoot family, the Cucurbitaceae. This large group of families is characterized by an early change from hypogyny to epigyny. Following this there is a division into two sublines, one being marked by a lack of connation and a retention of regular flowers. In the other subline, connate and irregular flowers both occur in advanced forms.

The Rosaceae (rose family) is one of the more primitive families in this line of evolutionary development. There exists a considerable variety of flower structure within the family. The more primitive members are regular, perfect, with numerous parts and with such a slight degree of perigyny as to be recognizable only with careful observation. This is exemplified by the boysenberry (Fig.

petal
stamens
stigma
style
ovary
receptacle
B

stigma
stamens

locule
seed

C

Figure 16.6 Solanaceae. *A*, branch of *Lycopersicon* showing leaf, flower, and buds; *B,* lontitudinal section of flower of *Lycopersicon; C,* cross section of tomato fruit.

16.9). Changes within the family involve a reduction in the number of floral parts, a shift from slight perigyny to true perigyny and to epigyny. Connation also occurs. Members of the family generally have regular flowers.

Fabaceae (pea family) are considered to be a more advanced family than Rosaceae, even though their flowers are hypogynous. There is a reduction in the number of parts, the gynoecium having only one carpel; connation of stamens and of some petals occurs, and the corolla parts are irregular. Apiaceae represent a further advance over Rosaceae in that the flowers are epigynous and exhibit syncarpy (Fig. 16.9). This line of ascent finds its climax in the Asteraceae, whose flowers all possess advanced characters, such as reduction in number of parts, modification of parts (pappus), irregular flowers, and some imperfect flowers, connation in each whorl, and adnation.

The Cucurbitaceae culminate a branch line from the rose order. Flowers of Cucurbitaceae have such advanced characters as loss of floral parts, imperfect flowers, epigyny, and connation.

Rosaceae

Whereas grasses and legumes supply bread and vegetables of basic importance, the rose family supplies fruit for dessert and roses for decoration. There are more than 2000 species in this family and over 100 genera, not counting the almost numberless cultivated forms of roses, peaches, apples, cherries, boysenberries (Fig. 16.9), almonds, and so on. The family is of worldwide distribution and somewhat heterogeneous. Many of the principles of angiosperm evolution can be demonstrated with its members. There are trees, shrubs, and herbs in the family. Leaves are usually alternate and bear stipules.

Cucurbitaceae

The gourd or melon family is of worldwide distribution in at

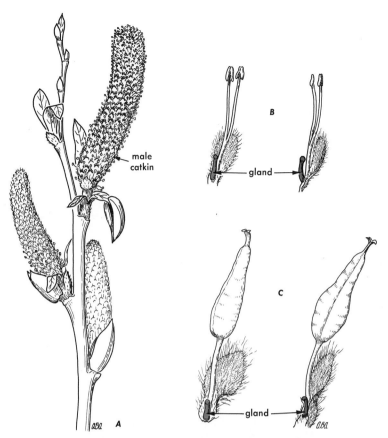

Figure 16.7 Willow *(Salix)* inflorescence and individual flowers; *A,* male catkin, ×1; *B,* single male flowers, X15; *C,* single female flowers, X15.

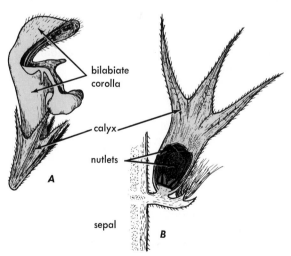

Figure 16.8 Lamiaceae. Irregular hypogynous flower of mint, showing union of sepals and petals.

least warmer regions of the world, and various peoples have selected different forms for domestication (Fig. 16.10). It is also probably the only group in which fruits are highly prized for ornamental purposes and for use as containers of various types. Pumpkins and squashes are

of Western Hemisphere origin. Cucumbers and melons have probably come from Africa and central Asia. Other species appear to have been first cultivated in the tropics of Asia, Polynesia, and India. They frequently use tendrils to climb and are annual or perennial vines. Most are rapid-growing and frost-tender. There are about 110 genera with 640 species. Stems are usually soft and hairy or prickly. The generally simple leaves are large and sometimes deeply cut. Flowers are usually unisexual.

Fabaceae (Leguminosae)

The legume family has such distinctive characteristics that its members can frequently be recognized with only little experience (Fig. 16.11). It is one of the three largest families, with 600 genera and 12,000 species. All major growth forms are represented: herbs, both annuals and perennials, shrubs, vines, and trees. It is of worldwide distribution. While less heterogeneous than the rose family, considerable variation is found among its various members. It is a family of considerable importance in supplying food for humans and their animals. Many legumes are used for ornaments, from shade trees to cut flowers. Some lumber is obtained from the black locust, and some of the tropical species furnish wood for fine cabinet work. Association of the nitrogen-fixing bacteria with roots of legumes places this family in a unique

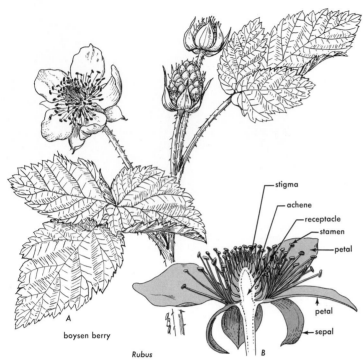

boysen berry

Rubus

- stigma
- achene
- receptacle
- stamen
- petal
- petal
- sepal

A

B

Figure 16.9 *Rubus* sp. (boysenberry) flowering branch with one flower split open to show the fleshy receptacle.

position relative to maintenance of soil fertility. Peas, beans, peanuts, clovers, and lupines are common herbaceous legumes; wisteria is a vine, brooms and redbuds are shrubs or low trees, and locusts and acacias are trees. The *Mimosa* genus alone has some 450 species, including the sensitive plant of the florist shop; there are others of varying habit from tall trees to low herbs. Leaves of legumes are prevailingly pinnately compound and quite generally with stipules. Sometimes a leaflet will be modified to a tendril.

Asteraceae (Compositae)

The Asteraceae is the largest family of angiosperms (Fig. 16.12). Only a few of these plants are woody. In this family of herbaceous plants there are around 900 genera and about 13,000 known species. The family is not only of worldwide distribution but in most places is relatively abundant. The family is not noted for its food plants; endive *(Cichorium)*, artichokes *(Cynara)*, chicory, lettuce, and sunflower form most of its contribution in this respect. Neither is it famous as a producer of drugs or other commercial products. One species, safflower *(Carthamus)*, is now being grown for oil and members of the genera *Taraxacum* and *Parthenium* are being considered as possible sources of latex for rubber. Paging through a seed catalog or a visit to a florist shop will show members of this family in their full grandeur. *Dahlia, Chrysanthemum, Aster, Zinnia, Tagetes, Gaillardia, Ageratum* and many others are representatives of this family. Dandelions color lawns, and various wild species add to fall infest of meadows and fields. There are 400 species of the genus *Artemisia,* that cover arid portions of the world and supply, among other things, tarragon for

fancy vinegar and nectar for desert honey. Leaves are of various shapes.

Monocotyledonous Line from Magnoliaceae to Orchidaceae

There are many obvious differences between the dicotyledonous plants such as the Magnolias and the monocotyledonous line of evolution. These have been discussed and are briefly reviewed in Table 16.3.

Table 16.3
Tabulation of Differences between Monocotyledons and Dicotyledons

Monocotyledons
1. One cotyledon or seed leaf.
2. Generally marked parallel leaf venation.
3. Flower parts typically in groups of three or multiples of three.
4. Vascular bundles of stems scattered throughout a cylindrical mass of ground tissue.
5. Vascular cambium lacking in most forms.

Dicotyledons
1. Two cotyledons or seed leaves.
2. Generally marked netted venation of leaves.
3. Flower parts typically in groups of four or five.
4. Vascular bundles of stems usually arranged in the form of a cylinder.
5. Vascular cambium present in forms having secondary growth.

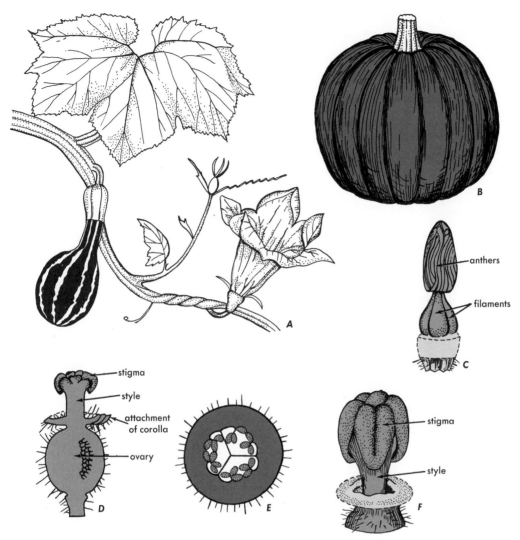

Figure 16.10 Cucurbitaceae. *A, Cucurbita pepo*, flowering and fruiting stem; *B, C. pepo*, fruit; *D, C. maxima*, staminal column; *D, C. maxima* stigma, longitudinal section through pistil, corolla tube removed; *E, C. maxima*, cross section through ovary; *F, C. maxima*, style and stigma.

However, some primitive monocotyledons, particularly certain water weeds such as arrowheads and water plantains, have much in common with buttercups and marsh marigolds. The families selected to represent the monocots are Liliaceae, Poaceae, Iridaceae, and Orchidaceae. Lilies differ from magnolias in all distinctive monocot characteristics. In addition, they show a reduction in the number of floral parts and a connation of carpels. Irises show a single advance over lilies in that they are epigynous. Many irises are regular, perfect, and with separate perianth parts. Stamens show no connation. Syncarpy occurs as it does in lilies. Orchids show evolutionary advance in that they have irregular flowers very highly specialized for insect pollination. Stamens have been reduced in number to one or two.

The order to which grasses belong may have arisen from the order to which lilies belong. The primitive characters of grasses include hypogyny, regular flowers, and only connation of carpels. The loss of perianth parts and the reduction in number of stamens and carpels mark them as a more advanced family than the Liliaceae.

Liliaceae

Unlike grasses, lilies are grown largely for ornamental purposes, although two genera, *Allium* (e.g., onions, garlic, and leeks) and *Asparagus,* are grown extensively for food. Some species yield drugs, and others have poisonous properties that may cause trouble in pastures or on ranges of western states (Fig. 16.13). There are about 4100 species in some 280 genera. Most lilies grow from a bulb or a bulblike organ and flower in a single growing season, after which the shoot dies down. A few, such as Joshua trees, are woody perennials. Tulips, hyacinths, day lilies, and aloes are other examples of the lily genera.

Selected Families of Angiosperms

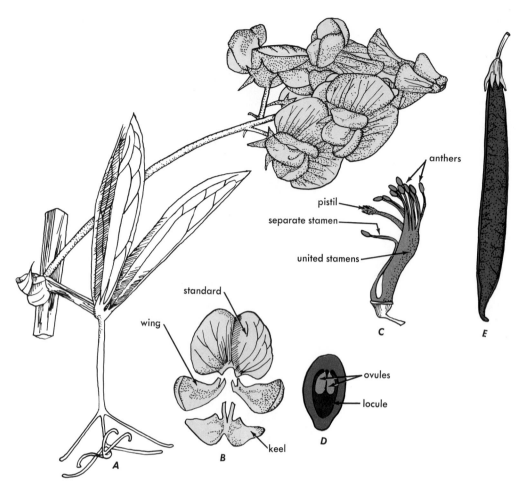

Figure 16.11   Fabaceae. *A,* inflorescence and leaf of *Lathyrus;*
B, exploded view of *Lathyrus* corolla; *C,* essential organs of
Lathyrus flower; *D,* cross section of ovary; *E,* pod of *Lathyrus.*

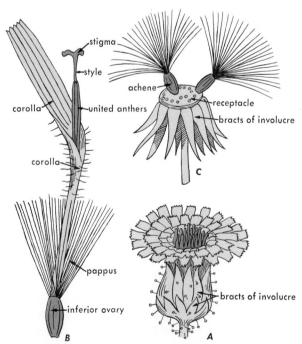

Figure 16.12   Composite flower and fruit of sow thistle
(Sonchus). A, flowering head; *B,* single ray flower; *C,* mature
achenes.

Iridaceae

The iris family contains about 60 genera and 800 species.
They are all herbaceous, largely perennial forms, usually
with rhizomes, bulbs, or corms, as in lilies. Leaves and
flowering stalks last for only one season. Some of the
choicest florists' plants occur in this family—*Iris* (Fig.
16.14), *Gladiolus, Freesia, Crocus, Watsonia,* and so forth.

Orchidaceae

It is agreed by all that orchids represent the most
advanced family in the Monocotyledoneae (Fig. 16.15).
They are all herbaceous, occurring throughout the world
largely in the tropics, but with a few genera extending into
colder temperate regions. There are some 17,000
recognized species in several hundred genera, making this
one of the three largest families of angiosperms. All forms
are perennial, with tuberous, bulbous, or otherwise
thickened roots. They may be erect, prostrate, or climbing.
A few are saprophytic and lack chlorophyll; others are
epiphytes, growing on trees without benefit of soil.

Poaceae (Gramineae)

Humans and other mammals have been associated with
the grass family for far more years than those of recorded
history. It is probably not an overstatement to say that

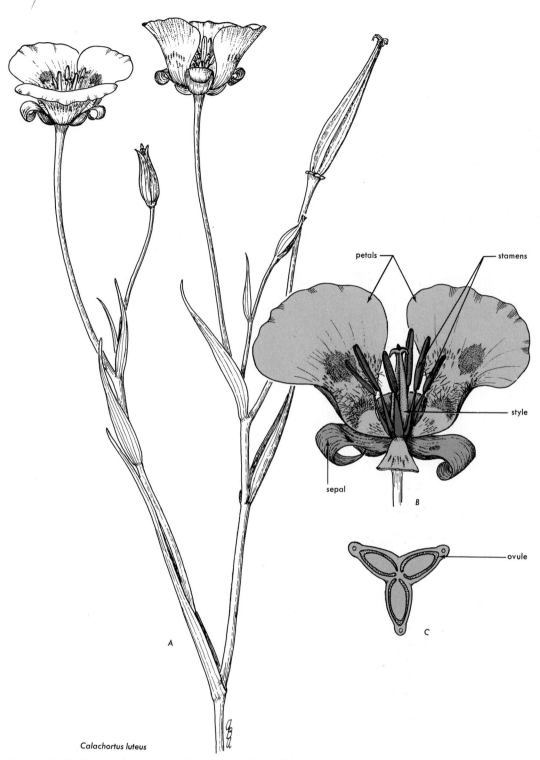

Figure 16.13 *Calachortus luteus. A*, flowering branch; *B*, flower;
C, cross section through ovary.

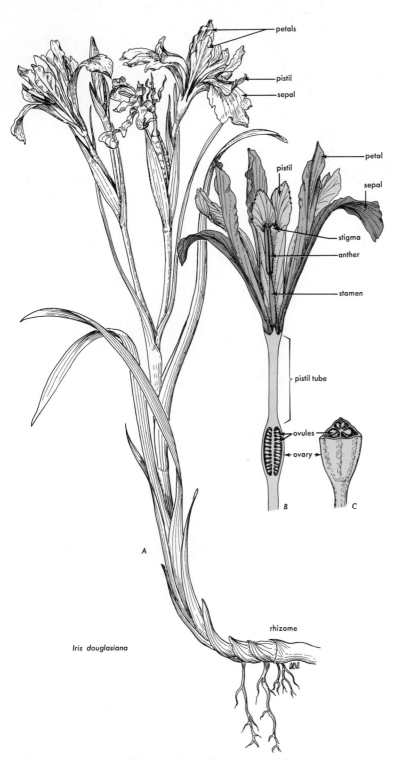

petals

pistil

sepal

petal

pistil

sepal

stigma

anther

stamen

pistil tube

ovules

ovary

B

C

A

rhizome

Iris douglasiana

Figure 16.14 *Iris douglasiana. A,* entire flowering plant; *B,* flower showing interior ovary; *C,* section through ovary.

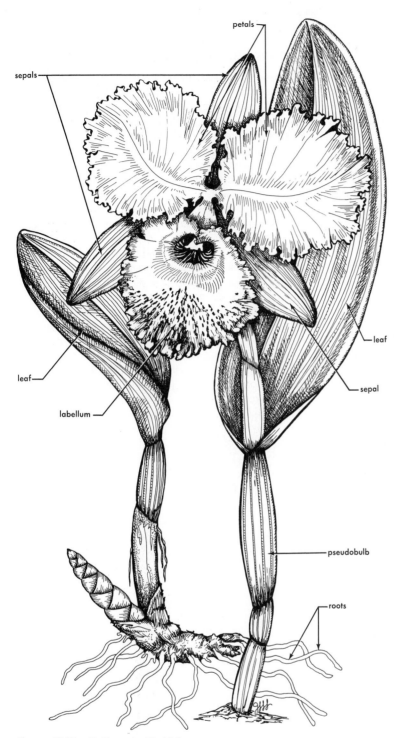

Figure 16.15 *Cattleya* sp. (Orchid).

without the grass family human civilization, as we know it, could not have developed. Different grasses were domesticated by the three centers of civilization; wheat, rye, barley, and oats by peoples of the Mediterranean region and Southwest Asia; corn by natives of the Western Hemisphere; and rice and millet *(Eleusine)* by peoples in the Orient. Grasses are grown for human and animal consumption, for ornament, and for uses in arts and industry. There are over 620 genera and about 10,000 species. Most grasses are herbaceous, either annuals or perennials; a few bamboos are climbers and some are woody and up to 20 m tall. Grasses of one kind or another grow in all kinds of soil and situations. Grasses may be conveniently divided into six groups according to their use. (a) Bamboos, mostly evergreen, stout perennials, are used for construction in many parts of the Orient, and some have been introduced into warmer parts of the United States as ornamentals. (b) Cereals and some other annual grasses supply grain and forage for both human beings and animals. (c) Sugar-producing species, such as sugar cane, are strong, upright perennials growing only in the tropics. (d) Sod-forming grasses, perennials that cover many square miles of the earth's surface, are used in lawns and meadows and for forage. (e) Bunch grasses, common in semi-arid regions, are perennials that do not form rhizomes and do not produce a turf as sod grasses do. (f) There is a large group of grasses grown for ornamental purposes, such as pampas grass. Flowers and vegetative characteristics of grasses have been described in some detail in Chapter 7 and shown in Fig. 7.6.

Summary of Angiosperm Evolution

1. Angiosperm families, according to the Besseyan system of classification, are all derived from the primitive order Ranales, to which buttercups and magnolias belong.
2. Important changes in floral evolution, according to the Besseyan system, are as follows: (a) from a spiral to a whorled arrangement of floral parts; (b) from many parts to few or even to a loss of parts; (c) from separate floral parts to connate parts; (d) from a regular flower to an irregular flower; (e) from hypogyny to epigyny.
3. According to the Besseyan system of classification there are three branched lines of ascent from the Ranales.
4. The first line of ascent following the Ranales includes: Malvaceae (mallow family), Brassicaceae (mustard family), Solanaceae (potato family), Salicaceae (willow family), and Lamiaceae (mint family).
5. The second line comprises: Rosaceae (rose family), Fabaceae (pea family), Apiaceae (carrot family), and Asteraceae, with Cucurbitaceae (melon family) as an offshoot from Rosaceae.
6. The third line comprises the Monocotyledoneae in the following order: Liliaceae (lily family), Iridiaceae (iris family), and Orchadaceae (orchid family). The Poaceae (grass family) is an offshoot from the Liliaceae.

A APPENDIX

basic ideas of chemistry

The living body is built of very small units of matter called **molecules**. There are thousands of kinds of molecules, which differ in size, shape, behavior, and role in the body. The largest molecules carry hereditary information and are slender threads that may reach 1 mm in length. But closer to the average size are molecules of table sugar (sucrose): about three million molecules of sucrose would have to be placed end to end, to span the printed word "cube." The living body builds nearly all of its own molecules, by rearranging the parts of simpler molecules taken from the environment. The systems in the body that build new molecules are themselves collections of molecules.

What Are Molecules Made Of?

A molecule is built of still smaller units of matter called **atoms**. An atom in turn contains three kinds of material particles: **electrons, protons,** and **neutrons**. The electrons in an atom move separately around a group of protons and neutrons that are joined together to form a single unit called the **atomic nucleus**. There are many kinds of nuclei, which differ in the numbers of protons and neutrons they contain. With rare exceptions nuclei are extremely stable. They are never modified by the events that occur in the living body.

A molecule that contains nuclei of differing kinds is called a **chemical compound**. In contrast a unit of matter is called an **element** if all of its nuclei contain the same number of protons. Several elemental substances were obtained and given common names before the discovery of atomic structure: silver, gold, iron, sulfur, oxygen, and others. The nuclei of an element may differ in the number of neutrons they contain, but this difference does not influence the chemical behavior of the nucleus. The neutrons seem to be important chiefly in holding the nucleus together as a unit. (Nuclei with the same number of protons but differing numbers of neutrons are called **isotopes**. An isotope with a number of neutrons far from the average may be unstable and may spontaneously decompose, a process known as **radioactive decay**. Such a nucleus splits into parts, releasing much energy. Decay events are rare in the molecules of life and will not be considered further.)

Some of the elements most prominent in life are listed below, along with the letter symbols that chemists use to indicate their presence in a molecule:

Name of element	Symbol	Number of Protons
*Hydrogen	H	1
*Carbon	C	6
*Nitrogen	N	7
*Oxygen	O	8
Sodium	Na	11
Magnesium	Mg	12
*Phosphorus	P	15
Sulfur	S	16
Chlorine	Cl	17
Potassium	K	19
Calcium	Ca	20
Iron	Fe	26

The elements that are starred provide almost all the nuclei of biological molecules. These nuclei combine in many different combinations, along with electrons, to form thousands of compounds.

Chemists indicate the composition of a molecule by using the elemental symbols to show the kinds of nuclei present. Subscript numbers indicate how many nuclei there are of each kind. Thus the symbol CH_4 means a molecule that has four hydrogen atoms and one carbon atom (the subscript "1" is assumed if no other number is supplied).

What Holds the Parts of the Molecule Together?

Protons, neutrons, and electrons are far smaller than the atom as a whole. Protons and neutrons are about equal in weight, and both are about 1840 times heavier than an electron. Thus almost all the weight or mass of an atom is in the nucleus. Nevertheless, the nucleus is extremely tiny. If an atom were magnified to the size of a house, the nucleus would be about as large as a pinhead. Since the electrons are also very small, this means that the atom (or the molecule) is mostly empty space.

The forces that hold the nuclei and electrons together in a molecule are electrical. Each electron carries a unit of negative electrical charge, and each proton has an equal-sized unit of positive electrical charge. Neutrons carry no charge. Two particles attract one another if they carry opposite charges and repel or push one another away if they carry charges of the same sign. These forces grow stronger as the particles approach closer. (But the attraction between an electron and a nucleus turns into a repulsion if they approach too close. This keeps the electrons and nuclei from colliding.)

Because of these electrical forces, each electron and nucleus in the molecule experiences a combination of pulls and pushes from all the other particles at all times. The molecule holds together because the attraction between nuclei and electrons is enough to overcome the forces of repulsion.

A nucleus can hold in its vicinity about as many electrons as it has protons. Therefore most molecules have an equal number of protons and electrons. Such a molecule is said to be **electrically neutral**, because any force that its electrons exert on a distant object is just countered by an opposite force exerted by the protons.

Though most molecules are electrically neutral, there are many kinds of molecules that have more protons than electrons or vice-versa. These molecules are called **ions**. In symbolizing an ion, chemists indicate the number and sign of the excess charge with a superscript: for example, K^+ indicates the potassium ion, with one more proton than the number of electrons; or SO_4^{-2} for the sulfate ion, with two more electrons than protons. Ions with a positive charge are **cations**; those with a net negative charge, **anions**. Ions that have the same charge repel one another at a distance; those with opposite signs tend to approach and stay close together. When two opposite ions are in contact, chemists often say they are joined by an **ionic bond**.

What Determines the Shape of a Molecule?

Nuclei are heavy and take up definite positions in the molecule. Their locations can be shown accurately in drawings or models. But the electrons continually move at high speed around and between the nuclei. Chemists describe this situation by saying that the nuclei are embedded in a cloud of electrons. Thus molecules do not have sharp boundaries. But two molecules repel one another if they get too close, because the two electron clouds repel one another. The size of a molecule is the distance at which this repulsion becomes great enough to prevent an oncoming molecule from moving any closer.

Though electrons do not settle in definite positions, they have regions where they spend most of their time. The region where a given electron can usually be found is called an **orbital**. The orbitals do not have sharp boundaries but they can be shown in diagrams by means of shaded areas.

The nuclei act as centers about which the orbitals are

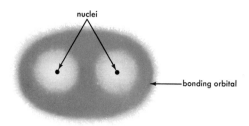

nuclei

bonding orbital

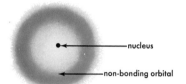

nucleus

non-bonding orbital

Figure A.2

oriented. Some of the electrons spend all their time near a single nucleus (Figure A.1). These electrons are said to be in a **nonbonding orbital**. Other electrons travel between two or more nuclei. These are **bonding electrons**. The region of space they used is called a **bonding orbital** (Figure A.2). The two nuclei that share the bonding electrons are said to be joined by a **covalent bond**. The sharing of electrons results in a powerful force holding the two nuclei at a stable distance from one another.

Covalent bonds within a molecule are shown by **structural formulas,** where the nuclei are represented by the elemental symbols and a covalent bonding orbital is shown by a line between the nuclei; for example, C–N. Electrons that are in nonbonding orbitals are usually not shown in the structural formula.

It is important to know how orbitals form around the nuclei of hydrogen (H), carbon (C), nitrogen (N), and oxygen (O), since these make up the bulk of the molecules of life. H is the simplest nucleus, consisting of just one proton. The charge on the proton is so weak that only one orbital forms around the H nucleus. C, N, and O nuclei have six, seven, and eight protons, respectively, and they exert a much stronger attraction for electrons. Each of these nuclei is the center for five orbitals. Since these three elements are almost alike with respect to the kinds of orbitals that form around them, they are considered as a group below.

One of the five orbitals is a spherical region that lies very close to the nucleus (Figure A.3). The electrons in this orbital are nonbonding electrons and are usually

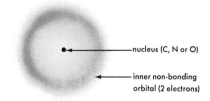

nucleus (C, N or O)

inner non-bonding orbital (2 electrons)

Figure A.3

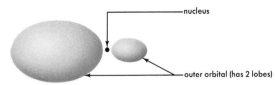

Figure A.4

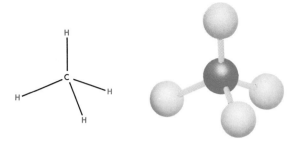

Figure A.6

omitted from diagrams of molecules. But in studying this nonbonding orbital we encounter a general rule that applies to all orbitals: *each orbital contains exactly two electrons* when the nucleus is part of a stable molecule. An orbital cannot hold more than two electrons and, if for some reason an orbital loses one of its electrons, there is a powerful tendency for the molecule to rearrange itself or to react with other molecules so that each orbital again has two electrons.

Most commonly, the four remaining orbitals all have the same shape: each is approximately dumbbell-shaped, with the nucleus at the narrow point and with one lobe much larger than the other (Figure A.4). The smaller lobe can be ignored. These four equal orbitals tend to be oriented in space so that their axes are as far from one another as possible (Figure A.5). This happens because the electrons of each orbital push those of neighboring orbitals away. (But actually the outer orbitals overlap more than the diagrams indicate, so that electrons shield the nucleus from external contact equally in all directions.)

In the case of carbon, all four of the outer orbitals are bonding orbitals. This is seen most simply with CH_4, or methane, which is shown in Figure A.6 with the structural formula and as it would appear in a ball-and-stick model. The C nucleus with its inner electrons forms the center of the molecule (Figure A.7). Each of the outer, oblong orbitals has an H nucleus embedded in one end, with the C nucleus at the other end. Therefore C forms four covalent bonds. Every H nucleus shares two electrons. These shared electrons might be considered as half-time electrons from the point of view of the hydrogen nucleus. Together, they add up to placing one full unit of negative charge near the H nucleus, to balance the proton's unit of positive charge. The C nucleus has the two electrons of the inner orbital to itself, and also shares four pairs of half-time electrons in the bonding orbitals. Therefore six electrons are likely to be near the C nucleus at any moment, balancing the six positive charges of the nucleus. Overall, the molecule is neutral; it has 10 electrons and 10 protons.

In summary, methane illustrates two important rules of molecular structure: (a) *each nucleus tends to control*

enough full-time (unshared) and half-time (shared) electrons to equal its own charge; and (b) *this is achieved in a way that places two electrons in each orbital.* The same rule applies to all other cases where C occurs in biological molecules, and it applies equally to H, N, and O.

Notice also that the four bonds made by a C atom tend to be rigidly spaced; the angle between any two of the bonds is about 109°. This consequence of orbital repulsion does much to fix the internal geometry of the molecule. No drawing of a molecule on a flat page can adequately show these angles, so chemists frequently resort to physical models by using parts such as balls to represent nuclei and springs or sticks to show covalent bonds.

The structural rules seen above can explain why nitrogen (N) nuclei commonly form three covalent bonds, as in ammonia (NH_3) (Figure A.8). The N nucleus has four electrons to itself: two in the inner nonbonding orbital and two in an outer orbital that are not shared with other nuclei (this is also a nonbonding orbital) (Figure A.9). Also, the N nucleus shares six electrons in the three bonding orbitals. Counting these half-time electrons, there are a total of seven negative charges near the N nucleus at any moment, matching the seven protons of the nucleus. The N nucleus forms only three covalent bonds rather than four because the extra proton in the N nucleus is sufficient to hold two electrons in one of the outer orbitals without sharing with another atom. As in methane, there are 10 protons and 10 electrons in the ammonia molecule. Notice that the usual repulsion between the orbitals occurs, so that the three bonds between N and H form a pyramidal arrangement.

Similar rules apply to the behavior of oxygen (O), as illustrated by water (H_2O) (Figure A.10). Oxygen forms only two bonds rather than four because the O nucleus has enough protons to hold pairs of electrons in two of the outer orbitals without having to share with other nuclei (Figure A.11). The eight protons of the O nucleus are

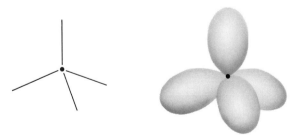

Figure A.5

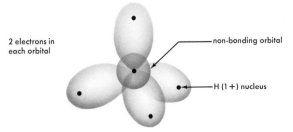

Figure A.7

Figure A.8

two non-bonding orbitals (each has 2 electrons)

Figure A.9

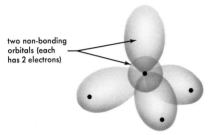

Figure A.10

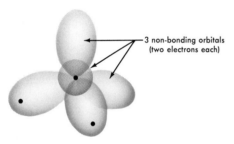

3 non-bonding orbitals (two electrons each)

Figure A.11

balanced by six full-time electrons in the three nonbonding orbitals, plus four half-time electrons that are shared in the bonds between O and H. Again, there are 10 protons and 10 electrons in the molecule.

A less common but still important orbital arrangement is seen in molecules like ethylene, C_2H_4 (Figure A.12). Each C nucleus has the usual pair of inner, nonbonding electrons and shares two pairs of electrons in bonds with H nuclei. But also, *four* electrons are shared between the two C nuclei. The sharing of four electrons is known as a **double bond**. Chemists represent a double bond by two lines between the nuclei. Double bonds such as C=N, C=C, and C=O are common in the molecules of life.

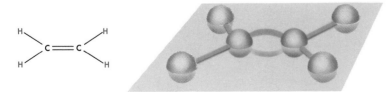

Figure A.12

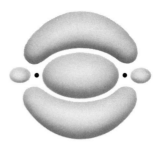

Figure A.13

Since an orbital can contain only two electrons, the double bond consists of two orbitals (Figure A.13). One of the orbitals has three parts: an oval region between the two C nuclei plus a small lobe on either end. The other orbital consists of two sausage-shaped regions. The two electrons in this orbital spend part of their time in one sausage-shaped region; part of the time in the other. Remember that the diagram of an orbital represents only the places where the electrons spend *most* of their time; they can pass between two lobes but they spend little time in the intervening space.

When two nuclei are joined by a double bond, all their other bonds lie in a single plane. This is a rigid arrangement that is maintained because the orbitals in the double bond resist being twisted. By contrast, a single bond forms an axis about which the joined parts of a molecule may freely rotate. Figure A.14 illustrates this arrangement with the substance ethane, C_2H_6. Rotation does not change the angle between two neighboring bonds, but rotation lends a good deal of variability to the shape of a molecule if there are several nuclei joined into chains by single bonds.

How Molecules Behave: Polarity, Hydrogen Bonds, and Solubility

Molecules may attract or repel one another because of their electrical charges. This is important in establishing how molecules will behave as a group in the body mass.

Even neutral molecules may have local regions of positive and negative charge. This happens because electrons that are shared between two unlike nuclei usually spend more of their time closer to one nucleus than the other. A measure called **electronegativity** expresses the tendency of a nucleus to dominate in sharing electrons: a higher value means a stronger tendency to hoard the electrons. Electronegativities for some common elements are:

Element	Electronegativity
O	3.5
N	3.0
C	2.5
S	2.5
H	2.1
P	2.1

Because of these differences in electronegativity, a molecule that contains a mixture of nuclei tends to be negative near the N and O nuclei, and positive near the C, H, P, and S nuclei. A molecule that has such internal charge separations is said to be **polar**. There is a degree of polarity in every bond between two unlike nuclei, but between C and H the polarity is insignificant whereas in bonds such as O–H, C–O, C–N, and N–H the polarity is high enough to affect the behavior of the molecule toward other molecules. Water is a strongly polar molecule. The local charge concentrations are less than the value of a whole electron, and are shown in diagrams by the symbols $\delta(-)$ and $\delta(+)$:

The H of one water molecule is attracted to the O of a neighboring water molecule, because of their opposite charges (Figure A.15). This attraction represents a weak bond called a **hydrogen bond**. It is only about 1/16 as strong as the covalent bond between the H and O within a water molecule. A hydrogen bond is indicated in diagrams by a dotted line.

Each of the two outer nonbonding orbitals in a water molecule can become involved in a hydrogen bond, and each H of the molecule can form another hydrogen bond. Therefore each water molecule can, and does, form as many as four hydrogen bonds with neighboring water molecules or other polar molecules that may be present. The force exerted by these bonds allows water molecules to move about only by breaking and re-forming hydrogen bonds one at a time.

Other molecules besides water can form hydrogen bonds. This happens when a hydrogen nucleus that is sharing electrons with O or N comes close to another O or N, with the three nuclei approximately in a straight line. Since most biological molecules contain O–H or N–H, they can make hydrogen bonds with one another or with water.

A mass of pure substance consists of many molecules of the same kind packed regularly (a crystal) or irregularly (an amorphous solid or liquid). On contact with water, the molecules may leave such a mass to become surrounded by water molecules. The mass is said to **dissolve**; the molecules that become embedded in the water are called **solute** molecules; the water is said to act as a **solvent**; and the resulting mixture of molecules is a **solution**. The number of solute molecules in a given volume of solvent is the **solute concentration**. Concentrations are usually expressed in terms of **molarity**. For example, a solution may be 0.01 molar, abbreviated as 0.01 *M*. A 1 *M* solution contains, by definition, 6.023×10^{23} molecules of solute (one mole) in one liter of solution. A 0.01 *M* solution contains 1/100 as much solute as a 1 *M* solution, and so on.

For most solutes, there is a limit to the concentration of solute molecules that a given solvent will accept. Excess solute molecules spontaneously aggregate to form masses such as crystals. A solution with the limiting solute concentration is said to be **saturated**. The **solubility** of a substance is the concentration at the point of saturation. Solubility is high when the solute molecules are attracted

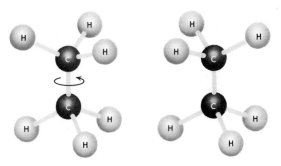

Figure A.14   The substance ethane, C_2H_6, as it would appear in a ball-and-stick model. One of the H atoms has been shaded to show how rotation around the axis of the central (C–C) bond can affect the internal geometry of the molecule.

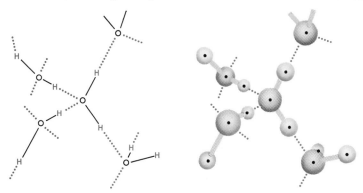

Figure A.15   Several water molecules joined by hydrogen bonds. (Dotted lines represent hydrogen bonds.) On the left, structural formulas; on the right, the arrangement as it would look using ball-and-stick models.

to solvent molecules more strongly than they are attracted to one another and more strongly than solvent molecules attract one another.

A general rule governing solubility is that *water will accept polar but not nonpolar molecules*. Polarity is a factor because the water molecules form a network, joined by hydrogen bonds between one another. A solute can move into this network if it can form hydrogen bonds with water; that is, if it is polar. If not, the network of water molecules resists being parted to make room for the solute molecules. Ions often are particularly soluble: solids composed of positive and negative ions, such as NaCl, often readily break up in contact with water because each ion acquires a blanket of oriented water molecules.

Masses of nonpolar molecules dissolve freely in solvents such as oil or gasoline, which are themselves made of nonpolar molecules. There are attractions between the polar solute molecules in a mass that prevent them from leaving the mass to be surrounded by nonpolar solvent molecules. These facts may be summarized by the dictum, "like dissolves like." But molecules differ widely in their degree of polarity and correspondingly they differ widely in the degree of solubility in various solvents.

How Molecules Behave: Chemical Reactions

A **chemical reaction** is an event in which new kinds of molecules are built by rearranging the parts of old molecules. Chemists symbolize a reaction by means of an arrow, with the initial molecules (**reactants**) at the tail and the final molecules (**products**) at the head.

Some common types of reactions are illustrated in Figure A.16. The simplest are **rearrangements** or **isomerizations**, in which a single molecule undergoes an internal reorganization. There are also **addition** reactions, where two molecules join to form a larger molecule. The reverse is a **decomposition** reaction. The most common kind of reaction is the **transfer**, in which two molecules meet and exchange or transfer parts to form two new molecules.

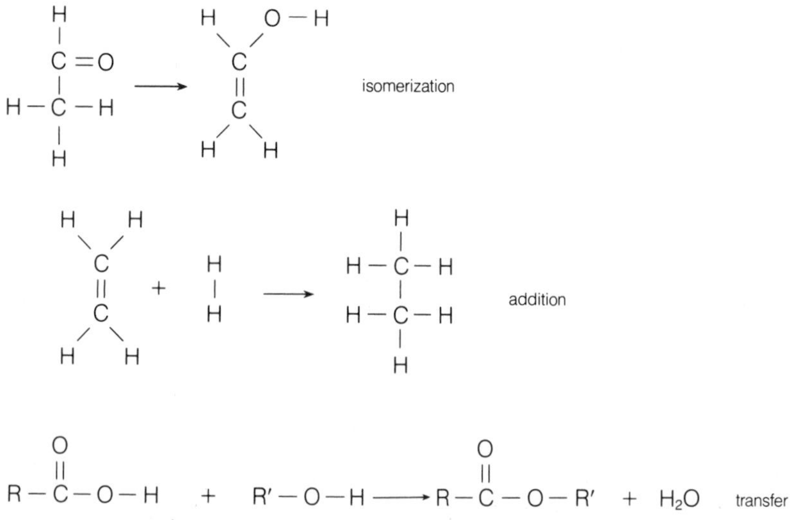

What governs the occurrence of chemical reactions? Most biological molecules are stable and resist change. This is because the nuclei and electrons fall into an arrangement that is the best compromise between the forces of attraction and repulsion. Any small change from this condition will create an imbalance of forces that tends to restore the initial condition. As illustrated in Figure A.17, there may be several different stable arrangements that a given collection of nuclei and electrons may fall into, given the chance. Some of these arrangements are more stable than others, but all have a degree of stability for the reasons given above. To go from one arrangement to another, the molecule would have to pass through an intermediate arrangement that can be reached only by working against electrical forces. The intermediate condition is a highly unstable arrangement that will spontaneously give way to any other arrangement where the forces are again balanced. A molecule that is in the process of changing is called an **activated intermediate**.

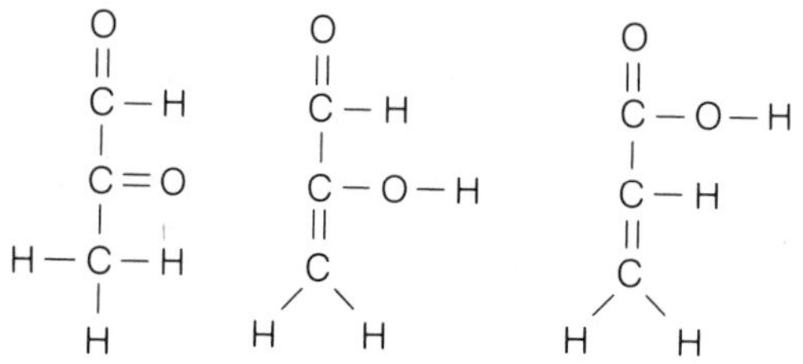

Figure A.17 Three possible structures that could occur with the formula $C_3H_4O_2$.

A collision between two molecules may set the stage for an exchange of parts, as illustrated in Figure A.18. But the exchange first requires the two molecules to form an unstable **activated complex**.

Activated complexes form only when outside forces push the molecules out of their stable arrangements. The required force is most commonly provided by the force of collision between two molecules. Momentum carries the

Figure A.16 Examples showing three general types of reactions.

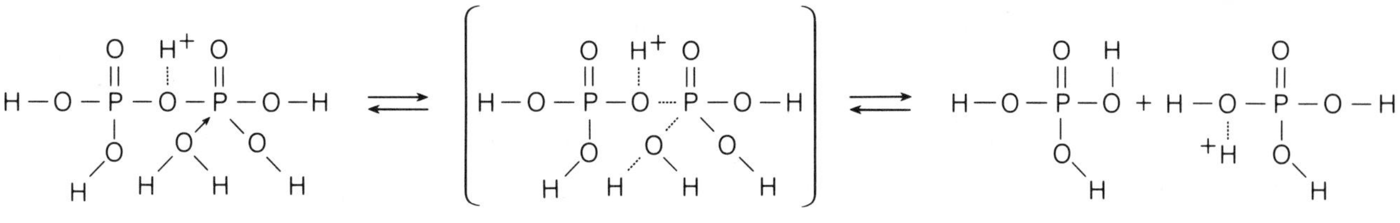

activated complex

Figure A.18 The reaction between pyrophosphoric acid and water, illustrating an activated complex. The bonds shown with dotted lines in the activated complex are those that will be affected by the reaction. As written, the reaction is a hydrolysis, one of several kinds of transfer reactions.

molecules together until their shapes become distorted, just as a basketball is flattened on one side at the moment of impact with the floor. Sufficient distortion will bring the molecules into the activated state enabling a chemical reaction to follow. A lesser degree of distortion allows the colliding molecules to rebound without reacting.

Chemists routinely use energy concepts to discuss chemical reactions. A typical graph of the energy relations in a reaction is shown in Figure A.19. The molecules contain a certain amount of **stored** or **potential energy** in their stable condition. The activated complex contains more potential energy. This reflects the fact that force must be applied (work must be done) to bring the stable molecule into the activated condition. Moving molecules have **kinetic energy,** or energy of motion, in addition to their stored potential energy. A collision between two molecules converts this kinetic energy into potential energy: the colliding molecules are momentarily at rest in a distorted state. When the molecules rebound from the collision, some of the potential energy is converted back into kinetic energy. In this process the total amount of energy remains constant. Energy cannot be created or destroyed by the reactions in life, though the form of the energy may be changed.

The products of a reaction may contain, between them, either more or less potential energy than the original reactants; the difference is made up by a compensating change in the amount of kinetic energy carried by the molecules. In Figure A.19 the products have less potential energy and therefore more kinetic energy than the reactants. The kinetic energy of the molecules is known as **heat.** Thus heat is produced as the reaction in Figure A.19 progresses.

The difference in energy between the activated intermediate and the separate molecules is termed the **activation energy.** This is the amount of energy that must be supplied in the collision if there is to be a reaction event. The higher the hill, the less likely is the collision to produce a reaction.

Whether the collision causes a reaction will depend on the nature of the molecules, their orientation when they hit, and how hard they hit. Most collisions at the temperatures of life fail to bring about chemical reactions because of defects in one of these factors.

In principle, all chemical reactions are **reversible.** Using Figure A.19 as an example, if A and B can collide to form an activated complex, so can C and D collide and form the same kind of activated complex. The latter can break apart to form either A and B, or C and D. The back reaction would look the same as if a film of the forward reaction were being run backward. As shown in Figure A.18, reversibility is shown in reaction diagrams by putting two opposed arrows at each step.

Which of the two opposed reactions will run the faster? The rate of a reaction depends on the frequency of collisions, hence on the concentrations of the reacting molecules. If we refer again to Figure A.19, the reaction from left to right depletes the supply of A and B so that the reaction rate declines with time. The products C and D accumulate and collide with one another increasingly often, raising the rate of the back reaction. Thus the overall rate, the difference between forward and back reactions, depends on the prevailing balance of concentrations. **Equilibrium** occurs when the concentrations of all the molecules are just sufficient for the forward and back reactions to run at the same rate. At equilibrium the overall rate is zero; the opposed reactions cancel.

What determines the balance of concentrations at equilibrium? In most reactions (Figure A.19) there is a difference in potential energy between the reactant and product molecules. This gives the back reaction a different activation energy requirement than the forward reaction: in the reaction of Figure A.6, the back reaction has a higher energy hill to climb. A higher activation energy requirement means that fewer collisions will have enough energy to form the activated complex. Therefore more collisions must occur per unit time if the back reaction rate is to equal the forward reaction rate. The rates will only be equal when there are more of the low-energy molecules than the high-energy molecules. In Figure A.19, C and D

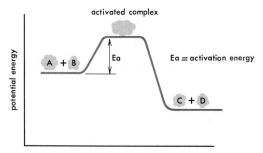

Figure A.19 Potential energy changes that occur during a reaction event in which molecules A and B collide, form an activated complex, and react to form C and D. Graph shows the potential energy of (A+B), the activated complex, and (C+D).

will be more abundant than A and B at equilibrium.

These ideas suggest that reactions should have the overall effect of replacing molecules that are rich in potential energy with molecules that are lower in potential energy. Nevertheless, living bodies do form many molecules that are higher in potential energy than the raw materials from which they are made. This is possible because the energy relations discussed above refer only to the *sum* of potential energy, taking both product molecules together. As long as this sum is greater for A and B than for C and D, the latter will tend to be the more abundant at equilibrium. If C is very low in potential energy, D may be richer in potential energy than A or B.

In biological systems, reactions rarely come to equilibrium because various materials are continually being withdrawn or supplied by other reactions. Also, back reactions are often insignificant because their activation energy requirements may be too high to meet in collisions at body temperature.

Special Reactions: Acids and Bases

Acidity results from a special type of transfer reaction, involving water and a hydrogen nucleus (proton). When two water molecules are joined by a hydrogen bond, occasionally the hydrogen nucleus jumps from one molecule to the other:

hydronium   hydroxyl

The products are H_3O^+, known as **hydronium**, and OH^-, called the **hydroxyl** ion. A meeting between hydronium and a hydroxyl ion usually results in a return to the two water molecules: the ionization of water strongly favors the two un-ionized molecules.

The concentration of hydronium ions determines the acidity of the solution. More hydronium means a more acidic solution. Chemists express the hydronium concentration by a measure known as the **pH**. The scale runs from 0 to 14. A change of 1 unit on the pH scale is a ten-fold change in hydronium concentration. The lower the pH, the higher is the hydronium concentration. Thus a solution with pH 2 has 10 times as much hydronium as a solution with pH 3. Pure water has a pH of 7. Most living material have a pH value near 7. Stomach juice in human beings is very acidic, with a pH around 1. Even at this strongly acidic pH, though, there are many more neutral water molecules than hydronium ions in the solution: pH 1 represents about 1 hydronium for every 180 neutral water molecules.

An **acid** is a compound that *raises* the hydronium concentration in water. Any compound that can donate a proton to a water molecule is an acid. If we use the symbol HA for a generalized acid, the ionization is:

$$HA + H_2O \longrightarrow H_3O^+ + A^-$$

Acids vary widely in strength. The very strong acid HCl (hydrochloric) completely breaks up into Cl^- and H_3O^+ when it contacts water. A weak acid remains mostly un-ionized, as HA. Water and acetic acid (which causes the sourness in vinegar) are both weak acids.

A **base** is a molecule that **reduces** the concentration of hydronium in water. For example, OH^- is a base. There are several ways in which a substance can decrease the hydronium concentration. Thus B^- can be an ion that reacts as:

$$B^- + H_3O^+ \longrightarrow BH + H_2O$$

The OH^- ion acts in this way. As another example, BOH may be a compound that ionizes in water to release OH^-. The OH^- then reacts as shown above.

Instead of talking about hydronium, chemists often abbreviate by limiting their attention to the H^+ ion itself, ignoring the part played by H_2O in carrying the proton (as H_3O^+). From this viewpoint, pH refers to the H^+ concentration. But free H^+ never occurs in the solution.

Functional Groups

The molecules of life are often large and complex. But they are easier to understand when we use the concept of **functional groups**. A functional group is a part of the molecule that is especially likely to participate in chemical reactions.

The functional groups **carboxyl**

and **phosphoryl**

act as acids. These can donate H^+ to water:

The carbon of carboxyl and the phosphorus of phosphoryl tend to have a partial positive charge; the attached oxygens have partially drawn away the shared electrons, leaving the C and P nuclei relatively exposed to attack by other molecules. Therefore these nuclei tend to be involved often in chemical reactions. By contrast, C nuclei that are only bonded to other C or H nuclei are less subject to chemical attack.

The **amino** group acts as a base. The

nonbonding orbital of the N atom can accept a proton from water or hydronium:

$$R-\underset{\overset{|}{H}}{\overset{H}{N}} \ + \ H_3O^+ \ \longrightarrow \ R-\underset{\overset{|}{H}}{\overset{H}{\underset{|}{N}}}{}^+\!\!-H \ + \ H_2O$$

Hydroxyl (—O—H) and **carbonyl** ($\overset{\backslash}{\underset{/}{C}}=O$) are also functional groups. They do not form ions appreciably unless they are joined together as carboxyl, but the oxygen atoms in these groups pull electrons from adjoining C nuclei, creating polarity and exposing the C nuclei to attack. A common reaction is that of a carboxyl with a hydroxyl group, to give a product called an **ester** plus a water molecule (See Figure A.16.) Such reactions, which give off a water molecule, are called **condensations**. The opposite, cleavage of a molecule by adding water, is a **hydrolysis** reaction. An example of a hydrolysis is shown in Figure A.18. Both condensations and hydrolysis reactions are transfer reactions.

Oxidation and Reduction

An **oxidation-reduction** is a reaction in which electrons are transferred from one group or molecule to another. The entity that loses the electrons is said to be **oxidized**; the one that gains the electrons is **reduced**. Often such a reaction also involves a transfer of atomic nuclei.

A simple example is the ionization of chlorophyll, an important event in photosynthesis. After absorbing a unit of light energy, the chlorophyll molecule passes an electron to a compound that acts as an electron acceptor:

$$\text{chlorophyll} + \text{acceptor}$$
$$\rightarrow \text{chlorophyll}^+ + \text{acceptor}^-$$

Here chlorophyll is oxidized and the acceptor is reduced.

Somewhat more complex is the formation of the compound NADPH from NADP$^+$, another event in photosynthesis:

$$2Z^- + NADP^+ + H_3O^+ \longrightarrow 2Z + NADPH + H_2O$$

In this reaction, which probably proceeds by several steps, NADP$^+$ receives two electrons, with Z as the electron donor. NADP$^+$ also gains a proton. The donor Z$^-$ is oxidized; NADP$^+$ becomes reduced.

Oxidation and reduction can also occur during an internal reorganization of a molecule. For example:

$$H-\underset{\overset{|}{\underset{R}{\overset{|}{C}=O}}}{\overset{\overset{OH}{|}}{C}}-H \quad \longrightarrow \quad \underset{H-\underset{\overset{|}{R}}{\overset{|}{C}}-OH}{\overset{\overset{O}{\|}}{C}-H}$$

Since O nuclei have a high affinity for electrons and will hoard the electrons that they share, the change in bonding that occurs in this reaction causes the uppermost C to lose a fraction of a negative charge (it is oxidized) while the lower C makes a matching gain in negative charge (it is reduced). The molecule as a whole is neither oxidized nor reduced since it has not lost or gained electrons.

As the last example illustrates, the electron-attracting capacity of O causes groups to be oxidized when O becomes attached to them, whereas a replacement of O by H is a reduction because H tends to donate electrons. Oxygen nuclei rarely are poorer in electrons than when two of them are competing for electrons in an O$_2$ molecule. With its high electronegativity, O will readily give up its position in O$_2$ to form bonds with C and H, which do not compete as well for electrons. The possibilities for drawing electrons from C and H have been maximally exploited by O in the molecules CO$_2$ and H$_2$O. Therefore these are very stable molecules. In them, C and H are highly oxidized and O is highly reduced. Compounds such as sugars,

$$H-\underset{\overset{|}{OH}}{\overset{\overset{H}{|}}{C}}-\underset{\overset{|}{OH}}{\overset{\overset{H}{|}}{C}}-\underset{\overset{|}{OH}}{\overset{\overset{H}{|}}{C}}-\underset{\overset{|}{H}}{\overset{\overset{OH}{|}}{C}}-\underset{\overset{|}{OH}}{\overset{\overset{H}{|}}{C}}-\underset{\overset{|}{H}}{\overset{\overset{}{}}{C}}=O$$

which have C—C and C—H bonds, contain C and H nuclei that are richer in electrons and therefore are more reduced than the C and H in CO$_2$ and H$_2$O. A compound such as sugar is highly reduced. The more bonds C and H have with one another rather than with O, the more reduced the compound is. Reduced compounds spontaneously react with O$_2$ (with a suitable catalyst) to give products in which the C and H are more fully combined with O and therefore are more highly oxidized. This occurs because it transfers electrons from C and H to O. The reverse, the removal of O, is unfavorable unless energy is supplied by the environment. Therefore in the presence of O$_2$, reduced compounds are rich in energy as compared to oxidized products. The stored energy is often called **redox energy** and may be defined as energy associated with an unfavorable distribution of electrons between the nuclei. This redox energy is released when a reaction allows the electrons to assume a more favorable arrangement.

genetics supplement

In many flowering plants, segments of a stem can be rooted to give a set of plants with identical heredity, a **clone.** Clones occur widely in nature because many plants can reproduce vegetatively (e.g., by runners or rhizomes). But plants of almost all species also reproduce sexually, by means of **matings.** In such a mating (Figure B.1), meiosis in one diploid parent produces haploid egg cells, while meiosis in the other parent produces haploid nuclei that are incorporated into pollen grains. The egg cells and the generative cells of pollen are **gametes.** Egg and pollen nuclei meet and fuse in the act of **fertilization,** to yield a diploid **zygote.*** The zygote is the first cell of the offspring or **progeny** plant. It contains all the hereditary information that will guide the young plant's development. Half of that information came from the pollen, and half from the egg.

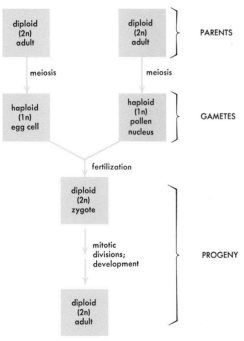

Figure B.1

With a clone, one knows exactly what to expect from one generation to another; each generation will be identical to the preceding one, except for variations that the environment causes. But matings make predictions more difficult. The progeny plant can be expected to differ from both of its parents in many ways, unless the two parents carried identical hereditary information. A major

* The details of meiosis are presented on pp. 38–43, while gamete formation and fertilization are described on pp. 130–140.

goal of genetics has been to deduce the laws that govern inheritance in matings. This is useful in plant breeding, where one must plan breeding programs to give predictable improvements in the plants; it is also useful in exploring the fundamental mechanisms of heredity.

A good breeding program requires plants whose matings can be controlled. The possibility of accidental pollination by wind or insects, or of self-fertilization, must be considered. In self-fertilization, pollen and eggs come from the same parent.

Success in a breeding program also depends on the choice of **traits** to be studied. Traits are observable characteristics such as color of petals and shape of leaves. Some traits are easier to study than others. For example, the weight of seeds produced by a plant is partly determined by heredity, but also it is strongly influenced by the environment. Even in a constant environment, though, seed weights show many intergrades between the heaviest and the lightest. Seed weight, oil or starch content, growth rates, and so on are **quantitative traits.** Such traits commonly show **continuous variation,** in which the differences between plants are distributed widely and evenly about a mean (Fig. B.13). Traits with continuous variation are hard to study because the progeny of matings are hard to classify and patterns of heredity are complex.

By contrast, some traits show **discontinuous variation.** Thus a population of garden pea plants might contain some individuals that form red flowers and some with purple flowers, but no intergrades. Traits that show discontinuous variation are easy to study because the progeny can be assigned to definite classes and the various types occur in simple ratios. Historically, studies of discontinuous traits gave us the key to understanding heredity. Today, it is still easiest to learn genetics by starting with discontinuous traits.

Traits That Show Discontinuous Variation

The first systematic breeding program was done in the mid-nineteenth century by Gregor Mendel. He chose the garden pea, partly because many varieties of peas were available and partly because their matings could be controlled. The stigmas mature before the anthers and before the flowers open. At this stage the flowers can be pried open, the immature anthers can be removed, and pollen from another plant can be applied to the stigma.

Fertilization occurs in the closed flower, with the stigma shielded from stray pollen.

Each variety of peas that Mendel studied had its own set of unique hereditary features. In a field sown with seeds of one variety, all the progeny for generation after generation would have the features typical of that variety. The plants in such a field are said to "breed true" for the traits under study; collectively, the plants form a **true-breeding strain**.

The Monohybrid Cross

The problem in genetics is to explore what will happen when plants of two true-breeding strains are mated. This is easiest to approach if the plants differ in only one trait. Matings of this kind are called **monohybrid crosses**. (Actually, the parental plants may differ in other characteristics as well, but so long as attention is confined to just one trait, such as flower color, the mating is a monohybrid cross.)

Figure B.2 shows the kind of results that Mendel obtained in monohybrid crosses with peas. The original parents are called the **P Generation**. One parental plant in Fig. B.2 came from a strain that always produces round seeds; the other from a strain that always forms wrinkled seeds. The progeny are the **First Filial or F₁ Generation**. In this cross, Mendel found that all of the F₁ plants formed round seeds. This seems surprising at first, because having received hereditary information for making both round and wrinkled seeds, one might expect some of the progeny to form wrinkled seeds. Either the hereditary information for wrinkled seeds has been suppressed, or it never entered the zygote during fertilization.

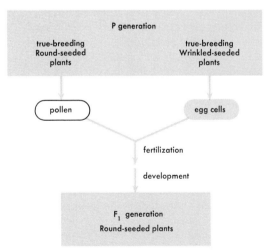

Figure B.2

An additional cross (Fig. B.3) proves that the information for making wrinkled seeds was indeed present in the F₁ plants. Here the F₁ plants were mated with one another. In such a cross, the F₁ plants are said to be **self-crossed** or **selfed**, and their progeny are the **F₂ Generation**. Although all the F₁ plants formed round seeds, some of the F₂ plants made wrinkled seeds. This

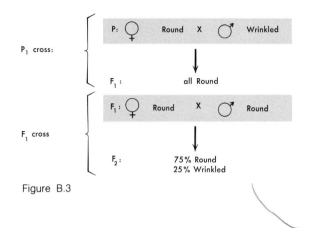

Figure B.3

shows that the F₁ plants did contain the information for wrinkled seeds, but the information was suppressed.

To explain the patterns of inheritance that he saw, Mendel proposed the scheme shown in Fig. B.4. He suggested that the hereditary information for seed shape is carried by particles. There were two versions of the particle. One kind (call it **R**) carries information for making round seeds. The other (**r**) carries information for making wrinkled seeds. Mendel proposed that each gamete carries one of these particles, so that the zygote contains two of them. To account for the suppression of the "wrinkled" trait, he proposed that the factor **R** is **dominant** over r when the two are both present in the same organism. That is, the presence of R prevents r from affecting the shape of seeds.

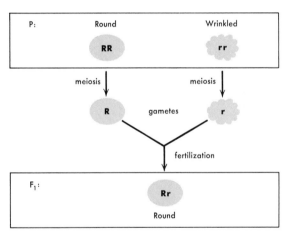

Figure B.4

Today, Mendel's particles are called **genes**. They are known to be made of the substance **DNA**. DNA can be modified in many ways, though new modifications (**mutations**) in any one gene occur only rarely (see Chapter 2 and pp. 215–217). The different versions of the gene proposed by Mendel are slightly different segments of DNA, carrying information for the same trait. Different versions of the same gene are now called **alleles**. Because there are many ways to mutate a gene, many different alleles are possible. But a diploid cell carries only two copies of each gene, so that only two alleles for that gene can be present. (There are exceptions to this rule, but they can be ignored in elementary studies.)

In modern language, we would rephrase Mendel's proposal by saying that allele R (round seeds) is dominant over allele r (wrinkled seeds); allele r is said to be **recessive** to R. This pair of alleles is said to show **classical dominance**. We have not explained *how* R suppresses r; we have merely developed a picture that helps to predict the progeny in matings. Classical dominance has been found between many pairs of alleles, affecting many traits in many plant and animal species.

The idea of dominance between alleles explains why some of the F_2 plants produced wrinkled seeds (Fig. B.3). The analysis is shown in Fig. B.5. Each nucleus in the parental plants contains the allele combination Rr. Meiosis separates the alleles, so that half of the gametes contain allele R and half contain allele r. Gametes with these alleles meet at random in matings, so that zygotes are formed with all the possible combinations. These are shown in the four boxes in Fig. B.5, along with the seed shapes that the progeny will form. Three of the four combinations result in round-seeded plants; the fourth combination, rr, yields plants with wrinkled seeds.

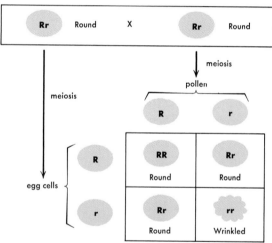

Figure B.5

To be useful, genetic analyses should predict not only the kinds of progeny that will occur, but also their relative abundance or **frequency**. Looking back to Fig. B.3, one can see that 25% of the progeny formed wrinkled seeds. Why 25%? To understand this, one must be aware that chance plays a major part in bringing the gametes together. The contained alleles do not affect a gamete's chances of meeting and fusing with another gamete. Since meiosis forms equal numbers of R eggs and r eggs, the laws of chance assure that the two types of eggs will be involved in matings equally often. Thus half the progeny should contain an r egg nucleus. By chance, half of these will combine with pollen that carry allele r, since meiosis also assures that r pollen are just as abundant as R pollen. Half of half is 25%, hence 25% of the F_2 plants are the rr type, which form wrinkled seeds. The frequency of a given progeny type is set partly by the mechanics of meiosis, which fixes the frequency of gamete types; and

partly by the laws of chance, which govern the random combinations of the various gamete types.

As anyone knows who has played games with dice or cards, the laws of chance leave room for variation. With many tosses of a coin, "heads" will come up very nearly 50% of the time. But with just a few tosses the percentage may be quite far from the ideal 50% figure. Likewise in counting the progeny of a mating, we do not expect the numbers to be exactly as predicted; deviation is likely to be greater, the smaller the number of progeny. Geneticists have adopted statistical methods that help to estimate how far the real results of a cross are likely to deviate from ideal predictions.

A note on terminology: the combination of alleles present in a plant is known as the **genotype**, while the observable trait itself—seed shape in this case—is the **phenotype**. True-breeding plants have two copies of the same allele, RR or rr. They are said to be **homozygous** for this gene. Plants that carry two different alleles, such as Rr, are **heterozygous**.

The Testcross

The dominance mechanism makes its easy to recognize plants that are homozygous for the recessive allele, because they are the only plants that show the recessive phenotype. But plants with genotypes RR and Rr are identical in appearance; both produce round seeds. How can we deduce the true genotype of a round-seeded plant, so as to make accurate predictions in matings that use the plant?

The easiest approach is to use a **testcross**. In a testcross, one parent has only recessive alleles for the trait being studied. The other parent has the dominant phenotype, but its genotype is unknown. The testcross for seed shape in peas is diagrammed in Fig. B.6. There it has been assumed that the unknown parent has the genotype Rr. The recessive parent forms only gametes with the allele r. The other parent produces gametes with allele R and an equal number with allele r. Only two progeny genotypes are possible; they are shown in the boxes. If the unknown parent is heterozygous, half the progeny form round seeds and half form wrinkled seeds. If the unknown parent were homozygous RR, all the progeny in the testcross would have formed round seeds. The testcross is useful because the alleles contributed by the recessive parent do not interfere with the expression of the other parent's alleles. The progeny phenotypes directly reveal the unknown parent's gamete types. From this information we can easily deduce the unknown parent's genotype.

Incomplete Dominance

Though many traits show classical dominance when pairs of alleles are brought together in matings, many others do not. One exception is shown in Fig. B.7. When plants from a red-flowered strain of snapdragon *(Antirrhinum)* are crossed with plants from a white-flowered strain, the F_1 progeny all have pink flowers. Here Mendel's idea of dominance does not apply. But the F_1 plants still do contain the information for making red and white flowers.

Traits That Show Discontinuous Variation

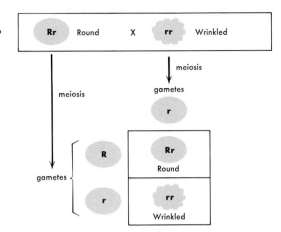

P

Figure B.6

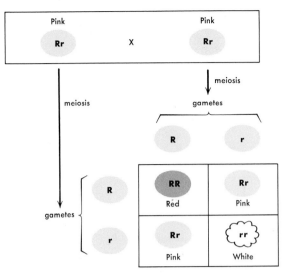

Figure B.8

When the F₁ snapdragons are selfed (Fig. B.7, bottom), half the progeny form pink flowers, but the rest produce red or white flowers.

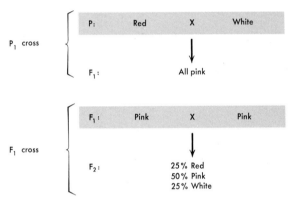

Figure B.7

Although dominance does not apply, these results can be explained easily with the allele theory of inheritance. In Fig. B.8, the red-flowered plants are assigned the genotype RR; white-flowered plants have the genotype rr. The F₁ progeny all have the genotype Rr, which is found by observation to give a pink-flowered phenotype. In the next cross, the pink-flowered parents (with genotype Rr) form equal numbers of R and r gametes. These unite at random so that four combinations result. One combination (RR) yields red-flowered plants; two combinations (Rr) give pink, and the fourth combination (rr) produces white-flowered plants. One would expect 25% red, 50% pink, and 25% white. This matches the observed progeny in the F₂ generation (Fig. B.7).

Pairs of alleles such as this one, where the heterozygous condition gives a different phenotype than either of the two homozygous phenotypes, are said to show **codominance** or **incomplete dominance**.

Dihybrid Crosses and Linkage

Many breeding programs are concerned with bringing two or more qualities together in the same plant. For example,

a strain of corn with high resistance to fungus attack but with poor grain yield may be crossed with another strain that has low fungus resistance but good yield. Such a cross, involving two traits, is a **dihybrid cross**. The goal would be to produce plants that yield well and are disease-resistant. But the progeny might also include plants that combine poor yield with poor resistance. How many of the progeny can be expected to have the desirable combination of features? The answer will determine the effectiveness of the breeding program.

In a monohybrid cross, we were dealing with only *one* gene that had two alleles. In dihybrid crosses, *two* genes are involved, each with two alleles. It is important to distinguish clearly between "genes" and "alleles" because confusion between these terms will make it impossible to understand inheritance in dihybrid crosses.

The genes are carried on chromosomes, with many genes per chromosome, governing many traits. Each kind of chromosome in the haploid set carries a particular group of genes, which are arranged in a definite sequence along the chromosome. Thus two genes on the same chromosome are physically coupled together. They are said to be **linked**. Two genes that are carried on different chromosomes are **unlinked**. Linkage affects the formation of gametes, which in turn limits the occurrence of certain progeny types. Therefore one must know whether two genes are linked or not, if good predictions are to be made about their behavior in matings.

In garden peas, the shape and color of the seeds are governed by a pair of unlinked genes. As to shape, round (R) is dominant over wrinkled (r). In seed color, yellow (Y) is dominant over green (y). Figure B.9 shows how a plant that is heterozygous for both genes produces gametes. Remember that meiosis is preceded by a duplication of the chromosomes, so that each chromosome entering meiosis has two identical chromatids.* Then in prophase of the first meiotic division, homologous chromosomes join together. The resulting **tetrad** is a group of four

---

* It may be worthwhile to review Fig. 3.32 and the associated text on pp. 42–43, where meiosis is shown in more detail).

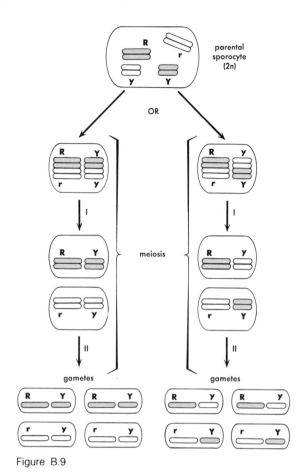

Figure B.9

Figure B.10

chromatids. In Fig. B.9 the long chromosomes form one tetrad; the short chromosomes form another. The tetrads then line up at the equator of the spindle. But tetrads move independently of one another. Therefore the two tetrads can assume two alternative positions relative to one another. In one arrangement (the left branch in Fig. B.9), meiosis gives two gametes with genotype RY and two with genotype ry. The other arrangement (right branch, Fig. B.9) results in two gametes with genotype Ry and two with genotype rY. Only one of the possible arrangements can occur in any one cell that undergoes meiosis. But meiosis occurs in many cells. Chance governs the positions that the tetrads take, so that the two arrangements will be equally common. *Therefore, if the two genes are unlinked, the four gamete types will be equally numerous.*

Next, let us consider the formation of gametes when the two genes are linked. This can be illustrated with another pair of genes in the garden pea: flower color, with purple (P) dominant over red (p); and pollen shape, with long (L) dominant over round (l). Figure B.10 shows the formation of gametes in a plant that is heterozygous for both genes. Here we assume that the parent has allele P on the same chromosome as allele L, and allele p is on the same chromosome as allele l.

Most cells that undergo meiosis will follow the sequence of events in the left branch of Fig. B.10. Pairing brings the homologous chromosomes together in a tetrad. The coupling between P and L is maintained, as is the

coupling between p and l. When meiosis is complete, half the gametes have the genotype PL and half have the genotype pl.

Some cells take the right branch of Fig. B.10. Here, an event called a **crossover** occurs in the tetrad: two of the homologous chromatids break at the same point. Then the broken ends rejoin in such a way that the chromatids have traded parts. Tracing through the rest of meiosis, one can see that the crossover has introduced two new classes of gametes. The chromatids that took part in the crossover will result in gametes with the genotypes Pl and pL. The chromatids that were *not* involved in the crossover will form gametes with genotypes PL and pl. Meiosis with a crossover has formed four kinds of gametes, whereas meiosis without a crossover gave only two kinds of gametes. When a large number of cells undergo meiosis, only a minority will have crossovers between the two genes. Therefore, *the effect of linkage is to reduce the frequency of two classes of gametes out of the four.*

As Figs. B.9 and B.10 show, linkage affects the frequency of gamete types. However, we cannot directly see the genotypes of gametes. How, then, can we determine whether the genes in a pair are linked or unlinked? The easiest method is to subject the doubly heterozygous plant to a testcross.

Figure B.11 shows what happens in the testcross when the genes are unlinked. As always in a testcross, one parent has only recessive alleles. This allows the other

---

Traits That Show Discontinuous Variation

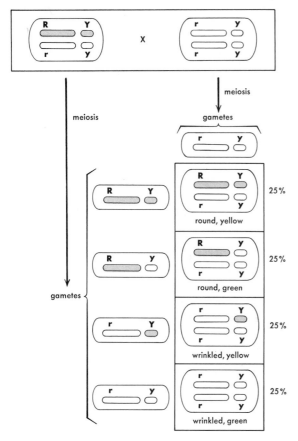

Figure B.11

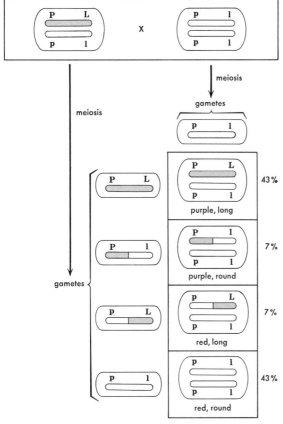

Figure B.12

parent's alleles to be fully expressed in the progeny. The genes are unlinked, so the heterozygous parent has produced four equally common kinds of gametes. Therefore the progeny also include four equally common phenotypes.

Contrast this with the behavior of linked genes in a testcross (Fig. B.12). Again there are four classes of progeny, but two classes are more common than the other two. The rare types are due to gametes that were formed by crossovers in meiosis.

One can see that the testcross of the double heterozygote can be useful in determining whether two genes are linked or not. But whatever the outcome, the testcross shows the relative frequencies of the various gamete types that a plant produces. This is useful information for breeding programs, because the gamete frequencies can be used to predict progeny genotype and phenotype frequencies. The mathematical methods for making these predictions cannot be described in the space we have here, but they are simple and can easily be learned from any elementary genetics text.

Though Fig. B.10 shows only one crossover between two chromatids, it should be mentioned that more than one crossover can occur in a given tetrad during a given meiosis. Also, even though a single crossover involves only two chromatids, additional crossovers may involve other chromatids. Chance plays a large part in setting the locations where crossovers will occur along the chromosome. Thus crossovers occur less often between two closely spaced genes than between two distant genes. Two or more crossovers may occur between genes that occur at opposite ends of the chromosome, so that in testcrosses the genes may appear to be unlinked. Geneticists have developed mathematical methods for taking these factors into account in planning crosses and predicting results. Such methods have also been used to map the chromosomes, showing the locations where genes occur.

Traits That Show Continuous Variation

The preceding discussion has concentrated on traits that show discontinuous variation. But continuous variation is the rule with many traits. For example, the seeds of corn (*Zea mays*) vary in protein content. In a study of this trait, plants that produced seeds high in protein were mated with plants that produce low-protein seeds. The F_1 contained progeny with a wide range of seed protein content (Fig. B.13). This pattern suggests that protein content might be controlled by environmental factors rather than heredity. Undoubtedly the environment does play a part. But another kind of experiment shows that heredity is involved too. A field of corn was planted and seeds from the crop were used to replant the field again the next year—but only seeds from plants that gave the best protein yields were used. Every year, a similar selection and replanting was done. Figure B.14 shows that the selection process gradually raised the average protein

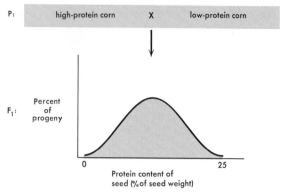

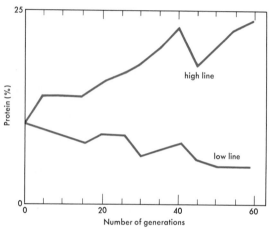

Figure B.13

Figure B.14. The effect of selection on seed protein content in *Zea mays*. Data from Woodworth, C. M., E. R. Leng, and R. W. Jugenheimer. 1952. Fifty generations of selection for protein and oil in corn. Agron. J. 44:60–65.

content of seeds produced in this field. In a parallel experiment, another field was replanted year after year, using only the seeds from plants that had the lowest protein yield in the preceding year's crop. In this experiment, the continued selection gradually decreased the average level of seed protein. These experiments show that protein content in the seeds depends on heredity: otherwise the protein content should not have been so affected by selection of the seeds.

If quantitative traits such as seed protein content are under hereditary control, why don't the progeny fall into discrete classes? Aside from environmental effects, the central explanation seems to be that several genes contribute to the control of the trait. Two or more alleles might be present in the population for each of these genes. In matings, then, gametes and zygotes with many combinations of alleles could result. Each combination might give a somewhat different phenotype with respect to the trait that is being studied. If many genes are involved, the effect of changing any one gene's alleles might be too small to detect easily against the background of environmental effects. For plant improvement in such cases, selection programs like the one in Fig. B.14 may be important, together with specific crosses between strains that have been selected for favorable features.

Maternal Inheritance

All the patterns of heredity mentioned so far have concerned genes that are carried in the nucleus, with a pollen grain and an egg cell making equal contributions. These include the great majority of cases. But anomalous patterns of heredity sometimes occur, in which the maternal (egg-producing) parent determines a trait in the progeny and the pollen seems to make no contribution. This can result from at least two causes. First, there are cases in which fertilization is bypassed and the flower forms seeds that start from a purely maternal cell (usually diploid) instead of a zygote. This process, called **apomixis**, occurs in the lawn dandelion. It is really a form of asexual reproduction and does not belong in a discussion of matings, but we mention it here because such cases can greatly confuse an attempt at controlled breeding. Second, **plastids** contain some genes that contribute to the development of the chloroplasts. These genes are passed from generation to generation by duplication and division of the plastids; their inheritance is distinct from that of nuclear genes. Since egg cells contribute plastids to the zygote while pollen cells usually do not, only the maternal parent has a part to play in plastid inheritance. Effects of the chloroplast genes are especially noted in the inheritance of variegated color patterns in the leaves of some plants (e.g., variegated *Mirabilis*). Here the egg cell contains both normal plastids and plastids that cannot develop into green chloroplasts because of defective genes. Some cells in the embryo come to have only the defective plastids. These cells divide further to produce the distinctive white or yellow patches on the leaves and stem of the mature plant.

table of metric equivalents

Exponents

Very large or very small numbers are often written in exponential form to save space. Thus, in the number 10^3, the **exponent** is the superscript 3. 10^3 is short for 1,000; or 1 followed by 3 zeroes. The number 10^{-2} represents 0.01; the digit 1 is placed 2 places to the right of the decimal.

Common Prefixes

nano = 10^{-9} (one billionth)
micro = 10^{-6} (one millionth)
milli = 10^{-3} (one thousandth)
kilo = 10^3 (one thousand times)
mega = 10^6 (one million times)

I. Units of Length

kilometer (km) = 10^3 m = 0.62 mile
meter (m) = 1.09 yd = 3.28 ft
decimeter (dm) = 10^{-1} m = 3.90 in.
centimeter (cm) = 10^{-2} m = 0.39 in.
millimeter (mm) = 10^{-3} m = 0.04 in.
micron, micrometer (μ) = 10^{-6} m
millimicron, nanometer (mμ or nm) = 10^{-9} m
angstrom (Å) = 10^{-10} m

II. Units of Weight

metric ton = 10^3 kg = 10^6 g = 1.10 short tons = 0.98 long ton = 2200 lb
kilogram (kg) = 10^3 g = 2.20 lb
gram (g) = 0.035 oz
milligram (mg) = 10^{-3} g = 0.015 grain

III. Units of Area

are (a) = 100 m^2 = 1075.84 ft^2
hectare (ha) = 100 are = 2.47 acres

IV. Units of Volume

liter (l) = 10^3 ml = 1.06 U.S. liquid qt = 0.91 U.S. dry qt = 0.88 British qt
milliliter, cubic centimeter (ml or cm^3) = 0.03 fluid oz = 0.06 in.^{3}

V. Miscellany; Units of Pressure, Light Intensity, Energy, Work, Force

bar = 0.99 atm of pressure = 14.54 lb/in.^{2}
lux (L) = 0.09 ft-c of light intensity
joule (j) = 0.735 ft-lb of work = 9.5×10^{-4} British Thermal Units (BTU)
erg = 10^{-7} joulc
watt (w) = 1.3×10^{-3} hp
calorie, gram-calorie (cal) = amount of heat needed to raise the temperature of 1 g of water 1°C from 14.5 to 15.5°C
kilocalorie, kilogram-calorie (kcal) = 1000 c = 3.97 BTU (1 BTU is the amount of heat needed to raise 1 lb of water 1°F)
solar constant (radiation at Earth's outer atmosphere edge) = 2 cal/cm^2/min = 13,400 ft-c = 4.06×10^7 BTU/m^2/yr
dyne = force required to accelerate 1 g at the rate of 1 cm/sec/sec = 1 g-cm/sec^2
newton = 10^5 dyne = 1 kg-m/sec^2

VI. Units of Temperature

one degree Celsius, Centigrade (1°C) = 1°K = 1.8°F
absolute zero = −273°C = 0°Kelvin = −460°F
boiling point of water = 100°C = 373°K = 212°F
freezing point of water = 0°C = 273°K = 32°F
to convert from °F to °C: (°F − 32) × 0.56 = °C

Abbreviation	Meaning
A.S.	Anglo-Saxon
D.	Dutch
dim.	diminutive
F.	French
Gr.	Greek
It.	Italian
L.	Latin
Lapp.	Lappland
M.E.	Medieval English
M.L.	Medieval Latin
N.L.	New Latin
O.F.	Old French
O.E.	Old English
R.	Russian
Sp.	Spanish

Å. Abbreviation for an **angstrom,** 0.0001 of a micron; there are 10,000,000 angstroms in a millimeter, 10 in a millimicron, and 10 in a nanometer.

Abscisic acid. A hormone variously inducing abscission, dormancy, stomatal closure, growth inhibition, and other responses in plants. Abbreviated **ABA.**

Abscission zone (L. *abscissus,* cut off). Zone of delicate thin-walled cells extending across the base of a petiole, the breakdown of which disjoins the leaf or fruit from the stem.

Absorb (L. *ab,* away + *sorbere,* to suck in). To suck up, to drink up, or to take in; in plant cells materials are taken in (absorbed) in solution.

Absorption spectrum. A graph relating the ability of a substance to absorb light of various wavelengths.

Accessory (M.L. *accessorius,* an additional appendage). Something aiding or contributing in a secondary way, such as buds in addition to the main axillary bud.

Achene. Simple, dry, one-seeded indehiscent fruit, with seed attached to ovary wall at one point only.

Acid. A substance that can donate a hydrogen ion.

Actinomorphic (Gr. *aktis,* ray + *morphe,* form). Said of flowers of a regular or star pattern, capable of bisection in two or more planes into similar halves.

Action spectrum (F. *acte,* a thing done). A graph relating the degree of physiological response (e.g., phototropism, photosynthesis) caused by different wavelengths of light.

Active solute absorption. The intake of dissolved materials by cells or organelles against an electrochemical potential gradient and requiring a supply of metabolic energy.

Adaptation (L. *ad,* to + *aptare,* to fit). Adjustment of an organism to the environment.

Adenine. A purine base present in nucleic acids and nucleotides.

Adhesion (L. *adhaerere,* to stick to). A sticking together of unlike things or materials.

Adnation (L. *adnasci,* to grow to). In flowers, the growing together of two or more whorls to a greater or less extent; compare **connation.**

ADP. Adenosine diphosphate.

Adsorption (L. *ad,* to + *sorbere,* to suck in). The concentration of molecules or ions of a substance at a surface or an interface (boundary) between two substances.

Advanced (M.E. *advaunce,* to forward). Said of a taxonomic trait thought to have evolved late in time from some more primitive trait.

Adventitious (L. *adventicius,* not properly belonging to). Referring to a structure arising from an unusual place: buds at other places than leaf axils, roots growing from stems or leaves.

Aeciospore (Gr. aikia, injury + spore). One of the dikaryotic asexual spores of rust fungi.

Aecium (plural, **aecia**) (Gr. *aikia,* injury). In rust, a sorus that produces aeciospores.

Aerate. To supply or impregnate with common air, such as by bubbling air through a culture solution.

Aerobe (Gr. *aer,* air + *bios,* life). An organism that uses molecular oxygen in its respiratory process.

Aerobic respiration. Respiration involving molecular oxygen.

Agar (Malay *agaragar*). A gelatinous substance obtained mainly from certain species of red algae.

Aggregate fruit (L. *ad*, to + *gregare*, to collect; to bring together). A fruit developing from the several separate carpels of a single flower; e.g., a strawberry; compare with **multiple fruit.**

Alcohol (M.L. from Arabic *al-kuhl*, a powder for painting eyelids; later applied, in Europe, to distilled spirits which were unknown in Arabia). A product of the distillation of wine or malt; any one of a class of compounds analogous to common alcohol; the ending **ol** designates a member of this class of compounds.

Aleurone layer (Gr. *aleuron*, flour). The outermost cell layer of the endosperm of wheat and other grains.

Alfisol. Modified podsol soil, typical of the northern part of the deciduous forest.

Alga (plural, **algae**) (L. *alga*, seaweed). A member of the large group of thallus plants containing chlorophyll and thus able to synthesize carbohydrates.

Algin. A long-chain polymer of mannuronic acid found in the cell walls of the brown algae.

Alkali (Arabic *alqili*, the ashes of the plant saltwort). A substance with marked basic properties.

Allele (Gr. *allelon*, of one another, mutually each other). Variant forms of a gene.

Alpine (L. *Alpes*, the Alps Mountains). Meadowlike vegetation at high elevation, above tree line.

Alternate. Referring to bud or leaf arrangement in which there is one bud or one leaf at a node.

Alternation of generations. The alternation of haploid (gametophytic) and diploid (sporophytic) phases in the life cycle of many organisms; the phases (generations) may be morphologically quite similar or very distinct, depending on the organism.

Amensalism (L. *a*, not + *mensa*, table). A form of biological interaction in which one organism is inhibited by another, by the addition of something to the environment.

Amino acid One of the building blocks of a protein.

Ammonification (*Ammon*, Egyptian sun god, near whose temple ammonium salts were first prepared from camel dung + L. *facere*, to make). Decomposition of amino acids, resulting in the production of ammonia.

Amyloplast (L. *amylum*, starch + *plastos*, formed). Cytoplasmic organelle specialized to store starch. Abundant in roots in storage organs such as tubers.

Anabolism (Gr. *ana*, up + metabolism). The constructive phase of metabolism, in which more complex molecules are built from simpler substances.

Anaerobe (Gr. *a*, without + *aer*, air + *bios*, life). An organism able to live in the absence of free oxygen, or in greatly reduced concentrations of free oxygen.

Anaphase (Gr. *ana*, up + *phais*, appearance). That stage in mitosis in which half chromosomes or sister chromatids move to opposite poles of the cell.

Androecium (Gr. *andros*, man + *oikos*, house). The aggregate of stamens in the flower of a seed plant.

Angiosperm (Gr. *angion*, a vessel + *sperma* from *speirein*, to sow, hence a seed or germ). Literally a seed borne in a vessel; thus a group of plants whose seeds are borne within a matured ovary.

Anisogamy (Gr. *an*, prefix meaning not + *isos*, equal + *gamete*, spouse). The condition in which the gametes, though similar in appearance, are not identical. Compare with **heterogamy.**

Annual (L. *annualis*, within a year). A plant that completes its life cycle within one year and then dies.

Annual ring. Ring of xylem in wood, which indicates an annual increment of growth in trees growing in temperate regions.

Annular vessels (L. *annularis*, a ring). Vessels with rings of secondary wall material.

Annulus (L. *anulus* or *annulus*, a ring). In ferns, a row of specialized cells in a sporangium, of importance in opening of the sporangium; in mosses, thick-walled cells along the rim of the sporangium to which the peristome teeth are attached.

Anther (M.L. *anthera*, from the Gr. *anthros*, meaning flower). Pollen-bearing portion of stamen.

Antheridiophore (Gr. *anthros*, flowers + Gr. *phoros*, to bear). In some liverworts, an elongate structure, bearing antheridia.

Antheridium (anther + *-idion*, a Gr. dim. ending, thus a little anther). Male gametangium or sperm-bearing organ with a sterile jacket of cells around the spermatocytes.

Anthocyanin (Gr. *anthros*, a flower + *kyanos*, dark blue). A blue, purple, or red vacuolar pigment.

Anticlinal cell division. Cell division where the newly formed cell wall is perpendicular to the axis of the organ surface.

Antipodal (Gr. *anti*, opposite + *pous*, foot). Referring to cells at the end of the embryo sac opposite that of the egg apparatus.

Apex (L. *apex*, a tip, point, or extremity). The tip, point, or angular summit of anything: the tip of a leaf; that portion of a root or shoot containing apical and primary meristems.

Apical dominance. The inhibition of lateral buds or meristems by the apical meristem.

Apical meristem. A mass of meristematic cells at the very tip of a shoot or root.

Aplanospore (Gr. *a*, not + *planetes*, wanderer + *sporos*, seed, spore). A nonmotile spore, one that is carried passively by wind, water, or other organisms.

Apomixis (Gr. *apo*, away from + *mixis*, a mingling). The production of offspring in the usual sexual structures without the mingling and segregation of chromosomes.

Apothecium (Gr. *apotheke*, a storehouse). A cup-shaped or saucer-shaped open ascocarp.

Arboretum (L. *arbor*, tree, also *arboretum*, a place grown with trees). A place, often outdoors, set aside for the display of living plants, including herbs and shrubs as well as trees, contrasts with an herbarium, which displays dead, preserved remains of plants.

Archegoniophore (Gr. *archgonos*, founder of a race + Gr. *phoros*, to bear). Elongate structure found on some liverworts that bears archegonia.

Archegonium (L. dim. of Gr. *archegonos*, literally a little founder of a race). Female gametangium or egg-bearing organ, in which the egg is protected by a jacket of sterile cells.

Aril (M.L. *arillus*, a wrapper for a seed). An accessory seed covering formed by an outgrowth at the base of the ovule in *Taxus*.

Ascus (plural, **asci**) (Gr. *askos*, a bag). A specialized cell, characteristic of the Ascomycetes, in which two haploid nuclei fuse, immediately after which three (generally) divisions occur, two of which constitute meiosis, resulting in eight ascospores still contained within the ascus.

-ase. A chemical suffix indicating an enzyme.

Asexual (Gr. *a*, without + L. *sexualis*, sexual). Any type of production not involving the union of gametes or meiosis.

Aspect (L. *aspectus*, appearance). The direction of slope of a surface, as a hillside with a south-facing aspect.

Assimilation (L. *assimilare*, to make like). The transformation of food into protoplasm.

Atom (Gr. *atomos*, indivisible). A unit of matter, consisting of a dense, central nucleus surrounded by a number of negatively charged electrons that are in constant motion. The nucleus consists of several positively charged protons and uncharged neutrons.

ATP. Adenosine triphosphate, a high-energy organic phosphate of great importance in energy transfer in cellular reactions.

Auricles (L. *auricula*, dim. of *auris*, ear). In grasses, small projections that grow out from the opposite side of the leaf sheath at its upper end where it joins the blade.

Autotrophic (Gr. *auto*, self + *trophein*, to nourish with food). Pertaining to an organism that is able to manufacture its own food.

Axial system (L. *axis*, axle). System of cells in secondary tissues that are oriented parallel to the long axis of the stem, forms from fusiform initials of the vascular cambium.

Axil (L. *axilla*, armpit). The upper angle between a petiole of a leaf and the stem from which it grows.

Axile placentation. Condition where ovules arise on the axis of an ovary with several locules.

Axillary bud. A bud formed in the axil of a leaf.

Auxin (Gr. *auxein*, to increase). A type of hormone that regulates many aspects of plant growth and development.

Banner (M.L. *bandum*, a standard). Large, broad, and conspicuous petal of legume type of flower.

Bark (Swedish *bark*, rind). The external group of tissues, from the cambium outward, of a woody stem or root.

Base. A substance that can accept a proton (H^+). Also, the purine and pyrimidine groups in nucleic acids and nucleotides are collectively called bases; they act as bases.

Basidiospore (M.L. *basidium*, a little pedestal + spore).

Type of meiospore borne by basidia in the Basidiomycetes.

Basidium (plural, **basidia**) (M.L. *basidium*, a little pedestal). A specialized reproductive cell of the Basidiomycetes in which nuclei fuse and meiosis occurs. It may be a special club-shaped cell, a short filamentous cell, or a short four-celled filament.

Benthon (Gr. *benthos*, the depths of the sea). Attached aquatic plants and animals, collectively.

Berry. A simple fleshy fruit, the ovary wall fleshy and including one or more carpels and seeds.

Biennial (L. *biennium*, a period of two years). A plant that requires two years to complete its life cycle. Flowering is normally delayed until the second year.

Bifacial leaf (L. *bis*, twice + *facies*, face). A leaf having upper and lower surfaces and anatomy distinctly different.

Binomial (L. *binominis*, two names). Two-named; in biology each species is generally indicated by two names, the genus to which it belongs and its own species name.

Bioassay (Gr. *bios*, life + L. *exagere*, to weigh or test). To test for the presence or quantity of a substance by using an organism's response as an indicator.

Biological barrier. A barrier to crossing (hybridization) of plants caused by differences in pollination vector or timing in flower opening, in contrast to physiological barriers (incompatability of pollen with stigma or style) or ecological barriers (habitats too far apart).

Biology (Gr. *bios*, life + *logos*, word, speech, discourse). The science that deals with living things.

Biosystematics (Gr. *bios*, life + *synistanai*, to place together). A field of taxonomy that emphasizes breeding behavior and chromosome characteristics.

Biotic (Gr. *biotikos*, relating to life). Relating to life.

Biotin. A vitamin of the B complex.

Bladder (O.E. *blaedre*, a blister). A gas-filled sac whose buoyancy keeps some aquatic plants upright.

Bloom (Gr. *blume*, flower). A flower; also the coloring of a body of fresh water by a high density of microscopic algal organisms.

Bordered pit. A pit in a tracheid or vessel member having a distinct rim of the cell wall overarching the pit membrane.

Botanic garden. See Arboretum.

Botany (Gr. *botane*, plant, herb). The science dealing with plant life.

Bract (L. *bractea*, a thin plate of precious metal). A modified leaf, from the axil of which arises a flower or an inflorescence.

Bud (M.E. *budde*, bud). An undeveloped shoot, largely meristematic tissue, generally protected by modified scale-leaves. Also a swelling on a yeast cell that will become a new yeast cell when released.

Bud scale. A modified protective leaf of a bud.

Bud scar. A scar left on a twig when the bud or bud scales fall away.

Bulb (L. *bulbus*, a modified bud, usually underground). A short, flattened, or disk-shaped underground stem, with many fleshy scale-leaves filled with stored food.

Bundle scar. Scar left where conducting strands passing out of the stem into the leaf stalk were broken off when the leaf fell.

Bundle sheath. Sheath of parenchyma cells that surround the vascular bundles of leaves, sometimes called border parenchyma.

C_3 cycle. The Calvin-Benson cycle of photosynthesis, in which the first stable products after CO_2 fixation are three-carbon molecules. The overall process builds sugar from CO_2, ATP, and NADPH.

C_4 cycle. The Hatch-Slack cycle of photosynthesis, in which the first stable products after CO_2 fixation are four-carbon molecules.

Callose (L. *callum,* thick skin + *ose,* a suffix indicating a carbohydrate). An amorphous polysaccharide deposited around pores in sieve tube members.

Callus (L. *callum,* thick skin). Mass of thin-walled cells, usually developed as the result of wounding or in tissue cultures.

Calorie (L. *calor,* heat). The amount of heat needed to raise the temperature of 1 g of water 1°C (usually from 14.5 to 15.5°C), also called gram-calorie; 1000 calories = 1 kilocalorie.

Calyptra (Gr. *kalyptra,* a veil, covering). In bryophytes, an envelope covering the developing sporophyte, formed by growth of the venter of the archegonium.

Calyx (Gr. *kalyx,* a husk, cup). Sepals collectively; outermost flower whorl.

Cambium (L. *cambium,* one of the alimentary body fluids supposed to nourish the body organs). A layer, usually regarded as one or two cells thick, of persistently meristematic tissue, giving rise to secondary tissues, resulting in growth in diameter.

Canopy (Gr. *kanopeion,* a cover over a bed to keep off gnats). The leafy portion of a tree or shrub.

Capillaries (L. *capillus,* hair). Very small spaces, or very fine bores in a tube.

Capsule (L. *capsula,* dim. of *capsa,* a case). A simple, dry, dehiscent fruit, with two or more carpels.

Carbohydrate (chemical combining forms, *carbo,* carbon + *hydrate,* containing water). Compounds with the general formula $C_n(H_2O)_n$ or $C_nH_{2n}O_n$

Carbon fixation. The enzymatic reaction in which CO_2 is attached to a receiver compound such as ribulose diphosphate, thereby adding to the supply of organic carbon. Occurs chiefly in photosynthesis.

Carotene (L. *carota,* carrot). A reddish-orange plastid pigment.

Carotenoids. A class of fat-soluble compounds (lipids) that includes carotenes as well as xanthophylls; most of them absorb light and appear yellow, orange or red.

Carpel (Gr. *karpos,* fruit). A floral leaf bearing ovules along the margins.

Carpogonium (Gr. *karpos,* fruit + *gonos,* producing). Female gametangium (in red algae).

Carpospore (Gr. *karpos,* fruit + spore). One of the spores produced in a carpogonium.

Carposporophyte (Gr. *karpos,* fruit + *sporos,* seed, spore + *phuton,* plant). One of two sporophyte generations in certain red algae; grows attached to the gametophyte generation, in contrast to the tetrasporophyte.

Caruncle (L. *caruncula,* dim. of *caro,* flesh, wart). A spongy outgrowth of the seed coat, especially prominent in the castor bean seed.

Caryopsis (Gr. *karyon,* a nut + *opsis,* appearance). A simple, dry, one-seeded, indehiscent fruit, with pericarp firmly united all around to the seed coat.

Casparian strip. Suberized strip that impregnates the radial and transverse walls of endodermal cells.

Catabolism (Gr. *kata,* down + metabolism). The phase of metabolism in which complex substances are broken down into simpler molecules, the chief role being to provide chemically reactive materials for use in anabolism and to provide ATP for cellular work.

Catalyst (Gr. *katelyein,* to dissolve). A substance that accelerates a chemical reaction but that is not used up in the reaction.

Cation exchange (Gr. *kata,* downward). The replacement of one positive ion (cation) by another, as on a negatively charged clay particle.

Catkin (literally a kitten, apparently first used in 1578 to describe the inflorescence of the pussy willow). A type of inflorescence, really a spike, generally bearing only pistillate flowers or only staminate flowers, which eventually fall from the plant entire.

Caulescent (Gr. *kaulos,* a plant stem). A plant whose stem bears leaves separated by visibly elongated internodes, as opposed to a rosette plant.

Cell (L. *cella,* small room). The smallest unit of material in the organism that is capable of self-reproduction. It is surrounded by a plasma membrane and contains a store of DNA together with a metabolic system.

Cell wall. A layer of material, chiefly elongated polymers, that is laid down outside the plasma membrane of most plant cells, and that serves to protect the protoplast and to limit its expansion.

Cellulose (cell + *ose,* a suffix indicating a carbohydrate). A polysaccharide occurring in the cell walls of the majority of plants; it is composed of hundreds of simple sugar molecules, glucose, linked together in a characteristic manner.

Cenozoic (Gr. *kainos,* recent + *zoe,* life). The geologic era extending f m 65 million years ago to the present.

Centromere (L. *centrum,* center + Gr. *meros,* part). Specialized part of chromosomes where spindle fibers attach and where two chromatids connect. Two **kinetochores,** one on each chromatid, compose one centromere.

Chalaza (Gr. *chalaza,* small tubercle). The region on a seed at the upper end of the raphe where the funiculus spreads out and unites with the base of the ovule.

Chaparral (Sp. *chaparro,* an evergreen oak). A vegetation type characterized by small leaved, evergreen shrubs growing together into a nearly impenetrable scrub; shrubby oaks are found in chaparral of California and the Mediterranean region, but other genera are typical of chaparral in Chile, South Africa, and Australia.

Chemotropism (chemo + Gr. *tropos,* a turning). Influence of a chemical substance on the direction of growth.

Chernozem (R. *cherny,* black + *zem,* earth). A soil characteristic of some grassland vegetation in warm areas with moderate rainfall; dark in color because of a high content of organic matter; a **mollisol.**

Chiasma (Gr. *chiasma,* two lines placed crosswise). The cross formed by breaking, during prophase I of meiosis, of two nonsister chromatids of homologous chromosomes and the rejoining of the broken ends of different chromatids.

Chitin (Gr. *chiton,* a coat of mail). A polymer in which the monomer unit is the modified sugar *N*-acetyl glucosamine; it is the principal stiffening material in the cell walls of most fungi and in the exoskeletons of insects and crustaceans.

Chlamydospore (Gr. *chlamys,* a horseman's or young man's coat + spore). A heavy-walled resting asexual spore.

Chlorenchyma (Gr. *chloros,* green + *-enchyma,* a suffix meaning tissue). Parenchyma tissue possessing chloroplasts.

Chlorophyll (Gr. *chloros,* green + *phyllon,* leaf). The green pigment found in the chloroplast, important in the absorption of light energy in photosynthesis.

Chloroplast (Gr. *chloros,* green + *plastos,* formed). Specialized cytoplasmic body, containing chlorophyll, in which occur important reactions of starch or sugar synthesis.

Chlorosis (Gr. *chloros,* green + *osis,* diseased state). Failure of chlorophyll development, because of a nutritional disturbance or because of an infection of virus, bacteria, or fungus.

Chromatid (chromosome + *-id,* L. suffix meaning daughters of). The half-chromosome during prophase and metaphase of mitosis, and between prophase I and anaphase II of meiosis.

Chromatin (Gr. *chroma,* color). Substance in the nucleus that readily takes artificial staining; also, that portion that bears the determiners of hereditary characters; made up of DNA and protein.

Chromoplast (Gr. *chroma,* color + *plastos,* formed). Specialized plastid containing yellow or orange pigments.

Chromosome (Gr. *chroma,* color + *soma,* body). A condensed mass of chromatin, visible during cell division.

Cilia (singular, **cilium**) (Fr. *cil,* an eyelash). Protoplasmic hairs which, by a whiplike motion, propel certain types of unicellular organisms, gametes, and zoospores through water.

Cisterna (plural, **cisternae**) (L. *cisterna,* a reservoir). A flattened sac, composed of a continuous surrounding membrane, together with the enclosed space.

Citric acid cycle. A system of reactions that contributes to the catabolic breakdown of fuels in respiration and that provides building materials for a number of important anabolic pathways. Also called the **Krebs cycle** and the **tricarboxylic acid (TCA) cycle.**

Cladode (Gr. *kladodes,* having many shoots). A branch resembling a foliage leaf.

Class (L. *classis,* one of the six divisions of Roman people). A taxonomic group below the division level but above the order level.

Clay. Soil particles less than 2 microns in diameter, composed mainly of aluminum (Al), oxygen (O), and silicon (S).

Cleistothecium (plura, **cleistothecia**) (Gr. *kleistos,* closed + *thekion,* a small receptacle). The closed, spherical ascocarp of the powdery mildews.

Climax community. The last stage of a natural succession; a community capable of maintaining itself as long as the climate does not change.

Clone (Gr. *klon,* a twig or slip). The aggregate of individual organisms produced asexually from one sexually produced individual.

Closed bundle. A vascular bundle lacking residual procambium.

Coal Age. The Carboniferous period, beginning 345 million years ago and ending 280 million years ago.

Coalescence (L. *coalescere,* to grow together). A condition in which there is union of separate parts of any one whorl of flower parts; synonyms are **connation** and **cohesion.**

Coenocyte (Gr. *koinos,* shared in common + *kytos,* a vessel). A plant or filament whose protoplasm is continuous and multinucleate and without any division by walls into separate protoplasts.

Coenzyme. A substance, usually nonprotein and of low molecular weight, necessary for the action of some enzymes.

Cohesion (L. *cohaerere,* to stick together). Union or holding together of parts of the same materials; the union of floral parts of the same whorl, as petals to petals.

Coleoptile (Gr. *koleos,* sheath + *ptilon,* down, feather). The first leaf in germination of grasses, which sheaths the succeeding leaves.

Coleorhiza (Gr. *koleos,* sheath + *rhiza,* root). Sheath that surrounds the radicle of the grass embryo and through which the young root bursts.

Collenchyma (Gr. *kolla,* glue + *-enchyma,* a suffix, derived from parenchyma and denoting a type of cell or tissue). A stem tissue composed of cells that fit rather closely together and with walls thickened at the angles of the cells.

Colloid (Gr. *kolla,* glue + *eidos,* form). Referring to matter composed of particles, ranging in size from 0.0001 to 0.000001 millimeter, dispersed in some medium, as clay particles in soil.

Colony (L. *colonia*, a settlement). A growth form characterized by a group of closely associated, but poorly differentiated, cells; sometimes filaments can be associated together in a colony (as in *Nostoc*), but more typically unicells are associated in a colony.

Community (L. *communitas*, a fellowship). All the populations within a given habitat; usually the populations are thought of as being interdependent.

Companion cells. Cell associated with sieve-tube members.

Compensation depth (L. *compensare*, to counterbalance). That depth, in a body of water, at which light intensity is too low for photosynthesis of floating or submerged plants to exceed respiration.

Competition (L. *competere*, to strive together). A form of biological interaction in which both organisms (at least initially) decline in growth or success because of the insufficient supply of some necessary factor(s).

Complete flower. A flower having four whorls of floral leaves: sepals, petals, stamens, and carpels.

Compound leaf. A leaf whose blade is divided into several distinct leaflets.

Conceptacle (L. *conceptaculum*, a receptacle). A cavity or chamber of a frond (of *Fucus*, for example) in which gametangia are borne.

Cone (Gr. *konos*, a pine cone). A fruiting structure composed of modified leaves or branches, which bear sporangia (microsporangia, megasporangia, pollen sacs, or ovules), and are frequently arranged in a spiral or four-ranked order; for example, a pine cone.

Cone scale. The flat, woody parts of pine cones that spiral out from the central axis and bear the ovules (and later seeds) on their upper surface; each is subtended by a sterile bract; strictly speaking, a cone scale is not equivalent to a megasporophyll but, instead, is thought to represent a modified branch system.

Conduction (L. *conducere*, to bring together). Act of moving or conveying water through the xylem in plant organs.

Conidia (singular, **conidium**) (Gr. *konis*, dust). Asexual reproductive cells of fungi, arising by the cutting off of terminal or lateral cells of special hyphae, or by being pushed out from a flask-shaped cell.

Conidiophore (conidia + Gr. *phoros*, bearing). Conidium-bearing branch of hypha.

Conifer (cone + L. *ferre*, to carry). A cone-bearing tree; in the coniferophyta.

Conjugation (L. *conjagatus*, united). Process of sexual reproduction involving the fusion of isogametes or of specialized cell extensions.

Connation (L. *connatus*, to be born together). Condition in flower where there is a union of similar parts of any one whorl of appendages; synonym of **coalescence.**

Conservative (L. *conservare*, to keep). Said of a taxonomic trait whose expression is not modified to any great extent by the external environment; a trait that is constant unless its genetic base is changed.

Convergent evolution. Process of successive progeny, originally of quite distinct parents, coming to appear more and more alike through time because of selection pressure.

Cork. An external, secondary tissue impermeable to water and gases.

Cork cambium. The cambium from which cork develops.

Corm (Gr. *kormos*, a trunk). A short, solid, vertical, enlarged underground stem in which food is stored.

Corolla (L. *corolla*, dim. of *corona*, a wreath, crown). Petals, collectively; usually the conspicuous colored flower whorl.

Cortex (L. *cortex*, bark). Primary tissue of a stem or root bounded externally by the epidermis and internally in the stem by the phloem and in the root by the pericycle.

Cotyledon (Gr. *kotyledon*, a cup-shaped hollow). Seed leaf; two, generally storing food in dicotyledons; one, generally a digestive organ in the monocotyledons.

Covalent bond. A binding force that holds two atoms together, due to the sharing of electrons.

Cristae (L. *crista*, a crest). Crests or ridges, used here to designate the infoldings of the inner mitochondrial membrane.

Cross-pollination. The transfer of pollen from a stamen to the stigma of a flower on another plant, except in clones.

Crossing-over. The exchange of corresponding segments between chromatids of homologous chromosomes.

Cultural eutrophication (Gr. *eu*, good, well + *trephein*, to nourish). Organic pollution of bodies of water resulting from mankind's activities; results in oxygen depletion and a change in the biota; occurs over a shorter period of time than natural eutrophication.

Cuticle (L. *cuticula*, dim. of *cutis*, the skin). Waxy layer on outer wall of epidermal cells.

Cutin (L. *cutis*, the skin). Waxy substance that is but slightly permeable to water, water vapor, and gases; a major part of the cuticle.

Cutinization. Impregnation of cell wall with a substance called cutin.

Cyme (Gr. *kyma*, a wave, a swelling). A type of inflorescence in which the apex of the main stalk or the axis of the inflorescence ceases to grow quite early, relative to the laterals.

Cystocarp (Gr. *kystos*, bladder + *karpos*, fruit). A peculiar diploid spore-bearing structure formed after fertilization in certain red algae.

Cytochrome (Gr. *kytos*, a receptacle or cell + *chroma*, color). A class of several electron-transport proteins serving as carriers in mitochondrial oxidations and in photosynthetic electron transport.

Cytokinesis (Gr. *kytos*, a hollow vessel + *kinesis*, motion). Division of cytoplasmic constituents at cell division.

Cytokinin. A class of hormones that participate in controlling many developmental processes in plants.

Cytology (Gr. *kytos*, a hollow vessel + *logos*, word, speech, discourse). The science dealing with the cell.

Cytoplasm (Gr. *kytos*, a hollow vessel + *plasma*, form). All the protoplasm of a protoplast outside the nucleus.

Cytosine. A pyrimidine base found in DNA and RNA.

Deciduous (L. *deciduus*, falling). Referring to trees and shrubs that lose their leaves in the fall.

Decomposer (L. *de*, from + *componere*, to put together). An organism that obtains food by breaking down dead organic matter into simpler molecules.

Decomposition (L. *de*, to denote an act undone + *componere*, to put together). A separation or dissolving into simpler compounds; rotting or decaying.

Dehiscent (L. *dehiscere*, to split open). Opening spontaneously when ripe, splitting into definite parts.

Deletion (L. *deletus*, to destroy, to wipe out). Used here to designate an area, or region, lacking from a chromosome.

Dendrogram (Gr. *dendron*, tree + *gramme*, what is written or drawn). A graph showing relationship between things at different levels of similarity; the graph resembles the limbs of a tree.

Denitrification (L. *de*, to denote an act undone + *nitrum*, nitro, a combining form indicating the presence of nitrogen + *facere*, to make). Conversion of nitrates into nitrites, or into gaseous oxides of nitrogen, or even into free nitrogen.

Deoxyribose nucleic acid (DNA). Hereditary material; the DNA molecule carries hereditary information.

Desert scrub (M.E., *schrubbe*, shrub). A vegetation type characterized by evergreen or drought-deciduous shrubs growing together rather openly generally in an area with annual precipitation below 25 cm.

Detritus (L. *detritus*, worn away). Particulate organic matter released in the process of decomposition of dead organisms or parts of organisms (such as plant litter).

Development (F. *developper*, to unfold). Changes in the plant body that result from controlled processes of cell division, cell growth, and cell differentiation.

Diatom (Gr. *diatomos*, cut in two). Member of a group of golden brown algae with silicious cell walls fitting together much as do the halves of a pill box.

Diatomite. Diatomaceous earth; that is, sedimentary deposits made up of the silica wall remains of diatoms.

Dicotyledon (Gr. *dis*, twice + *kotyledon*, a cup-shaped hollow). A plant whose embryo has two cotyledons.

Dictyosome (Gr. *diktyon*, a net + *soma*, body). One of the component parts of the Golgi apparatus; in plant cells a complex of flattened double lamellae.

Differentially permeable. See Selectively permeable.

Differentiation (L. *differre*, to carry different ways). Developmental change of a cell leading to the presence of features that equip the cell for performing specialized functions.

Diffuse porous. Wood with an equal and random distribution of large xylem vessel members throughout the growth season.

Diffuse secondary growth. Secondary growth, such as in palm trees, that is caused by a proliferation of parenchyma cells and not by vascular cambium.

Diffusion (L. *diffusus*, spread out). The movement of molecules from a region of higher concentration to a region of lower concentration.

Digestion (L. *digestio*, dividing, or tearing to pieces, an orderly distribution). The processes of rendering food available for metabolism by breaking it down into simpler compounds, chiefly through actions of enzymes.

Dikaryon (Gr. *di*, two + *karyon*, nut). A hypha or mycelium in which each cell contains two haploid nuclei, the two usually derived from different parent organisms. The dikaryotic condition is often abbreviated as the n+n condition.

Dinoflagellate (Gr. *dinein*, to whirl + L. *flagellum*, a whip). The common name for members of the algal division Pyrrhophyta; the organisms are typically unicellular and motile, with cell walls made up of overlapping plates.

Dioecious (Gr. *dis*, twice + *oikos*, house). Unisexual; having the male and female elements in different individuals.

Diploid (Gr. *diploos*, double + *oides*, like). Having a double set of chromosomes, or referring to an individual containing a double set of chromosomes per cell; usually a sporophyte generation.

Divergent evolution. Process of successive progeny, originally of quite similar parents, coming to appear more and more different through time because of isolation and selection pressure.

Division. A major portion of the plant kingdom; equivalent to phylum.

DNA. Deoxyribonucleic acid.

Dominant (L. *dominari*, to rule). Referring, in ecology, to species of a community that receive the full force of the macroenvironment; usually the most abundant of such species.

Dormant (L. *dormire*, to sleep). Being in a state of reduced physiological activity such as occurs in seeds, buds, etc.

Dorsiventral (L. *dorsum*, the back + *venter*, the belly). Having upper and lower surfaces distinctly different, as many leaves do.

Double bond. A covalent bond that involves four electrons.

Drupe (L. *drupa*, an overripe olive). A simple, fleshy fruit, derived from a single carpel, usually one-seeded, in which the exocarp is thin, the mesocarp fleshy, and the endocarp stony.

Early wood. That portion of an annual ring formed during spring, characterized by large cells and thin walls, also called **spring wood.**

Ecology (Gr. *oikos*, home + *logos*, discourse). The study of life in relation to environment.

Ecosystem (Gr. *oikos*, house + *synistanai*, to place together). An inclusive term for a living community and all the factors of its nonliving environment.

Ecotype (Gr. *oikos*, house + *typos*, the mark of a blow).

Genetic variant within a species that is adapted to a particular environment yet remains interfertile with all other members of the species.

Edaphic (Gr. *edaphos*, soil). Pertaining to soil conditions that influence plant growth.

Egg (A.S. *aeg*, egg). A female gamete.

Elater (Gr. *elater*, driver). An elongated, spindle-shaped, sterile, hygroscopic cell in the sporangium of a some Bryophyta sporophytes.

Electron (Gr. *elektron*, gleaming in the sun, from L. *electrum*, amber, from which electricity was first produced by friction). An elementary particle of matter bearing a unit of negative electrical charge. Low in mass and in constant rapid motion, electrons repel one another and are attracted to the positively charged atomic nucleus. The motion of the electrons defines the size and chemical properties of the atom or molecule.

Electron transport chain. A membrane-bound system that controls the flow of electrons from reduced to oxidized compounds, so that some of the energy carried by the electrons is used to form ATP. The chain consists of several compounds (carriers) that alternately accept and donate electrons. Found in mitochondria and chloroplasts.

Electron microscope. A microscope that uses a beam of electrons rather than light to produce a magnified image.

Element (L. *elementa*, the first principles). In modern chemistry, a substance that cannot be divided or reduced by any known chemical means to a simpler substance; 92 natural elements are known, of which gold, carbon, oxygen, and iron are examples; several, including plutonium, have been formed in atomic piles.

Embryo (Gr. *en*, in + *bryein*, to swell). A young sporophytic plant, while still retained in the gametophyte or in the seed.

Embryo sac. The female gametophyte of the angiosperms; generally a seven-celled structure; the seven cells are two synergids, one egg cell, three antipodal cells (each with a single haploid nucleus), and one endosperm mother cell with two haploid nuclei.

Endocarp (Gr. *endon*, within + *karpos*, fruit). Inner layer of fruit wall (pericarp).

Endodermis (Gr. *endon*, within + *derma*, skin). The layer of living cells, with various characteristically thickened walls and no intercellular spaces, which surrounds the vascular tissue in nearly all roots and certain stems and leaves.

Endogenous (Gr. *endon*, within + *genes*, born). Produced within the cell or organism.

Endoplasmic reticulum (Gr. *endon*, within + *plasma*, anything formed or molded; L. *reticulum*, a small net). A system of membrane-bound cysternae found in the cytoplasm.

Endosperm (Gr. *endon*, within + *sperma*, seed). The nutritive tissue formed within the embryo sac of seed plants; it is often consumed as the seed matures, but remains in the seeds of corn and other cereals.

Endosperm mother cell. One of the seven cells of the mature embryo sac, containing the two polar nuclei and, after reception of a sperm cell, giving rise to the primary endosperm cell from which the endosperm develops.

Environment (O.F., *environ*, around). The living and nonliving factors that surround a given organism or community of organisms.

Enzyme (Gr. *en*, in + *zyme*, yeast). A protein that acts as a catalyst to speed chemical reactions.

Epicotyl (Gr. *epi*, upon + *kotyledon*, a cup-shaped hollow). The upper portion of the axis of embryo or seedling, above the cotyledons.

Epidermis (Gr. *epi*, upon + *derma*, skin). A superficial layer of cells occurring on all parts of the primary plant body: stems, leaves, roots, flowers, fruits, and seeds; it is absent from the root cap and not differentiated on the apical meristems.

Epigyny (Gr. *epi*, upon + *gyne*, woman). The arrangement of floral parts in which the ovary is embedded in the receptacle so that the other parts appear to arise from the top of the ovary.

Epiphyte (L. *epi*, upon + *phyton*, a plant). A plant that grows upon another plant, but is not parasitic.

ER. Endoplasmic reticulum.

Ethylene. C_2H_4, a hormone that participates in the control of many developmental processes in plants.

Etiolation (F. *etioler*, to blanch). A condition involving increased stem elongation, poor leaf development, and lack of chlorophyll, found in plants growing in the absence, or in a greatly reduced amount, of light.

Eukaryote (L. *eu*, true + *karyon*. A nut, referring in modern biology to the nucleus); any organism characterized by having cellular organelles, including a nucleus, bounded by membranes.

Eutrophication (Gr. *eu*, good, well + *trephein*, to nourish). Pollution of bodies of water resulting from slow, natural, geological, or biological processes such as siltation or encroachment of vegetation or accumulation of detritus, also called natural eutrophication.

Evapotranspiration (L. *evaporare*, e, out of + *vapor*, vapor + F. *transpirer*, to perspire). The process of water loss in vapor form from a unit surface of land both directly and from leaf surfaces.

Evolution (L. *evolutio*, an unrolling). The development of a race, genus or other larger group of plants or animals.

Exine (L. *exterus*, outside). Outer coat of pollen.

Exocarp (Gr. *exo*, without, outside + *karpos*, fruit). Outermost layer of fruit wall (pericarp).

Exogenous (Gr. *exe*, out, beyond + *genos*, race, kind). Produced outside of, originating from, or due to external causes.

Facultative (L. *facultas*, capability). Referring to an organism having the power to live under a number of certain specific conditions, e.g., a facultative parasite may be either parasitic or saprophytic.

Family (L. *familia*, family). In plant taxonomy, a group of genera; families are grouped in orders.

Fascicle (L. *fasciculus*, a small bundle). A bundle of leaves arising from one point on the stem.

Fascicular cambium. Cambium within vascular bundles.

Fermentation (L. *fermentum*, a drink made from fermented barley, beer). Catabolic breakdown of fuels by a process that does not involve molecular oxygen.

Fern (O.E., *fearn*, fern). Common name for members of the division Pterophyta, part of the lower vascular plants.

Ferredoxin. An electron-transferring protein containing iron, involved in photosynthesis and in nitrogen fixation.

Fertilization (L. *fertilis*, capable of producing fruit). The union of egg and sperm.

Fiber (L. *fibra*, a fiber or filament). An elongated, tapering, thick-walled strengthening cell occurring in various parts of plant bodies.

Fiber-tracheid. Xylem elements found in pine that are structurally intermediate between tracheids and fibers.

Field capacity. The amount of water retained in a soil (generally expressed as percent by weight) after large capillary spaces have been drained by gravity.

Filament (L. *filum*, a thread). Stalk of stamen bearing the anther at its tip; also, a slender row of cells (certain algae).

Flagellum (plural, **flagella**) (L. *flagellum*, a whip). A long, slender whip of protoplasm.

Flora (L. *floris*, a flower). An enumeration of all the species that grow in a region; also, the collective term for all the species that grow in a region.

Floret (F. *fleurette*, a dim. of *fleur*, flower). One of the small flowers that make up the composite inflorescence (head) or the spike of the grasses.

Flower (F. *fleur*, L. *flos*, a flower). Floral leaves grouped together on a stem and adapted for sexual reproduction in the angiosperms.

Follicle (L. *folliculus*, dim. of *follis*, bag). A simple, dry, dehiscent fruit, with one carpel, splitting along one suture.

Food (A.S. *fōda*). Any organic substance that furnishes energy and building materials directly for vital processes.

Food chain. The path along which caloric energy is transferred within a community (from producers to consumers to decomposers).

Foot (O.E., *fot*, foot). That portion of the sporophyte of bryophytes and lower vascular plants which is sunk in gametophyte tissue and absorbs food parasitically from the gametophyte.

Fossil (L. *fossio*, a digging). Any impression, impregnated remains, or other trace of an animal or plant of past geological ages that has been preserved in the earth's crust.

Fossil fuel. Hydrocarbon deposits, currently mined and refined for use as fuel (coal, gas, oil), which were originally deposits of detritus of now-extinct plants.

Fret (O. F. *frette*, latticework). Flattened membrane sacs that connect grana in chloroplasts.

Frond (L. *frons*, branch, leaf). A synonym for a large divided leaf, especially for a fern leaf.

Fruit (L. *fructus*, that which is enjoyed, hence product of the soil, trees, cattle, etc.). A matured ovary; in some seed plants other parts of the flower may be included; also applied, as **fruiting body**, to reproductive structures of some fungi.

Frustule (L. *frustulum*, little piece). A diatom cell, composed of two overlapping halves (valves).

Fucoxanthin (Gr. *phykos*, seaweed + *xanthos*, yellowish brown). A brown pigment found in brown algae.

Fungus (plural **fungi**) (L. *fungus*, a mushroom). A eukaryotic organism that lacks plastids and that reproduces by means of spores.

Funiculus (L. *funiculus*, dim. of *funis*, rope or small cord). A stalk of the ovule, containing vascular tissue.

Fusiform initials (L. *fusus*, spindle + form). Meristematic cells in the vascular cambium that develop into xylem and phloem cells comprising an axial system.

Gametangium (Gr. *gametes*, a husband, *gamete*, a wife + *angeion*, a vessel). Organ bearing gametes.

Gamete (Gr. *gametes*, a husband, *gamete*, a wife). A protoplast that fuses with another protoplast to form the zygote in the process of sexual reproduction.

Gametic life cycle. A life cycle in which the haploid phase is limited to gametes, as in some diatoms and many animals.

Gametophyte (gamete + Gr. *phyton*, a plant). The gamete-producing plant.

Gel (L. *gelare*, to freeze). Jellylike colloidal mass.

Gemma (plural, gemmae) (L. *gemma*, a bud). A small mass of reproductive tissue in some fungi and bryophytes.

Gene (Gr. *genos*, race, offspring). A group of base pairs in the DNA molecule that determines one or more hereditary characters.

Gene recombination. The appearance of gene combinations in the progeny different from the combinations present in the parents.

Generation (L. *genus*, birth, race, kind). Any phase of a life cycle characterized by a particular chromosome number, as the gametophyte generation and the sporophyte generation.

Genetics (Gr. *genesis*, origin). The science of heredity.

Genotype (gene + type). The assemblage of genes in an organism.

Genus (plural, **genera**) (Gr. *genos*, race, stock). A group of structurally and phylogenetically related species.

Geotropism (Gr. *ge*, earth + *tropos*, turning). A growth curvature induced by gravity.

Germination (L. *germinare*, to sprout). The beginning of growth of a seed, spore, or other once-dormant structure.

Gibberellins. A class of hormones that participate in controlling many developmental processes in plants.

Girdle (O.E. *gyrdel*, enclosure, girdle). That region of a frustule where the two valves overlap.

Glucose (Gr. *glykys,* sweet + *ose,* a suffix indicating a carbohydrate). A common hexose sugar.

Glume (L. *gluma,* husk). An outer and lowermost bract of a grass spikelet.

Glycogen (Gr. *glykys,* sweet + *genes,* born). A polysaccharide built of glucose, akin to starch, serving as a food reserve in animals and fungi and some prokaryotes.

Glycolysis (Gr. *glykys,* sweet + *lysis,* a loosening). Decomposition of sugar compounds without involving free oxygen; early steps of respiration.

Golgi body (Italian cytologist Camillo Golgi 1844–1926, who first described the organelle). In animal cells, a complex perinuclear region thought to be associated with secretion; in plant cells a series of flattened plates, more properly called **dictyosomes.**

Grana (singular, **granum**) (L. *granum* a seed). Structures within chloroplasts, seen as a series of parallel lamellae with the electron microscope.

Ground cover. The area of ground covered by a plant when its canopy edge is projected perpendicularly down.

Ground meristem (Gr. *meristos,* divisible). A primary meristem that gives rise to cortex and pith.

Ground tissues. Category of primary tissues (parenchyma, collenchyma, and sclerenchyma) that provide such basic functions as storage, support, and secretion.

Growth (A.S. *growan,* probably from Old Teutonic *gro,* from which grass is also derived). An irreversible increase in size.

Growth retardant. A chemical (such as cycocel, CCC) that selectively interferes with normal hormonal promotion of growth—but without appreciable toxic effects.

Guanine. A purine base found in DNA and RNA.

Guttation (L. *gutta,* drop, exudation of drops). Exudation of water from plants, in liquid form.

Gynoecium (Gr. *gyne,* woman + *oikos,* house). The aggregate of carpels in the flower of a seed plant.

Gymnosperm (Gr. *gumnos,* naked + *sperma,* seed, sperm). The common name for a group of divisions, including the Coniferophyta, which bear exposed seeds, in contrast to the flowering plants that bear the seeds enclosed in a fruit (mature ovary).

Haploid (Gr. *haploos,* single + *oides,* like). Having a single complete set of chromosomes, or referring to an individual or generation containing such a single set of chromosomes per cell; usually a gametophyte generation.

Hapteron (Gr. *haptein,* to fasten). An individual branch of the hold fast organ of a kelp; also called haptere (plural, hapteres).

Haustorium (plural, **haustoria**) (M.L. *haustrum,* a pump). A projection that acts as a penetrating and absorbing organ.

Head. An inflorescence, typical of the composite family, in which flowers are sessile and grouped closely on a receptacle.

Heartwood. Wood in the center of old secondary stems that is plugged with resins and tyloses and is not active.

Helix (Gr. *helix,* anything twisted). Anything having a spiral form.

Hemicellulose. A class of polysaccharides of the cell wall, built of several different kinds of simple sugars linked in various combinations.

Herb (L. *herba,* grass, green blades). A seed plant that does not develop woody tissues.

Herbaceous (L. *herbaceus,* grassy). Referring to plants having the characteristics of herbs.

Herbal (L. *herba,* grass). A book that contains the names and descriptions of plants, especially those which are thought to have medicinal uses.

Herbarium (L. *herba,* grass). A collection of dried and pressed plant specimens.

Herbicide (L. *herba,* grass or herb + *cidere,* to kill). A chemical used to kill plants, frequently chemically related to a hormone.

Heredity (L. *hereditas,* being a heir). The transmission of morphological and physiological characters of parents to their offspring.

Heterobasidiomycetidae (Gr. *heteros,* other + Basidiomycete). A subclass of Basidiomycetes with variable basidia, never club-shaped cells.

Heterocyst (Gr. *heteros,* different + *cytis,* a bag). An enlarged colorless cell that may occur in the filaments of certain blue-green algae; associated with nitrogen fixation.

Heteroecious (Gr. *heteros,* different + *oikos,* house). Referring to fungi that cannot carry through their complete life cycle unless two different host species are present.

Heterogametes (Gr. *heteros,* different + gamete). Gametes dissimilar from each other in size and behavior, like egg and sperm.

Heterogamy (Gr. *heteros,* different + *gamos,* union or reproduction). Reproduction involving two types of gametes.

Heterospory (Gr. *heteros,* different + spore). The condition of producing microspores and megaspores.

Heterothallic (Gr. *heteros,* different + thallus). Referring to species in which male gametangia and female gametangia are produced by different individual plant bodies.

Heterotrichy (Gr. *heteros,* different + *trichos,* a hair). In the algae, the occurrence of two types of filaments, erect and prostrate.

Heterotrophic (Gr. *heteros,* different + *trophein,* to nourish with food). Referring to a plant obtaining nourishment from outside sources.

Heterozygous (Gr. *heteros,* different + *zygon,* yoke). Having two different alleles for a given gene in the diploid cell.

Hexose (Gr. *hexa,* six + *-ose,* suffix indicating carbohydrate). A carbohydrate with six carbon atoms (e.g., $C_6H_{12}O_6$).

Higher vascular plant. Common name for extant seed-producing plants; hence now includes only gymnosperm and angiosperm divisions although in the past other groups did produce seeds.

Hilum (L. *hilum,* a trifle). Scar on seed, which marks the place where the seed broke from the stalk.

Homobasidiomycetidae (Gr. *homo,* the same + Basidiomycete). A subclass of Basidiomycetes with a typical club-shaped cell as a basidium.

Homologous chromosomes (Gr. *homologos,* the same). Members of a chromosome pair; they may be heterozygous or homozygous.

Homospory (Gr. *homos,* one and the same + spore). The condition of producing one sort of spore only.

Homothallic (Gr. *homos,* one and the same + thallus). Referring to species in which male gametangia and female gametangia are produced. By the same individual plant body.

Homozygous (Gr. *homos,* one and the same + *zygon,* yoke). Having two identical alleles for a given gene in the diploid cell.

Hormogonia (singular, **hormogonium**) (Gr. *hormos,* necklace + *gonos,* offspring). Short filaments, the result of a breaking apart of filaments of certain blue-green algae at the heterocysts.

Hormone (Gr. *hormaein,* to excite). A compound that is normally produced by a plant and whose sole function is to act as a signal in controlling development.

Humidity, relative (L. *humidus,* moist). The ratio of the weight of water vapor in a given quantity of air, to the total weight of water vapor that quantity of air is capable of holding at the temperature in question, expressed as percent.

Humus (L. *humus,* the ground). Decomposing organic matter in the soil.

Hybrid (L. *hybrida,* offspring of a tame sow and a wild boar, a mongrel). The offspring of two plants or animals differing in at least one Mendelian character; or the offspring formed by mating two plants or animals that differ genetically.

Hydathode (Gr. *hydro,* water + O.E. *thoden,* stem or *thyddan,* to thrust). A structure, usually on leaves, which releases liquid water during guttation.

Hydrogen acceptor. A substance capable of accepting hydrogen atoms or electrons in the oxidation-reduction reactions of metabolism.

Hydrogen bond. A weak bond that occurs between a hydrogen atom and an oxygen or nitrogen atom when the hydrogen is already covalently bonded to another oxygen or nitrogen atom. The hydrogen bond is not covalent.

Hydrolysis (Gr. *hydro,* water + *lysis,* loosening). Reaction of a compound with water, attended by decomposition into less complex compounds.

Hydrophyte (Gr. *hudor,* water + *phuton,* plant). A plant that normally grows in a wet habitat.

Hymenium (Gr. *hymen,* a membrane). Spore-bearing tissue in various fungi.

Hypanthium (L. *hypo,* under + Gr. *anthos,* flower). Fusion of calyx and corolla part way up their length to form a cup, as in many members of the rose family.

Hypertrophy (Gr. *hyper,* over + *trophein,* to nourish with food). A condition of overgrowth or excessive development of an organ or part.

Hypha (plural, **hyphae**) (Gr. *hyphe,* a web). A slender, elongated, threadlike cell or filament of cells of a fungus.

Hypocotyl (Gr. *hypo,* under + *kotyledon,* a cup-shaped hollow). That portion of an embryo or seedling between the cotyledons and the radicle or young root.

Hypogyny (Gr. *hypo,* under + *gyne,* female). A condition in which the receptacle is convex or conical, and the flower parts are situated one above another in the following order, beginning with the lowest: sepals, petals, stamens, carpels.

Hypothesis (Gr. *hypothesis,* foundation). A tentative theory or supposition provisionally adopted to explain certain facts and to guide in the investigation of other facts.

IAA. Indoleacetic acid.

Imbibition (L. *imbibere,* to drink). The absorption of liquids or vapors into the ultramicroscopic spaces or pores found in such materials as cellulose or a block of gelatine; an adsorption phenomenon.

Imperfect flower. A flower lacking either stamens or pistils.

Imperfect fungi. Fungi reproducing only by asexual means.

Incomplete flower. A flower lacking one or more of the four kinds of flower parts.

Indehiscent (L. *in,* not + *dehiscere,* to divide). Not opening by valves or along regular lines.

Indicator species. A species that has a narrow range of tolerance for one or more environmental factors so that, from its occurrence at a site, one can predict these factors at that site (e.g., nutrient availability or summer temperatures).

Indusium. Membranous growth of the epidermis of a fern leaf that covers a sorus.

Infect (L. *infectus,* to put into, to taint with morbid matter). Specifically to produce disease by such agents as bacteria or viruses.

Inferior ovary. An ovary more or less, (sometimes completely) attached to the calyx and corolla.

Inflorescence (L. *inflorescere,* to begin to bloom). A flower cluster.

Inheritance (O.F. *enheritance,* inheritance). The reception or acquisition of characters or qualities by transmission of parent to offspring.

Inorganic. Referring to compounds that do not contain *both* carbon and hydrogen.

Integuments (L. *integumentum,* covering). Cell layers around ovule that develop into the seed coat.

Interphase (L. *inter,* between + phase). Period between mitotic divisions, consists of G1, pre-DNA synthesis phase; S, DNA synthesis; and G2, post-DNA synthesis phase.

Intercalary (L. *intercalare,* to insert). Descriptive or meristematic tissue or growth not restricted to the apex of an organ, i.e., growth at nodes.

Intercellular (L. *inter*, between + cells). Lying between cells.

Interfascicular cambium (L. *inter*, between + *fasciculus*, small bundle). Cambium that develops between vascular bundles.

Internode (L. *inter*, between + *nodus*, a knot). The region of a stem between two successive nodes.

Intine (L. *intus*, within). The innermost coat of a pollen grain.

Intracellular (L. *intra*, within + cell). Lying within cells.

Introgressive hybridization (L. *intro*, to the inside + *gress*, walk + *hybrida*, halfbreed). Back-crossing between complete or partial hybrids and the original parental stock.

Involucre (L. *involucrum*, a wrapper). A whorl or rosette of bracts surrounding an inflorescence.

Ion. An atom or molecule that has a net negative or positive electrical charge because the number of electrons does not equal the number of protons.

Ionic compound. A substance whose molecules readily break up into ions when placed in contact with water.

Irregular flower. A flower in which one or more members of at least one whorl are of different form from other members of the same whorl; zygomorphic flower.

Isobilateral leaf (Gr. *isos*, equal + L. *bis*, twice, twofold + *lateralis*, pertaining to the side). A leaf having the upper and lower surfaces and anatomy essentially similar.

Isodiametric (Gr. *isos*, equal + diameter). Having diameters equal in all directions, as a ball.

Isogametes (Gr. *isos*, equal + gametes). Gametes similar in size and behavior.

Isogamy (Gr. *isos*, equal + *gamete*, spouse). The condition in which the gametes are identical.

Isolating barriers. Any barrier to the crossing (hybridization) of plants; includes biological, physiological, and ecological categories.

Isomers (Gr. *isos*, equal + *meros*, part). Two or more compounds having the same molecular formula but different internal structure; for example, glucose and fructose, both having the formula $C_6H_{12}O_6$.

K selection. Natural selection that favors long-lived, late-maturing individuals that devote a small fraction of their resources into reproduction; tree species are K strategists.

Karyogamy (Gr. *karyon*, nut + *gamos*, marriage). The fusion of two nuclei.

Keel (A.S. *ceol*, ship). A structure of the legume type of flower, made up of two petals loosely united along their edges.

Kelp. (M.E. *culp*, seaweed). A collective name for any of the large brown marine algae.

Kinetochore. (Gr. *kinein*, to move + *chorein*, to move apart). Specialized portion of chromosome, marks point of spindle fiber attachment, related to mechanism of chromosome movement.

Krebs cycle. See **Citric acid cycle.**

Lamella (plural, **lamellae**) (Gr. *lamin*, a thin blade).

Cellular membranes, frequently those seen in choroloplasts.

Lamina (L. *lamina*, a thin plate). Blade or expanded part of a leaf.

Late wood. That portion of an annual ring which formed during summer and fall, characterized by small diameter cells with thick walls, also called **summer wood.**

Lateral bud. A bud that grows out of the side of a stem.

Laterite (L. *later*, a brick). A soil characteristic of rain forest vegetation; color is red from oxidized iron in the A horizon; in the oxisol soil order.

Latex (L. *latex*, juice). A milky secretion.

Leach (O.E. *leccan*, to moisten). To extract a soluble or moveable substance (as ions or clay particles or bits of organic matter) by causing water to filter down through a material (as a soil horizon).

Leaf. Lateral outgrowth of stem axis, which is the usual primary photosynthetic organ, and in the axil of which may be bud.

Leaf axil. Angle formed by the leaf stalk and the stem.

Leaf scar. Characteristic scar on stem axis made after leaf abscission.

Leaflet. Separate part of the blade of a compound leaf.

Leaf primordium (L. *primordium*, a beginning). A lateral outgrowth from the apical meristem, which will become a leaf.

Legume (L. *legumen*, any leguminous plant, particularly bean). A simple, dry dehiscent fruit with one carpel, splitting along two sutures.

Lemma (Gr. *lemma*, a husk). Lower bract that subtends a grass flower, but above the glumes.

Lenticel (M.L. *lenticella*, a small lens). A structure of the bark that permits the passage of gas inward and outward.

Leucoplast (Gr. *leukos*, white + *plastos*, formed). A colorless plastid.

Liana (F. *liane* from *lier*, to bind). A plant that climbs upon other plants, depending upon them for mechanical support; a plant with climbing shoots.

Lichen (Gr. *leichen*, thallus plants growing on rocks and trees). A composite plant consisting of a fungus living symbiotically with an alga.

Lignification (L. *lignum*, wood + *facere*, to make). Impregnation of a cell wall with lignin.

Lignin (L. *lignum*, wood). An organic substance or group of substances impregnating the cellulose framework of certain plant cell walls.

Ligule (L. *ligula*, dim. of *lingua*, tongue). In grass leaves, an outgrowth from the upper and inner side of the leaf blade where it joins the sheath.

Line transect. A method of sampling vegetation by stretching a tape along a straight line and measuring the canopy cover of plants beneath that line or which cut through a vertical plane described by that line.

Linkage. The grouping of genes on the same chromosome.

Linked characters. Characters of a plant or animal

controlled by genes grouped together on the same chromosome.

Lipase (Gr. *lipos*, fat + -*ase*, suffix indicating an enzyme). Any enzyme that breaks fats into glycerin and fatty acids.

Lipids (Gr. *lipos*, fat + L. *ides*, suffix meaning son of; now used in sense of having the quality of). An artificial grouping of compounds consisting of substances that are insoluble in water and soluble in fat solvents.

Liverwort (liver + M.E. *wort*, a plant, literally, a liver plant, so named in medieval times because of its fancied resemblance to the lobes of the liver). Common name for the Class Hepaticae of the Bryophyta

Loam (O.E. *lam* or Old Teutonic *lai*, to be sticky, clayey). A particular soil texture class, referring to a soil having 30–50% sand, 30–40% silt, and 10–25% clay.

Lobed leaf (Gr. *lobos*, lower part of the ear). A leaf divided by clefts or sinuses.

Locule (L. *loculus*, dim. of *locus*, a place). A cavity of the ovary in which ovules occur.

Lodicules (L. *lodicula*, a small coverlet). Two scalelike structures that lie at the base of the ovary of a grass flower.

Lower vascular plant. Common name for extant vascular plants that do not produce seeds (even though in the past some of their members did produce seeds); the Psilophyta, Lycophyta, Sphenophyta, and Pterophyta divisions.

Lumen (L. *lumen*, light, an opening for light). The cavity of the cell within the cell walls.

Lysis (Gr. *lysis*, a loosening). A process of disintegration or cell destruction.

Macroenvironment (Gr. *makros*, large + O.F. *environ*, about). The environment due to the general, regional climate; traditionally measured some 1.5 m above the ground and away from large obstructions.

Macronutrient (Gr. *makros*, large + L. *nutrire*, to nourish). An essential element required by plants in relatively large quantities.

Mating types. A term applied to organisms that show no visible male-female differentiation. Two individuals that can mate belong to different mating types. Mating types are often designated as (+) and (−).

Medulla (L. *medulla*, marrow). The filamentous center of certain lichens and kelp blades and stipes.

Megaphyll (Gr. *megas*, great + *phyllon*, leaf). A leaf whose trace is marked with a gap in the stem's vascular system.

Megasporangiate cone. In gymnosperms, a cone that produces megaspores and, ultimately, seeds; synonyms include female cone, ovulate cone, seed cone.

Megasporangium (Gr. *megas*, large + sporangium). Sporangium that bears megaspores.

Megaspore (Gr. *megas*, large + spore). The meiospore of vascular plants, which gives rise to a female gametophyte.

Megasporocyte (Gr. *megas*, large + *spora*, seed + L. *cyta*, vessel). A diploid cell in which meiosis will occur,

resulting in four megaspores; synonymous with megaspore mother cell of other texts.

Megasporophyll (Gr. *megas*, large + spore + Gr. *phyllon*, leaf). A leaf bearing megasporangia.

Meiocyte (meiosis + Gr. *kytos*, currently meaning a cell). Any cell in which meiosis occurs.

Meiosis (Gr. *meioun*, to make smaller). Two special cell divisions occurring in the life cycle of every sexually reproducing plant and animal, halving the chromosome number and effecting a segregation of genetic determiners.

Meiospore (meiosis + spore). Any spore resulting from the meiotic divisions.

Membrane (L. *membrana*, skin covering the separate members of the body). A limiting protoplasmic surface, consisting of protein and lipid, which surrounds cellular organelles.

Meristem (Gr. *meristos*, divisible). Undifferentiated tissue, the cells of which are capable of active cell division and differentiation into specialized tissues.

Meristoderm (meristem + epidermis). The outer meristematic cell layer (epidermis) of some Phaeophyta.

Mesocarp (Gr. *mesos*, middle + *karpos*, fruit). Middle layer of fruit wall (pericarp).

Mesophyll (Gr. *mesos*, middle + *phyllon*, leaf). Parenchyma tissue of leaf between epidermal layers.

Mesophyte (Gr. *mesos*, middle + *phuton*, plant). A plant that normally grows in moist habitats.

Mesozoic (Gr. *mesos*, middle + zoe, life). A geologic era beginning 225 million years ago and ending 65 million years ago.

Metabolism (M.L. from Gr. *metabolos*, to change). The overall set of chemical reactions occurring in an organism or cell.

Metabolite (Gr. *metabolos*, changeable + *ites*, one of a group). A chemical that is a normal cell constituent capable of entering into the biochemical transformations within living cells

Metamorphic rock (Gr. *meta*, change + *morphe*, shape or form). One of three major categories of rock; rocks whose original structure or mineral composition has been changed by pressures or temperatures in the earth's crust.

Metaphase (Gr. *meta*, after + *phsis*, appearance). Stage of mitosis during which the chromosomes, or at least the kinetochores, lie in the central plane of the spindle.

Metaxylem. Last formed primary xylem.

Microbody (Gr. *mikros*, small + body). A cellular organelle, always bound by a single membrane, frequently spherical, from 20 to 60 nanometers in diameter, containing a variety of enzymes.

Microcapillary space. Exceedingly small spaces, such as those found between microfibrils of cellulose.

Microenvironment (Gr. *mikros*, small + O.F. *environ*, about). The environment close enough to the surface of a living or nonliving object to be influenced by it.

Microfibrils (Gr. *mikros*, small + *fibrils*, dim. of fiber).

Very small fibers of the cell wall.

Microfossil (Gr. *mikros*, small + L. *fossilis*, dug up). Fossils of microscopic organisms, only visible when thin sections of rock are examined.

Micrometer (Gr. *mikros*, small + *metron*, measure). One millionth (10^{-6}) of a meter, or 0.001 millimeter; also called a **micron,** and abbreviated μm.

Micronutrient (Gr. *mikros*, small + L. *nutrire*, to nourish). An essential element required by plants in relatively small quantities.

Microphyll (Gr. *mikros*, small + *phyllon*, leaf). A leaf whose trace is not marked with a gap in the stem's vascular system; microphylls are thought to represent epidermal outgrowths.

Micropylar chamber. The space between the micropyle and the nucellus; sealed off from the outside when the micropyle closes after pollination.

Micropyle (Gr. *mikros*, small + *pulon*, orifice, gate). A pore leading from the outer surface of the ovule between the edges of the integuments down to the surface of the nucellus.

Microsporangiate cone. In gymnosperms, a cone that produces microspores and, ultimately, pollen; synonyms include male cone, pollen cone, and staminate cone.

Microsporangium (plural, **microsporangia**) (Gr. *mikros*, iittle + *sporangium*). A sporangium that bears microspores.

Microspore (Gr. *mikros*, small + *spore*). A spore which, in vascular plants, gives rise to a male gametophyte.

Microsporocyte (Gr. *mikros*, small + *spora*, seed + L. *cyta*, vessel). A diploid cell in which meiosis will occur, resulting in four microspores; synonymous with **microspore mother cell.**

Microsporophyll (Gr. *mikros*, little + *spore* + Gr. *phyllon*, leaf). A leaf bearing microsporangia

Microtubule (Gr. *mikros*, small + *tubule*, dim. of tube). A tubule 25 nm in diameter and of indefinite length, occurring in the cytoplasm of many types of cells.

Middle lamella (L. *lamella*, a thin plate or scale). Thin layer separating two adjacent protoplasts and remaining as a distinct cementing layer between adjacent cell walls.

Millimeter. The 0.001 part of a meter, equal to 0.0394 inch.

Mitochondrion (plural, **mitochondria**) (Gr. *mitos*, thread + *chondrion*, a grain). A membrane-bounded organelle associated with intracellular respiration.

Mitosis (plural, **mitoses**) (Gr. *mitos*, a thread). Nuclear division, involving coiling and equal distribution of duplicate chromosomes into derivative nuclei; the chromosome number remains constant. Compare with **meiosis.**

Mitospore (mitosis + spore). A spore forming after mitosis.

Mixed bud. A bud containing both rudimentary leaves and flowers.

Molecular biology. A field of biology that emphasizes the interaction of biochemistry and genetics in the life of an organism.

Molecule (F. *môle*, + *clue*, a dim.; literally, a little mass). A unit of matter, the smallest portion of an element or a compound that retains chemical identity with the substance in mass; the molecule usually consists of a union of two or more atoms, some organic molecules containing a very large number of atoms.

Mollisol (L. *mollis*, soft + *solum*, soil, solid). One of the ten world soil orders, characterized by containing more than 1% organic matter in the top 17.5 cm and associated with grassland vegetation; synonymous with chernozem.

Monocotyledon (Gr. *monos*, solitary + *kotyledon*, a cup-shaped hollow). A plant whose embryo has one cotyledon.

Monoecious (Gr. *monos*, solitary + *oikos*, house). Having the reproductive organs in separate structures, but borne on the same individual.

Monophyletic (Gr. *mono*, single + *phyle*, tribe). Said of organisms having a common (but sometimes quite ancient) ancestor.

Morphogenesis (Gr. *morphe*, form + L. *genitus*, to produce). The changes in body form that occur during development of an organism.

Morphology (Gr. *morphe*, form + *logos*, discourse). The study of form and its development.

Moss (L. *muscus*, moss). A bryophytic plant.

Multiciliate (L. *multus*, many + F. *cil*, an eyelash). Having many cilia present on a sperm or spore or other type of ciliated cell.

Multiple fruit. A cluster of matured ovaries produced by separate flowers; e.g., a pineapple.

Mutation (L. *mutare*, to change). A sudden, heritable change appearing in an individual as the result of a change in genes or chromosomes.

Mutualism (L. *mutuus*, reciprocal). A form of biological interaction in which both organisms must associate together for continued success of both.

Mycelium (Gr. *mykes*, mushroom). The mass of hyphae forming the body of the fungus.

Mycology (Gr. *mykes*, mushroom + *logos*, discourse). The branch of botany dealing with fungi.

Mycorrhiza (Gr. *mykos*, fungus + *riza*, root). A symbiotic association between a fungus and usually the root of a higher plant.

NAD. Nicotinamide adenine dinucleotide, a coenzyme capable of being reduced.

NADH. Reduced NAD.

NADP. Nicotinamide adenine dinucleotide phosphate, a coenzyme capable of being reduced.

NADPH. Reduced NADP.

Naked bud. A bud not protected by bud scales.

Nanometer (Gr. *nanos*, small). One millionth (10^{-6}) of a millimeter, equals 10 angstroms; abbreviated nm.

Natural biotic unit. A species defined by isolating barriers rather than by morphological features; the species of biosystematists.

Natural classification. A classification scheme that is

based on the phylogenetic nature of the organisms classified; contrasts with an artificial classification, which separates organisms on the basis of convenient traits, but fails to show the evolutionary relationships among the organisms.

Natural selection. The effect of the environment in channeling the genetic variation of organisms down particular pathways.

Nectar (Gr. *nektar,* drink of the gods). A fluid rich in sugars secreted by nectaries, which are often located near or in flowers.

Nectar guide. A mark of contrasting color or texture that may serve to guide pollinators to nectaries within the flower.

Nectary (Gr. *nektar,* the drink of the gods). A nectar-secreting gland.

Net productivity. The arithmetic difference between calories produced in photosynthesis and calories lost in respiration.

Net radiation. The arithmetic difference between incoming solar radiation and outgoing terrestrial radiation.

Net venation. Veins of leaf blade visible to the unaided eye, branching frequently and joining again, forming a network.

Neutron (L. *neuter,* neither). An uncharged particle found in the atomic nucleus of all elements except hydrogen.

Niche (It. *nicchia,* a recess in a wall). The functional relationship of an organism to its ecosystem.

Nitrification (L. *nitrum,* nitro, a combining form indicating the presence of nitrogen + *facere,* to make). Change of ammonium salts into nitrates through the activites of certain bacteria.

Nitrogen fixation. The process of reducing N_2 gas into ammonia and incorporating it into the protoplast; accomplished only by certain prokaryotes.

nm. See **nanometer.**

n + n. See **Dikaryon.**

Node (L. *nodus,* a knot). Slightly enlarged portion of the stem where leaves and buds arise, and where branches originate.

Nonseptate. Descriptive of hyphae or algal filaments lacking crosswalls.

Nucellus (L. *nucella,* a small nut). Tissue composing the chief part of the young ovule, in which the embryo sac develops; megasporangium.

Nucleic acid. A polymeric molecule consisting of subunits called nucleotides, linked together in a chain.

Nucleolus (L. *nucleolus,* a small nucleus). Dense protoplasmic body in the nucleus.

Nucleosides. Components of nucleic acids consisting of a base and a sugar; in DNA, the sugar is deoxyribose, and in RNA, ribose; the bases adenine, guanine, and cytosine occur in both DNA and RNA; the base thymine occurs in DNA; the base uracil occurs in RNA.

Nucleotide. A nucleoside to which a phosphate unit is attached.

Nucleus (L. *nucleus,* kernel of a nut). A membrane-bounded organelle that contains most of the DNA in eukaryotic cells.

Numerical taxonomy. A field of taxonomy that does not place subjective weight on any particular type of evidence that shows relationships between taxa.

Nut (L. *nux,* nut). A dry, indehiscent, hard, one-seeded fruit, generally produced from a compound ovary.

Obligate anaerobe. An organism that cannot live in the presence of oxygen.

Obligate parasite. An organism that can only live as a parasite.

Obligate saprophyte. An organism that can only live as a saprophyte.

Oogamy (Gr. *oion,* egg + *gamete,* spouse). The condition in which the gametes are different in form and activity, i.e., sperms and eggs.

Oogonium (L. dim. of Gr. *oogonos,* literally, a little egg layer). Female gametangium of egg-bearing organ not protected by a jacket of sterile cells, characteristic of the thallophytes.

Oospore (Gr. *oion,* an egg + spore). A resistant spore developing from a zygote resulting from the fusion of heterogametes.

Open bundle. A vascular bundle with residual procambium.

Operculum (L. *operculum,* a lid). In mosses, cap of sporangium.

Opposite. Referring to leaf arrangement in which there are two leaves opposite each other at a node.

Order (L. *ordo,* a row of threads in a loom). A taxonomic category below class and above family.

Organ (L. *organum,* an instrument or engine of any kind). A part or member of a plant body adapted for a particular function.

Organelle. A membrane-bound specialized region within a cell such as the mitochondrion or dictyosome.

Organic. Referring to compounds that contain both carbon and hydrogen; also generally referring to the material products of living organisms.

Organic evolution. See **Evolution.**

Organism. An individual living body.

-ose. A chemical suffix indicating a carbohydrate.

Osmosis (Gr. *osmos,* a pushing). Diffusion of a solvent through a differentially permeable membrane.

Ovary (L. *ovum,* an egg). Enlarged basal portion of the pistil, which becomes the fruit.

Ovulate. Referring to a cone, scale, or other structure bearing ovules.

Ovulate cone. See **Megasporangiate cone.**

Ovule (F. *ovule,* from L. *ovulum,* dim. of *ovum,* egg). A rudimentary seed, containing, before fertilization, the female gametophyte, with egg cell, all being surrounded by the nucellus and one or two integuments.

Ovuliferous (ovule + L. *ferre,* to bear). Referring to a scale or sporophyll bearing ovules.

Oxidation. The removal of electrons or hydrogen, or the addition of oxygen to a compound.

Oxisol (Gr. *oxus*, sharp, sour + L. *solum*, soil, solid). One of the ten world soil orders, characterized by a soil horizon at least 30 cm thick that is highly weathered, rich in clay, but low in nutrients, and red in color, and associated with tropical vegetation; synonymous with laterite.

P_{fr} and P_r. abbreviations for the far-red (FR) or red (R) absorbing forms of phytochrome (P).

P-protein. Proteinaceous contents of phloem sieve tube members, sometimes called slime.

Palea (L. *palea*, chaff). Upper bract that subtends a grass flower.

Paleoecology (Gr. *palaios*, ancient). A field of ecology that reconstructs past vegetation and climate from fossil evidence.

Paleozoic (Gr. *palaios*, ancient + *zoe*, life). A geologic era beginning 570 million years ago and ending 225 million years ago.

Palisade parenchyma. Elongated cells, containing many chloroplasts, found just beneath the upper epidermis of leaves.

Palmately veined (L. *palma*, palm of the hand). Descriptive of a leaf blade with several principal veins spreading out from the upper end of the petiole.

Panicle (L. *panicula*, a tuft). An inflorescence, the main axis of which is branched, and whose branches bear loose racemose flower clusters.

Pappus (L. *pappus*, woolly, hairy seed or fruit of certain plants). Scales or bristles representing a reduced calyx in composite flowers.

Parallel venation. Type of venation in which veins of a leaf blade that are clearly visible to the unaided eye are parallel to each other.

Paraphysis (plural, **paraphyses**) (Gr. *para*, beside + *physis*, growth). A sterile, slender, multicellular hair growing beside fertile cells in certain thallophytes and bryophytes.

Parasite (Gr. *parasitos*, one who eats at the table of another). An organism deriving its food from the living body of another plant or an animal.

Parenchyma (Gr. *parenchein*, an ancient Greek medical term meaning to pour beside and expressing the ancient concept that the liver and other internal organs were formed by blood diffusing through the blood vessels and coagulating, thus designating ground tissue). A tissue composed of cells that usually have thin walls; site of most essential processes such as photosynthesis, secretion, and storage.

Parent material. The original rock or depositional matter from which the soil of a region has been formed.

Parietal (F. *pariétal*, attached to the wall, from L. *paries*, wall). Belonging to, connected with, or attached to the wall of a hollow organ or structure, especially of the ovary or cell.

Parietal placentation. A type of placentation in which placentae are on the ovary wall.

Parthenocarpy (Gr. *parthenos*, virgin + *karpos*, fruit). The development of fruit without fertilization.

Parthenogenesis (Gr. *parthenos*, virgin + *genesis*, origin). The development of a gamete into a new individual without fertilization.

Passive solute absorption. Absorption due only to forces of simple diffusion.

Pathogen (Gr. *pathos*, suffering + *genesis*, beginning). An organism that causes a disease.

Pathology (Gr. *pathos*, suffering + *logos*, account). The study of diseases, their effects on plants or animals, and their treatment.

Pathway. A sequence of chemical reactions, each governed by an enzyme, that gradually transform a starting molecule into some final product.

Peat (M.E. *pete*, of Celtic origin, a piece of turf used as fuel). Any mass of semicarbonized vegetable tissue, such as *Sphagnum*, formed by partial decomposition in water.

Pectin (Gr. *pektos*, congealed). A class of polymers of the cell wall that are built chiefly of partially oxidized sugars.

Pedicel (L. *pediculus*, a little foot). Stalk or stem of the individual flowers of an inflorescence.

Peduncle (L. *pedunculus*, a late form of *pediculus*, a little foot). Stalk or stem of a flower that is borne singly; or the main stem of an inflorescence.

Penicillin. An antibiotic derived from the mold *Penicillium*.

Pentose (Gr. *pente*, five + OSE). A five-carbon sugar, $C_5H_{10}O_5$.

PEP. Phosphoenolpyruvic acid, or phosphoenolpyruvate.

Perennial (L. *perennis*, lasting the whole year through). A plant that lives from year to year.

Perfect flower. A flower having both stamens and pistils.

Perianth (Gr. *peri*, around, about + *anthos*, flower). The petals and sepals taken together.

Pericarp (Gr. *peri* + *karpos*, fruit). Fruit wall, developed from ovary wall.

Periclinal cell division. Cell division where the newly formed cell wall is parallel to the axis of the organ.

Pericycle (Gr. *peri* + *kyklos*, circle). Tissue, generally of root, bound externally by the endodermis and internally by the phloem.

Periderm (Gr. *peri* + Gr. *derma*, skin). Protective tissue that replaces the epidermis after secondary growth is initiated. Consists of cork, cork cambium, and phelloderm.

Peridium (plural, **peridia**) (Gr. *peridion*, a little pouch). External covering of the hymenium of certain fungi; in Myxomycetes, the hardened envelope that covers the sporangium.

Perigyny (Gr. *peri* + *gyne*, a female). A condition in which the receptacle is more or less concave, at the margin of which the sepals, petals, and stamens have their origin, so that these parts seem to be attached around the ovary; also called half-inferior.

Perisperm. Nutritive tissue in some seeds that forms from the nucellus.

Peristome (Gr. *peri* + *stoma*, a mouth). In mosses, a fringe of teeth about the opening of the sporangium.

Perithecium (Gr. *peri* + *theke*, a box). A spherical or flask-shaped ascocarp having a small opening.

Permafrost (L. *permanere*, to remain + A.S. *freosan*, to freeze). Soil that is permanently frozen; usually found some distance below a surface layer that thaws during warm weather.

Permeable (L. *permeabilis*, that which can be penetrated). Said of a membrane, cell, or cell system through which substances may diffuse.

Peroxysome. An organelle of the microbody class that contains enzymes capable of destroying peroxides.

Petal (Gr. *petalon*, a flower leaf). One of the flower parts, usually conspicuously colored.

Petiole (L. *petiolus*, a little foot or leg). Stalk of leaf.

PGA. 3-Phosphoglyceric acid, a three-carbon compound formed by the interaction of carbon dioxide and a five-carbon compound, ribulose diphosphate; the reaction yields two molecules of PGA for each molecule of ribulose diphosphate; PGA is the first stable product of carbon fixation in the C_3 cycle of photosynthesis.

Phelloderm (Gr. *phellos*, cork + *derma*, skin). A layer of cells formed in the stems of some plants from the inner cells of the cork cambium.

Phellogen (Gr. *phellos*, cork + *genesis*, birth). Cork cambium, a cambium giving rise externally to cork and in some plants internally to phelloderm.

Phenotype (Gr. *phaneros*, showing + type). The bodily characteristics of an organism.

Phloem (Gr. *phloos*, bark). Food-conducting tissue, consisting of sieve tubes members or sieve cells, companion cells, parenchyma, and fibers.

Phosphoenolpyruvate (PEP) carboxylase. The enzyme responsible for the fixation of inorganic CO_2 into oxaloacetic acid in a dark reaction of the C_4 photosynthesis cycle.

Phosphorylation. A reaction in which phosphate is added to a compound, e.g., the formation of ATP from ADP and inorganic phosphate.

Photon. A quantum of light; the energy of a photon is proportional to its frequency: $E = h\nu$ where E is energy; h, Planck's constant, 6.62×10^{-27} erg-second; and ν is the frequency.

Photoperiod (Gr. *photos*, light + period). The optimum length of day or period of daily illumination required for the normal growth and maturity of a plant.

Photophosphorylation. A reaction in which light energy is converted into chemical energy in the form of ATP produced from ADP and inorganic phosphate.

Photoreceptor (Gr. *photos*, light + L. *receptor*, a receiver). A light-absorbing molecule involved in converting light into some metabolic (chemical energy) form, e.g., chlorophyll and phytochrome.

Photosynthesis (Gr. *photos*, light + *syn*, together + *tithenai*, to place). A process in which light energy is used to drive the formation of organic compounds.

Phototropism (Gr. *photos*, light + Gr. *tropos*, a turning). Influence of light on the direction of plant growth.

Phycobilins (Gr. phucos, seaweed). Red and blue accessory pigments characteristic of blue-green and red algae.

Phycobiliproteins. Pigments found in the red and blue-green algae; a phycobilin associated with a protein.

Phycocyanin (Gr. *phykos*, seaweed + *kyanos*, blue). A blue phycobilin pigment occurring in blue-green algae.

Phycoerythrin (Gr. *phykos*, seaweed + *erythros*, red). A red phycobilin pigment occurring in red algae.

Phylogenetic classification. See **Natural classification.**

Phylogeny (Gr. *phylon*, race or tribe + *genesis*, beginning). The evolution of a group of related individuals.

Phylum (Gr. *phylon*, race or tribe). A primary division of the animal or plant kingdom.

Physiology (Gr. *physis*, nature + *logos*, discourse). The science of the functions and activities of living organisms.

Phytobenthon (Gr. *phyton*, a plant + *benthos*, depths of the sea). Attached aquatic plants, collectively.

Phytochrome. A pigment found in green plants; it is associated with the control of development in response to light stimuli.

Phytoplankton (Gr. *phyton*, a plant + *planktos*, wandering). Free-floating plants, collectively.

Pigment. A substance that absorbs visible light, hence, appears colored.

Pileus (L. *pileus*, a cap). Umbrella-shaped cap of fleshy fungi.

Pinna (plural, *pinnae*) (L. *pinna*, a feather). Leaflet or division of a compound leaf (frond).

Pinnately veined (L. *pinna*, a feather + *vena*, a vein). Descriptive of a leaf blade with single midrib from which smaller veins branch off, somewhat like the divisions of a feather.

Pioneer community. The first stage of a succession.

Pistil (L. *pistillum*, a pestle). Central organ of the flower, typically consisting of ovary, style, and stigma.

Pistillate flower. A flower having pistils but no stamens.

Pit. A thin area of a secondary cell wall.

Pith. The parenchymatous tissue occupying the central portion of a stem.

Placenta (plural, **placentae**) (L. *placenta*, a cake). The tissue within the ovary to which the ovules are attached.

Placentation (L. *placenta*, a cake + *-tion*, state of). Manner in which the placentae are distributed in the ovary.

Plankton (Gr. *planktos*, wandering). Free-floating aquatic plants and animals, collectively; generally microscopic.

Plasma membrane. The membrane that separates the living protoplast from the external environment; found in all cells.

Plasmalemma (Gr. *plasma,* anything formed + *lemma,* a husk of a fruit). A synonym for **plasma membrane.**

Plasmodesma (plural, **plasmodesmata**) (Gr. *plasma,* something formed + *desmos,* a bond, a band). Fine protoplasmic thread passing through the wall that separates two protoplasts.

Plasmodium (Gr. *plasma,* something former + mod. L. *odium,* something of the nature of). In Myxomycetes, a coenocytic mass of protoplasm, with no surrounding wall.

Plasmogamy (Gr. *plasma,* anything molded or formed + *gamos,* marriage). The fusion of protoplasts, not accompanied by nuclear fusion.

Plasmolysis (Gr. *plasma,* something formed + *lysis,* a loosening). The separation of the cytoplasm from the cell wall due to removal of water from the protoplast.

Plastid (Gr. *plastis,* a builder). A class of organelles, including the chloroplast and several related kinds of bodies; the latter kinds are associated with storage of food materials and some of them (chromoplasts) are highly pigmented.

Plastoquinone. A quinone, one of a group of compounds involved in the transport of electrons during photosynthesis in chloroplasts.

Plumule (L. *plumula,* a small feather). The first bud of an embryo or that portion of the young shoot above the cotyledons.

Podzol (R. *pod,* under + *zola,* ashes). A soil characteristic of taiga vegetation; color of the A horizon is gray because of excessive leaching; in the spodosol soil order.

Polar transport. The directed movement within plants of compounds (usually hormones) predominantly in one direction; polar transport overcomes the tendency for diffusion in all directions.

Polarity (Gr. *pol,* an axis). The observed differentiation of an organism, tissue, or cell into parts having opposed or contrasted properties or form.

Pollen (L. *pollen,* fine flour). The germinated microspores or partially developed male gametophytes of seed plants.

Pollen mother cell. See **Microsporocyte.**

Pollen profile. A diagrammatic summary of the sequence and abundance of pollen types that have been chronologically trapped in sediments.

Pollen tube. The parasitic, complete male gametophyte of seed plants, which grows through the nucellus of a gymnosperm ovule or through the pistil of an angiosperm.

Pollination. The transfer of pollen from a stamen or staminate cone to a stigma or ovulate cone.

Pollinium (L. *pollentis,* powerful or *pollinis,* fine flour + *ium,* group). A mass of pollen that sticks together and is transported by pollinators as a mass; present in orchids and milkweeds.

Pollutant (L. polluere, to dirty + lutum, mud). An unnatural, human-related substance that is introduced to the environment; may be gas, liquid, or solid, easily broken down or long-lasting.

Polymer. A molecule that is made by coupling together many small molecules (monomers) that are similar to one another.

Polymerization. The chemical union of monomers to produce a polymer.

Polynomial (Gr. *polys,* many + L. *nomen,* name). Scientific name for an organism composed of more than two words; compare to **binomial.**

Polynucleotides (Gr. *polys,* much, many). Long-chain molecules composed of units (monomers) called nucleotides; nucleic acid is a polynucleotide.

Polyphyletic (Gr. *polys,* many + *phyle,* tribe). Referring to organisms that did not have an ancestor in common.

Polyploid (Gr. *polys,* many + *ploos,* fold). Referring to a plant or tissue with more than two complete sets of chromosomes per cell.

Polyribosome (Gr. *polys,* many + ribosomes). An aggregation of ribosomes; frequently simply *polysome.*

Polysaccharides (Gr. *polys,* much, many + *sakcharon,* sugar). Polymeric molecules composed of units (monomers) of a sugar; starch and cellulose are polysaccharides.

Pome (Gr. *pomme,* apple). A simple fleshy fruit, the outer portion of which is formed by the floral parts that surround the ovary.

Population (L. *populus,* people). A group of closely related, interbreeding organisms.

Pore spaces. Spaces between soil particles that may be filled with air or water and into which root hairs may penetrate; the larger the soil particles, the larger the pore space.

Prairie (L. *pratum,* meadow). Grassland vegetation, with trees essentially absent; often considered to have more rainfall than does the steppe.

Predation (L. *predatio,* plundering). A form of biological interaction in which one organism is destroyed (by ingestion); parasitism is a form of predation.

Primary (L. *primus,* first). First in order of time or development.

Primary pitfield. Thin areas of primary cell walls.

Primary endosperm cell. A cell of the embryo sac after fertilization, generally containing a nucleus resulting from fusion of the two polar nuclei with a sperm nucleus; the endosperm develops from this cell.

Primary meristems. Meristems of the shoot or root tip giving rise to the tissues of the primary plant body.

Primary tissues. Those tissues, epidermis, xylem, phloem, and ground tissues, which form from primary meristems.

Primitive (L. *primus,* first). Referring to a taxonomic trait thought to have evolved early in time.

Primordium (L. *primus,* first + *ordiri,* to begin to weave; literally beginning to weave, or to put things in order). The beginning or origin of any part of an organ.

Procambium (L. *pro,* before + cambium). A primary meristem that gives rise to primary vascular tissues and, in most woody plants, to the vascular cambium.

Producer (L. *producere,* to draw forward). An organism

that produces organic matter for itself and other organisms (consumers and decomposers) by photosynthesis.

Proembryo (L. *pro*, before + *embryon*, embryo). A group of cells arising from the division of the fertilized egg cell before the cells that are to become the embryo are recognizable.

Prokaryotes (L. *pro*, before + Gr. *karyon*, a nut, referring in modern biology to the nucleus). Primitive organisms, bacteria, and blue-green algae, which do not have the DNA separated from the cytoplasm by an envelope.

Prophase (Gr. *pro*, before + *phasis*, appearance). An early stage in nuclear division, characterized by coiling of the chromosomes and formation of the mitotic spindle.

Proplastid (Gr. *pro*, before + plastid). A type of plastid, occurring generally in meristematic cells, which will develop into a chloroplast.

Protease (protein + *-ase*, a suffix indicating an enzyme). An enzyme breaking down a protein.

Protein (Gr. *proteios*, holding first place). A polymeric molecule made of subunits called amino acids which are linked together in a chain by a type of bond called the peptide bond.

Proterozoic (Gr. *protero*, before in time + *zoe* life). The earliest geologic era, beginning about 4.5–5 billion years ago and ending 570 million years ago; also called the Precambrian era.

Protochlorophyll (Gr. *protos*, first + *chloros*, green + *phyllos*, leaf). One of the precursors of chlorophyll.

Prothallus (L. *pro*, before + Gr. *thallos*, young shoot, sprout). A synonym for the gametophyte generation of ferns; also called prothallium; the prothallial cells of some male gametophytes are sterile and thought to represent a much reduced vegetative body.

Protoxylem. First formed primary xylem.

Protoderm (Gr. *protos*, first + *derma*, skin). A primary meristem that gives rise to epidermis.

Proton (Gr. *proton*, first). The nucleus of a hydrogen atom is a single positively charged particle, the proton; the nucleus of all other elements consists of protons and neutrons; the mass of a proton is 1.67×10^{-24} gram.

Protonema (plural, **protonemata**) (Gr. *protos*, first + *nema*, a thread). An algal-like filamentous growth; an early stage in development of the gametophyte of mosses.

Protoplasm (Gr. *protos*, first + *plasma*, something formed). Living substance.

Protoplast (Gr. *protoplastos*, formed first). The organized living unit of a single cell.

Pseudopodium (Gr. *pseudes*, false + *podion*, a foot). In Myxomycetes, an armlike projection from the body by which the plant creeps over the surface.

Purines. A group of compounds in which carbon and nitrogen atoms form a double ring structure, one ring with six atoms, the other with five atoms; includes the compounds adenine and guanine.

Pyrenoid (Gr. *pyren*, the stone of a fruit + L. *oïdes*, like). A denser body occurring within the chloroplasts of certain algae and liverworts and apparently associated with starch deposition.

Pyrimidines. A group of compounds in which carbon and nitrogen atoms form a ring structure; includes the compounds cytosine, thymine, and uracil.

Quadrat (L. *quadrus*, a square). A frame of any shape that, when placed over vegetation, defines a unit sample area within which the plants may be counted or measured.

Quantum (L. *quantum*, how much). An elemental unit of energy.

Quiescent center (L. *quiescere*, to rest). Disk-shaped region of root apex containing slowly dividing cells.

r selection. Natural selection that favors short-lived, early-maturing individuals which devote a large fraction of their resources into reproduction; annual herbs are r strategists.

Raceme (L. *racemus*, a bunch of grapes). An inflorescence in which the main axis is elongated and the flowers are born on pedicels that are about equal in length.

Rachilla (Gr. *rhachis*, a backbone + L. dim, ending *-illa*). Shortened axis of spikelet.

Rachis (Gr. *rhachis*, a backbone). Main axis of spike; axis of fern leaf (frond) from which pinnae arise; in compound leaves, the extension of the petiole corresponding to the midrib of an entire leaf.

Radicle (L. *radix*, root). Portion of the plant embryo that develops into the primary or seed root.

Random plant distribution. A distribution of a plant species within an area such that the probability of finding an individual at one point is the same for all points.

Raphe (Gr. *rhaphe*, seam). Ridge on seeds, formed by the stalk of the ovule, in those seeds in which the funiculus is sharply bent at the base of the ovule.

Raphides (Gr. *rhaphis*, a needle). Fine, sharp, needlelike crystals.

Ray initials. Meristematic cells in the vascular cambium that develop into xylem and phloem cells comprising the ray system.

Ray system (L. *radius*, a beam or ray). System of cells in secondary tissues that are oriented perpendicular to the long axis of the stem, form from ray initials of the vascular cambium.

Receptacle (L. *receptaculum*, a reservoir). Enlarged end of the pedicel or peduncle to which other flower parts are attached.

Recombination (L. *re*, repeatedly + *combinatus*, joined). The mixing of genotypes that results from sexual reproduction.

Red tide. The coloring of nearshore, marine water by a high density of mircroscopic algal organisms that may additionally release toxic byproducts into the water.

Reduction (F. *reduction*, L. *reductio*, a bringing back). Any chemical reaction involving the removal of oxygen from or the addition of hydrogen or an electron to a substance.

Regular flower. A flower in which the corolla is made up of similarly shaped petals equally spaced and radiating from the center of the flower.

Reproduction (L. *re*, repeatedly + *producere*, to give birth to). The process by which plants and animals give rise to offspring.

Reproductive isolation. The separation of populations in time or space so that genetic flow between them is cut off.

Residual meristem. Meristematic region near the tip of the shoot apex that remains after differentiation of the pith and cortex.

Resin duct. Resin canal; in conifers, continuous tubes lined with secretory cells that run through the sapwood; they function as repositories for metabolic byproducts, but may have an ecological use as deterrents to wood-boring insects and they have economic value as a source of turpentine and other naval stores.

Respiration (L. *re*, repeatedly, + *spirare*, to breathe). In the cell, the catabolic process by which sugars and other fuels are oxidized and broken down, with some of the energy captured in the formation of ATP.

Rhizoid (Gr. *rhiza*, root + L. *oïdes*, like). One of the cellular filaments that perform the functions of roots.

Rhizome (Gr. *rhiza*, root). An elongated, underground, horizontal stem.

Rhizophores (Gr. *rhiza*, root + *phoros*, bearing). Leafless branches that grow downward from the leafy stems of certain Lycophyta and give rise to roots when they come into contact with the soil.

Ribose. A pentose sugar.

Ribosomes (*ribo*, from RNA + Gr. *somatos*, body). Small particles 10–20 nm in diameter, containing RNA and protein; active in protein synthesis.

Ring porous. Wood with large xylem vessel members mostly in early wood; compare with **diffuse porous.**

Ripening (A.S. *rifi*, perhaps related to reap). Changes in a fruit that follow seed maturation and that prepare the fruit for its function of seed dispersal.

RNA. Ribonucleic acid.

Root (A.S. *rōt*). The descending axis of a plant, normally below ground, serving to anchor the plant and absorb and conduct water and mineral nutrients.

Root cap. A thimblelike mass of living cells covering and protecting the apical meristems of a root; site of perception of gravity in geotropism.

Root hairs. Epidermal projections of root cells in region of maturation, provide means to increase the absorptive surface of root.

Root pressure. Pressure in the xylem arising as a result of osmosis in the root.

Rootstock. An elongated, underground, horizontal stem.

Rosette. A shoot with a very short stem, composed of several unelongated internodes with fully expanded leaves.

Runner. A stem that grows horizontally along the ground surface.

Samara (L. *samara*, the fruit of the elm). Simple, dry, one- or two-seeded indehiscent fruit with pericarp bearing a winglike outgrowth.

Sand. Soil particles between 50 and 2000 microns in diameter.

Saprophyte (Gr. *sapros*, rotten + *phyton*, a plant). Ar. organism deriving its food from the dead body or the nonliving products of another plant or animal.

Sapwood. Peripheral wood that actively transports.

Savannah (Sp. *sabana*, a large plain). Vegetation of scattered trees in a grassland matrix.

Schizocarp (Gr. *schizein*, to split + *karpos*, fruit). Dry fruit with two or more united carpels that split apart at maturity.

Sclereids (Gr. *skleros*, hard). Sclerenchyma cells having variably shaped, heavily lignified cell walls.

Sclerenchyma (Gr. *skleros*, hard + *-echyma*, a suffix denoting tissue). A strengthening tissue composed of cells with heavily lignified cell walls.

Scrub (A.S. *scrob*, a shrub). Vegetation dominated by shrubs; described as thorn forest in areas with moderate rainfall, or as chaparral or desert in areas with low rainfall.

Scutellum (L. *scutella*, a dim. of *scutum*, shield). Single cotyledon of grass embryo.

Seaweed. Any large, marine alga; usually a red or brown alga; includes the kelps.

Secondary tissues. Those tissues, xylem, phloem, and periderm, that form from secondary meristems.

Secretory structures. Any of a number of specialized plant structures such as nectaries, glands, and hydathodes that secrete secondary plant substances.

Sedimentary rock (L. *sedere*, to sit). Rock formed from material deposited as sediment, then physically or chemically changed by compaction and hardening while buried in the earth crust.

Seed (A.S. *sed*, anything which may be sown). Popularly as originally used, anything that may be sown; i.e., ''seed'' potatoes, ''seeds'' of corn, sunflower, etc.; botanically, a seed is the matured ovule without accessory parts.

Seed coat. A hardened, outer layer of the seed, derived from the integument(s) of the ovule, and which functions to prevent mechanical disturbance to and water loss from the embryo; it may also regulate germination in several ways.

Seed plant. Common name for members of the gymnosperm and angiosperm groups, and for extinct members of other groups that produced seeds.

Selectively permeable. Referring to a membrane that permits some kinds of molecules to pass through while not allowing other kinds of molecules to pass through.

Self-pollination. Transfer of pollen from the stamens to the stigma of either the same flower or flowers on the same plant.

Seminal root. The root or roots forming from primordia present in the seed.

Sepals (M.L. *sepalum*, a covering). Whorl of sterile, leaf-like structures that usually enclose the other flower parts.

Septate (L. *septum*, fence). Divided by crosswalls into cells or compartments.

Septum (L. *septum*, fence). Any dividing wall or partition; frequently a crosswall in a fungal or algal filament.

Serpentine (L. *serpens*, a serpent). Referring to soil derived from metamorphic parent material characterized (among other things) by low calcium (Ca), high magnesium (Mg), and a greenish-gray color.

Sessile (L. *sessilis*, low, dwarf, from *sedere*, to sit). Sitting, referring to a leaf lacking a petiole or a flower or fruit lacking a pedicel.

Seta (plural, **setae**) (L. *seta*, a bristle). In bryophytes, a short stalk of the sporophyte, which connects the foot and the capsule.

Sexual reproduction. Reproduction that requires meiosis and fertilization for a complete life cycle.

Shade tolerance. The ability to grow slowly in the shade of an overstory canopy; an essential characteristic for a climax species.

Sheath. Part of leaf that wraps around the stem, as in grasses.

Shoot (derivation uncertain, but early referring to new plant growth; 1450, "Take a feyr schoyt of blake thorne"). A young branch that grows out from the main stock of a tree, or the young main portion of a plant growing above ground.

Shoot tip. Terminal portion of the shoot containing apical and primary meristems and cells in early stages of differentiation.

Sibling species. Species morphologically nearly identical but incapable of producing fertile hybrids.

Side-chain. A part of a polymer that extends laterally from the main chain.

Sieve cell. A long and slender sieve element without a companion cell, with relatively unspecialized sieve areas, and with tapering end walls that lack sieve plates; found in gymnosperms and ferns.

Sieve plate. Perforated wall area in a sieve-tube member through which pass strands connecting sieve-tube protoplasts.

Sieve tube. A series of sieve-tube members forming a long cellular tube specialized for the conduction of food materials.

Sieve tube members. An enucleate phloem cell primarily responsible for photosynthate transport; separated from other sieve tube members by sieve plates.

Silique (L. *siliqua*, pod). The fruit characteristic of Brassicaceae (mustards); two-celled, the valves splitting from the bottom and leaving the placentae with a partition stretched between.

Silt. Soil particle between 2 and 50 μm in diameter.

Simple pit. Pit not surrounded by an overarching border; in contrast to bordered pit.

Single bond. A covalent bond that involves the sharing of two electrons.

Siphonous line. A line of evolutionary development in the algae in which mitosis is not followed by cytokinesis; this results in an elongated multinucleate, coenocytic filament.

Soil (L. *solum*, soil, solid). The uppermost stratum of the earth's crust, which has been modified by weathering and organic activity into (typically) three horizons: an upper A horizon that is leached, a middle B horizon in which the leached material accumulates, and a lower C horizon that is unweathered parent material.

Soil texture. Refers to the amounts of sand, silt, and clay in a soil, as a sandy loam, loam, or clay texture.

Solute (L. *solutus*, from *solvere*, to loosen). A dissolved substance.

Solution. A homogeneous mixture, the molecules of the dissolved substance (e.g., sugar), the solute, being dispersed between the molecules of the solvent (e.g., water).

Solvent. A substance, usually a liquid, having the properties of dissolving other substances.

Soredium (plural, **soredia**) (Gr. *soros*, a heap). A sexual reproductive body of lichens, consisting of a few algal cells surrounded by fungal hyphae.

Sorus (plural, **sori**) (Gr. *soros*, a heap). A cluster of sporangia in ferns.

Species (L. *species*, appearance, form, kind). A group of individuals usually interbreeding freely and having many characteristics in common.

Sperm (Gr. *sperma*, the generative substance or seed of a male animal). A male gamete.

Spermagonium (plural, **spermagonia**) (Gr. *sperma*, sperm + *gonos*, offspring). Flask-shaped structure characteristic of the sexual phase of the rust fungi; bearing receptive hyphae and spermatia.

Spermatophyte (Gr. *sperma*, seed + *phyton*, plant). A seed plant.

Spike (L. *spica*, an ear of grain). An inflorescence in which the main axis is elongated and the flowers are sessile.

Spikelet (L. *spica*, an ear of grain + dim. ending *-let*). The unit of inflorescence in grasses; a small group of grass flowers.

Spindle (A.S. *spinel*, and instrument employed in spinning thread by hand). Referring in mitosis and meiosis to the spindle-shaped intracellular aggregate of microtubules involved in chromosome movement.

Spodosol (Gr. *spodos*, wood ashes; R. *pod*, under + *zola*, ashes). One of the ten world soil orders, characterized by an ashy, sandy, bleached, acidic A_2 horizon and associated mainly with coniferous forest vegetation; synonomous with podzol.

Sporangiophore (sporangium + Gr. *-phore*, a root of *phorein*, to bear). A branch bearing one or more sporangia.

Sporangium (spore + Gr. *angeion*, a vessel). Spore case.

Spore (Gr. *spora*, seed). A reproductive cell that develops into a plant without union with other cells.

Sporic life cycle. A life cycle that includes alternation of generations, the sporophyte and gametophyte being more than zygote and gametes, respectively.

Sporocyte (spore + L. *cyta,* vessel). A diploid or haploid cell that will undergo mitosis or meiosis to produce spores.

Sporophore (spore + Gr. *phorein,* to bear). The fruiting body of fleshy and woody fungi, which produces spores.

Sporophyll (spore + Gr. *phyllon,* leaf). A spore-bearing leaf.

Sporophyte (spore + Gr. *phyton,* a plant). In alternation of generations, the plant in which meiosis occurs and which thus produces meiospores.

Spring wood. See Early wood.

Stamen (L. *stamen,* the standing-up things or a tuft of thready things). Flower structure made up of an anther (pollen-bearing portion) and a filament.

Staminate cone. See Microsporangiate cone.

Staminate flower. A flower having stamens but no pistils.

Starch (M.E. *sterchen,* to stiffen). A polysaccharide composed of glucose; the chief food storage material of many plants.

Statolith (Gr. *statos,* standing + *lithos,* stone). An organelle that moves to its position in a cell as a result of gravity, thus providing an initial sensing of, or orientation to, gravity by a cell.

Stele (Gr. *stele,* a post). The central cylinder, inside the cortex, of roots and stems of vascular plants.

Stem (O.E. *stemn*). The main body of the portion above ground of tree, shrub, herb, or other plant; the ascending axis, whether above or below ground, of a plant, in contradistinction to the descending axis or root.

Steppe (R. *step,* a lowland). An arid grassland vegetation.

Sterigma (plural, **sterigmata**) (Gr. *sterigma,* a prop). A slender, pointed protuberance at the end of a basidium, which bears a basidiospore.

Stigma (L. *stigma,* a prick, a spot, a mark). Receptive portion of the style to which pollen adheres.

Stipe (L. *stipes,* post, tree trunk). The stem portion of a kelp, to which are attached bladders and blades.

Stipule (L. *stipula,* dim of *stipes,* a stock or trunk). A leaflike structure from either side of the leaf base.

Stolon (L. *stolo,* a shoot). A stem that grows horizontally along the ground surface.

Stoma (plural, **stomata**) (Gr. *stoma,* mouth). Epidermal structure on stems and leaves composed of two guard cells plus the small pore between them, through which gases pass.

Strobilus (Gr. *strobilos,* a cone). A number of modified, spore-producing scales or leaves (sporophylls) grouped together on an axis.

Stroma (Gr. *stroma,* a bed or covering). A mass of protecting vegetative filaments; the background substance of chloroplasts, probably the location of the carbon cycle of photosynthesis.

Style (Gr. *stylos,* a column). Slender column of tissue that arises from the top of the ovary and through which the pollen tube grows.

Suberin (L. *suber,* the cork oak). A waxy material found in cell walls of cork tissue and endodermis.

Substrate. A molecule that engages in a reaction catalyzed by an enzyme.

Succession (L. *successio,* a coming into the place of another). A sequence of changes in time of the species that inhabit an area, from an initial pioneer community to a final climax community.

Succulent. A plant with fleshy, water-storing parts.

Sucrose. Table sugar, a disaccharide made of a molecule of glucose linked to a molecule of fructose.

Sugar. A simple carbohydrate such as glucose.

Summer wood. See Late wood.

Superior ovary. An ovary completely separate and free from the calyx.

Suspensor (L. *suspendere,* to hang). A cell or chain of cells developed from a zygote whose function is to place the embryo cells in an advantageous position to receive food.

Suture (L. *sutura,* a sewing together; originally the sewing together of flesh or bone wounds). The junction, or line of junction, of contiguous parts.

Symbiosis (Gr. *syn,* with + *bios,* life). An association of two different kinds of living organisms.

Sympetaly (Gr. *syn,* with + *petalon,* leaf). A condition in which petals are united.

Synandry (Gr. *syn,* with + *andros,* a man). A condition in which stamens are united.

Syncarpy (Gr. *syn,* with + *karpos,* fruit). A condition in which carpels are united.

Synergids (Gr. *synergos,* toiling together). The two nuclei at one end of the embryo sac, which, with the third (the egg), constitute the egg appartus.

Synsepaly (Gr. *syn,* with + sepals). A condition in which sepals are united.

Taiga (Teleut *taiga,* rocky mountainous terrain). A broad northern belt of vegetation dominated by conifers; also, a similar belt in mountains just below alpine vegetation.

Tannin. A substance that has an astringent, bitter taste.

Tapetum (l. *tapete,* carpet). The tissue that lines developing pollen sacs (microsporangia) of seed plants; it degenerates and provides nutrition to the tissue within.

Taxon (plural, **taxa**) (Gr. *taxis,* order). A general term for any taxonomic rank, from subspecific to divisional.

Taxonomy (Gr. *taxis,* arrangement + *nomos,* law). Systematic botany; the science dealing with the describing, naming, and classifying of plants

TCA cycle. See Citric acid cycle.

Teliospore (Gr. *telos,* completion + spore). Resistant spore characteristic of the Heterobasidiomycetidae, in which karyogamy and meiosis occur and from which a basidium develops.

Telium (plural, *telia*) (Gr. *telos,* completion). A sorus of teliospores.

Telophase (Gr. *telos,* completion + phase). The last stage of mitosis, in which daughter nuclei are reorganized.

Tendril (L. *tendere,* to stretch out, to extend). A slender coiling organ that aids in the support of stems.

Terminal bud. A bud at the end of a stem.

Testa (L. *testa*, brick, shell). The outer coat of the seed.

Tetrad (Gr. *tetradeion*, a set of four). A group of four, usually referring to the meiospores immediately after meiosis.

Tetraploid (Gr. *tetra*, four + *ploos*, fold). Having four sets of chromosomes per nucleus.

Tetraspores (Gr. *tetra*, four + spores). Four spores formed by division of the sporocyte.

Tetrasporophyte (Gr. *tetra*, four + *sporos*, seed, spore + *phuton*, plant). One of two sporophyte generations in certain red algae; produces meiospores in tetrad clusters.

Tetrasporine line (tetraspore + L. suffix *-ine*, like). A line of evolutionary development in the algae in which mitosis is directly followed by cytokinesis, resulting in a filament, thallus, or complex plant body of varied form.

Thallophytes (Gr. *thallos*, a sprout + *phyton*, plant). A division of plants whose body is a thallus, i.e., lacking roots, stems, and leaves.

Thallus (Gr. *thallos*, a sprout). Plant body without true roots, stems, or leaves.

Thermoperiod (Gr. *therme*, heat + *periods*, a cycle). A difference in temperature between day and night.

Thymidine. A nucleoside incorporated in DNA, but not in RNA.

Thymidine ³H. Tritiated or radioactive thymidine.

Thymine. A pyrimidine occurring in DNA, but not in RNA.

Tiller (O.E. *telga*, a branch). A grass stem arising from a lateral bud at a basal node; tillering is the process of tiller formation.

Tissue. A group of cells that perform a collective function.

Tonoplast (Gr. *tonos*, stretching tension + *plastos*, molded, formed). The cytoplasmic membrane bordering the vacuole; so-called by de Vries, as he thought it regulated the pressure exerted by the cell sap.

Toxin (L. *toxicum*, poison). A poisonous secretion of a plant or animal.

Tracheid (Gr. *tracheia*, windpipe). An elongated, tapering xylem cell, with lignified pitted walls, adapted for conduction and support.

Tracheophytes (Gr. *tracheia*, windpipe + *phyton*, plant). Vascular plants.

Trait. A distinctive definable characteristic; a mark of individuality.

Transfer cells. Specialized cells modified by their cell wall projections for efficient short distance transport.

Transfusion tissue (L. *trans*, across + *fundere*, to pour, to melt). In pine needles, the tissue surrounding the central veins, which may serve to transfer water, nutrients, and food between the vascular tissue and the mesophyll.

Translocation (L. *trans*, across + *locare*, to place). The transfer of food materials or products of metabolism.

Transmit. To pass or convey something from one person, organism, or place to another person, organism, or place.

Transpiration (F. *transpirer*, to perspire). The giving off of water vapor from the surface of leaves.

Trichogyne (Gr. *trichos*, a hair + *gyne*, female). Receptive hairlike extension of the female gametangium in the Rhodophyta and Ascomycetes.

Trichome (Gr. *trichoma*, a growth of hair). A short filament of cells.

Triose (Gr. *treis*, three + *-ose*, suffix indicating a carbohydrate). Any three-carbon sugar.

Tritium. A hydrogen atom, the nucleus of which contains one proton and two neutrons; it is written as ³H; the more common hydrogen nucleus consists only of a proton.

Tropical rain forest. Vegetation with several tree strata, characteristic of tropical lowland regions.

Tropism (Gr. *trope*, a turning). An orientation of the direction of growth in an organ, guided by an external stimulus such as light or gravity.

Tuber (L. *tuber*, a bump, swelling). A much-enlarged, short, fleshy underground stem tip.

Tundra (Lapp. *tundar*, hill). Meadowlike vegetation at low elevation in cold regions that do not experience a single month with average daily maximum temperatures above 50°F.

Turgid (L. *trugidus*, swollen, inflated). Swollen, distended; referring to a cell that is firm due to water uptake.

Turgor pressure (L. *turgor*, a swelling). The pressure within the cell resulting from the absorption of water into the vacuole and the imbibition of water by the protoplasm.

Tylosis (plural, **tyloses**) (Gr. *tylos*, a lump or knot). A growth of one cell into the cavity of another.

Type specimen. The herbarium specimen selected by a taxonomist to serve as a basis for the naming and descriptions of a new species.

Ultisol. A modified podsol soil, with red or yellow B horizon, representative of the southern deciduous forest.

Umbel (L. *umbella*, a sunshade). An inflorescence, the individual pedicels of which all arise from the apex of the peduncle.

Unavailable water. Water held by the soil so strongly that root hairs cannot readily absorb it.

Unicell (L. *unus*, one + cell). An organism consisting of a single cell; generally used in describing algae.

Uniseriate (L. *unus*, one + M.L. *seriatus*, to arrange in a series). Said of a filament having a single row of cells.

Uracil. A pyrimidine found in RNA but not in DNA.

Uredium (plural, **uredia**) (L. *uredo*, a blight). A sorus of uredospores.

Uredospore (L. *uredo*, a blight + spore). A red, one-celled summer spore in the life cycle of the rust fungi.

Vacuole (L. dim. of *vacuus*, empty). A watery solution of various substances forming a portion of the protoplast distinct from the protoplasm.

Vascular (L. *vasculum*, a small vessel). Referring to any plant tissue or region consisting of or giving rise to conducting tissue, e.g., bundle, cambium, ray.

Vascular bundle. A strand of tissue containing primary xylem and primary phloem (and procambium if present) and sometimes enclosed by a bundle sheath of parenchyma or fibers.

Vascular cambium. Cambium giving rise to secondary phloem and secondary xylem.

Vascular plant (L. *vasculum*, small vessel). The common name for any plant that has xylem and phloem; includes the higher and lower vascular plants but not the kelps (which possess sieve tubelike cells) or the bryophytes (some of which contain cells resembling tracheids and sieve tubes).

Vegetation (L. *vegetare*, to quicken). The plant cover that clothes a region; it is formed of the species that make up the flora, but is characterized by the abundance and life form (tree, shrub, herb, evergreen, deciduous plant, etc.) of certain of them.

Venation (L. *vena*, a vein). Arrangement of veins in leaf blade.

Venter (L. *venter*, the belly). Enlarged basal portion of an archegonium in which the egg cell is borne.

Ventral canal cell. The cell just above the egg cell in the archegonium.

Ventral suture (L. *ventralis*, pertaining to the belly). The line of union of the two edges of a carpel.

Vernalization (L. *vernalis*, belonging to spring + *izare*, to make). The promotion of flowering by naturally or artificially applied periods of extended low temperature; seeds, bulbs, or entire plants may be so treated.

Vessel (L. *vasculum*, a small vessel). A series of xylem elements whose function it is to conduct water and mineral nutrients.

Vessel element. A xylem cell derived from the vascular cambium or procambium; a portion of a vessel.

Vitamins (L. *vita*, life + amine). Naturally occurring organic substances, akin to enzymes, necessary in small amounts for the normal metabolism of plants and animals.

Volvocine line (*Volvox* + L. suffix -*ine*, like). A line of evolutionary development in the algae in which the cells remain separate, as in *Volvox*, and are never connected to form a filament or flattened thallus.

Water potential. Refers to the difference between the activity of water molecules in pure distilled water at standard temperature and pressure, and the activity of water molecules in any other system; the activity of these water molecules may be greater (positive) or less (negative) than the activity of the water molecules under standard conditions.

Weathering. Physical and chemical change in parent material that leads to soil formation.

Weed (A.S. *wēod*, used at least since 888 in its present meaning). Generally a herbaceous plant or shrub not valued for use or beauty, growing where unwanted, and regarded as using ground or hindering the growth of more desirable plants.

Whorl. A circle of three or more flower parts, or of leaves.

Whorled. Referring to bud or leaf arrangement in which there are three or more buds or three or more leaves at a node.

Wings. Lateral petals of legume type of flower.

Wood (M.E., *wode*, *wude*, a tree). Secondary xylem.

Xanthophyll (Gr. *xanthos*, yellowish brown + *phyllon*, leaf). A yellow chloroplast pigment.

Xerophyte (Gr. *xeros*, dry + *phuton*, plant). A plant that normally grows in dry habitats.

Xylem (Gr. *xylon*, wood). A plant tissue consisting of tracheids, vessel elements, parenchyma cells, and fibers; wood.

Zoosporangium (Gr. *zoon*, an animal + sporangium). A sporangium bearing zoospores.

Zoospore (Gr. *zoon*, an animal + spore). A motile spore.

Zygomorphic (Gr. *zygon* a yoke + *morphe*, form). Referring to bilateral symmetry; said of organisms, or a flower, capable of being divided into two symmetrical halves only by a single longitudinal plane passing through the axis.

Zygospore (Gr. *zygon*, a yoke + spore). A thick-walled resistant spore developing from a zygote resulting from the fusion of isogametes.

Zygote (Gr. *zygon*, a yoke). A protoplast resulting from the fusion of gametes.

Zygotic life cycle. A life cycle in which the diploid phase is represented only by the zygote.

E APPENDIX

illustration credits

Noncredited figures are either original to this book or are used with permission from T. E. Weier *et al.*, *Botany: An Introduction to Plant Science*, 5th Ed., Wiley and Sons, New York. Figure numbers are given in boldface type.

Chapter 1: Introduction
1.1 Photo courtesy of R. M. Liebaert.
1.2 Photo courtesy of Dr. N. J. Lang.
1.3 Photo courtesy of Dr. N. J. Lang.
1.4 Photo courtesy of D. Dreyfus.
1.5 Photo courtesy of R. M. Liebaert.
1.6 Photo courtesy of R. M. Liebaert.
1.7 Photo courtesy of D. Brown.

Chapter 2: Metabolism
2.6 From *Scientific American*, April 1972. Copyright 1972 by Richard E. Dickerson and Irving Geis.
2.8 From V. A. Greulach and J. E. Adams, *Plants: An Introduction to Modern Botany*, 3rd Ed., John Wiley & Sons, Inc., 1976. Used with permission.

Chapter 3: The Plant Cell
3.1 From Hooke, *Micrographia* (1664). The Council of the Royal Society of London for Improving Natural Knowledge.
3.4 Courtesy of Prof. Harry T. Horner, Jr.
3.5 Courtesy of Prof. James P. Braselton.
3.6 Courtesy of Prof. Harry T. Horner, Jr.
3.7 Courtesy of Prof. Harry T. Horner, Jr.
3.9 Courtesy of Prof. Chin Ho Lin.
3.10 Courtesy of J. B. Pantastico.
3.11 Courtesy of Prof. Harry T. Horner, Jr.
3.12 Courtesy of Prof. Harry T. Horner, Jr.
3.15 Courtesy of E. G. Cutter.
3.17 Courtesy of Prof. Harry T. Horner, Jr.
3.18 *A* and *B* from E. H. Newcomb, *Protoplasma 84*, 3, 11. Copyright Springer-Verlag.
3.19 Courtesy of D. Branton.
3.25 Courtesy of A. Bajer, *Chromosoma 25*, 249. Copyright Springer-Verlag.
3.27 Courtesy of A. Bajer.
3.28 Courtesy of P. K. Hepler and W. T. Jackson, *J. Cell Biol. 38*, 437.
3.30 Courtesy of M. C. Ledbetter.
3.31 *B*, courtesy of Gankin.

Chapter 4: The Plant Body
4.1 *C*, courtesy of E. B. Risley.
4.9 *A* and *B* redrawn from Carl C. Forsaith. *The Technology of New York State Timbers*, New York State College of Forestry Publications 18; *F*, after A. S. Foster.
4.10 From J. S. Pate and B. E. S. Gunning, *Protoplasma 68*, 140.

4.11 Courtesy of A. S. Crafts.
4.12 *A*, courtesy of J. Cronshaw and K. Esau, *J. Cell Biol. 38*, 25; *B*, as above, p. 292.
4.13 *A* through *D* redrawn from Carl C. Forsaith, The Technology of New York State Timbers, New York State College of Forestry Publication 18.
4.14 *B*, courtesy of L. M. Srivastava.
4.16 After McArthur.
4.23 Slides courtesy of Triarch Products.
4.25 Slides courtesy of D. Graham; stereoscan courtesy of D. Hess.
4.30 *A*, courtesy of Richard Parker, National Audubon Society; *B*, courtesy of Diamon T. Smithers, National Audubon Society.
4.31 From K. Esau, *Plant Anatomy*, 2nd Ed., Wiley, New York.
4.42 Courtesy of G. Breckon.
4.43 *A*, from R. M. Holman and W. W. Robbins, *A Textbook of General Botany*, Wiley, New York; *B*, courtesy of C. H. Lin; *C*, courtesy of W. W. Thomson.
4.44 *A* and *B*, courtesy of S. Lynch.
4.48 Redrawn after G. S. Avery, *Amer. J. Bot. 20*, 565.
4.50 *A* and *B*, courtesy of Tomato Genetics Cooperative; *E*, drawing by Peggy-Ann Kessler Duke in V. A. Greulach, *Plant Function and Structure*, Macmillan, New York.
4.51 *D*, courtesy of W. Russell; *E*, courtesy of P. Jones.
4.53 From H. J. Fuller and O. Tippo, *College Botany*, Holt, Rinehart, and Winston, New York.
4.55 *E*, from L. Kutschera, Wurzelatlas, Geschäftsführer des DLG–Verlages, Frankfurt.
4.56 From Rogers and Head *in* W. J. Whittington (Ed.), *Root Growth*, Plenum Press, New York. Also courtesy of East Malling Research Station.
4.58 Drawn by Sue MacLeod.
4.62 *A* through *D*, after K. Esau.

Chapter 5: The Absorption and Transport Systems
5.9 Photo courtesy of E. L. Proebsting.
5.10 Photo courtesy of D. R. Hoagland.
5.11 Photo courtesy of D. R. Hoagland.
5.12 Redrawn from McDougall, *Plant Ecology*, 4th Ed., Lea and Febiger, 1949, in V. A. Greulach and J. E. Adams, *Plants: An Introduction to Modern Botany*, 3rd Ed., John Wiley and Sons, Inc., 1976. Used with permission.
5.13 Photo courtesy of D. R. Hoagland.
5.14 Photo courtesy of U. S. Forest Service.

Chapter 7: Flowers, Fruits and Seeds

7.3 Redrawn from A. S. Foster and E. M. Gifford, Jr., *Comparative Morphology of Vascular Plants*, W. H. Freeman, San Francisco; also from I. W. Bailey and A. C. Smith, *J. Arnold Arboretum 23*, 256–265, 1942; also from J. E. Canright, *Amer. J. Botany 39*, 484–497, 1952.

7.13 Courtesy of D. Hess.

7.16 *A* and *B*, redrawn from Priestly and Scott, *An Introduction to Botany*, 4th Ed., Longmans Group Ltd.

7.24 *B*, from W. W. Robbins, *The Botany of Crop Plants*, McGraw-Hill, New York. Used with permission of McGraw-Hill Book Company.

7.27 Redrawn after Wilhelm Troll.

7.28 Courtesy of D. Hess and H. Drever.

7.35 *A* and *B*, redrawn after Wilhelm Troll.

Chapter 8: The Control of Growth and Development

8.9 Photo courtesy of L. Rappaport.

8.12 Photo courtesy of J. Goeschl and H. Pratt.

8.13 Photo courtesy of U.S. Department of Agriculture.

8.14 Photo courtesy of Plant Industry Station, Crops Research Division, Agricultural Research Service, U.S. Department of Agriculture.

8.15 Photo courtesy of Plant Industry Station, Crops Research Division, Agricultural Research Service, U.S. Department of Agriculture.

8.17 Photo courtesy of A. Lang.

Chapter 9: Plant Ecology

9.1 From *Silvics of Forest Trees of the United States*, U.S. Department of Agriculture Handbook No. 271, Washington, D.C.

9.2 Redrawn from W. D. Billings, *Plants and the Ecosystem*, Wadsworth Publishing Company, Belmont, California.

9.3 Redrawn from D. M. Gates, *Ecology 46*:1–13. Copyright, 1965, by the Ecological Society of America.

9.4 Courtesy of the New York Botanical Garden.

9.6 Redrawn from D. M. Gates, *Energy Exchange in the Biosphere*, Harper and Row, New York.

9.12 Courtesy of H. Biswell.

9.14 Courtesy of C. H. Muller.

9.15 Redrawn from H. A. Mooney and W. D. Billings, *Ecological Monographs 31*:1–29. Copyright, 1961, by the Ecological Society of America.

9.17 Courtesy of U. S. Forest Service.

9.18 Redrawn from P. W. Richards, *Tropical Rain Forest*, Cambridge University Press.

9.21 Courtesy of U. S. Forest Service.

9.22 Courtesy of J. Major.

9.23 Courtesy of J. Major.

9.24 Courtesy of W. D. Billings.

9.26 Courtesy of G. Webster.

9.27 Courtesy of O. A. Leonard.

9.28 Courtesy of J. Major.

9.30 Courtesy of Ansel Adams.

9.31 From R. M. Love, "The rangelands of the western United States," *Scientific American 222*, 94. Copyright, 1970, by Scientific American, Inc. All rights reserved.

9.34 From J. R. McBride, *et al., California Agriculture 29*(12):8–9.

Chapter 10: Plant Taxonomy and Evolution

10.2 Courtesy of the Library of the New York Botanical Garden.

10.8 With the assistance of D. Brown.

10.17 Courtesy of L. D. Gottlieb.

10.18 Courtesy of W. M. Hiesey.

10.19 Courtesy of the Field Museum of Natural History.

10.21 Courtesy of E. S. Barghoorn.

10.22 Courtesy of J. W. Schopf.

10.23 From J. W. Schopf, *Journal of Paleontology 42*:651–688, and courtesy of the Society of Economic Paleontologists and Mineralogists.

10.24 Redrawn from H. P. Banks, *Evolution and Plants of the Past*, Wadsworth, Belmont, California. Reprinted by permission of the publisher.

10.25 *A*, redrawn from W. Goldring, *Scientific Monthly 24*, 515–527; *B*, redrawn from C. B. Beck, *American Journal of Botany 49*, 376, 1962.

10.26 *A*, courtesy of Field museum of Natural History; *B*, cartoon courtesy of Sidney Harris.

10.27 Redrawn from M. Hirmer, *Handbuch der Paläobotanik 1*, 182–232.

10.28 Redrawn from M. Hirmer, *Handbuch der Paläobotanik 1*, 409–452, 1927.

10.29 Redrawn from E. Dorf, *American Scientist 48*, 341–364, 1960. Reprinted by permission, American Scientist, Journal of Sigma Xi, the Scientific Research Society of North America.

10.30 *A* and *B*, courtesy of C. Heiser; *C*, courtesy of F. Smith.

10.31 Redrawn from J. M. Savage, *Evolution*, Holt, Rinehart and Winston, New York.

10.32 Adapted from G. L. Stebbins, *Processes of Organic Evolution*, 2nd Ed., Prentice-Hall, Englewood Cliffs, New Jersey. Copyright 1971, p. 89, by permission of Prentice-Hall, Inc.

Chapter 11: Algae

11.4 Redrawn from M. Neushul, *Ecology 48*, 90. Copyright 1967 by the Ecological Society of America.

11.5 Redrawn from M. N. Hill, *BioScience 18*(10), 965, published by the American Institute of Biological Sciences.

11.7 Redrawn from F. T. Haxo and L. R. Blinks, *Journal of General Physiology 33*, 389–442, The Rockefeller University Press.

11.8 All courtesy of N. J. Lang.

11.9 Courtesy of I. Abbott.

11.10 Courtesy of Johns-Manville Corporation.

11.13 Courtesy of N. J. Lang.

11.14 *A* and *B*, courtesy of D. Brandon; *C*, *D*, and *E*, courtesy of N. J. Lang.

11.16 *B*, redrawn from R. F. Scagel *et al.*, *An Evolutionary Survey of the Plant Kingdom*, 1965, Wadsworth Publishing Co., Inc., Belmont, California, 94002. Reprinted by permission of the publisher.

11.17 Courtesy of K. Schmitz.

11.18 *A* and *C*, redrawn after G. M. Smith, *Freshwater Algae of the United States*, 1933, McGraw-Hill, New York. Used with permission of McGraw-Hill Book Co.

11.19 Redrawn after F. Moewns.

11.23 Redrawn from R. F. Scagel, *et al., An Evolutionary Survey of the Plant Kingdom*, 1965, Wadsworth Publishing Co., Inc., Belmont, California, 94002. Reprinted by permission of the publisher.

11.25 *F* through *P*, redrawn from R. F. Scagel, *et al., An Evolutionary Survey of the Plant Kingdom*, 1965, Wadsworth Publishing Co., Inc., Belmont, California, 94002. Reprinted by permission of the publisher.

Chapter 12: The Mycota

12.1 Redrawn from C. J. Alexopoulous, *Introductory Mycology*, John Wiley & Sons, Inc., New York, 1962.

12.4 From *The Saprolegniaceae: With Notes On Other Water Molds* by William C. Coker, The University of North Carolina Press, 1923. By permission of the publisher.

12.5 From *The Saprolegniaceae: With Notes On Other Water

Molds by William C. Coker, The University of North Carolina Press, 1923. By permission of the publisher.

12.6 Based on L. R. Jones, N. J. Giddins and B. F. Lutman, "Investigations of the Potato Fungus," Vermont Exp. Sta. Bull. 168, 9, 1912.

12.7 Photo courtesy of Plant Pathology Department, University of California, Davis.

12.8 From R. M. Holman and W. W. Robbins, *A Textbook of General Botany*, John Wiley & Sons, New York.

12.9 Photo courtesy of R. N. Campbell.

12.12 Photo courtesy of C. E. Bracker and E. E. Butler.

12.15 *A* through *C; E, F:* from *Comparative Morphology of Fungi*, by E. A. Gäuman, Copyright 1928, McGraw-Hill Book Co. Used with permission of McGraw-Hill Book Co. *D*, redrawn from E. A. Gäuman, *The Fungi: A Description of Their Morphological and Evolutionary Development*, Hafner Publishing Co., New York, 1952.

12.17 Redrawn from *Selecta Fungorum Carpologia of the Brothers Tulasne*, L. R. and C., Vol. I, 1861–65. Translated into English by W. B. Grove; A. H. R. Buller and C. L. Shear, Eds., Oxford (Clarendon Press) 1931. Copyright Commonwealth Mycological Institute, Ferry Lane, Kew, Richmond, Surrey, TW9 3AF, U. K. Used with permission.

12.19 Photo courtesy of J. Ogawa.

12.23 Redrawn from A. H. R. Buller, *Researches on Fungi*, Vol. III, *The Production and Liberation of Spores in Hymenomycetes and Uredineae*. Copyright Longman Group Ltd., Harlow, Essex (U. K.). Used with permission.

12.24 Photo courtesy of Roche.

12.25 Photo courtesy of Brownell.

12.27 *A*, from A. H. R. Buller, *Researches on Fungi*, Vol. VII, *The Sexual Process in the Uredinales*. Royal Society of Canada, University of Toronto Press, 1922. Used with permission of the Royal Society of Canada and University of Toronto Press. *D*, photo courtesy of E. E. Butler.

12.30 *A*, after K. B. Raper and D. F. Alexander, J. Elisha Mitchell Scientific Society, Vol. 61, August 1945. Used with permission.

12.31 Redrawn from E. A. Gäuman, *Fungi: A Description of Their Morphological Features and Morphological Development*, Hafner Publishing Co., New York, 1950.

12.32 Redrawn from Darbishire.

Chapter 13: The Bryophyta

13.1 Redrawn from W. P. Schimper, *Recherches sur les mousses*, Strasbourg, 1948.

13.2 Redrawn from R. F. Scagel *et al., An Evolutionary Survey of the Plant Kingdom,* 1965, Wadsworth Publishing Co., Inc., Belmont, California, 94002. Reprinted by permission of the publisher.

13.3 *B*, courtesy of C. Laning.

13.4 C. Hebant, 1967, *Naturalia Monspeliensia Ser. Bot.* 18, 301.

13.8 Courtesy of W. Russell and D. Hess.

13.14 Redrawn from G. M. Smith, F. M. Gilbert, G. S. Bryan, R. I. Evans, and J. F. Stauffer. *A Textbook of General Botany*, Macmillan, New York. Reprinted with permission of Macmillan Publishing Co., Inc. Copyright 1949 by L. H. Bailey, renewed 1977 by E. Z. Bailey.

13.15 Redrawn from R. M. Holman and W. W. Robbins, *A Textbook of General Botany*, John Wiley & Sons, New York.

Chapter 14: Lower Vascular Plants

14.3 Courtesy of D. Brandon.

14.4 Courtesy of D. Brandon.

14.8 *B*, redrawn from M. G. Sykes, *Annals of Botany* (London) 22, 63; *C*, redrawn from E. Pritzel in A. Engel and K. Prantl, *Die nätürlichen pflanzenfamilien; E, G,* and *H,* redrawn from H. Bruchmann, *Flora 101,* 220, with permission of Gustav Fischer Verlag, New York; *J,* redrawn from material supplied by A. J. Eames.

14.10 *E, H, I, J, K,* and *L,* redrawn from R. A. Slagg, *American Journal of Botany 19,* 106–7; *I, J,* and *O* redrawn from H. Bruchmann, *Flora 104,* 180, with permission of Gustav Fischer Verlag, New York.

14.13 Redrawn from E. R. Walker, *The Botanical Gazette 92,* 7, copyright by the University of Chicago Press.

14.18 *E, F,* and *G,* redrawn from M. E. Hartmann, The Botanical Gazette 91, 259, copyright by the University of Chicago Press; *H* and *I,* redrawn from D. H. Campbell, 1905, *Mosses and Ferns,* Macmillan, New York; *J* and *K,* redrawn from R. M. Holman and W. W. Robbins, *A Textbook of General Botany,* Wiley, New York.

14.19 *A,* courtesy of D. Brandon.

Chapter 15: Gymnosperms

15.4 *A,* courtesy of the American Museum of Natural History; *B,* redrawn from L. H. Bailey, *The Cultivated Conifers in North America,* Macmillan, New York.

15.12 *G, H,* and *I,* redrawn from Coulter and Chamberlain, *Morphology of the Gymnosperms,* University of Chicago Press, with permission from the University of Chicago Press.

15.14 Slide courtesy of Triarch Products.

15.16 Courtesy of the Field Museum of Natural History.

15.18 Courtesy of A. Addicott.

Chapter 16: The Angiosperms

16.1 *C* and *D*, redrawn from Bonnier and Sablon, *Cours de Botanique*, Librairie Generale de l'Enseignement.

16.2 Courtesy of McMinn.

16.6 Redrawn from L. H. Bailey, *Manual of Cultivated Plants*, Macmillan, New York. Copyright 1949 by L. H. Bailey, renewed 1977 by E. Z. Bailey.

16.10 Redrawn from L. H. Bailey, *Manual of Cultivated Plants*, Macmillan, New York. Copyright 1949 by L. H. Bailey, renewed 1977 by E. Z. Bailey.

16.11 Redrawn from L. H. Bailey, *Manual of Cultivated Plants*, Macmillan, New York. Copyright 1949 by L. H. Bailey, renewed 1977 by E. Z. Bailey.

16.12 Redrawn from E. Korsmo, *Unkrauter Im Ackerbau Der Neuzeit*, 1930, Springer.

Illustration Credits

E3

INDEX TO GENERA

This index includes all the genera mentioned in the text. Species are not shown, thus a given set of entries for a genus may refer to more than one species. In parentheses following the genus is the family (for genera in the Anthophyta) or the class (for genera in the Mycota) or the division (for all other genera). Common names are listed, but they give only cross-references to the scientific name, and all page entries appear after the scientific name. Common names selected are those preferred by Bailey *(Manual of Cultivated Plants)* or Munz and Keck *(A California Flora).*

Index to Genera

SUBJECT INDEX

collenchyma, 77
cork, 63
cortical, 85, 89
culture, **161,** 163, 164, 167
endodermal, **89**
epidermal, 47, 88
parenchyma, 78, 288
phloem, 51, 288
primary, 48–57, 89–91
sclerenchyma, 53
secondary, **48**
sporogenous, 279, 281, 286
vascular, 53–57, 89, 96, 155, 161, 198, 204, 224, 288, 296, 324, 338
Tonoplast, **19,** 31, 32
Torus, 55
Toxin, 182, 271, 227
TPN, *see* NADP
Traces, 57
Tracheids, 55–57, 77, 91, 96, 100, 208, 274, 288, 317, 323
Transcription, **13,** 14, 15
Transfer cells, **53**
Transfer reaction, 374, 375
Translation, **13,** 16
Translocation, 107–109, 110, 157, 239, 252, 276
Transpiration, 69, 96, 99, 100, 102, 109, 120, 175, 191
and absorption, 102
conditions affecting rate, 101–102
cuticular, 102
osmotic factors, 96–102
pull, 99–100
rate, 96, 174
regulation by guard cells, 101–102, 166
stomatal, 101–102
Transport, **96**–110
active, 101, 102, 106, 157
polar, 157, 158, 159, 167
Triassic, 209, 212
Trichogyne, **246,** 248
Trichomes, **76**
Triose, **4**
phosphate, 119
Tropical rain forest, *see* Forest
Tropism, **159**
Tryptophan, **157**
Tube cell, **131,** 328
Tuber, **68**–70, 169, 257
Tumor, 270
Tundra, **187,** 194
alpine, 184, 188
arctic, 184, 187–188, 192, 272
Turgor pressure, **97**–99, 102, 109, 156, 157, 166, 254

Tyloses, 60, 63, 84

Ultisol **179,** 189
Ultraviolet radiation, 210, 237
Umbel, **129,** 140
Umbelliferae, *see* Apiaceae
Unicell, 222–223, 234, 237, 242, 244, 252
Unisexual, 327, 328. *See also* Flower, *and* pistillate staminate
Units of measurement, 378
Uracil, **13,** 14
Uredinia, **270**
Uredospores, 269, **270**
Uridine, 13

Vacuole, **18,** 19, 21, 30–32, 99, 156, 234, 254
Vascular bundle, *see* Bundle
Vascular cambium, *see* Cambium
Vascular plants, *see* Plant
Vascular rays, 58
Vascular system, 76–77, 155. *See also* Tissue
Vegetation, world types, 187–192, 194, 216
Veins, 73, 76, 78, 84, 108, 305, 316
Venation, 72, 73, 338
Venter, **276,** 280, 281, 285
Vernalization, **170,** 172
Vertisol, **179**
Vessels, **56,** 57, 77, 96, 100, 309
elements, **55**–57, 91, 205
evolution, 203
lower vascular plants and gymnosperms, 291, 315, 324
Vines, 189, 297
Vitamin, 164, 255
Vitamin B₁, 255

Wall, cell, pressure, 98
primary, 56
secondary patterns, **56,** 57, 96, 160
see also Cell, wall
Water, 3, 9, 10, 11, 15, 96–102, 121–122, 155, 371, 373
absorption, 96–101, 109, 166
adhesion, 99, 100
bound, 99
cohesion, 99–100, 373
loss, 102
movement, mechanism, 99–100, 101–102, 109
photosynthesis, 111, 112, 176
potential, **96**–101, 102, 108, 109
quantity transpired, 96
reactions, 374, 375, 376, 377
respiration, 9–11

role in plant, 96, 168
storage, 102
stress, 166, 172
transport, 96–101
see also, Environmental factors
Waxes, 4, 217
Weathering, 177, 194
Weed, 101, 182, 183, **215**
Whorled leaves, 46
Wilting, 98, 102, 166
Wind, 166
dispersal agent, 137, 162
see also Environmental factors
Wood, 255, 265
anatomy, 62
diffuse porous, **60,** 62
early, 60
heart, 60
late, 60
petrified, 208, 211–212
ring porous, **60**
sap, **60**
spring, 60
summer, 60
Wound healing, 160, 161

Xanthophylls, 204
Xanthophyta, 236
Xerophytes, 102, **174**
Xylem, 100, 102, 107, 108, 109, 155, 157, 159, 164, 210, 239, 252, 276, 288, 308, 309, 315, 317, 323, 324, 331
differentiation, 160, 161
leaf, 77
lower vascular plants, 288, 289–290, 291, 293, 297
primary, 51, 55–56, 69
ray, 315
root, 85, 93, 94, 96
sap, **96,** 100
secondary, 59, 66, 316, 324
Xylose, 234

Yeast, 11, 163, 253, 265, 273

Zeatin, 157
Zinc, 104, 152
Zoosporangium, 256
Zoospores, **255,** 256, 258, 246, **242**
Zygomycetes, 253, 260, 261, 273
Zygospore, **257,** 258
Zygote, **135,** 136, 139, 147, 156, 220, 232, 243, 253, 244–252, 255–273, 274, 276, 280, 285, 288–309, 319, 323, 326, 328, 330, 331
Zygotic life cycle, 244, 252